Practice Problems *for the* Environmental Engineering PE Exam

A Companion to the *Environmental Engineering Reference Manual*

Michael R. Lindeburg, PE

Professional Publications, Inc. • Belmont, CA

Where do I find help solving these Practice Problems?

Practice Problems for the Environmental Engineering PE Exam presents complete, step-by-step solutions for 440 problems like those found on the Environmental PE exam. These problems appear at the end of chapters in the *Environmental Engineering Reference Manual*. You can find all the background information, including charts and tables of data, that you need to solve these problems in the *Environmental Engineering Reference Manual*.

The *Environmental Engineering Reference Manual* may be purchased from Professional Publications (800-426-1178 or **www.ppi2pass.com**) or from your favorite bookstore.

PRACTICE PROBLEMS FOR THE ENVIRONMENTAL ENGINEERING PE EXAM

Printed in the United States of America

Professional Publications, Inc.
1250 Fifth Avenue, Belmont, CA 94002
(650) 593-9119
www.ppi2pass.com

Current printing of this edition: 1

ISBN 1-888577-55-X

Topics

How to Get Online Updates for This Book

I wish I could claim that this book is 100% perfect, but 25 years of publishing have taught me that engineering textbooks seldom are. Even if you only took one engineering course in college, you are familiar with the issue of mistakes in textbooks.

I am inviting you to log on to Professional Publications' website at **www.ppi2pass.com** to obtain a current listing of known errata in this book. From the website home page, click on "Errata." Every significant known update to this book will be listed as fast as we can say "HTML." Suggestions from readers (such as yourself) will be added as they are received, so check in regularly.

PPI and I have gone to great lengths to ensure that we have brought you a high-quality book. Now, we want to provide you with high-quality after-publication support. Please visit us at **www.ppi2pass.com**.

Michael R. Lindeburg, PE
Publisher, Professional Publications, Inc.

Table of Contents

Preface

The label "environmentalist" has always encompassed people from all walks of life; and the title "environmental engineer" can now apply to all engineers (if it doesn't already, it should). All engineers have the ability to affect the environment, from the materials and processes we implicitly design into our creations, through the residuals left behind in their manufacture, to the resultant products of their use. For that reason, the breadth of this book is wide.

Practice Problems for the Environmental Engineering PE Exam is a convenient collection of practice problems covering familiar but perhaps faintly remembered (or impossible-to-remember) subjects. Although many people will use it initially for PE exam preparation, the intent of this book is to provide you with a resource for your daily work long after you have taken the exam.

This book also has sections that will be useful to practicing civil, mechanical, and chemical engineers, as well as to prospective engineering licensees in those branches.

Since this is the first edition, there isn't much to say about previous editions or changes made to them to get to this one. So, here are some of the conventions that I tried to follow while writing this book:

- As much as possible, the chapters use common nomenclature.
- The nomenclature used in the various source documents has been adopted.
- The most current references available at the time of writing were used.
- Some environmental subjects are traditionally evaluated in traditional U.S. (English, British, etc.) units, while others use a mix of metric systems. In this book, SI units, equations, data, and examples are presented in parallel with their customary U.S. counterparts.
- Many problems are in multiple-choice (objective) format, which is consistent with the exam format.
- Careful attention has been given to the scenario statements in all problems, with an eye toward what you are expected to know or assume. To eliminate the need to make assumptions that, though logical, might lead to incorrect answers, many items are explicitly stated.

I admit that the scope of this book slightly exceeds that of the PE exam, and that some of the problems and questions are marginally more difficult than their exam counterparts. As I wrote the book, I included the subjects that I thought you should know something about. If you use this book as a PE exam reference, you will end up being somewhat overprepared for the exam.

The choice of subjects to be included in this book, though greatly influenced by hundreds of engineers, was ultimately mine, as was the choice of degree and depth of coverage of these subjects. After all these years, I think that I have a pretty good feel for the academic needs of a working engineer. But time marches on. You can help me keep this book relevant by doing what thousands of engineers have already done: stay in touch. The next edition will be result of your input.

Thanks!

Michael R. Lindeburg, P.E.
mlindeburg@ppi2pass.com

Acknowledgments

I know enough about environmental engineering to know how little I know. So, for the most critical chapters, I asked several environmental experts from industry and academia to fill in holes in my outline. Their contributions ranged from small sections to entire chapters. When their contributions were submitted to me, I edited every sentence and checked every calculation. However, the genius of the chapters remains theirs. Of the 58 chapters in this book, 9 were created in this manner. The following experts were invaluable in helping prepare this edition in an accurate and timely manner.

Jeffrey S. Mueller, MS, PE, CHP: Biological Effects of Radiation, Nuclear Shielding (Chs. 48, 49)

R. Wane Schneiter, Ph.D., PE, DEE: Air Quality (Ch. 38) plus technical review of Protection of Wetlands (Ch. 45)

James R. Sheetz, PE, DEE: Risk Analysis, Emergency Response, Protection of Wetlands, Toxicology, Industrial Hygiene, Health and Safety (Chs. 11 (part), 28, 45, 46, 47, 51)

You should see the original manuscript for this book, particularly after the copyediting, calculation checking, and developmental editing stages were completed. Talk about a sow's ear! The task of sewing the silk purse again fell to the Professional Publications production team. Cathy Schrott, as Production Department manager, gave short shrift to delays despite an unpredictable schedule. Jessica Holden, to whom I am beholden, didn't relax her copyediting standards, even while bescolding me to keep to my beschedule. Yvonne Sartain illustrated the illustrations and cover art. Kate Hayes and Miriam Hanes apparently swapped letters in their last names while typesetting this book, until Kate eventually disappeared altogether. After about a year, Miriam got tired of taking care of Kate's cat, Shammah, and Kate reappeared. Tracey Berdichevskiy got farther and farther away from her desk as she proofed this book, but David Nielsen won the distance-proofing contest when he moved his chair outside in order to keep his tan. Once, I almost had Tamer Hadi, PPIs staff engineer who did the developmental edit and calculation check. But it turned out Tamer was right, and I was wrong, so Hadi had me after all. Aline Magee became Mama Magee while organizing the contributions to and requesting all the permissions for this book.

This production team is an incredibly competent, dedicated, driven, talented, and forgiving group, and after all the books it has produced for me, I am running out of clever things to say about its members. I was, nevertheless, very fortunate to have had all of these people work with me.

Thanks to you all!

Michael R. Lindeburg, P.E.
mlindeburg@ppi2pass.com

Codes Referenced in This Manual

PPI lists on its website the dates of the codes, standards, and regulations on which NCEES has announced the exams are based. However, these postings do not contain any announcements specifically about the environmental exam. The conclusion you and I must reach from such an omission is that the exam is not sensitive to changes in standards, regulations, or announcements in the Federal Register.

How to Use This Book

This book is primarily a companion to the *Environmental Engineering Reference Manual*, eighth edition. As a collection of solved problems, there are a few, but not very many, ways to use it.

Before it grew to behemoth size, I envisioned this book (with its included problem statements) being taken to work, on the bus or train, on business trips, and on weekend pleasure getaways to the beach. I figured that it would "carry" a lot easier than the big *Environmental Engineering Reference Manual*, which I knew before I started to write would be a big book. Though I don't think you'll be taking this book on any backpacking trips, you might still find yourself working problems in the cafeteria during your lunch break.

The really big issue is whether you really work the practice problems or not. Some people think they can read a problem statement, think about it for about ten seconds, read the solution, and then say "Yes, that's what I was thinking of, and that's what I would have done." Sadly, these people find out too late that the human brain doesn't learn very efficiently that way. Under pressure, they find they know and remember little. For real learning, you have to spend some time with a stubby pencil.

There are so many places where you can get messed up solving a problem. Maybe it's in the use of your calculator, like pushing log instead of ln, or forgetting to set the angle to radians instead of degrees, and so on. Maybe it's rusty math. What is $\ln(e^x)$, anyway? How do you factor a polynomial? Maybe it's in finding the data needed or the proper unit conversion. Maybe it's just trying to find out if that funky building code equation expects L to be in feet or inches. These things take time. And you have to make the mistakes once so that you don't make them again.

If you do decide to get your hands dirty and actually work these problems, you'll have to decide how much reliance you place on the solutions. It's tempting to turn to a solution when you get slowed down by details or stumped by the subject material. You'll probably want to maximize the number of problems you solve by spending as little time as possible on each one. I want you to struggle a little bit more than that.

Studying a new subject is analogous to using a machete to cut a path through a dense jungle. By doing the work, you develop pathways that weren't there before. It's a lot different from just looking at the route on a map. You actually get nowhere by looking at a map. But cut that path once, and you're in business.

So do the problems. All of them. Don't look at the answers until you've sweated a little. And let's not have any whining. Please.

1 Systems of Units

PRACTICE PROBLEMS

1. Convert 250°F to degrees Celsius.

(A) 115°C
(B) 121°C
(C) 124°C
(D) 420°C

2. Convert the Stefan-Boltzmann constant (0.1713×10^{-8} Btu/ft^2-hr-°R^4) from English to SI units.

(A) 5.14×10^{-10} W/m^2·K^4
(B) 0.95×10^{-8} W/m^2·K^4
(C) 5.67×10^{-8} W/m^2·K^4
(D) 7.33×10^{-6} W/m^2·K^4

3. How many U.S. tons (2000 lbm per ton) of coal with a heating value of 13,000 Btu/lbm must be burned to provide as much energy as a complete nuclear conversion of 1 g of its mass? (Hint: Use Einstein's equation: $E = mc^2$.)

(A) 1.7 tons
(B) 14 tons
(C) 779 tons
(D) 3300 tons

SOLUTIONS

1. The conversion to degrees Celsius is given by

$$\begin{aligned} ^\circ\text{C} &= \left(\tfrac{5}{9}\right)(^\circ\text{F} - 32^\circ\text{F}) \\ &= \left(\tfrac{5}{9}\right)(250^\circ\text{F} - 32^\circ\text{F}) \\ &= \left(\tfrac{5}{9}\right)(218^\circ\text{F}) \\ &= \boxed{121.1^\circ\text{C}\ (121^\circ\text{C})} \end{aligned}$$

The answer is (B).

2. In U.S. customary units, the Stefan-Boltzmann constant, σ, is 0.1713×10^{-8} Btu/hr-ft^2-°R^4.

Use the following conversion factors.

$$\begin{aligned} 1\ \text{Btu/hr} &= 0.2931\ \text{W} \\ 1\ \text{ft} &= 0.3048\ \text{m} \\ T_{^\circ\text{R}} &= \frac{9}{5}T_\text{K} \end{aligned}$$

Performing the conversion gives

$$\begin{aligned} \sigma &= \left(0.1713 \times 10^{-8}\ \frac{\text{Btu}}{\text{hr-ft}^2\text{-}^\circ\text{R}^4}\right)\left(0.2931\ \frac{\text{W}}{\frac{\text{Btu}}{\text{hr}}}\right) \\ &\quad \times \left(\frac{1\ \text{ft}}{0.3048\ \text{m}}\right)^2 \left(\frac{1^\circ\text{R}}{\frac{5}{9}\text{K}}\right)^4 \\ &= \boxed{5.67 \times 10^{-8}\ \text{W/m}^2\cdot\text{K}^4} \end{aligned}$$

The answer is (C).

3. The energy produced from the nuclear conversion of any quantity of mass is given as

$$E = mc^2$$

The speed of light, c, is 3×10^8 m/s.

For a mass of 1 g (0.001 kg),

$$\begin{aligned} E &= mc^2 \\ &= (0.001\ \text{kg})\left(3 \times 10^8\ \frac{\text{m}}{\text{s}}\right)^2 \\ &= 9 \times 10^{13}\ \text{J} \end{aligned}$$

Convert to U.S. customary units with the conversion 1 Btu = 1055 J.

$$\begin{aligned} E &= (9 \times 10^{13}\ \text{J}) \left(\frac{1\ \text{Btu}}{1055\ \text{J}} \right) \\ &= 8.53 \times 10^{10}\ \text{Btu} \end{aligned}$$

The number of tons of 13,000 Btu/lbm coal is

$$\frac{8.53 \times 10^{10}\ \text{Btu}}{\left(13{,}000\ \frac{\text{Btu}}{\text{lbm}}\right)\left(2000\ \frac{\text{lbm}}{\text{ton}}\right)} = \boxed{3281\ \text{tons}\ (3300\ \text{tons})}$$

The answer is (D).

2 Properties of Areas

PRACTICE PROBLEMS

1. Locate the centroid of the area.

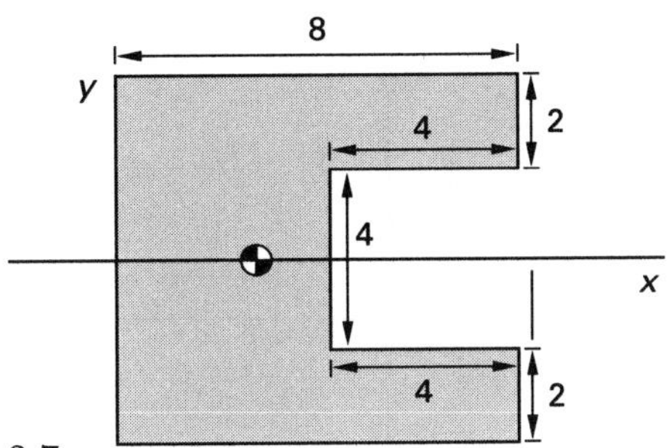

(A) 2.7
(B) 2.9
(C) 3.1
(D) 3.3

2. Replace the distributed load with three concentrated loads, and indicate the points of application.

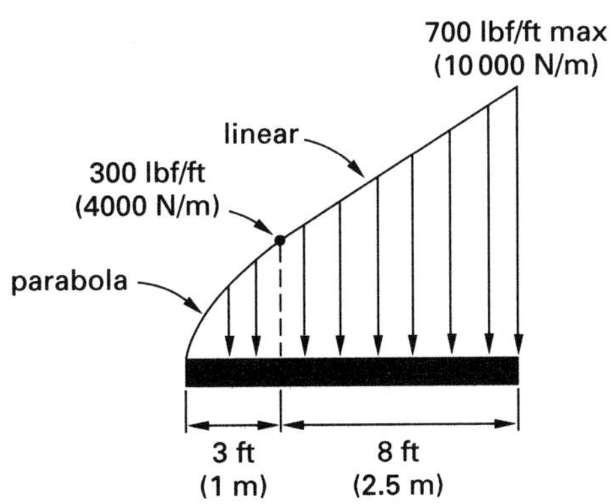

3. Find the centroidal moment of inertia about an axis parallel to the x-axis.

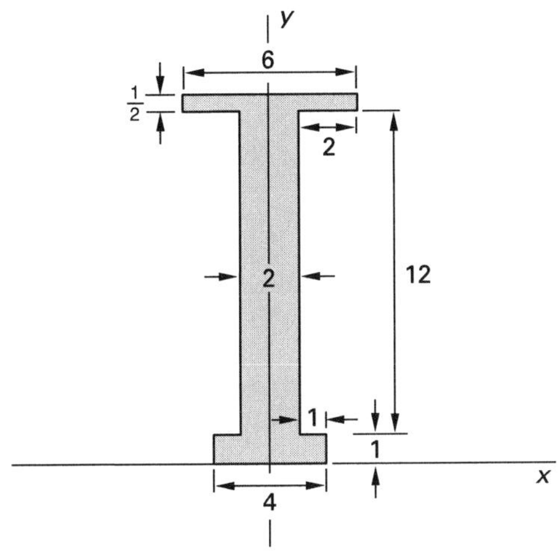

(A) 160 units^4
(B) 290 units^4
(C) 570 units^4
(D) 740 units^4

SOLUTIONS

1. The area is divided into three basic shapes.

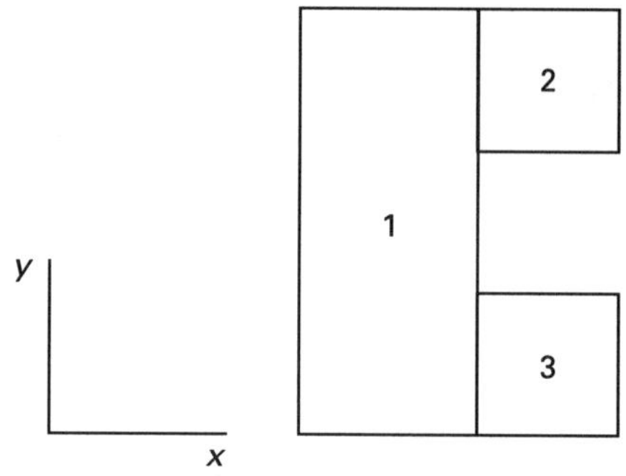

First, calculate the areas of the basic shapes.

$$A_1 = (4)(8) = 32 \text{ units}^2$$
$$A_2 = (4)(2) = 8 \text{ units}^2$$
$$A_3 = (4)(2) = 8 \text{ units}^2$$

Next, find the x-components of the centroids of the basic shapes.

$$x_{c1} = 2 \text{ units}$$
$$x_{c2} = 6 \text{ units}$$
$$x_{c3} = 6 \text{ units}$$

Finally, use Eq. 2.5.

$$x_c = \frac{\sum A_i x_{ci}}{\sum A_i} = \frac{(32)(2) + (8)(6) + (8)(6)}{32 + 8 + 8} = \boxed{3.33 \text{ units}}$$

The answer is (D).

2. *Customary U.S. Solution*

The parabolic shape is

$$f(x) = 300 \ \frac{\text{lbf}}{\text{ft}} \sqrt{\frac{x}{3}}$$

First, use Eq. 2.3 to find the concentrated load given by the area.

$$A = \int f(x)dx = \int_0^{3 \text{ ft}} 300\sqrt{\frac{x}{3}}dx = \left(\frac{300 \ \frac{\text{lbf}}{\text{ft}}}{\sqrt{3}}\right)\left[\frac{x^{\frac{3}{2}}}{\frac{3}{2}}\right]_0^{3 \text{ ft}} = \left(\frac{300 \ \frac{\text{lbf}}{\text{ft}}}{\left(\frac{3}{2}\right)\sqrt{3}}\right)\left(3^{\frac{3}{2}} - 0^{\frac{3}{2}}\right) = \boxed{600 \text{ lbf}}$$

From Eq. 2.4,

$$dA = f(x)dx = 300\sqrt{\frac{x}{3}}$$

Finally, use Eq. 2.1 to find the location x_c of the concentrated load from the left end.

$$x_c = \frac{\int x dx}{A} = \frac{1}{600 \text{ lbf}} \int_0^3 300x\sqrt{\frac{x}{3}}dx = \frac{300}{600\sqrt{3}} \int_0^3 x^{\frac{3}{2}} dx = \left(\frac{300 \ \frac{\text{lbf}}{\text{ft}}}{600\sqrt{3}}\right)\left[\frac{x^{\frac{5}{2}}}{\frac{5}{2}}\right]_0^{3 \text{ ft}} = \left(\frac{300 \ \frac{\text{lbf}}{\text{ft}}}{600\sqrt{3}\left(\frac{5}{2}\right)}\right)\left(3^{\frac{5}{2}} - 0^{\frac{5}{2}}\right) = \boxed{1.8 \text{ ft}}$$

Alternative solution for the parabola:

Use App. 2.A.

$$A = \frac{2bh}{3} = \frac{(2)\left(300 \ \frac{\text{lbf}}{\text{ft}}\right)(3 \text{ ft})}{3} = \boxed{600 \text{ lbf}}$$

The centroid is located at

$$\frac{3h}{5} = \frac{(3)(3 \text{ ft})}{5} = \boxed{1.8 \text{ ft}}$$

The concentrated load for the triangular shape is the area from App. 2.A.

$$A = \frac{bh}{2} = \frac{\left(700 \ \frac{\text{lbf}}{\text{ft}} - 300 \ \frac{\text{lbf}}{\text{ft}}\right)(8 \text{ ft})}{2} = \boxed{1600 \text{ lbf}}$$

From App. 2.A, the location of the concentrated load from the right end is

$$\frac{h}{3} = \frac{8 \text{ ft}}{3} = \boxed{2.67 \text{ ft}}$$

The concentrated load for the rectangular shape is the area from App. 2.A.

$$A = bh = \left(300 \ \frac{\text{lbf}}{\text{ft}}\right)(8 \text{ ft}) = \boxed{2400 \text{ lbf}}$$

From App. 2.A, the location of the concentrated load from the right end is

$$\frac{h}{2} = \frac{8\text{ ft}}{2} = \boxed{4\text{ ft}}$$

SI Solution

The parabolic shape is

$$f(x) = 4000\ \frac{\text{N}}{\text{m}}\sqrt{x}$$

First, use Eq. 2.3 to find the concentrated load given by the area.

$$\begin{aligned} A &= \int f(x)dx = \int_0^{1\text{ m}} 4000\ \frac{\text{N}}{\text{m}}\sqrt{x}\,dx \\ &= \left[\frac{4000\ \frac{\text{N}}{\text{m}}x^{\frac{3}{2}}}{\frac{3}{2}}\right]_0^{1\text{ m}} = \left(\frac{4000\ \frac{\text{N}}{\text{m}}}{\frac{3}{2}}\right)\left(1^{\frac{3}{2}} - 0^{\frac{3}{2}}\right) \\ &= \boxed{2666.7\text{ N}} \quad \text{[first concentrated load]} \end{aligned}$$

From Eq. 2.4,

$$dA = f(x)dx = 4000\sqrt{x}\,dx$$

Finally, use Eq. 2.1 to find the location, x_c, of the concentrated load from the left end.

$$\begin{aligned} x_c &= \frac{\int x dA}{A} = \frac{1}{2666.7\text{ N}}\int_0^{1\text{ m}} 4000\ \frac{\text{N}}{\text{m}}x\sqrt{x}\,dx \\ &= \frac{4000\ \frac{\text{N}}{\text{m}}}{2666.7}\int_0^{1\text{ m}} x^{\frac{3}{2}}dx = \left(\frac{4000\ \frac{\text{N}}{\text{m}}}{2666.7}\right)\left[\frac{x^{\frac{5}{2}}}{\frac{5}{2}}\right]_0^{1\text{ m}} \\ &= \left(\frac{4000\ \frac{\text{N}}{\text{m}}}{2666.7}\right)\left(\frac{1^{\frac{5}{2}} - 0^{\frac{5}{2}}}{\frac{5}{2}}\right) = \boxed{0.60\text{ m}} \quad \text{[location]} \end{aligned}$$

Alternative solution for the parabola:

Use App. 2.A.

$$\begin{aligned} \text{A} &= \frac{2bh}{3} = \frac{(2)\left(4000\ \frac{\text{N}}{\text{m}}\right)(1\text{ m})}{3} \\ &= \boxed{2666.7\text{ N}} \end{aligned}$$

The centroid is located at

$$\frac{3h}{5} = \frac{(3)(1\text{ m})}{5} = \boxed{0.6\text{ m}}$$

The concentrated load for the triangular shape is the area from App. 2.A.

$$\begin{aligned} A &= \frac{bh}{2} = \frac{\left(10\,000\ \frac{\text{N}}{\text{m}} - 4000\ \frac{\text{N}}{\text{m}}\right)(2.5\text{ m})}{2} \\ &= \boxed{7500\text{ N}} \quad \text{[second concentrated load]} \end{aligned}$$

From App. 2.A, the location of the concentrated load from the right end is

$$\frac{h}{3} = \frac{2.5\text{ m}}{3} = \boxed{0.83\text{ m}}$$

The concentrated load for the rectangular shape is the area from App. 2.A.

$$\begin{aligned} A &= bh = \left(4000\ \frac{\text{N}}{\text{m}}\right)(2.5\text{ m}) \\ &= \boxed{10\,000\text{ N}} \quad \text{[third concentrated load]} \end{aligned}$$

From App. 2.A, the location of the concentrated load from the right end is

$$\frac{h}{2} = \frac{2.5\text{ m}}{2} = \boxed{1.25\text{ m}}$$

3. The area is divided into three basic shapes.

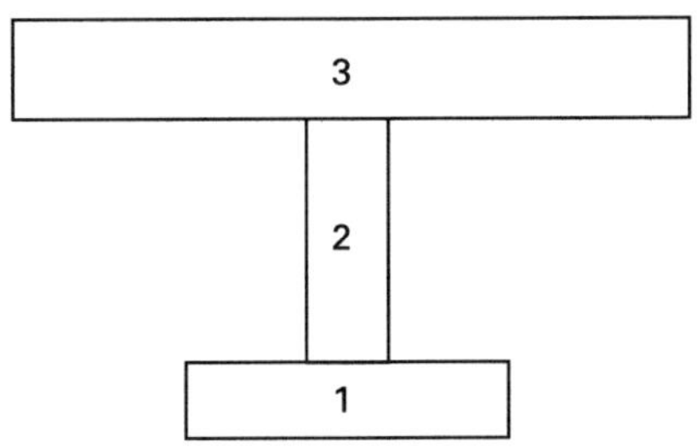

First, calculate the areas of the basic shapes.

$$\begin{aligned} A_1 &= (4)(1) = 4\text{ units}^2 \\ A_2 &= (2)(12) = 24\text{ units}^2 \\ A_3 &= (6)(0.5) = 3\text{ units}^2 \end{aligned}$$

Next, find the y-components of the centroids of the basic shapes.

$$\begin{aligned} y_{c1} &= 0.5\text{ units} \\ y_{c2} &= 7\text{ units} \\ y_{c3} &= 13.25\text{ units} \end{aligned}$$

From Eq. 2.6, the centroid of the area is

$$\begin{aligned} y_c &= \frac{\sum A_i y_{ci}}{\sum A_i} = \frac{(4)(0.5) + (24)(7) + (3)(13.25)}{4 + 24 + 3} \\ &= 6.77 \text{ units} \end{aligned}$$

From App. 2.A, the moment of inertia of basic shape 1 about its own centroid is

$$I_{cx1} = \frac{bh^3}{12} = \frac{(4)(1)^3}{12} = 0.33 \text{ units}^4$$

The moment of inertia of basic shape 2 about its own centroid is

$$I_{cx2} = \frac{bh^3}{12} = \frac{(2)(12)^3}{12} = 288 \text{ units}^4$$

The moment of inertia of basic shape 3 about its own centroid is

$$I_{cx3} = \frac{bh^3}{12} = \frac{(6)(0.5)^3}{12} = 0.063 \text{ units}^4$$

From the parallel axis theorem, Eq. 2.20, the moment of inertia of basic shape 1 about the centroidal axis of the section is

$$\begin{aligned} I_{x1} &= I_{cx1} + A_1 d_1^2 = 0.33 + (4)(6.77 - 0.5)^2 \\ &= 157.6 \text{ units}^4 \end{aligned}$$

The moment of inertia of basic shape 2 about the centroidal axis of the section is

$$\begin{aligned} I_{x2} &= I_{cx2} + A_2 d_2^2 = 288 + (24)(7.0 - 6.77)^2 \\ &= 289.3 \text{ units}^4 \end{aligned}$$

The moment of inertia of basic shape 3 about the centroidal axis of the section is

$$\begin{aligned} I_{x3} &= I_{cx3} + A_3 d_3^2 = 0.063 + (3)(13.25 - 6.77)^2 \\ &= 126.0 \text{ units}^4 \end{aligned}$$

The total moment of inertia about the centroidal axis of the section is

$$\begin{aligned} I_x &= I_{x1} + I_{x2} + I_{x3} \\ &= 157.6 \text{ units}^4 + 289.3 \text{ units}^4 + 126.0 \text{ units}^4 \\ &= \boxed{572.9 \text{ units}^4} \end{aligned}$$

The answer is (C).

3 Algebra

PRACTICE PROBLEMS

Series

1. Calculate the following sum.

$$\sum_{j=1}^{5}[(j+1)^2-1]$$

(A) 15
(B) 24
(C) 35
(D) 85

Logarithms

2. If every 0.1 sec a quantity increases by 0.1% of its current value, calculate the doubling time.

(A) 14 sec
(B) 70 sec
(C) 690 sec
(D) 69,000 sec

SOLUTIONS

1. Let $S_n = (j+1)^2 - 1$.

For $j = 1$,

$$S_1 = (1+1)^2 - 1 = 3$$

For $j = 2$,

$$S_2 = (2+1)^2 - 1 = 8$$

For $j = 3$,

$$S_3 = (3+1)^2 - 1 = 15$$

For $j = 4$,

$$S_4 = (4+1)^2 - 1 = 24$$

For $j = 5$,

$$S_5 = (5+1)^2 - 1 = 35$$

Substituting the above expressions gives

$$\begin{aligned}\sum_{j=1}^{5}\left((j+1)^2-1\right) &= \sum_{j=1}^{5} S_j \\ &= S_1 + S_2 + S_3 + S_4 + S_5 \\ &= 3 + 8 + 15 + 24 + 35 \\ &= \boxed{85}\end{aligned}$$

The answer is (D).

2. Let n represent the number of elapsed periods of 0.1 sec, and let y_n represent the amount present after n periods.

y_0 represents the initial quantity.

$$\begin{aligned}y_1 &= 1.001y_0 \\ y_2 &= 1.001y_1 = (1.001)\big((1.001)y_0\big) = (1.001)^2 y_0\end{aligned}$$

Therefore, by deduction,

$$y_n = (1.001)^n y_0$$

The expression for a doubling of the original quantity is

$$2y_0 = y_n$$

Substitute for y_n.

$$\begin{aligned} 2y_0 &= (1.001)^n y_0 \\ 2 &= (1.001)^n \end{aligned}$$

Take the logarithm of both sides.

$$\begin{aligned} \log(2) &= \log(1.001)^n \\ &= n \log(1.001) \end{aligned}$$

Solve for n.

$$n = \frac{\log(2)}{\log(1.001)} = 693.5$$

Since each period is 0.1 sec, the time is given by

$$\begin{aligned} t &= n(0.1\ \text{sec}) \\ &= (693.5)(0.1\ \text{sec}) = \boxed{69.35\ \text{sec}} \end{aligned}$$

The answer is (B).

4 Linear Algebra

PRACTICE PROBLEMS

Determinants

1. What is the determinant of matrix **A**?

$$\mathbf{A} = \begin{bmatrix} 8 & 2 & 0 & 0 \\ 2 & 8 & 2 & 0 \\ 0 & 2 & 8 & 2 \\ 0 & 0 & 2 & 4 \end{bmatrix}$$

(A) 459
(B) 832
(C) 1552
(D) 1776

Simultaneous Linear Equations

2. Use Cramer's rule to solve for the values of $x, y,$ and z that simultaneously satisfy the following equations.

$$x + y = -4$$
$$x + z - 1 = 0$$
$$2z - y + 3x = 4$$

(A) $(x, y, z) = (3, 2, 1)$
(B) $(x, y, z) = (-3, -1, 2)$
(C) $(x, y, z) = (3, -1, -3)$
(D) $(x, y, z) = (-1, -3, 2)$

SOLUTIONS

1. Expand by cofactors of the first row since there are two zeros in that row.

$$D = 8\begin{vmatrix} 8 & 2 & 0 \\ 2 & 8 & 2 \\ 0 & 2 & 4 \end{vmatrix} - 2\begin{vmatrix} 2 & 0 & 0 \\ 2 & 8 & 2 \\ 0 & 2 & 4 \end{vmatrix} + 0 - 0$$

by first row:

$$\begin{vmatrix} 8 & 2 & 0 \\ 2 & 8 & 2 \\ 0 & 2 & 4 \end{vmatrix} = (8)[(8)(4) - (2)(2)] - (2)[(2)(4) - (2)(0)] = (8)(28) - (2)(8) = 208$$

by first row:

$$\begin{vmatrix} 2 & 0 & 0 \\ 2 & 8 & 2 \\ 0 & 2 & 4 \end{vmatrix} = (2)[(8)(4) - (2)(2)] = 56$$

$$D = (8)(208) - (2)(56) = \boxed{1552}$$

The answer is (C).

2. Rearrange the equations.

$$\begin{aligned} x + y \quad &= -4 \\ x \quad + z &= 1 \\ 3x - y + 2z &= 4 \end{aligned}$$

Write the set of equations in matrix form: $\mathbf{AX} = \mathbf{B}$.

$$\begin{bmatrix} 1 & 1 & 0 \\ 1 & 0 & 1 \\ 3 & -1 & 2 \end{bmatrix} \begin{bmatrix} x \\ y \\ z \end{bmatrix} = \begin{bmatrix} -4 \\ 1 \\ 4 \end{bmatrix}$$

Find the determinant of the matrix **A**.

$$\begin{aligned} |\mathbf{A}| &= \begin{vmatrix} 1 & 1 & 0 \\ 1 & 0 & 1 \\ 3 & -1 & 2 \end{vmatrix} \\ &= 1\begin{vmatrix} 0 & 1 \\ -1 & 2 \end{vmatrix} - 1\begin{vmatrix} 1 & 0 \\ -1 & 2 \end{vmatrix} + 3\begin{vmatrix} 1 & 0 \\ 0 & 1 \end{vmatrix} \\ &= (1)\big((0)(2) - (1)(-1)\big) \\ &\quad - (1)\big((1)(2) - (-1)(0)\big) \\ &\quad + (3)\big((1)(1) - (0)(0)\big) \\ &= (1)(1) - (1)(2) + (3)(1) \\ &= 1 - 2 + 3 \\ &= 2 \end{aligned}$$

Find the determinant of the substitutional matrix $\mathbf{A_1}$.

$$
\begin{aligned}
|\mathbf{A_1}| &= \begin{vmatrix} -4 & 1 & 0 \\ 1 & 0 & 1 \\ 4 & -1 & 2 \end{vmatrix} \\
&= -4\begin{vmatrix} 0 & 1 \\ -1 & 2 \end{vmatrix} - 1\begin{vmatrix} 1 & 0 \\ -1 & 2 \end{vmatrix} + 4\begin{vmatrix} 1 & 0 \\ 0 & 1 \end{vmatrix} \\
&= (-4)\big((0)(2) - (1)(-1)\big) \\
&\quad - (1)\big((1)(2) - (-1)(0)\big) \\
&\quad + (4)\big((1)(1) - (0)(0)\big) \\
&= (-4)(1) - (1)(2) + (4)(1) \\
&= -4 - 2 + 4 \\
&= -2
\end{aligned}
$$

Find the determinant of the substitutional matrix $\mathbf{A_2}$.

$$
\begin{aligned}
|\mathbf{A_2}| &= \begin{vmatrix} 1 & -4 & 0 \\ 1 & 1 & 1 \\ 3 & 4 & 2 \end{vmatrix} \\
&= 1\begin{vmatrix} 1 & 1 \\ 4 & 2 \end{vmatrix} - 1\begin{vmatrix} -4 & 0 \\ 4 & 2 \end{vmatrix} + 3\begin{vmatrix} -4 & 0 \\ 1 & 1 \end{vmatrix} \\
&= (1)\big((1)(2) - (4)(1)\big) \\
&\quad - (1)\big((-4)(2) - (4)(0)\big) \\
&\quad + (3)\big((-4)(1) - (1)(0)\big) \\
&= (1)(-2) - (1)(-8) + (3)(-4) \\
&= -2 + 8 - 12 \\
&= -6
\end{aligned}
$$

Find the determinant of the substitutional matrix $\mathbf{A_3}$.

$$
\begin{aligned}
|\mathbf{A_3}| &= \begin{vmatrix} 1 & 1 & -4 \\ 1 & 0 & 1 \\ 3 & -1 & 4 \end{vmatrix} \\
&= 1\begin{vmatrix} 0 & 1 \\ -1 & 4 \end{vmatrix} - 1\begin{vmatrix} 1 & -4 \\ -1 & 4 \end{vmatrix} + 3\begin{vmatrix} 1 & -4 \\ 0 & 1 \end{vmatrix} \\
&= (1)\big((0)(4) - (-1)(1)\big) \\
&\quad - (1)\big((1)(4) - (-1)(-4)\big) \\
&\quad + (3)\big((1)(1) - (0)(-4)\big) \\
&= (1)(1) - (1)(0) + (3)(1) \\
&= 1 - 0 + 3 \\
&= 4
\end{aligned}
$$

Use Cramer's rule.

$$x = \frac{|\mathbf{A_1}|}{|\mathbf{A}|} = \frac{-2}{2} = \boxed{-1}$$

$$y = \frac{|\mathbf{A_2}|}{|\mathbf{A}|} = \frac{-6}{2} = \boxed{-3}$$

$$z = \frac{|\mathbf{A_3}|}{|\mathbf{A}|} = \frac{4}{2} = \boxed{2}$$

The answer is (D).

5 Vectors

PRACTICE PROBLEMS

Chapter 5 of the *Environmental Engineering Reference Manual* contains no practice problems.

6 Trigonometry

PRACTICE PROBLEMS

1. A 5 lbm (5 kg) block sits on a 20° incline without slipping. (a) Draw the freebody with respect to axes parallel and perpendicular to the surface of the incline. (b) Determine the magnitude of the frictional force on the block.

(A) 1.71 lbf (16.8 N)
(B) 3.35 lbf (32.9 N)
(C) 4.70 lbf (46.1 N)
(D) 5.00 lbf (49.1 N)

SOLUTIONS

1. (a)

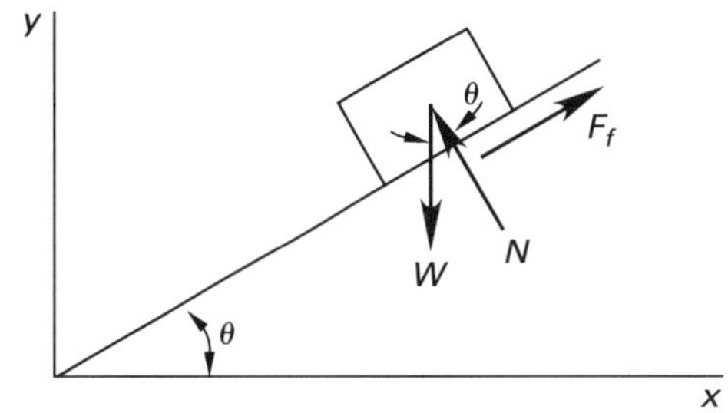

Customary U.S. Solution

(b) The mass of the block is $m = 5$ lbm.

The angle of inclination is $\theta = 20°$. The weight is

$$W = \frac{mg}{g_c} = \frac{(5 \text{ lbm})\left(32.2 \ \frac{\text{ft}}{\text{sec}^2}\right)}{32.2 \ \frac{\text{lbm-ft}}{\text{lbf-sec}^2}} = 5 \text{ lbf}$$

The frictional force is

$$F_f = W\sin\theta = (5 \text{ lbf})(\sin 20°) = \boxed{1.71 \text{ lbf}}$$

The answer is (A).

SI Solution

(b) The mass of the block is $m = 5$ kg.

The angle of inclination is $\theta = 20°$. The gravitational force is

$$W = mg = (5 \text{ kg})\left(9.81 \ \frac{\text{m}}{\text{s}^2}\right) = 49.1 \text{ N}$$

The frictional force is

$$F_f = W\sin\theta = (49.1 \text{ N})(\sin 20°) = \boxed{16.8 \text{ N}}$$

The answer is (A).

7 Analytic Geometry

PRACTICE PROBLEMS

1. The diameter of a sphere and the base of a cone are equal. What percentage of that diameter must the cone's height be so that both volumes are equal?

(A) 133%
(B) 150%
(C) 166%
(D) 200%

SOLUTIONS

1. Let d be the diameter of the sphere and the base of the cone.

The volume of the sphere is given by

$$\begin{aligned} V_{\text{sphere}} &= \left(\tfrac{4}{3}\right)\pi r^3 = \left(\tfrac{4}{3}\right)\pi\left(\frac{d}{2}\right)^3 \\ &= \left(\frac{\pi}{6}\right)d^3 \end{aligned}$$

The volume of the circular cone is given by

$$\begin{aligned} V_{\text{cone}} &= \left(\tfrac{1}{3}\right)\pi r^2 h = \left(\tfrac{1}{3}\right)\pi\left(\frac{d}{2}\right)^2 h \\ &= \left(\frac{\pi}{12}\right)d^2 h \end{aligned}$$

Since the volume of the sphere and cone are equal,

$$\begin{aligned} V_{\text{cone}} &= V_{\text{sphere}} \\ \left(\frac{\pi}{12}\right)d^2 h &= \left(\frac{\pi}{6}\right)d^3 \\ h &= 2d \end{aligned}$$

The height of the cone must be 200% of the diameter.

The answer is (D).

8 Differential Calculus

PRACTICE PROBLEMS

1. What are the values of a, b, and c in the following expression such that $n(\infty) = 100$, $n(0) = 10$, and $dn(0)/dt = 0.5$?

$$n(t) = \frac{a}{1 + be^{ct}}$$

(A) $a = 10,\ b = 9,\ c = 1$
(B) $a = 100,\ b = 10,\ c = 1.5$
(C) $a = 100,\ b = 9,\ c = -0.056$
(D) $a = 1000,\ b = 10,\ c = 0.056$

2. Find all minima, maxima, and inflection points for

$$y = x^3 - 9x^2 - 3$$

(A) maximum at $x = 0$
inflection at $x = 3$
minimum at $x = 6$

(B) maximum at $x = 0$
inflection at $x = -3$
minimum at $x = -6$

(C) minimum at $x = 0$
inflection at $x = 3$
maximum at $x = 6$

(D) minimum at $x = 3$
inflection at $x = 0$
maximum at $x = -3$

SOLUTIONS

1. If c is positive, then $n(\infty) = \infty$, which is contrary to the data given. Therefore, $c \leq 0$. If $c = 0$, then $n(\infty) = a/(1 + b) = 100$, which is possible depending on a and b. However, $n(0) = a/(1+b)$ would also equal 100, which is contrary to the given data. Therefore, $c \neq 0$.

c must be less than 0.

Since $c < 0$, then $n(\infty) = a$, so $\boxed{a = 100.}$

Applying the condition $t = 0$ gives

$$n(0) = \frac{a}{1+b} = 10$$

Since $a = 100$,

$$\begin{aligned} n(0) &= \frac{100}{1+b} \\ 100 &= (10)(1+b) \\ 10 &= 1+b \\ &\boxed{b = 9} \end{aligned}$$

Substitute the results for a and b into the expression.

$$n(t) = \frac{100}{1 + 9e^{ct}}$$

Take the first derivative.

$$\frac{d}{dt}n(t) = \left(\frac{100}{(1 + 9e^{ct})^2}\right)(-9ce^{ct})$$

Apply the initial condition.

$$\begin{aligned} \frac{d}{dt}n(0) = \left(\frac{100}{(1 + 9e^{c(0)})^2}\right)\left(- 9ce^{c(0)}\right) &= 0.5 \\ \left(\frac{100}{(1+9)^2}\right)(-9c) &= 0.5 \\ (1)(-9c) &= 0.5 \\ c &= \frac{-0.5}{9} \\ &= \boxed{-0.0556} \end{aligned}$$

The answer is (C).

Substitute the terms a, b, and c into the expression.

$$n(t) = \frac{100}{1 + 9e^{-0.0556t}}$$

2. Determine the critical points by taking the first derivative of the function and setting it equal to zero.

$$\begin{aligned} \frac{dy}{dx} &= 3x^2 - 18x = 3x(x-6) \\ 3x(x-6) &= 0 \\ x(x-6) &= 0 \end{aligned}$$

The critical points are located at $x = 0$ and $x = 6$.

Determine the inflection points by setting the second derivative equal to zero. Take the second derivative.

$$\begin{aligned} \frac{d^2y}{dx^2} &= \left(\frac{d}{dx}\right)\left(\frac{dy}{dx}\right) = \frac{d}{dx}(3x^2 - 18x) \\ &= 6x - 18 \end{aligned}$$

Set the second derivative equal to zero.

$$\begin{aligned} \frac{d^2y}{dx^2} &= 0 = 6x - 18 = (6)(x-3) \\ (6)(x-3) &= 0 \\ x - 3 &= 0 \\ x &= 3 \end{aligned}$$

This inflection point is at $x = 3$.

Determine the local maximum and minimum by substituting the critical points into the expression for the second derivative.

At the critical point $x = 0$,

$$\begin{aligned} \left.\frac{d^2y}{dx^2}\right|_{x=0} &= (6)(x-3) = (6)(0-3) \\ &= -18 \end{aligned}$$

Since $-18 < 0$, $x = 0$ is a local maximum.

At the critical point $x = 6$,

$$\begin{aligned} \left.\frac{d^2y}{dx^2}\right|_{x=6} &= (6)(x-3) = (6)(6-3) \\ &= 18 \end{aligned}$$

Since $18 > 0$, $x = 6$ is a local minimum.

The answer is (A).

9 Integral Calculus

PRACTICE PROBLEMS

Chapter 9 of the *Environmental Engineering Reference Manual* contains no practice problems.

10 Differential Equations

PRACTICE PROBLEMS

1. Solve the following differential equation for y.

$$y'' - 4y' - 12y = 0$$

(A) $A_1e^{6x} + A_2e^{-2x}$
(B) $A_1e^{-6x} + A_2e^{2x}$
(C) $A_1e^{6x} + A_2e^{2x}$
(D) $A_1e^{-6x} + A_2e^{-2x}$

2. Solve the following differential equation for y.

$$y' - y = 2xe^{2x} \qquad y(0) = 1$$

(A) $y = 2e^{-2x}(x-1) + 3e^{-x}$
(B) $y = 2e^{2x}(x-1) + 3e^{x}$
(C) $y = -2e^{-2x}(x-1) + 3e^{-x}$
(D) $y = 2e^{2x}(x-1) + 3e^{-x}$

3. The oscillation exhibited by the top story of a certain building in free motion is given by the following differential equation.

$$x'' + 2x' + 2x = 0 \qquad x(0) = 0 \qquad x'(0) = 1$$

(a) What is x as a function of time?

(A) $e^{-2t}\sin t$
(B) $e^{t}\sin t$
(C) $e^{-t}\sin t$
(D) $e^{-t}\sin t + e^{-t}\cos t$

(b) What is the building's fundamental natural frequency of vibration?

(A) $1/2$
(B) 1
(C) $\sqrt{2}$
(D) 2

(c) What is the amplitude of oscillation?

(A) 0.32
(B) 0.54
(C) 1.7
(D) 6.6

(d) What is x as a function of time if a lateral wind load is applied with a form of $\sin t$?

(A) $\frac{6}{5}e^{-t}\sin t + \frac{2}{5}e^{-t}\cos t$
(B) $\frac{6}{5}e^{t}\sin t + \frac{2}{5}e^{t}\cos t$
(C) $\frac{2}{5}e^{-t}\sin t + \frac{6}{5}e^{-t}\cos t + \frac{2}{5}\sin t - \frac{1}{5}\cos t$
(D) $\frac{6}{5}e^{-t}\sin t + \frac{2}{5}e^{-t}\cos t + \frac{1}{5}\sin t - \frac{2}{5}\cos t$

4. (*Time limit: one hour*) A 90 lbm (40 kg) bag of a chemical is accidentally dropped in an aerating lagoon. The chemical is water soluble and nonreacting. The lagoon is 120 ft (35 m) in diameter and filled to a depth of 10 ft (3 m). The aerators circulate and distribute the chemical evenly throughout the lagoon.

Water enters the lagoon at the rate of 30 gal/min (115 L/min). Fully mixed water is pumped into a reservoir at a rate of 30 gal/min (115 L/min).

The established safe concentration of this chemical is 1 ppb (part per billion). How many days will it take for the concentration of the discharge water to reach this level?

(A) 25 days
(B) 50 days
(C) 100 days
(D) 200 days

5. A tank contains 100 gal (100 L) of brine made by dissolving 60 lbm (60 kg) of salt in pure water. Salt water with a concentration of 1 lbm/gal (1 kg/L) enters the tank at a rate of 2 gal/min (2 L/min). A well-stirred mixture is drawn from the tank at a rate of 3 gal/min (3 L/min). Find the mass of salt in the tank after 1 hr.

(A) 13 lbm (13 kg)
(B) 37 lbm (37 kg)
(C) 43 lbm (43 kg)
(D) 51 lbm (51 kg)

SOLUTIONS

1. Obtain the characteristic equation by replacing each derivative with a polynomial term of equal degree.

$$r^2 - 4r - 12 = 0$$

Factor the characteristic equation.

$$(r-6)(r+2) = 0$$

The roots are $r_1 = 6$ and $r_2 = -2$.

Since the roots are real and distinct, the solution is

$$\begin{aligned} y &= A_1 e^{r_1 x} + A_2 e^{r_2 x} \\ &= \boxed{A_1 e^{6x} + A_2 e^{-2x}} \end{aligned}$$

The answer is (A).

2. The equation is a first-order linear differential equation of the form

$$\begin{aligned} y' + p(x)y &= g(x) \\ p(x) &= -1 \\ g(x) &= 2xe^{2x} \end{aligned}$$

The integration factor $u(x)$ is given by

$$\begin{aligned} u(x) &= \exp\left(\int p(x)dx\right) \\ &= \exp\left(\int (-1)dx\right) \\ &= e^{-x} \end{aligned}$$

The closed form of the solution is given by

$$\begin{aligned} y &= \left(\frac{1}{u(x)}\right)\left(\int u(x)g(x)dx + C\right) \\ &= \left(\frac{1}{e^{-x}}\right)\left(\int (e^{-x})(2xe^{2x})dx + C\right) \\ &= e^x\left(2(xe^x - e^x) + C\right) \\ &= e^x\left(2e^x(x-1) + C\right) \\ &= 2e^{2x}(x-1) + Ce^x \end{aligned}$$

Apply the initial condition $y(0) = 1$ to obtain the integration constant C.

$$\begin{aligned} y(0) = 2e^{(2)(0)}(0-1) + Ce^0 &= 1 \\ (2)(1)(-1) + C(1) &= 1 \\ -2 + C &= 1 \\ C &= 3 \end{aligned}$$

Substituting in the value for the integration constant C, the solution is

$$y = \boxed{2e^{2x}(x-1) + 3e^x}$$

The answer is (B).

3. (a) The differential equation is a homogeneous second-order linear differential equation with constant coefficients. Write the characteristic equation.

$$r^2 + 2r + 2 = 0$$

This is a quadratic equation of the form $ar^2 + br + c = 0$ where $a = 1$, $b = 2$, and $c = 2$.

Solve for r.

$$\begin{aligned} r &= \frac{-b \pm \sqrt{b^2 - 4ac}}{2a} \\ &= \frac{-2 \pm \sqrt{(2)^2 - (4)(1)(2)}}{(2)(1)} \\ &= \frac{-2 \pm \sqrt{4-8}}{2} \\ &= \frac{-2 \pm \sqrt{-4}}{2} \\ &= \frac{-2 \pm 2\sqrt{-1}}{2} \\ &= -1 \pm \sqrt{-1} \\ &= -1 \pm i \\ r_1 &= -1 + i \text{ and } r_2 = -1 - i \end{aligned}$$

Since the roots are imaginary and of the form $\alpha + i\omega$ and $\alpha - i\omega$ where $\alpha = -1$ and $\omega = 1$, the general form of the solution is given by

$$\begin{aligned} x(t) &= A_1 e^{\alpha t}\cos \omega t + A_2 e^{\alpha t}\sin \omega t \\ &= A_1 e^{-1t}\cos(1t) + A_2 e^{-1t}\sin(1t) \\ &= A_1 e^{-t}\cos t + A_2 e^{-t}\sin t \end{aligned}$$

Apply the initial conditions $x(0) = 0$ and $x'(0) = 1$ to solve for A_1 and A_2.

First, apply the initial condition $x(0) = 0$.

$$\begin{aligned} x(t) = A_1 e^0 \cos 0 + A_2 e^0 \sin 0 &= 0 \\ A_1(1)(1) + A_2(1)(0) &= 0 \\ A_1 &= 0 \end{aligned}$$

Substituting, the solution of the differential equation becomes

$$x(t) = A_2 e^{-t}\sin t$$

To apply the second initial condition, take the first derivative.

$$\begin{aligned} x'(t) &= \frac{d}{dt}(A_2e^{-t}\sin t) \\ &= A_2\frac{d}{dt}(e^{-t}\sin t) \\ &= A_2\left(\sin t\frac{d}{dt}(e^{-t}) + e^{-t}\frac{d}{dt}\sin t\right) \\ &= A_2\Big(\sin t(-e^{-t}) + e^{-t}(\cos t)\Big) \\ &= A_2(e^{-t})(-\sin t + \cos t) \end{aligned}$$

Apply the initial condition, $x'(0) = 1$.

$$\begin{aligned} x(0) = A_2(e^0)(-\sin 0 + \cos 0) &= 1 \\ A_2(1)(0+1) &= 1 \\ A_2 &= 1 \end{aligned}$$

The solution is

$$\begin{aligned} x(t) &= A_2e^{-t}\sin t \\ &= (1)e^{-t}\sin t \\ &= \boxed{e^{-t}\sin t} \end{aligned}$$

The answer is (C).

(b) To determine the natural frequency, set the damping term to zero. The equation has the form

$$x'' + 2x = 0$$

This equation has a general solution of the form

$$x(t) = x_0\cos\omega t + \left(\frac{\mathrm{v}_0}{\omega}\right)\sin\omega t$$

ω is the natural frequency. Given the equation $x''+2x = 0$, the characteristic equation is

$$\begin{aligned} r^2 + 2 &= 0 \\ r &= \sqrt{-2} \\ &= \pm\sqrt{2}i \end{aligned}$$

Since the roots are imaginary and of the form $\alpha + i\omega$ and $\alpha - i\omega$ where $\alpha = 0$ and $\omega = \sqrt{2}$, the general form of the solution is given by

$$\begin{aligned} x(t) &= A_1e^{\alpha t}\cos\omega t + A_2e^{\alpha t}\sin\omega t \\ &= A_1e^{0t}\cos\sqrt{2}t + A_2e^{0t}\sin\sqrt{2}t \\ &= A_1(1)\cos\sqrt{2}t + A_2(1)\sin\sqrt{2}t \\ &= A_1\cos\sqrt{2}t + A_2\sin\sqrt{2}t \end{aligned}$$

Apply the initial conditions, $x(0) = 0$ and $x'(0) = 1$ to solve for A_1 and A_2. Applying the initial condition $x(0) = 0$ gives

$$\begin{aligned} x(0) = A_1\cos\left((\sqrt{2})(0)\right) + A_2\sin(\sqrt{2})(0) &= 0 \\ A_1\cos 0 + A_2\sin 0 &= 0 \\ A_1(1) + A_2(0) &= 0 \\ A_1 &= 0 \end{aligned}$$

Substituting, the solution of the different equation becomes

$$x(t) = A_2\sin\sqrt{2}t$$

To apply the second initial condition, take the first derivative.

$$\begin{aligned} x'(t) &= \frac{d}{dt}(A_2\sin\sqrt{2}t) \\ &= A_2\sqrt{2}\cos\sqrt{2}t \end{aligned}$$

Apply the second initial condition, $x'(0) = 1$.

$$\begin{aligned} x'(0) = A_2\sqrt{2}\cos(\sqrt{2})(0) &= 1 \\ A_2\sqrt{2}\cos(0) &= 1 \\ A_2(\sqrt{2})(1) &= 1 \\ A_2\sqrt{2} &= 1 \\ A_2 = \frac{1}{\sqrt{2}} &= \frac{\sqrt{2}}{2} \end{aligned}$$

Substituting, the undamped solution becomes

$$x(t) = \left(\frac{\sqrt{2}}{2}\right)\sin\sqrt{2}t$$

Therefore, the undamped natural frequency is $\boxed{\omega = \sqrt{2}.}$

The answer is (C).

(c) The amplitude of the oscillation is the maximum displacement.

Take the derivative of the solution, $x(t) = e^{-t}\sin t$.

$$\begin{aligned} x'(t) &= \frac{d}{dt}(e^{-t}\sin t) \\ &= \sin t\frac{d}{dt}(e^{-t}) + e^{-t}\frac{d}{dt}\sin t \\ &= \sin(t)(-e^{-t}) + e^{-t}\cos t \\ &= e^{-t}\cos t - \sin t \end{aligned}$$

The maximum displacement occurs at $x'(t) = 0$.

Since $e^{-t} \neq 0$ except as t approaches infinity,

$$\begin{aligned} \cos t - \sin t &= 0 \\ \tan t &= 1 \\ t &= \tan^{-1}(1) \\ &= 0.785 \text{ rad} \end{aligned}$$

At $t = 0.785$ rad, the displacement is maximum. Substitute into the orginal solution to obtain a value for the maximum displacement.

$$\begin{aligned} x(0.785) &= e^{-0.785}\sin(0.785) \\ &= 0.322 \end{aligned}$$

The amplitude is $\boxed{0.322.}$

The answer is (A).

(d) (An alternative solution using Laplace transforms follows this solution.) The application of a lateral wind load with the form $\sin t$ revises the differential equation to the form

$$x'' + 2x' + 2x = \sin t$$

Express the solution as the sum of the complementary x_c and particular x_p solutions.

$$x(t) = x_c(t) + x_p(t)$$

From part (a),

$$x_c(t) = A_1 e^{-t}\cos t + A_2 e^{-t}\sin t$$

The general form of the particular solution is given by

$$x_p(t) = x^s(A_3\cos t + A_4\sin t)$$

Determine the value of s; check to see if the terms of the particular solution solve the homogeneous equation.

Examine the term $A_3\cos(t)$.

Take the first derivative.

$$\frac{d}{dx}(A_3\cos t) = -A_3\sin t$$

Take the second derivative.

$$\begin{aligned} \frac{d}{dx}\left(\frac{d}{dx}(A_3\cos t)\right) &= \frac{d}{dx}(-A_3\sin t) \\ &= -A_3\cos t \end{aligned}$$

Substitute the terms into the homogeneous equation.

$$\begin{aligned} x'' + 2x' + 2x &= -A_3\cos t + (2)(-A_3\sin t) \\ &\quad + (2)(-A_3\cos t) \\ &= A_3\cos t - 2A_3\sin t \\ &\neq 0 \end{aligned}$$

Except for the trival solution $A_3 = 0$, the term $A_3\cos t$ does not solve the homogeneous equation.

Examine the second term $A_4\sin t$.

Take the first derivative.

$$\frac{d}{dx}(A_4\sin t) = A_4\cos t$$

Take the second derivative.

$$\begin{aligned} \frac{d}{dx}\left(\frac{d}{dx}(A_4\sin t)\right) &= \frac{d}{dx}(A_4\cos t) \\ &= -A_4\sin t \end{aligned}$$

Substitute the terms into the homogeneous equation.

$$\begin{aligned} x'' + 2x' + 2x &= -A_4\sin t + (2)(A_4\cos t) \\ &\quad + (2)(A_4\sin t) \\ &= A_4\sin t + 2A_4\cos t \\ &\neq 0 \end{aligned}$$

Except for the trival solution $A_4 = 0$, the term $A_4\sin t$ does not solve the homogeneous equation.

Neither of the terms satisfies the homogeneous equation $s = 0$; therefore, the particular solution is of the form

$$x_p(t) = A_3\cos t + A_4\sin t$$

Use the method of undetermined coefficients to solve for A_3 and A_4. Take the first derivative.

$$\begin{aligned} x_p'(t) &= \frac{d}{dx}(A_3\cos t + A_4\sin t) \\ &= -A_3\sin t + A_4\cos t \end{aligned}$$

Take the second derivative.

$$\begin{aligned} x_p''(t) &= \frac{d}{dx}\left(\frac{d}{dx}(A_3\cos t + A_4\sin t)\right) \\ &= \frac{d}{dx}(-A_3\sin t + A_4\cos t) \\ &= -A_3\cos t - A_4\sin t \end{aligned}$$

Substitute the expressions for the derivatives into the differential equation.

$$\begin{aligned} x'' + 2x' + 2x &= (-A_3\cos t - A_4\sin t) \\ &\quad + (2)(-A_3\sin t + A_4\cos t) \\ &\quad + (2)(A_3\cos t + A_4\sin t) \\ &= \sin t \end{aligned}$$

Rearranging terms gives

$$\begin{aligned} &(-A_3 + 2A_4 + 2A_3)\cos t \\ &\qquad +(-A_4 - 2A_3 + 2A_4)\sin t = \sin t \\ &(A_3 + 2A_4)\cos t + (-2A_3 + A_4)\sin t = \sin t \end{aligned}$$

Equating coefficients gives

$$\begin{aligned} A_3 + 2A_4 &= 0 \\ -2A_3 + A_4 &= 1 \end{aligned}$$

Multiplying the first equation by 2 and adding equations gives

$$\begin{aligned} A_3 + 2A_4 &= 0 \\ +(-2A_3 + A_4) &= 1 \\ \hline 5A_4 = 1 \text{ or } A_4 &= \tfrac{1}{5} \end{aligned}$$

From the first equation for $A_4 = {}^1\!/_5$, $A_3 + (2)({}^1\!/_5) = 0$ and $A_3 = -{}^2\!/_5$.

Substituting for the coefficients, the particular solution becomes

$$x_p(t) = -\tfrac{2}{5}\cos t + \tfrac{1}{5}\sin t$$

Combining the complementary and particular solutions gives

$$\begin{aligned} x(t) &= x_c(t) + x_p(t) \\ &= A_1 e^{-t}\cos t + A_2 e^{-t}\sin t - \tfrac{2}{5}\cos t + \tfrac{1}{5}\sin t \end{aligned}$$

Apply the initial conditions to solve for the coefficients A_1 and A_2; then apply the first initial condition, $x(0) = 0$.

$$\begin{aligned} x(t) = A_1 e^0 \cos 0 + A_2 e^0 \sin 0 & \\ - \tfrac{2}{5}\cos 0 + \tfrac{1}{5}\sin 0 &= 0 \\ A_1(1)(1) + A_2(1)(0) + \left(-\tfrac{2}{5}\right)(1) + \left(\tfrac{1}{5}\right)(0) &= 0 \\ A_1 - \tfrac{2}{5} &= 0 \\ A_1 &= \tfrac{2}{5} \end{aligned}$$

Substituting for A_1, the solution becomes

$$x(t) = \tfrac{2}{5}e^{-t}\cos t + A_2 e^{-t}\sin t - \tfrac{2}{5}\cos t + \tfrac{1}{5}\sin t$$

Take the first derivative.

$$\begin{aligned} x'(t) &= \frac{d}{dx}\left(\tfrac{2}{5}e^{-t}\cos t + A_2 e^{-t}\sin t\right) \\ &\quad + \left(\left(-\tfrac{2}{5}\right)\cos t + \tfrac{1}{5}\sin t\right) \\ &= \left(\tfrac{2}{5}\right)\left(-e^{-t}\cos t - e^{-t}\sin t\right) \\ &\quad + A_2\left(-e^{-t}\sin t + e^{-t}\cos t\right) \\ &\quad + \left(-\tfrac{2}{5}\right)(-\sin t) + \tfrac{1}{5}\cos t \end{aligned}$$

Apply the second initial condition, $x'(0) = 1$.

$$\begin{aligned} x'(0) &= \left(\tfrac{2}{5}\right)\left(-e^0\cos 0 - e^0 \sin 0\right) \\ &\quad + A_2\left(-e^0\sin 0 + e^0\cos 0\right) \\ &\quad + \left(-\tfrac{2}{5}\right)(-\sin 0) + \tfrac{1}{5}\cos 0 \\ &= 1 \end{aligned}$$

$$\begin{aligned} \left(\tfrac{2}{5}\right)\left(-(1)(1) - (1)(0)\right) + A_2\big(-(1)(0) & \\ +(1)(1)\big) + \left(-\tfrac{2}{5}\right)(0) + \left(\tfrac{1}{5}\right)(1) &= 1 \\ \left(\tfrac{2}{5}\right)(-1) + A_2(1) + \left(\tfrac{1}{5}\right) &= 1 \\ A_2 &= \tfrac{6}{5} \end{aligned}$$

Substituting for A_2, the solution becomes

$$x(t) = \boxed{\begin{array}{l} \tfrac{2}{5}e^{-t}\cos t + \tfrac{6}{5}e^{-t}\sin t \\ -\tfrac{2}{5}\cos t + \tfrac{1}{5}\sin t \end{array}}$$

The answer is (D).

(d) *Alternative solution:*

Use the Laplace transform method.

$$\begin{aligned} x'' + 2x' + 2x &= \sin t \\ \mathcal{L}(x'') + 2\mathcal{L}(x') + 2\mathcal{L}(x) &= \mathcal{L}(\sin t) \\ s^2\mathcal{L}(x) - 1 + 2s\mathcal{L}(x) + 2\mathcal{L}(x) &= \frac{1}{s^2+1} \\ \mathcal{L}(x)(s^2 + 2s + 2) - 1 &= \frac{1}{s^2+1} \end{aligned}$$

$$\begin{aligned} \mathcal{L}(x) &= \frac{1}{s^2+2s+2} + \frac{1}{(s^2+1)(s^2+2s+2)} \\ &= \frac{1}{(s+1)^2+1} + \frac{1}{(s^2+1)(s^2+2s+2)} \end{aligned}$$

Use partial fractions to expand the second term.

$$\frac{1}{(s^2+1)(s^2+2s+2)} = \frac{A_1 + B_1 s}{s^2+1} + \frac{A_2 + B_2 s}{s^2+2s+2}$$

Cross multiply.

$$= \frac{\begin{array}{r} A_1 s^2 + 2A_1 s + 2A_1 + B_1 s^3 + 2B_1 s^2 \\ +2B_1 s + A_2 s^2 + A_2 + B_2 s^3 + B_2 s \end{array}}{(s^2+1)(s^2+2s+2)}$$

$$= \frac{\begin{array}{r} s^3(B_1 + B_2) + s^2(A_1 + A_2 + 2B_1) \\ +s(2A_1 + 2B_1 + B_2) + 2A_1 + A_2 \end{array}}{(s^2+1)(s^2+2s+2)}$$

Compare numerators to obtain the following four simultaneous equations.

$$\begin{array}{rcrcrcrcl} & & & & B_1 & + & B_2 & = & 0 \\ A_1 & + & A_2 & + & 2B_1 & & & = & 0 \\ 2A_1 & & & + & 2B_1 & + & B_2 & = & 0 \\ 2A_1 & + & A_2 & & & & & = & 1 \end{array}$$

Use Cramer's rule to find A_1.

$$A_1 = \frac{\begin{vmatrix} 0 & 0 & 1 & 1 \\ 0 & 1 & 2 & 0 \\ 0 & 0 & 2 & 1 \\ 1 & 1 & 0 & 0 \end{vmatrix}}{\begin{vmatrix} 0 & 0 & 1 & 1 \\ 1 & 1 & 2 & 0 \\ 2 & 0 & 2 & 1 \\ 2 & 1 & 0 & 0 \end{vmatrix}} = \frac{-1}{-5} = \frac{1}{5}$$

The rest of the coefficients are found similarly.

$$A_1 = \tfrac{1}{5}$$
$$A_2 = \tfrac{3}{5}$$
$$B_1 = -\tfrac{2}{5}$$
$$B_2 = \tfrac{2}{5}$$

Then,

$$\mathcal{L}(x) = \frac{1}{(s+1)^2+1} + \frac{\frac{1}{5}}{s^2+1} + \frac{-\frac{2}{5}s}{s^2+1} + \frac{\frac{3}{5}}{s^2+2s+2} + \frac{\frac{2}{5}s}{s^2+2s+2}$$

Take the inverse transform.

$$\begin{aligned} x(t) &= \mathcal{L}^{-1}\{\mathcal{L}(x)\} \\ &= e^{-t}\sin t + \tfrac{1}{5}\sin t - \tfrac{2}{5}\cos t + \tfrac{3}{5}e^{-t}\sin t \\ &\quad + \tfrac{2}{5}(e^{-t}\cos t - e^{-t}\sin t) \\ &= \boxed{\tfrac{6}{5}e^{-t}\sin t + \tfrac{2}{5}e^{-t}\cos t + \tfrac{1}{5}\sin t - \tfrac{2}{5}\cos t} \end{aligned}$$

The answer is (D).

4. *Customary U.S. Solution*

The differential equation is given as

$$m'(t) = a(t) - \frac{m(t)o(t)}{V(t)}$$

$a(t)$ = rate of addition of chemical
$m(t)$ = mass of chemical at time t
$o(t)$ = volumetric flow out of the lagoon (= 30 gal/min)
$V(t)$ = volume in the lagoon at time t

Water flows into the lagoon at a rate of 30 gal/min, and a water-chemical mix flows out of the lagoon at a rate of 30 gal/min. Therefore, the volume of the lagoon at time t is equal to the initial volume.

$$\begin{aligned} V(t) &= \left(\frac{\pi}{4}\right)(\text{diameter of lagoon})^2(\text{depth of lagoon}) \\ &= \left(\frac{\pi}{4}\right)(120\ \text{ft})^2(10\ \text{ft}) \\ &= 113{,}097\ \text{ft}^3 \end{aligned}$$

Use a conversion factor of 7.48 gal/ft^3.

$$\begin{aligned} o(t) &= \frac{30\ \frac{\text{gal}}{\text{min}}}{7.48\ \frac{\text{gal}}{\text{ft}^3}} \\ &= 4.01\ \text{ft}^3/\text{min} \end{aligned}$$

Substituting into the general form of the differential equation gives

$$\begin{aligned} m'(t) &= a(t) - \frac{m(t)o(t)}{V(t)} \\ &= (0) - m(t)\left(\frac{4.01\ \frac{\text{ft}^3}{\text{min}}}{113{,}097\ \text{ft}^3}\right) \\ &= -\left(\frac{3.55\times 10^{-5}}{\text{min}}\right)m(t) \end{aligned}$$

$$m'(t) + \left(\frac{3.55\times 10^{-5}}{\text{min}}\right)m(t) = 0$$

The differential equation of the problem has a characteristic equation.

$$\begin{aligned} r + \frac{3.55\times 10^{-5}}{\text{min}} &= 0 \\ r &= -3.55\times 10^{-5}/\text{min} \end{aligned}$$

The general form of the solution is given by

$$m(t) = Ae^{rt}$$

Substituting for the root, r, gives

$$m(t) = Ae^{\left(\frac{-3.55\times 10^{-5}}{\text{min}}\right)t}$$

Apply the initial condition $m(0) = 90$ lbm at time $t = 0$.

$$\begin{aligned} m(0) = Ae^{\left(\frac{-3.55\times 10^{-5}}{\text{min}}\right)(0)} &= 90\ \text{lbm} \\ Ae^0 &= 90\ \text{lbm} \\ A &= 90\ \text{lbm} \end{aligned}$$

Therefore,

$$m(t) = (90\ \text{lbm})\,e^{\left(\frac{-3.55\times 10^{-5}}{\text{min}}\right)t}$$

Solve for t.

$$\begin{aligned} \frac{m(t)}{90\ \text{lbm}} &= e^{\left(\frac{-3.55\times 10^{-5}}{\text{min}}\right)t} \\ \ln\left(\frac{m(t)}{90\ \text{lbm}}\right) &= \ln\left(e^{\left(\frac{-3.55\times 10^{-5}}{\text{min}}\right)t}\right) \\ &= \left(\frac{-3.55\times 10^{-5}}{\text{min}}\right)t \\ t &= \frac{\ln\left(\frac{m(t)}{90\ \text{lbm}}\right)}{\frac{-3.55\times 10^{-5}}{\text{min}}} \end{aligned}$$

The initial mass of the water in the lagoon is given by

$$\begin{aligned} m_i &= V\rho \\ &= (113{,}097\ \text{ft}^3)\left(62.4\ \frac{\text{lbm}}{\text{ft}^3}\right) \\ &= 7.05\times 10^6\ \text{lbm} \end{aligned}$$

The final mass of chemicals is achieved at a concentration of 1 ppb or

$$\begin{aligned} m_f &= \frac{7.06\times 10^6\ \text{lbm}}{1\times 10^9} \\ &= 7.06\times 10^{-3}\ \text{lbm} \end{aligned}$$

Find the time required to achieve a mass of 7.06×10^{-3} lbm.

$$\begin{aligned} t &= \left(\frac{\ln\left(\frac{m(t)}{90\ \text{lbm}}\right)}{\frac{-3.55\times 10^{-5}}{\text{min}}}\right)\left(\frac{1\ \text{hr}}{60\ \text{min}}\right)\left(\frac{1\ \text{day}}{24\ \text{hr}}\right) \\ &= \left(\frac{\ln\left(\frac{7.06\times 10^{-3}\ \text{lbm}}{90\ \text{lbm}}\right)}{\frac{-3.55\times 10^{-5}}{\text{min}}}\right)\left(\frac{1\ \text{hr}}{60\ \text{min}}\right)\left(\frac{1\ \text{day}}{24\ \text{hr}}\right) \\ &= \boxed{185\ \text{days}} \end{aligned}$$

The answer is (D).

SI Solution

The differential equation is given as

$$m'(t) = a(t) - \frac{m(t)o(t)}{V(t)}$$

$a(t)$ = rate of addition of chemical
$m(t)$ = mass of chemical at time t
$o(t)$ = volumetric flow out of the lagoon (= 115 L/min)
$V(t)$ = volume in the lagoon at time t

Water flows into the lagoon at a rate of 115 L/min, and a water-chemical mix flows out of the lagoon at a rate of 115 L/min. Therefore, the volume of the lagoon at time t is equal to the initial volume.

$$\begin{aligned} V(t) &= \left(\frac{\pi}{4}\right)(\text{diameter of lagoon})^2(\text{depth of lagoon}) \\ &= \left(\frac{\pi}{4}\right)(35\ \text{m})^2(3\ \text{m}) \\ &= 2886\ \text{m}^3 \end{aligned}$$

Using a conversion factor of 1 m^3/1000 L gives

$$\begin{aligned} o(t) &= \left(115\ \frac{\text{L}}{\text{min}}\right)\left(\frac{1\ \text{m}^3}{1000\ \text{L}}\right) \\ &= 0.115\ \text{m}^3/\text{min} \end{aligned}$$

Substitute into the general form of the differential equation.

$$\begin{aligned} m'(t) &= a(t) - \frac{m(t)o(t)}{V(t)} \\ &= 0 - m(t)\left(\frac{0.115\ \frac{\text{m}^3}{\text{min}}}{2886\ \text{m}^3}\right) \\ &= -\left(\frac{3.985\times 10^{-5}}{\text{min}}\right)m(t) \end{aligned}$$

$$m'(t) + \left(\frac{3.985\times 10^{-5}}{\text{min}}\right)m(t) = 0$$

The differential equation of the problem has the following characteristic equation.

$$\begin{aligned} r + \frac{3.985\times 10^{-5}}{\text{min}} &= 0 \\ r &= -3.985\times 10^{-5}/\text{min} \end{aligned}$$

The general form of the solution is given by

$$m(t) = Ae^{rt}$$

Substituting for the root, r, gives

$$m(t) = Ae^{\left(\frac{-3.985\times 10^{-5}}{\text{min}}\right)t}$$

Apply the initial condition $m(0) = 40$ kg at time $t = 0$.

$$\begin{aligned} m(0) = Ae^{\left(\frac{-3.985\times 10^{-5}}{\text{min}}\right)(0)} &= 40\ \text{kg} \\ Ae^0 &= 40\ \text{kg} \\ A &= 40\ \text{kg} \end{aligned}$$

Therefore,

$$m(t) = (40\ \text{kg})e^{\left(\frac{-3.985\times 10^{-5}}{\text{min}}\right)t}$$

Solve for t.

$$\begin{aligned} \frac{m(t)}{40\ \text{kg}} &= e^{\left(\frac{-3.985\times 10^{-5}}{\text{min}}\right)t} \\ \ln\left(\frac{m(t)}{40\ \text{kg}}\right) &= \ln\left(e^{\left(\frac{-3.985\times 10^{-5}}{\text{min}}\right)t}\right) \\ &= \left(\frac{-3.985\times 10^{-5}}{\text{min}}\right)t \\ t &= \ln\left(\frac{\frac{m(t)}{40\ \text{kg}}}{\frac{-3.985\times 10^{-5}}{\text{min}}}\right) \end{aligned}$$

The initial mass of water in the lagoon is given by

$$\begin{aligned} m_i &= V\rho \\ &= (2886\ \text{m}^3)\left(1000\ \frac{\text{kg}}{\text{m}^3}\right) \\ &= 2.886 \times 10^6\ \text{kg} \end{aligned}$$

The final mass of chemicals is achieved at a concentration of 1 ppb or

$$\begin{aligned} m_f &= \frac{2.886 \times 10^6\ \text{kg}}{1 \times 10^9} \\ &= 2.886 \times 10^{-3}\ \text{kg} \end{aligned}$$

Find the time required to achieve a mass of 2.886×10^{-3} kg.

$$\begin{aligned} t &= \left(\frac{\ln \dfrac{m(t)}{40\ \text{kg}}}{\dfrac{-3.985 \times 10^{-5}}{\text{min}}}\right)\left(\frac{1\ \text{h}}{60\ \text{min}}\right)\left(\frac{1\ \text{day}}{24\ \text{h}}\right) \\ &= \left(\frac{\ln\left(\dfrac{2.886 \times 10^{-3}\ \text{kg}}{40\ \text{kg}}\right)}{\dfrac{-3.985 \times 10^{-5}}{\text{min}}}\right)\left(\frac{1\ \text{h}}{60\ \text{min}}\right)\left(\frac{1\ \text{day}}{24\ \text{h}}\right) \\ &= \boxed{166\ \text{days}} \end{aligned}$$

The answer is (D).

5. Let

$$\begin{aligned} m(t) &= \text{mass of salt in tank at time } t \\ m_0 &= 60 \text{ mass units} \\ m'(t) &= \text{rate at which salt content is changing} \end{aligned}$$

2 mass units of salt enter each minute, and 3 volumes leave each minute. The amount of salt leaving each minute is

$$\begin{aligned} &\left(3\ \frac{\text{vol}}{\text{min}}\right)\left(\text{concentration in } \frac{\text{mass}}{\text{vol}}\right) \\ &\qquad = \left(3\ \frac{\text{vol}}{\text{min}}\right)\left(\frac{\text{salt content}}{\text{volume}}\right) \\ &\qquad = \left(3\ \frac{\text{vol}}{\text{min}}\right)\left(\frac{m(t)}{100 - t}\right) \end{aligned}$$

$$\begin{aligned} m'(t) &= 2 - (3)\left(\frac{m(t)}{100 - t}\right) \quad \text{or } m'(t) + \frac{3m(t)}{100 - t} \\ &= 2\ \text{mass/min} \end{aligned}$$

This is a first-order linear differential equation. The integrating factor is

$$\begin{aligned} m &= \exp\left[3\int \frac{dt}{100 - t}\right] \\ &= \exp\left[(3)\left(-\ln(100 - t)\right)\right] \\ &= (100 - t)^{-3} \end{aligned}$$

$$\begin{aligned} m(t) &= (100 - t)^3\left[2\int \frac{dt}{(100 - t)^3} + k\right] \\ &= 100 - t + (k)(100 - t)^3 \end{aligned}$$

But $m = 60$ mass units at $t = 0$, so $k = -0.00004$.

$$m(t) = 100 - t - (0.00004)(100 - t)^3$$

At $t = 60$ min,

$$\begin{aligned} m &= 100 - 60\ \text{min} - (0.00004)(100 - 60\ \text{min})^3 \\ &= \boxed{37.44\ \text{mass units}} \end{aligned}$$

The answer is (B).

11 Probability, Statistical Analysis of Data, and Risk Analysis

PRACTICE PROBLEMS

Probability

1. Four military recruits whose respective shoe sizes are 7, 8, 9, and 10 report to the supply clerk to be issued boots. The supply clerk selects one pair of boots in each of the four required sizes and hands them at random to the recruits.

(a) What is the probability that all recruits will receive boots of an incorrect size?

(A) 0.25
(B) 0.38
(C) 0.45
(D) 0.61

(b) What is the probability that exactly three recruits will receive boots of the correct size?

(A) 0
(B) 0.063
(C) 0.17
(D) 0.25

Probability Distributions

2. The time taken by a toll taker to collect the toll from vehicles crossing a bridge is an exponential distribution with a mean of 23 sec. What is the probability that a random vehicle will be processed in 25 sec or more (i.e., will take longer than 25 sec)?

(A) 0.17
(B) 0.25
(C) 0.34
(D) 0.52

3. The number of cars entering a toll plaza on a bridge during the hour after midnight follows a Poisson distribution with a mean of 20.

(a) What is the probability that 17 cars will pass through the toll plaza during that hour on any given night?

(A) 0.076
(B) 0.12
(C) 0.16
(D) 0.23

(b) What is the probability that three or fewer cars will pass through the toll plaza at that hour on any given night?

(A) 0.00032
(B) 0.0019
(C) 0.079
(D) 0.11

4. A mechanical component exhibits a negative exponential failure distribution with a mean time to failure of 1000 hr. What is the maximum operating time such that the reliability remains above 99%?

(A) 3.3 hr
(B) 5.6 hr
(C) 8.1 hr
(D) 10 hr

5. (*Time limit: one hour*) A survey field crew measures one leg of a traverse four times. The following results are obtained.

repetition	measurement	direction
1	1249.529	forward
2	1249.494	backward
3	1249.384	forward
4	1249.348	backward

The crew chief is under orders to obtain readings with confidence limits of 90%.

(a) Which readings are acceptable?

(A) No readings are acceptable.
(B) Two readings are acceptable.
(C) Three readings are acceptable.
(D) All four readings are acceptable.

(b) Which readings are not acceptable?

(A) No readings are unacceptable.
(B) One reading is unacceptable.
(C) Two readings are unacceptable.
(D) All four readings are unacceptable.

(c) Explain how to determine which readings are not acceptable.

(A) Readings inside the 90% confidence limits are unacceptable.

(B) Readings outside the 90% confidence limits are unacceptable.

(C) Readings outside the upper 90% confidence limit are unacceptable.

(D) Readings outside the lower 90% confidence limit are unacceptable.

(d) What is the most probable value of the distance?

(A) 1249.399

(B) 1249.410

(C) 1249.439

(D) 1249.452

(e) What is the error in the most probable value (at 90% confidence)?

(A) 0.08

(B) 0.11

(C) 0.14

(D) 0.19

(f) If the distance is one side of a square traverse whose sides are all equal, what is the most probable closure error?

(A) 0.14

(B) 0.20

(C) 0.28

(D) 0.35

(g) What is the probable error of part (f) expressed as a fraction?

(A) 1:17,600

(B) 1:14,200

(C) 1:12,500

(D) 1:10,900

(h) What is the order of accuracy of the closure?

(A) first-order

(B) second-order

(C) third-order

(D) fourth-order

(i) Define accuracy and distinguish it from precision.

(A) If an experiment can be repeated with identical results, the results are considered accurate.

(B) If an experiment has a small bias, the results are considered precise.

(C) If an experiment is precise, it cannot also be accurate.

(D) If an experiment is unaffected by experimental error, the results are accurate.

(j) Give an example of systematic error.

(A) measuring river depth as a motorized ski boat passes by

(B) using a steel tape that is too short to measure consecutive distances

(C) locating magnetic north near a large iron ore deposit along an overland route

(D) determining local wastewater BOD after a toxic spill

Statistical Analysis

6. *(Time limit: one hour)* California law requires a statistical analysis of the average speed driven by motorists on a road prior to the use of radar speed control. The following speeds (all in mi/hr) were observed in a random sample of 40 cars.

44, 48, 26, 25, 20, 43, 40, 42, 29, 39, 23, 26, 24, 47, 45, 28, 29, 41, 38, 36, 27, 44, 42, 43, 29, 37, 34, 31, 33, 30, 42, 43, 28, 41, 29, 36, 35, 30, 32, 31

(a) Tabulate the frequency distribution of the data.

(b) Draw the frequency histogram.

(c) Draw the frequency polygon.

(d) Tabulate the cumulative frequency distribution.

(e) Draw the cumulative frequency graph.

(f) What is the upper quartile speed?

(A) 30 mph

(B) 35 mph

(C) 40 mph

(D) 45 mph

(g) What is the median speed?

(A) 31 mph

(B) 33 mph

(C) 35 mph

(D) 37 mph

(h) What is the standard deviation of the sample data?

(A) 2.1 mph

(B) 6.1 mph

(C) 6.8 mph

(D) 7.4 mph

(i) What is the sample standard deviation?

(A) 7.5 mph

(B) 18 mph

(C) 35 mph

(D) 56 mph

(j) What is the sample variance?

(A) 56 mi^2/hr^2

(B) 324 mi^2/hr^2

(C) 1225 mi^2/hr^2

(D) 3136 mi^2/hr^2

7. A spot speed study is conducted for a stretch of roadway. During a normal day, the speeds were found to be normally distributed with a mean of 46 and a standard deviation of 3.

(a) What is the 50th percentile speed?

(A) 39
(B) 43
(C) 46
(D) 49

(b) What is the 85th percentile speed?

(A) 47.1
(B) 48.3
(C) 49.1
(D) 52.7

(c) What is the upper two standard deviation speed?

(A) 47.2
(B) 49.3
(C) 51.1
(D) 52.0

(d) The daily average speeds for the same stretch of roadway on consecutive normal days were determined by sampling 25 vehicles each day. What is the upper two standard deviation average speed?

(A) 46.6
(B) 47.2
(C) 52.0
(D) 54.7

8. The diameters of bolt holes drilled in structural steel members are normally distributed with a mean of 0.502 in and a standard deviation of 0.005 in. Holes are out of specification if their diameters are less than 0.497 in or more than 0.507 in.

(a) What is the probability that a hole chosen at random will be out of specification?

(A) 0.16
(B) 0.22
(C) 0.32
(D) 0.68

(b) What is the probability that 2 holes out of a sample of 15 will be out of specification?

(A) 0.074
(B) 0.12
(C) 0.15
(D) 0.32

9. Two resistances, the meter resistance and a shunt resistor, are connected in parallel in an ammeter. Most of the current passing through the meter goes through the shunt resistor. In order to determine the accuracy of the resistance of shunt resistors being manufactured for a line of ammeters, a manufacturer tests a sample of 100 shunt resistors. The numbers of shunt resistors with the resistance indicated (to the nearest hundredth of an ohm) are as follows.

0.200 Ω, 1; 0.210 Ω, 3; 0.220 Ω, 5; 0.230 Ω, 10; 0.240 Ω, 17; 0.250 Ω, 40; 0.260 Ω, 13; 0.270 Ω, 6; 0.280 Ω, 3; 0.290 Ω, 2

(a) What is the mean resistance?

(A) 0.235 Ω
(B) 0.247 Ω
(C) 0.251 Ω
(D) 0.259 Ω

(b) What is the sample standard deviation?

(A) 0.0003
(B) 0.010
(C) 0.016
(D) 0.24

(c) What is the median resistance?

(A) 0.221 Ω
(B) 0.244 Ω
(C) 0.252 Ω
(D) 0.259 Ω

(d) What is the sample variance?

(A) 0.00027
(B) 0.0083
(C) 0.0114
(D) 0.0163

Hypothesis Testing

10. 100 bearings were tested to failure. The average life was 1520 hr, and the standard deviation was 120 hr. The manufacturer claims a 1600 hr life. Evaluate using confidence limits of 95% and 99%.

(A) The claim is accurate at both 95% and 99% confidence.
(B) The claim is inaccurate only at 95%.
(C) The claim is inaccurate only at 99%.
(D) The claim is inaccurate at both 95% and 99% confidence.

Curve Fitting

11. (a) Find the best equation for a line passing through the points given. (b) Find the correlation coefficient.

x	y
400	370
800	780
1250	1210
1600	1560
2000	1980
2500	2450
4000	3950

12. Find the best equation for a line passing through the points given.

s	t
20	43
18	141
16	385
14	1099

13. The number of vehicles lining up behind a flashing railroad crossing has been observed for five trains of different lengths, as given. What is the mathematical formula that relates the two variables?

no. of cars in train	no. of vehicles
2	14.8
5	18.0
8	20.4
12	23.0
27	29.9

14. The following yield data are obtained from five identical treatment plants. (a) Develop a mathematical equation to correlate the yield and average temperature. (b) What is the correlation coefficient?

treatment plant	average temperature (T)	average yield (Y)
1	207.1	92.30
2	210.3	92.58
3	200.4	91.56
4	201.1	91.63
5	203.4	91.83

15. The following data are obtained from a soil compaction test. What is the mathematical formula that relates the two variables?

x	y
−1	0
0	1
1	1.4
2	1.7
3	2
4	2.2
5	2.4
6	2.6
7	2.8
8	3

Quantitative Risk Analysis

16. What is the difference between risk analysis and risk management?

(A) Risk analysis is typically performed by engineers, and risk management is typically performed by risk managers.

(B) Risk analysis is primarily a technical evaluation, while risk management is primarily a political (or management) decision.

(C) Risk analysis is a fact-based process and has no uncertainties associated with it, while risk management involves a great many uncertainties and judgmental factors.

(D) Risk analysis can be applied to many different situations and conditions, but risk management is limited to industrial processes.

17. What is the sequence of steps in the risk analysis process?

I. hazard identification
II. toxicity assessment
III. risk characterization
IV. exposure assessment
V. fate and transport modeling
VI. process hazard analysis

(A) I, III, IV, V
(B) II, IV, V, VI
(C) I, II, IV, III
(D) I, II, V, VI

18. The objective of an initial screening of hazardous chemicals at a remediation site is to

(A) identify the carcinogenic and noncarcinogenic chemicals

(B) identify the potential release mechanisms at the site

(C) identify the potential pathways of exposure

(D) identify the chemicals or their surrogates that will account for 99% of the risk

19. A child has been exposed for 5 years to a drinking water well that is contaminated with heptachlor at a concentration of 1.3 μg/L. The slope factor (oral) for heptachlor is 4.5 (mg/kg·day)$^{-1}$. For a standard factor of 1.0 L/day water consumption and 10 kg body mass, and assuming heptachlor is carcinogenic, the excess risk of cancer over the child's lifetime is most nearly

(A) 3.4×10^{-4}
(B) 4.2×10^{-5}
(C) 3.6×10^{-6}
(D) 9.7×10^{-6}

20. An adult has been continuously exposed to two noncarcinogens for 2 years by the inhalation route, as shown below.

chemical	exposure (mg/m^3)	RfD, inhalation (mg/kg·day)
toluene	0.02	1.4
propylene glycol	0.33	0.572

Standard factors are 20 m^3/day of air breathed and 70 kg body mass. The hazard index for exposure to these two chemicals is most nearly

(A) 0.17
(B) 0.34
(C) 0.68
(D) 1.24

SOLUTIONS

1. (a) There are 4! = 24 different possible outcomes. By enumeration, there are 9 completely wrong combinations.

$$p\{\text{all wrong}\} = \frac{9}{24} = \boxed{0.375}$$

The answer is (B).

(b) If three recruits get the correct size, the fourth recruit will also since there will be only one pair remaining.

$$p\{\text{exactly 3}\} = \boxed{0}$$

The answer is (A).

2. For an exponential distribution function, the mean is given as

$$\mu = \frac{1}{\lambda}$$

For a mean of 23,

$$\mu = 23 = \frac{1}{\lambda}$$
$$\lambda = 0.0435$$

For an exponential distribution function,

$$\begin{aligned} p\{X < x\} &= F(x) = 1 - e^{-\lambda x} \\ p\{x > X\} &= 1 - p\{X < x\} \\ &= 1 - F(x) \\ &= 1 - \left(1 - e^{-\lambda x}\right) \\ &= e^{-\lambda x} \end{aligned}$$

The probability of a random vehicle being processed in 25 sec or more is given by

$$\begin{aligned} p\{x > 25\} &= e^{-(0.0435)(25)} \\ &= e^{-1.0875} \\ &= \boxed{0.337} \end{aligned}$$

The answer is (C).

3. (a) The distribution is a Poisson distribution with an average of $\lambda = 20$.

The probability for a Poisson distribution is given by

$$p\{x\} = f(x) = \frac{e^{-\lambda}\lambda^x}{x!}$$

The probability of 17 cars is

$$p\{x = 17\} = f(17) = \frac{e^{-20} \times 20^{17}}{17!}$$
$$= \boxed{0.076 \ \ (7.6\%)}$$

The answer is (A).

(b) The probability of three or fewer cars is given by

$$\begin{aligned} p\{x \leq 3\} &= p\{x = 0\} + p\{x = 1\} + p\{x = 2\} \\ &\quad + p\{x = 3\} \\ &= f(0) + f(1) + f(2) + f(3) \\ &= \frac{e^{-20} \times 20^0}{0!} + \frac{e^{-20} \times 20^1}{1!} \\ &\quad + \frac{e^{-20} \times 20^2}{2!} + \frac{e^{-20} \times 20^3}{3!} \\ &= 2 \times 10^{-9} + 4.1 \times 10^{-8} \\ &\quad + 4.12 \times 10^{-7} + 2.75 \times 10^{-6} \\ &= 3.2 \times 10^{-6} \ \ (3.2 \times 10^{-4}\%) \\ &= \boxed{0.00032} \end{aligned}$$

The answer is (A).

4. $\lambda = \frac{1}{\text{MTTF}}$

$$= \frac{1}{1000} = 0.001$$

The reliability function is

$$R\{t\} = e^{-\lambda t} = e^{-0.001t}$$

Since the reliability is greater than 99%,

$$\begin{aligned} e^{-0.001t} &> 0.99 \\ \ln(e^{-0.001t}) &> \ln(0.99) \\ -0.001t &> \ln(0.99) \\ t &< -1000 \ln(0.99) \\ &\boxed{t < 10.05} \end{aligned}$$

The maximum operating time such that the reliability remains above 99% is 10.05 hr.

The answer is (D).

5. Find the average.

$$\begin{aligned} \overline{x} &= \frac{\sum x_i}{n} \\ &= \frac{1249.529 + 1249.494 + 1249.384 + 1249.348}{4} \\ &= 1249.439 \end{aligned}$$

Since the sample population is small, use the sample standard deviation.

$$\begin{aligned} s &= \sqrt{\frac{\sum(x_i - \overline{x})^2}{n-1}} \\ &= \sqrt{\frac{\begin{array}{c}(1249.529-1249.439)^2 + (1249.494-1249.439)^2 \\ +(1249.384-1249.439)^2 + (1249.348-1249.439)^2\end{array}}{4-1}} \\ &= 0.08647 \end{aligned}$$

From the standard deviation table, a 90% confidence limit falls within $1.645s$ of $\overline{x}$.

$$1249.439 \pm (1.645)(0.08647) = 1249.439 \pm 0.142$$

Therefore, (1249.297, 1249.581) is the 90% confidence range.

(a) By observation, all the readings fall within the 90% confidence range.

The answer is (D).

(b) No readings are unacceptable.

The answer is (A).

(c) Readings outside the 90% confidence limits are unacceptable.

The answer is (B).

(d) The unbiased estimate of the most probable distance is 1249.439.

The answer is (C).

(e) The error for the 90% confidence range is 0.142.

The answer is (C).

(f) If the surveying crew places a marker, measures a distance x, places a second marker, and then measures the same distance x back to the original marker, the ending point should coincide with the original marker. If, due to measurement errors, the ending and starting points do not coincide, the difference is the closure error.

In this example, the survey crew moves around the four sides of a square, so there are two measurements in the x-direction and two measurements in the y-direction. If the errors E_1 and E_2 are known for two measurements, x_1 and x_2, the error associated with the sum or difference $x_1 \pm x_2$ is

$$E\{x_1 \pm x_2\} = \sqrt{E_1^2 + E_2^2}$$

In this case, the error in the x-direction is

$$\begin{aligned} E_x &= \sqrt{(0.1422)^2 + (0.1422)^2} \\ &= 0.2011 \end{aligned}$$

The error in the y-direction is calculated the same way and is also 0.2011. E_x and E_y are combined by the Pythagorean theorem to yield

$$E_{\text{closure}} = \sqrt{(0.2011)^2 + (0.2011)^2} = \boxed{0.2844}$$

The answer is (C).

(g) In surveying, error may be expressed as a fraction of one or more legs of the traverse. Assume that the total of all four legs is to be used as the basis.

$$\frac{0.2844}{(4)(1249)} = \boxed{\frac{1}{17{,}567}}$$

The answer is (A).

(h) In surveying, a class 1 third-order error is smaller than 1/10,000. The error of 1/17,567 is smaller than the third-order error; therefore, the error is within the third-order accuracy.

The answer is (C).

(i) An experiment is accurate if it is unchanged by experimental error. Precision is concerned with the repeatability of the experimental results. If an experiment is repeated with identical results, the experiment is said to be precise. However, it is possible to have a highly precise experiment with a large bias.

The answer is (D).

(j) A systematic error is one that is always present and is unchanged from sample to sample. For example, a steel tape that is 0.02 ft short introduces a systematic error.

The answer is (B).

6. (a) and (d) Tabulate the frequency distribution data.

(Note that the lowest speed is 20 mi/hr and the highest speed is 48 mi/hr; therefore, the range is 28 mi/hr. Choose 10 cells with a width of 3 mi/hr.)

midpoint	interval (mi/hr)	frequency	cumulative frequency	cumulative percent
21	20–22	1	1	3
24	23–25	3	4	10
27	26–28	5	9	23
30	29–31	8	17	43
33	32–34	3	20	50
36	35–37	4	24	60
39	38–40	3	27	68
42	41–43	8	35	88
45	44–46	3	38	95
48	47–49	2	40	100

(b)

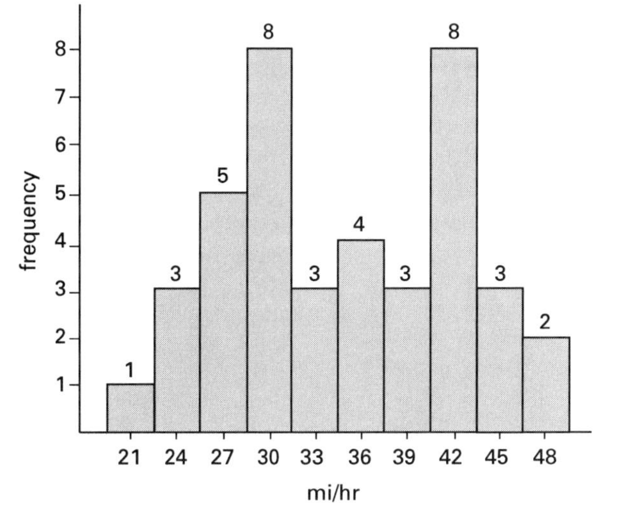

(c)

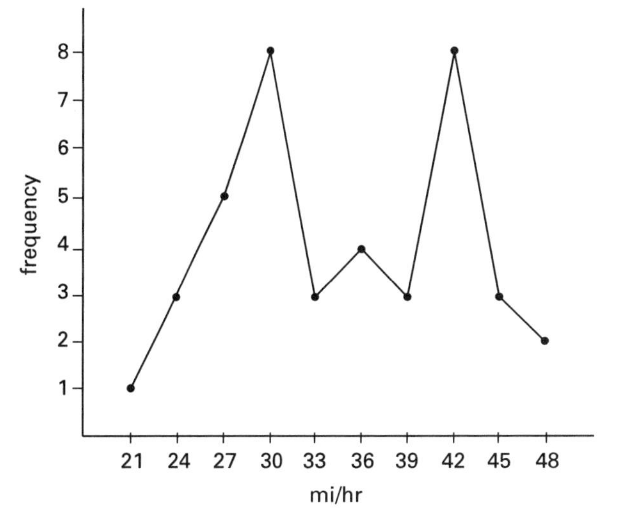

(e)

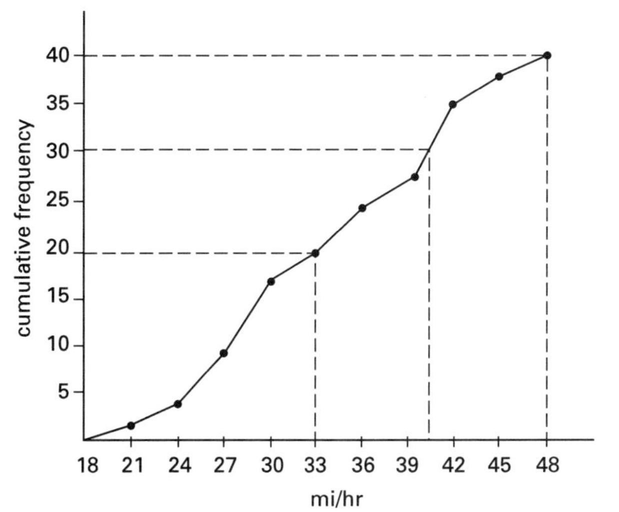

(f) From the cumulative frequency graph in part (e), the upper quartile speed occurs at 30 cars or 75%, which corresponds to approximately 40 mi/hr.

The answer is (C).

(g) The mode occurs at two speeds (the frequency at each of the two speeds is 8), which are 30 mi/hr and 42 mi/hr.

From the cumulative frequency chart, the median occurs at 50% or 20 cars and corresponds to 33 mi/hr.

$$\sum x_i = 1390 \text{ mi/hr}$$
$$n = 40$$

The mean is computed as

$$\overline{x} = \frac{\sum x_i}{n} = \frac{1390 \ \frac{\text{mi}}{\text{hr}}}{40} = \boxed{34.75 \text{ mi/hr}}$$

The answer is (B).

(h) The standard deviation of the sample data is given as

$$\sigma = \sqrt{\frac{\sum x^2}{n} - \mu^2}$$
$$\sum x^2 = 50{,}496 \text{ mi}^2/\text{hr}^2$$

Use the sample mean as an unbiased estimator of the population mean, μ.

$$\begin{aligned}\sigma &= \sqrt{\frac{\sum x^2}{n} - \mu^2} \\ &= \sqrt{\frac{50{,}496 \ \frac{\text{mi}^2}{\text{hr}^2}}{40} - \left(34.75 \ \frac{\text{mi}}{\text{hr}}\right)^2} \\ &= \boxed{7.405 \text{ mi/hr}}\end{aligned}$$

The answer is (D).

(i) The sample standard deviation is given by

$$\begin{aligned}s &= \sqrt{\frac{\sum x^2 - \frac{\left(\sum x\right)^2}{n}}{n-1}} \\ &= \sqrt{\frac{50{,}496 \ \frac{\text{mi}^2}{\text{hr}^2} - \frac{\left(1390 \ \frac{\text{mi}}{\text{hr}}\right)^2}{40}}{40-1}} \\ &= \boxed{7.500 \text{ mi/hr}}\end{aligned}$$

The answer is (A).

(j) The sample variance is given by the square of the sample standard deviation.

$$s^2 = \left(7.500 \ \frac{\text{mi}}{\text{hr}}\right)^2 = \boxed{56.25 \text{ mi}^2/\text{hr}^2}$$

The answer is (A).

7. (a) The 50th percentile speed is the mean speed, $\boxed{46.}$

The answer is (C).

(b) The 85th percentile speed is the speed that is exceeded by only 15% of the measurements. Since this is a normal distribution, App. 11.A can be used. 15% in the upper tail corresponds to 35% between the mean and the 85th percentile. This occurs at approximately 1.04σ. The 85th percentile speed is

$$x_{85\%} = \mu + 1.04\sigma = 46 + (1.04)(3) = \boxed{49.12}$$

The answer is (C).

(c) The upper 2σ speed is

$$x_{2\sigma} = \mu + 2\sigma = 46 + (2)(3) = \boxed{52}$$

The answer is (D).

(d) According to the central limit theorem, the mean of the average speeds is the same as the distribution mean, and the standard deviation of sample means is

$$\sigma_{\overline{x}} = \frac{\sigma_x}{\sqrt{K}} = \frac{3}{\sqrt{25}} = 0.6$$
$$\overline{x}_{2\sigma} = \mu + 2\sigma_{\overline{x}} = 46 + (2)(0.6) = \boxed{47.2}$$

The answer is (B).

8. (a) From Eq. 11.43,

$$z_{\text{upper}} = \frac{0.507 \text{ in} - 0.502 \text{ in}}{0.005 \text{ in}} = +1$$

From App. 11.A, the area outside $z = +1$ is

$$0.5 - 0.3413 = 0.1587$$

Since these are symmetrical limits, $z_{\text{lower}} = -1$.

$$\text{total fraction defective} = (2)(0.1587) = \boxed{0.3174}$$

The answer is (C).

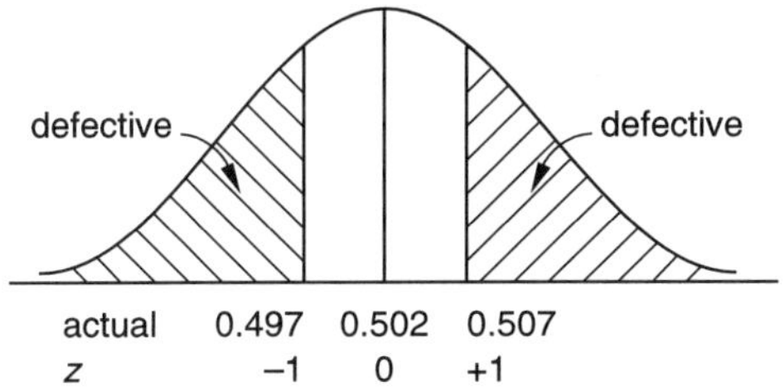

(b) This is a binomial problem.

$$p = p\{\text{defective}\} = 0.3174$$
$$q = 1 - p = 0.6826$$

From Eq. 11.28,

$$f(2) = \binom{15}{2}(0.3174)^2(0.6826)^{13}$$
$$= \left(\frac{15!}{13!2!}\right)(0.3174)^2(0.6826)^{13} = \boxed{0.0739}$$

The answer is (A).

9. (a)

R	f	fR	fR^2
0.200	1	0.200	0.0400
0.210	3	0.630	0.1323
0.220	5	1.100	0.2420
0.230	10	2.300	0.5290
0.240	17	4.080	0.9792
0.250	40	10.000	2.5000
0.260	13	3.380	0.8788
0.270	6	1.620	0.4374
0.280	3	0.840	0.2352
0.290	2	0.580	0.1682
	100	24.730	6.1421

$$\overline{R} = \frac{\sum fR}{\sum f} = \frac{24.730\ \Omega}{100} = \boxed{0.2473\ \Omega}$$

The answer is (B).

(b) The sample standard deviation is given by Eq. 11.61.

$$s = \sqrt{\frac{\sum fR^2 - \frac{(\sum fR)^2}{n}}{n-1}}$$
$$= \sqrt{\frac{6.1421\ \Omega - \frac{(24.73\ \Omega)^2}{100}}{99}}$$
$$= \boxed{0.0163\ \Omega}$$

The answer is (C).

(c) 50% of the observations are below the median. 36 observations are below 0.240. 14 more are needed.

$$0.240\ \Omega + \left(\frac{14}{40}\right)(0.250\ \Omega - 0.240\ \Omega) = \boxed{0.2435\ \Omega}$$

The answer is (B).

(d) $s^2 = (0.0163\ \Omega)^2 = \boxed{0.0002656\ \Omega^2}$

The answer is (A).

10. This is a typical hypothesis test of two sample population means. The two populations are the original population the manufacturer used to determine the 1600 hr average life value and the new population the sample was taken from. The mean ($\overline{x} = 1520$ hr) of the sample and its standard deviation ($s = 120$ hr) are known, but the mean and standard deviation of a population of average lifetimes are unknown.

Assume that the average lifetime population mean and the sample mean are identical.

$$\overline{x} = \mu = 1520 \text{ hr}$$

The standard deviation of the average lifetime population is

$$\sigma_{\overline{x}} = \frac{s}{\sqrt{n}} = \frac{120 \text{ hr}}{\sqrt{100}} = 12 \text{ hr}$$

The manufacturer can be reasonably sure that the claim of a 1600 hr average life is justified if the average test life is near 1600 hr. "Reasonably sure" must be evaluated based on acceptable probability of being incorrect. If the manufacturer is willing to be wrong with a 5% probability, then a 95% confidence level is required.

Since the direction of bias is known, a one-tail test is required. To determine if the mean has shifted downward, test the hypothesis that 1600 hr is within the 95% limit of a distribution with a mean of 1520 hr and a standard deviation of 12 hr. From a standard normal table, 5% of a standard normal distribution is outside of $z = 1.645$. Therefore, the 95% confidence limit is

$$1520 \text{ hr} + (1.645)(12 \text{ hr}) = 1540 \text{ hr}$$

The manufacturer can be 95% certain that the average lifetime of the bearings is less than 1600 hr.

If the manufacturer is willing to be wrong with a probability of only 1%, then a 99% confidence limit is required. From the normal table, $z = 2.33$ and the 99% confidence limit is

$$1520 \text{ hr} + (2.33)(12 \text{ hr}) = 1548 \text{ hr}$$

The manufacturer can be 99% certain that the average bearing life is less than 1600 hr.

The answer is (D).

11. (a) Plot the data points to determine if the relationship is linear.

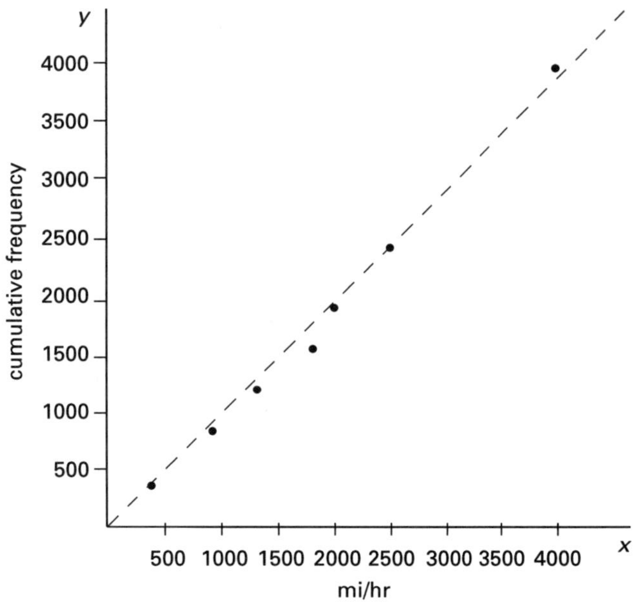

The data appear to be essentially linear. The slope, m, and the y-intercept, b, can be determined using linear regression.

The individual terms are

$$n = 7$$

$$\begin{aligned}\textstyle\sum x_i &= 400 + 800 + 1250 + 1600 + 2000 + 2500 \\ &\quad + 4000 \\ &= 12{,}550\end{aligned}$$

$$\begin{aligned}\left(\textstyle\sum x_i\right)^2 &= (12{,}550)^2 \\ &= 1.575 \times 10^8\end{aligned}$$

$$\begin{aligned}\overline{x} &= \frac{\sum x_i}{n} \\ &= \frac{12{,}550}{7} \\ &= 1792.9\end{aligned}$$

$$\begin{aligned}\textstyle\sum x_i^2 &= (400)^2 + (800)^2 + (1250)^2 + (1600)^2 \\ &\quad + (2000)^2 + (2500)^2 + (4000)^2 \\ &= 3.117 \times 10^7\end{aligned}$$

Similarly,

$$\begin{aligned}\textstyle\sum y_i &= 370 + 780 + 1210 + 1560 + 1980 \\ &\quad + 2450 + 3950 \\ &= 12{,}300\end{aligned}$$

$$\left(\textstyle\sum y_i\right)^2 = (12{,}300)^2 = 1.513 \times 10^8$$

$$\overline{y} = \frac{\sum y_i}{n} = \frac{12{,}300}{7} = 1757.1$$

$$\begin{aligned}\textstyle\sum y_i^2 &= (370)^2 + (780)^2 + (1210)^2 + (1560)^2 \\ &\quad + (1980)^2 + (2450)^2 + (3950)^2 \\ &= 3.017 \times 10^7\end{aligned}$$

Also,

$$\begin{aligned}\textstyle\sum x_i y_i &= (400)(370) + (800)(780) + (1250)(1210) \\ &\quad + (1600)(1560) + (2000)(1980) \\ &\quad + (2500)(2450) + (4000)(3950) \\ &= 3.067 \times 10^7\end{aligned}$$

The slope is

$$\begin{aligned}m &= \frac{n\sum x_i y_i - \sum x_i \sum y_i}{n\sum x_i^2 - \left(\sum x_i\right)^2} \\ &= \frac{(7)(3.067 \times 10^7) - (12{,}550)(12{,}300)}{(7)(3.117 \times 10^7) - (12{,}550)^2} \\ &= 0.994\end{aligned}$$

The y-intercept is

$$\begin{aligned}b &= \overline{y} - m\overline{x} \\ &= 1757.1 - (0.994)(1792.9) \\ &= -25.0\end{aligned}$$

The least squares equation of the line is

$$\begin{aligned}y &= mx + b \\ &= \boxed{0.994x - 25.0}\end{aligned}$$

(b) The correlation coefficient is

$$\begin{aligned}r &= \frac{n\sum(x_i y_i) - (\sum x_i)(\sum y_i)}{\sqrt{[n\sum x_i^2 - (\sum x_i)^2][n\sum y_i^2 - (\sum y_i)^2]}} \\ &= \frac{(7)(3.067 \times 10^7) - (12{,}500)(12{,}300)}{\sqrt{\begin{array}{l}[(7)(3.117 \times 10^7) - (12{,}500)^2] \\ \quad \times [(7)(3.017 \times 10^7) - (12{,}300)^2]\end{array}}} \\ &\approx \boxed{1.00}\end{aligned}$$

12. Plotting the data shows that the relationship is nonlinear.

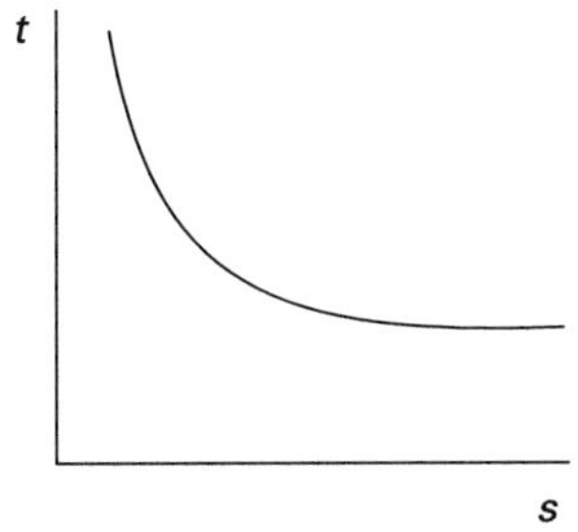

This appears to be an exponential with the form

$$t = ae^{bs}$$

Take the natural log of both sides.

$$\begin{aligned}\ln t &= \ln(ae^{bs})\\ &= \ln a + \ln(e^{bs})\\ &= \ln a + bs\end{aligned}$$

But, ln a is just a constant, c.

$$\ln t = c + bs$$

Make the transformation $R = \ln t$.

$$R = c + bs$$

s	R
20	3.76
18	4.95
16	5.95
14	7.00

This is linear.

$$\begin{aligned}n &= 4\\ \sum s_i &= 20 + 18 + 16 + 14 = 68\\ \overline{s} &= \frac{\sum s}{n} = \frac{68}{4} = 17\\ \sum s_i^2 &= (20)^2 + (18)^2 + (16)^2 + (14)^2 = 1176\\ \left(\sum s_i\right)^2 &= (68)^2 = 4624\\ \sum R_i &= 3.76 + 4.95 + 5.95 + 7.00 = 21.66\\ \overline{R} &= \frac{\sum R_i}{n} = \frac{21.66}{4} = 5.415\\ \sum R_i^2 &= (3.76)^2 + (4.95)^2 + (5.95)^2 + (7.00)^2\\ &= 123.04\\ \left(\sum R_i\right)^2 &= (21.66)^2 = 469.16\\ \sum s_i R_i &= (20)(3.76) + (18)(4.95) + (16)(5.95)\\ &\quad + (14)(7.00)\\ &= 357.5\end{aligned}$$

The slope, b, of the transformed line is

$$\begin{aligned}b &= \frac{n\sum s_i R_i - \sum s_i \sum R_i}{n\sum s_i^2 - \left(\sum s_i\right)^2}\\ &= \frac{(4)(357.5) - (68)(21.66)}{(4)(1176) - (68)^2} = -0.536\end{aligned}$$

The intercept is

$$\begin{aligned}c &= \overline{R} - b\overline{s} = 5.415 - (-0.536)(17)\\ &= 14.527\end{aligned}$$

The transformed equation is

$$\begin{aligned}R &= c + bs\\ &= 14.527 - 0.536s\end{aligned}$$

$$\boxed{\ln t = 14.527 - 0.536s}$$

13. The first step is to graph the data.

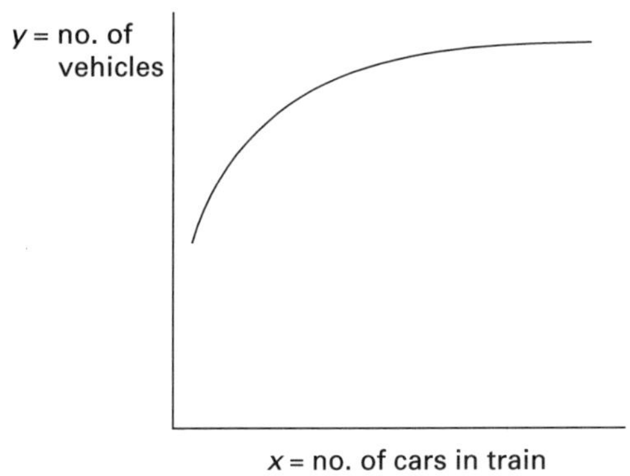

It is assumed that the relationship between the variables has the form $y = a + b \log x$. Therefore, the variable change $z = \log x$ is made, resulting in the following set of data.

z	y
0.301	14.8
0.699	18.0
0.903	20.4
1.079	23.0
1.431	29.9

$$\begin{aligned}\sum z_i &= 4.413\\ \sum y_i &= 106.1\\ \sum z_i^2 &= 4.6082\\ \sum y_i^2 &= 2382.2\end{aligned}$$

$$\left(\sum z_i\right)^2 = 19.475$$
$$\left(\sum y_i\right)^2 = 11{,}257.2$$
$$\overline{z} = 0.8826$$
$$\overline{y} = 21.22$$
$$\sum z_i y_i = 103.06$$
$$n = 5$$

The slope is

$$\begin{aligned} m &= \frac{n\sum z_i y_i - \sum z_i \sum y_i}{n\sum z_i^2 - (\sum z_i)^2} \\ &= \frac{(5)(103.06) - (4.413)(106.1)}{(5)(4.6082) - 19.475} \\ &= 13.20 \end{aligned}$$

The y-intercept is

$$\begin{aligned} b &= \overline{y} - m\overline{z} \\ &= 21.22 - (13.20)(0.8826) \\ &= 9.570 \end{aligned}$$

The resulting equation is

$$y = 9.570 + 13.20z$$

The relationship between x and y is approximately

$$\boxed{y = 9.570 + 13.20 \log x}$$

This is not an optimal correlation, as better correlation coefficients can be obtained if other assumptions about the form of the equation are made. For example, $y = 9.1 + 4\sqrt{x}$ has a better correlation coefficient.

14. (a) Plot the data to verify that they are linear.

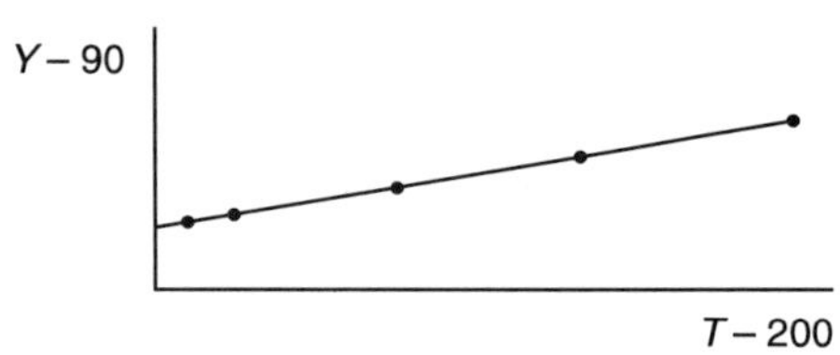

x $T - 200$	y $Y - 90$
7.1	2.30
10.3	2.58
0.4	1.56
1.1	1.63
3.4	1.83

step 1:

$$\begin{aligned} \sum x_i &= 22.3 & \sum y_i &= 9.9 \\ \sum x_i^2 &= 169.43 & \sum y_i^2 &= 20.39 \\ (\sum x_i)^2 &= 497.29 & (\sum y_i)^2 &= 98.01 \\ \overline{x} &= \frac{22.3}{5} = 4.46 & \overline{y} &= 1.98 \end{aligned}$$

$$\sum x_i y_i = 51.54$$

step 2: From Eq. 11.70, the slope is

$$m = \frac{(5)(51.54) - (22.3)(9.9)}{(5)(169.43) - 497.29} = 0.1055$$

step 3: From Eq. 11.71, the y-intercept is

$$b = 1.98 - (0.1055)(4.46) = 1.509$$

The equation of the line is

$$y = 0.1055x + 1.509$$
$$Y - 90 = (0.1055)(T - 200) + 1.509$$
$$Y = \boxed{0.1055T + 70.409}$$

(b) *step 4:* Use Eq. 11.72 to get the correlation coefficient.

$$\begin{aligned} r &= \frac{(5)(51.54) - (22.3)(9.9)}{\sqrt{[(5)(169.43) - 497.29][(5)(20.39) - 98.01]}} \\ &= \boxed{0.995} \end{aligned}$$

15. Plot the data to see if they are linear.

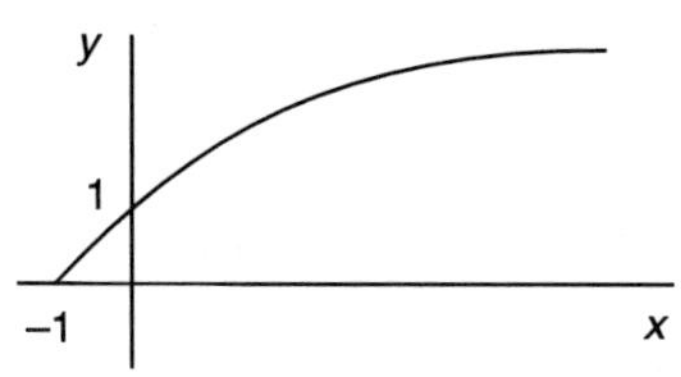

This looks like it could be of the form

$$y = a + b\sqrt{x}$$

However, when x is negative (as in the first point), the function is imaginary. Try shifting the curve to the right, replacing x with $x + 1$.

$$y = a + bz$$
$$z = \sqrt{x + 1}$$

z	y
0	0
1	1
1.414	1.4
1.732	1.7
2	2
2.236	2.2
2.45	2.4
2.65	2.6
2.83	2.8
3	3

Since $y \approx z$, the relationship is

$$\boxed{y = \sqrt{x+1}}$$

In this problem, the answer was found accidentally. Usually, regression would be necessary.

16. ***The answer is (B).***

17. ***The answer is (C).***

18. ***The answer is (D).***

19. Use Eq. 11.75.

$$\text{CDI}_{a/w} = \frac{C(\text{CR})(\text{EF})(\text{ED})}{(\text{BW})(\text{AT})}$$

$$\begin{aligned}\text{CDI}_w &= \frac{\left(1.3\ \frac{\mu\text{g}}{\text{L}}\right)\left(1.0\ \frac{\text{L}}{\text{day}}\right)\left(365\ \frac{\text{days}}{\text{yr}}\right)(5\ \text{yr})}{(10\ \text{kg})(70\ \text{yr})\left(365\ \frac{\text{days}}{\text{yr}}\right)\left(1000\ \frac{\mu\text{g}}{\text{mg}}\right)} \\ &= 9.2857 \times 10^{-6}\ \text{mg/kg·day}\end{aligned}$$

Use Eq. 11.78.

$$\begin{aligned}R_c &= (\text{CDI})(\text{SF}) \\ &= \left(9.2857 \times 10^{-6}\ \frac{\text{mg}}{\text{kg·day}}\right)\left(4.5\ \frac{\text{kg·day}}{\text{mg}}\right) \\ &= \boxed{4.17 \times 10^{-5}\quad (4.2 \times 10^{-5})}\end{aligned}$$

The answer is (B).

20. Use Eq. 11.75.

$$\text{CDI}_{a/w} = \frac{C(\text{CR})(\text{EF})(\text{ED})}{(\text{BW})(\text{AT})}$$

For toluene,

$$\begin{aligned}\text{CDI}_a &= \frac{\left(0.02\ \frac{\text{mg}}{\text{m}^3}\right)\left(20\ \frac{\text{m}^3}{\text{day}}\right)\left(365\ \frac{\text{days}}{\text{yr}}\right)(2\ \text{yr})}{(70\ \text{kg})\left(365\ \frac{\text{days}}{\text{yr}}\right)(2\ \text{yr})} \\ &= 5.714 \times 10^{-3}\ \text{mg/kg·day}\end{aligned}$$

Use Eq. 11.77.

$$\text{HI} = \frac{\text{CDI}}{\text{RfD}} = \frac{5.714 \times 10^{-3}\ \frac{\text{mg}}{\text{kg·day}}}{1.4\ \frac{\text{mg}}{\text{kg·day}}} = 4.08 \times 10^{-3}$$

For propylene glycol,

$$\begin{aligned}\text{CDI}_a &= \frac{\left(0.33\ \frac{\text{mg}}{\text{m}^3}\right)\left(20\ \frac{\text{m}^3}{\text{day}}\right)\left(365\ \frac{\text{days}}{\text{yr}}\right)(2\ \text{yr})}{(70\ \text{kg})\left(365\ \frac{\text{days}}{\text{yr}}\right)(2\ \text{yr})} \\ &= 9.429 \times 10^{-2}\ \frac{\text{mg}}{\text{kg·day}}\end{aligned}$$

$$\text{HI} = \frac{\text{CDI}}{\text{RfD}} = \frac{9.429 \times 10^{-2}\ \frac{\text{mg}}{\text{kg·day}}}{0.572\ \frac{\text{mg}}{\text{kg·day}}} = 0.165$$

The total HI is

$$4.08 \times 10^{-3} + 0.165 = \boxed{0.169\quad (0.17)}$$

The answer is (A).

12 Numerical Analysis

PRACTICE PROBLEMS

1. A function is given as $y = 3x^{0.93} + 4.2$. What is the percent error if the value of y at $x = 2.7$ is found by using straight-line interpolation between $x = 2$ and $x = 3$?

2. Given the following data points, find y by straight-line interpolation for $x = 2.75$.

x	y
1	4
2	6
3	2
4	−14

SOLUTIONS

1. The actual value at $x = 2.7$ is given by

$$\begin{aligned} y(x) &= 3x^{0.93} + 4.2 \\ y(2.7) &= (3)(2.7)^{0.93} + 4.2 \\ &= 11.756 \end{aligned}$$

At $x = 3$,

$$\begin{aligned} y(3) &= (3)(3)^{0.93} + 4.2 \\ &= 12.534 \end{aligned}$$

At $x = 2$,

$$\begin{aligned} y(2) &= (3)(2)^{0.93} + 4.2 \\ &= 9.916 \end{aligned}$$

Use straight-line interpolation.

$$\frac{x_2 - x}{x_2 - x_1} = \frac{y_2 - y}{y_2 - y_1}$$

$$\frac{3 - 2.7}{3 - 2} = \frac{12.534 - y}{12.534 - 9.916}$$

$$y = 11.749$$

The relative error is given by

$$\frac{\text{actual value} - \text{predicted value}}{\text{actual value}} = \frac{11.756 - 11.749}{11.756}$$

$$= \boxed{0.0006 \ \ (0.06\%)}$$

2. Let $x_1 = 2$; therefore, from the table of data points, $y_1 = 6$. Let $x_2 = 3$; therefore, from the table of data points, $y_2 = 2$.

Let $x = 2.75$. By straight-line interpolation,

$$\frac{x_2 - x}{x_2 - x_1} = \frac{y_2 - y}{y_2 - y_1}$$

$$\frac{3 - 2.75}{3 - 2} = \frac{2 - y}{2 - 6}$$

$$\boxed{y = 3}$$

13 Energy, Work, and Power

PRACTICE PROBLEMS

Chapter 13 of the *Environmental Engineering Reference Manual* contains no practice problems.

14 Fluid Properties

PRACTICE PROBLEMS

(Use $g = 32.2$ ft/sec^2 or 9.81 m/s^2 unless told to do otherwise in the problem.)

Pressure

1. What is the absolute pressure if a gauge reads 8.7 psi (60 kPa) vacuum?

(A) 4 psi (27 kPa)
(B) 6 psi (41 kPa)
(C) 8 psi (55 kPa)
(D) 10 psi (68 kPa)

Viscosity

2. Calculate the kinematic viscosity of air at 80°F (27°C) and 70 psia (480 kPa).

(A) 3.54×10^{-5} ft^2/sec (3.30×10^{-6} m^2/s)
(B) 4.25×10^{-5} ft^2/sec (3.96×10^{-6} m^2/s)
(C) 4.96×10^{-5} ft^2/sec (4.62×10^{-6} m^2/s)
(D) 6.37×10^{-5} ft^2/sec (5.94×10^{-6} m^2/s)

Solutions

3. Volumes of an 8% solution, a 10% solution, and a 20% solution of nitric acid are to be mixed in order to get 100 mL of a 12% solution. If the 8% solution contributes half of the total volume of nitric acid contributed by the 10% and 20% solutions, what volume of 10% acid solution is required?

(A) 20 mL
(B) 30 mL
(C) 50 mL
(D) 80 mL

SOLUTIONS

1. *Customary U.S. Solution*

$$p_{\text{gage}} = -8.7 \text{ lbf/in}^2$$
$$p_{\text{atmospheric}} = 14.7 \text{ lbf/in}^2$$

The relationship between absolute, gage, and atmospheric pressure is given by

$$\begin{aligned} p_{\text{absolute}} &= p_{\text{gage}} + p_{\text{atmospheric}} \\ &= -8.7 \ \frac{\text{lbf}}{\text{in}^2} + 14.7 \ \frac{\text{lbf}}{\text{in}^2} \\ &= \boxed{6 \text{ lbf/in}^2 \ \ (6 \text{ psi})} \end{aligned}$$

The answer is (B).

SI Solution

$$p_{\text{gage}} = -60 \text{ kPa}$$
$$p_{\text{atmospheric}} = 101.3 \text{ kPa}$$

The relationship between absolute, gage, and atmospheric pressure is given by

$$\begin{aligned} p_{\text{absolute}} &= p_{\text{gage}} + p_{\text{atmospheric}} \\ &= -60 \text{ kPa} + 101.3 \text{ kPa} \\ &= \boxed{41.3 \text{ kPa}} \end{aligned}$$

The answer is (B).

2. *Customary U.S. Solution*

For air at 14.7 psia and 80°F, the absolute viscosity independent of pressure is $\mu = 3.85 \times 10^{-7}$ lbf-sec/ft^2.

Determine the density of air at 70 psia and 80°F. (Assume an ideal gas.)

$$\rho = \frac{p}{RT}$$

For air, $R = 53.3$ lbf-ft/lbm-°R.

Substituting gives

$$\rho = \frac{\left(70\ \frac{\text{lbf}}{\text{in}^2}\right)\left(144\ \frac{\text{in}^2}{\text{ft}^2}\right)}{\left(53.3\ \frac{\text{lbf-ft}}{\text{lbm-°R}}\right)(80°\text{F} + 460)}$$
$$= 0.350\ \text{lbm/ft}^3$$

The kinematic viscosity, ν, is related to the absolute viscosity by

$$\nu = \frac{\mu g_c}{\rho}$$
$$= \frac{\left(3.85 \times 10^{-7}\ \frac{\text{lbf-sec}}{\text{ft}^2}\right)\left(32.2\ \frac{\text{lbm-ft}}{\text{lbf-sec}^2}\right)}{0.350\ \frac{\text{lbm}}{\text{ft}^3}}$$
$$= \boxed{3.54 \times 10^{-5}\ \text{ft}^2/\text{sec}}$$

The answer is (A).

SI Solution

For air at 480 kPa and 27°C, the absolute viscosity independent of pressure is $\mu = 1.84 \times 10^{-5}$ Pa·s.

Determine the density of air at 480 kPa and 27°C. (Assume an ideal gas.)

$$\rho = \frac{p}{RT}$$

For air, $R = 287$ J/kg·K.

Substituting gives

$$\rho = \frac{(480\ \text{kPa})\left(1000\ \frac{\text{Pa}}{\text{kPa}}\right)}{\left(287\ \frac{\text{J}}{\text{kg·K}}\right)(27°\text{C} + 273)}$$
$$= 5.575\ \text{kg/m}^3$$

The kinematic viscosity, ν, is related to the absolute viscosity by

$$\nu = \frac{\mu}{\rho}$$
$$= \frac{1.84 \times 10^{-5}\ \text{Pa·s}}{5.575\ \frac{\text{kg}}{\text{m}^3}}$$
$$= \boxed{3.30 \times 10^{-6}\ \text{m}^2/\text{s}}$$

The answer is (A).

3. Let

$$x = \text{volume of 8\% solution}$$
$$y = \text{volume of 10\% solution}$$
$$z = \text{volume of 20\% solution}$$

The three conditions that must be satisfied are

$$x + y + z = 100\ \text{mL}$$
$$0.08x + 0.10y + 0.20z = (0.12)(100\ \text{mL}) = 12\ \text{mL}$$
$$0.08x = \left(\tfrac{1}{2}\right)(0.10y + 0.20z)$$

Simplifying these equations,

$$\begin{aligned} x + y + z &= 100 \\ 4x + 5y + 10z &= 600 \\ 8x - 5y - 10z &= 0 \end{aligned}$$

Adding the second and third equations gives

$$12x = 600$$
$$x = \boxed{50\ \text{mL}}$$

Work with the first two equations to get

$$y + z = 100 - 50 = 50$$
$$5y + 10z = 600 - (4)(50) = 400$$

Multiplying the top equation by -5 and adding to the bottom equation,

$$5z = 150$$
$$z = 30\ \text{mL}$$

From the first equation,

$$y = 20\ \text{mL}$$

The answer is (A).

15 Fluid Statics

PRACTICE PROBLEMS

(Use g = 32.2 ft/sec^2 or 9.81 m/s^2 unless told to do otherwise in the problem.)

Buoyancy

1. A blimp contains 10,000 lbm (4500 kg) of hydrogen (specific gas constant = 766.5 ft-lbf/lbm-°R (4124 J/kg·K)) at 56°F (13°C) and 30.2 in Hg (770 mm Hg). What is its lift if the hydrogen and air are in thermal and pressure equilibrium?

(A) 7.6×10^3 lbf (3.4×10^4 N)
(B) 1.2×10^4 lbf (5.3×10^4 N)
(C) 1.3×10^5 lbf (5.9×10^5 N)
(D) 1.7×10^5 lbf (7.7×10^5 N)

2. A hollow 6 ft (1.8 m) diameter sphere floats half-submerged in seawater. What mass of concrete is required as an external anchor to just submerge the sphere completely?

(A) 2700 lbm (1200 kg)
(B) 4200 lbm (1900 kg)
(C) 5500 lbm (2500 kg)
(D) 6300 lbm (2700 kg)

SOLUTIONS

1. *Customary U.S. Solution*

Assume the weight of the blimp structure is small (negligible) compared with the weight of the hydrogen.

The lift of the hydrogen-filled blimp (F_{lift}) is equal to the difference between the buoyant force (F_b) and the weight of the hydrogen contained in the blimp (W_{H}).

$$F_{\text{lift}} = F_b - W_{\text{H}}$$

The weight of the hydrogen is calculated from the mass of hydrogen by

$$\begin{aligned} W_{\text{H}} &= \frac{mg}{g_c} \\ &= \frac{(10{,}000 \text{ lbm})\left(32.2 \frac{\text{ft}}{\text{sec}^2}\right)}{32.2 \frac{\text{lbm-ft}}{\text{lbf-sec}^2}} \\ &= 10{,}000 \text{ lbf} \end{aligned}$$

The buoyant force is equal to the weight of the displaced air. The weight of the displaced air is calculated by knowing that the volume of the air displaced is equal to the volume of hydrogen enclosed in the blimp. Compute the volume of the hydrogen contained in the blimp by assuming the hydrogen behaves like an ideal gas.

$$V_{\text{H}} = \frac{mRT}{p}$$

For hydrogen, R = 766.5 ft-lbf/lbm-°R.

The temperature of the hydrogen is given as 56°F. Convert to absolute temperature (°R).

$$T = 56°\text{F} + 460 = 516°\text{R}$$

The pressure of the hydrogen is given as 30.2 in Hg. Convert the pressure to units of pounds per square foot.

$$\begin{aligned} p &= (30.2 \text{ in Hg})\left(\frac{1 \frac{\text{lbf}}{\text{in}^2}}{2.036 \text{ in Hg}}\right)\left(144 \frac{\text{in}^2}{\text{ft}^2}\right) \\ &= 2136 \text{ lbf/ft}^2 \end{aligned}$$

Compute the volume of hydrogen.

$$V_{\text{H}} = \frac{mRT}{p}$$
$$= \frac{(10{,}000\ \text{lbm})\left(766.5\ \frac{\text{ft-lbf}}{\text{lbm-°R}}\right)(516°\text{R})}{2136\ \frac{\text{lbf}}{\text{ft}^2}}$$
$$= 1.85 \times 10^6\ \text{ft}^3$$

Since the volume of the hydrogen contained in the blimp is equal to the air displaced, the air displaced can be computed from the ideal gas equation by assuming the air behaves like an ideal gas.

$$m = \frac{pV_{\text{H}}}{RT}$$

Since the air and hydrogen are assumed to be in thermal and pressure equilibrium, the temperature and pressure are equal to the value given for the hydrogen.

For air, $R = 53.35$ ft-lbf/lbm-°R.

Substituting gives

$$m_{\text{air}} = \frac{pV_{\text{H}}}{RT}$$
$$= \frac{\left(2136\ \frac{\text{lbf}}{\text{ft}^2}\right)(1.85 \times 10^6\ \text{ft}^3)}{\left(53.35\ \frac{\text{ft-lbf}}{\text{lbm-°R}}\right)(516°\text{R})}$$
$$= 1.435 \times 10^5\ \text{lbm}$$

Recall that the buoyant force is equal to the weight of the air.

$$F_b = W_{\text{air}} = \frac{mg}{g_c}$$
$$= \frac{(1.435 \times 10^5\ \text{lbm})\left(32.2\ \frac{\text{ft}}{\text{sec}^2}\right)}{32.2\ \frac{\text{lbm-ft}}{\text{lbf-sec}^2}}$$
$$= 1.435 \times 10^5\ \text{lbf}$$

Therefore, the lift can be calculated as

$$F_{\text{lift}} = F_b - W_{\text{H}}$$
$$= 1.435 \times 10^5\ \text{lbf} - 10{,}000\ \text{lbf}$$
$$= \boxed{1.335 \times 10^5\ \text{lbf}}$$

The answer is (C).

SI Solution

Assume the mass of the blimp structure is small (negligible) compared with the mass of the hydrogen.

The lift of the hydrogen-filled blimp (F_{lift}) is equal to the difference between the buoyant force (F_b) and the weight of the hydrogen contained in the blimp (W_{H}).

$$F_{\text{lift}} = F_b - W_{\text{H}}$$

The weight of the hydrogen is calculated from the mass of hydrogen by

$$W_{\text{H}} = mg$$
$$= (4500\ \text{kg})\left(9.81\ \frac{\text{m}}{\text{s}^2}\right)$$
$$= 44\,145\ \text{N}$$

The buoyant force is equal to the weight of the displaced air. The weight of the displaced air is calculated by knowing that the volume of the air displaced is equal to the volume of hydrogen enclosed in the blimp. Compute the volume of the hydrogen contained in the blimp by assuming the hydrogen behaves like an ideal gas.

$$V_{\text{H}} = \frac{mRT}{p}$$

For hydrogen, $R = 4124$ J/kg·K.

The temperature of the hydrogen is given as 13°C. Convert to absolute temperature (K).

$$T = 13°\text{C} + 273 = 286\text{K}$$

The pressure of the hydrogen is given as 770 mm Hg. Convert the pressure to units of pascals.

$$p = \frac{(770\ \text{mm Hg})\left(133.4\ \frac{\text{kPa}}{\text{m}}\right)}{1000\ \frac{\text{mm}}{\text{m}}}$$
$$= 102.7\ \text{kPa}$$

The volume of hydrogen is

$$V_{\text{H}} = \frac{mRT}{p}$$
$$= \frac{(4500\ \text{kg})\left(4124\ \frac{\text{J}}{\text{kg·K}}\right)(286\text{K})}{(102.7\ \text{kPa})\left(1000\ \frac{\text{Pa}}{\text{kPa}}\right)}$$
$$= 5.168 \times 10^4\ \text{m}^3$$

Since the volume of the hydrogen contained in the blimp is equal to the air displaced, the air displaced can be computed from the ideal gas equation by assuming the air behaves like an ideal gas.

$$m = \frac{pV_{\text{H}}}{RT}$$

Since the air and hydrogen are assumed to be in thermal and pressure equilibrium, the temperature and pressure are equal to the value given for the hydrogen.

For air, $R = 287$ J/kg·K.

Substituting gives

$$m_{\text{air}} = \frac{pV_{\text{H}}}{RT} = \frac{(102.7\ \text{kPa})\left(1000\ \frac{\text{Pa}}{\text{kPa}}\right)(5.168 \times 10^4\ \text{m}^3)}{\left(287\ \frac{\text{J}}{\text{kg·K}}\right)(286\text{K})} = 6.466 \times 10^4\ \text{kg}$$

The buoyant force is equal to the weight of the air, so

$$F_b = W_{\text{air}} = mg = (6.466 \times 10^4\ \text{kg})\left(9.81\ \frac{\text{m}}{\text{s}^2}\right) = 6.34 \times 10^5\ \text{N}$$

Therefore, the lift can be calculated as

$$F_{\text{lift}} = F_b - w_{\text{H}} = 6.34 \times 10^5\ \text{N} - 44\,145\ \text{N} = \boxed{5.90 \times 10^5\ \text{N}}$$

The answer is (C).

2. *Customary U.S. Solution*

The weight of the sphere is equal to the difference in the buoyant force of the fully submerged sphere (F_{fs}) and the buoyant force of the half-submerged sphere (F_{hs}).

The buoyant force is given by

$$F_b = \frac{\rho g V_{\text{displaced}}}{g_c}$$

Since the volume of the half-submerged sphere is equal to half that of the fully submerged sphere,

$$W_{\text{sphere}} = F_{\text{fs}} - F_{\text{hs}} = \left(\tfrac{1}{2}\right)\left(\frac{\rho g V_{\text{sphere}}}{g_c}\right)$$

For seawater, $\rho = 64.0\ \text{lbm/ft}^3$.

The volume of the sphere is given by

$$V_{\text{sphere}} = \left(\frac{\pi}{6}\right)d^3 = \left(\frac{\pi}{6}\right)(6\ \text{ft})^3 = 113.1\ \text{ft}^3$$

The weight of the sphere is

$$W_{\text{sphere}} = \left(\tfrac{1}{2}\right)\left(\frac{\rho g V_{\text{sphere}}}{g_c}\right) = \left(\tfrac{1}{2}\right)\left(\frac{\left(64.0\ \frac{\text{lbm}}{\text{ft}^3}\right)\left(32.2\ \frac{\text{ft}}{\text{sec}^2}\right) \times (113.1\ \text{ft}^3)}{32.2\ \frac{\text{lbm-ft}}{\text{lbf-sec}^2}}\right) = 3619\ \text{lbf}$$

The buoyant force equation for a fully submerged sphere and anchor can be solved for the concrete volume.

$$W_{\text{sphere}} + W_{\text{concrete}} = (V_{\text{sphere}} + V_{\text{concrete}})\rho_{\text{water}}$$

$$W_{\text{sphere}} + \rho_{\text{concrete}}V_{\text{concrete}}\left(\frac{g}{g_c}\right) = (V_{\text{sphere}} + V_{\text{concrete}})\rho_{\text{water}} \times \left(\frac{g}{g_c}\right)$$

$$3619\ \text{lbf} + \left(150\ \frac{\text{lbm}}{\text{ft}^3}\right)(V_{\text{concrete}})\left(\frac{32.2\ \frac{\text{ft}}{\text{sec}^2}}{32.2\ \frac{\text{ft-lbm}}{\text{lbf-sec}^2}}\right) = (113.1\ \text{ft}^3 + V_{\text{concrete}}) \times \left(64.0\ \frac{\text{lbm}}{\text{ft}^3}\right) \times \left(\frac{32.2\ \frac{\text{ft}}{\text{sec}^2}}{32.2\ \frac{\text{ft-lbm}}{\text{lbf-sec}^2}}\right)$$

$$V_{\text{concrete}} = 42.09\ \text{ft}^3$$

$$m_{\text{concrete}} = \rho_{\text{concrete}}V_{\text{concrete}} = \left(150\ \frac{\text{lbm}}{\text{ft}^3}\right)(42.09\ \text{ft}^3) = \boxed{6314\ \text{lbm}}$$

The answer is (D).

SI Solution

The weight of the sphere is equal to the difference in the buoyant force of the fully submerged sphere (F_{fs}) and the buoyant force of the half-submerged sphere (F_{hs}).

The buoyant force is given by

$$F_b = \rho g V_{\text{displaced}}$$

Since the volume of the half-submerged sphere is equal to half that of the fully submerged sphere,

$$W_{\text{sphere}} = F_{\text{fs}} - F_{\text{hs}} = \tfrac{1}{2}\rho g V_{\text{sphere}}$$

For seawater, $\rho = 1024\ \text{kg/m}^3$.

The volume of the sphere is given by

$$\begin{aligned} V_{\text{sphere}} &= \left(\frac{\pi}{6}\right) d^3 \\ &= \left(\frac{\pi}{6}\right)(1.8\ \text{m})^3 \\ &= 3.054\ \text{m}^3 \end{aligned}$$

The weight of the sphere required is

$$\begin{aligned} W_{\text{sphere}} &= \tfrac{1}{2}\rho g V_{\text{sphere}} \\ &= \left(\tfrac{1}{2}\right)\left(1024\ \frac{\text{kg}}{\text{m}^3}\right)\left(9.81\ \frac{\text{m}}{\text{s}^2}\right)(3.054\ \text{m}^3) \\ &= 15\,339\ \text{N} \end{aligned}$$

The buoyant force equation for a fully submerged sphere and anchor can be solved for the concrete volume.

$$\begin{aligned} W_{\text{sphere}} + W_{\text{concrete}} &= (V_{\text{sphere}} + V_{\text{concrete}})\rho_{\text{water}} \\ W_{\text{sphere}} + \rho_{\text{concrete}} g V_{\text{concrete}} &= g(V_{\text{sphere}} + V_{\text{concrete}})\rho_{\text{water}} \\ 15\,339\ \text{N} + \left(2400\ \frac{\text{kg}}{\text{m}^3}\right)\left(9.81\ \frac{\text{m}}{\text{s}^2}\right)(V_{\text{concrete}}) &= (3.054\ \text{m}^3 + V_{\text{concrete}}) \\ &\quad \times \left(1024\ \frac{\text{kg}}{\text{m}^3}\right)\left(9.81\ \frac{\text{m}}{\text{s}^2}\right) \\ V_{\text{concrete}} &= 1.136\ \text{m}^3 \\ m_{\text{concrete}} &= \rho_{\text{concrete}} V_{\text{concrete}} \\ &= \left(2400\ \frac{\text{kg}}{\text{m}^3}\right)(1.136\ \text{m}^3) \\ &= \boxed{2726\ \text{kg}} \end{aligned}$$

The answer is (D).

16 Fluid Flow Parameters

PRACTICE PROBLEMS

Use the following values unless told to do otherwise in the problem:

$$g = 32.2 \text{ ft/sec}^2 \ (9.81 \text{ m/s}^2)$$
$$\rho_{\text{water}} = 62.4 \text{ lbm/ft}^3 \ (1000 \text{ kg/m}^3)$$
$$p_{\text{atmospheric}} = 14.7 \text{ psia} \ (101.3 \text{ kPa})$$

Hydraulic Radius

1. A 10 in (25 cm) composition pipe is compressed by a tree root until its inside height is only 7.2 in (18 cm). What is its approximate hydraulic radius when flowing half full? State your assumptions.

(A) 2.2 in (5.5 cm)
(B) 2.7 in (6.9 cm)
(C) 3.2 in (8.1 cm)
(D) 4.5 in (11.4 cm)

2. A pipe with an inside diameter of 18.812 in contains water to a depth of 15.7 in. What is the hydraulic radius? (Work in customary U.S. units only.)

(A) 4.39 in
(B) 5.08 in
(C) 5.72 in
(D) 6.51 in

SOLUTIONS

1. *Customary U.S. Solution*

The perimeter of the pipe is

$$\begin{aligned} p &= \pi d = \pi(10 \text{ in}) \\ &= 31.42 \text{ in} \end{aligned}$$

If the pipe is flowing half full, the wetted perimeter becomes

$$\begin{aligned} \text{wetted perimeter} &= \tfrac{1}{2}p = \left(\tfrac{1}{2}\right)(31.42 \text{ in}) \\ &= 15.71 \text{ in} \end{aligned}$$

Assume the compressed pipe is an elliptical cross section. The ellipse will have a minor axis, b, equal to one-half the height of the compressed pipe or

$$b = \frac{7.2 \text{ in}}{2} = 3.6 \text{ in}$$

When the pipe is compressed, the perimeter of the pipe will remain constant. The perimeter of an ellipse is given by

$$p = 2\pi\sqrt{\tfrac{1}{2}(a^2 + b^2)}$$

Solve for the major axis.

$$\begin{aligned} a &= \sqrt{(2)\left(\frac{p}{2\pi}\right)^2 - b^2} \\ &= \sqrt{(2)\left(\frac{31.42 \text{ in}}{2\pi}\right)^2 - (3.6 \text{ in})^2} \\ &= 6.09 \text{ in} \end{aligned}$$

The flow area or area of the ellipse is given by

$$\begin{aligned} \text{flow area} &= \tfrac{1}{2}\pi ab \\ &= \tfrac{1}{2}\pi(6.09 \text{ in})(3.6 \text{ in}) \\ &= 34.4 \text{ in}^2 \end{aligned}$$

The hydraulic radius is

$$\begin{aligned} r_h &= \frac{\text{area in flow}}{\text{wetted perimeter}} = \frac{34.4 \text{ in}^2}{15.7 \text{ in}} \\ &= \boxed{2.19 \text{ in}} \end{aligned}$$

The answer is (A).

SI Solution

The perimeter of the pipe is

$$p = \pi d = \pi(25\text{ cm}) = 78.54\text{ cm}$$

If the pipe is flowing half full, the wetted perimeter becomes

$$\text{wetted perimeter} = \tfrac{1}{2}p = \left(\tfrac{1}{2}\right)(78.54\text{ cm}) = 39.27\text{ cm}$$

Assume the compressed pipe is an elliptical cross section. The ellipse will have a minor axis, b, equal to one-half the height of the compressed pipe or

$$b = \frac{18\text{ cm}}{2} = 9\text{ cm}$$

When the pipe is compressed, the perimeter of the pipe will remain constant. The perimeter of an ellipse is given by

$$p = 2\pi\sqrt{\left(\tfrac{1}{2}\right)(a^2+b^2)}$$

Solve for the major axis.

$$a = \sqrt{(2)\left(\frac{p}{2\pi}\right)^2 - b^2} = \sqrt{(2)\left(\frac{78.54\text{ cm}}{2\pi}\right)^2 - (9\text{ cm})^2} = 15.2\text{ cm}$$

The flow area or area of the ellipse is given by

$$\text{flow area} = \tfrac{1}{2}\pi ab = \tfrac{1}{2}\pi(15.2\text{ cm})(9\text{ cm}) = 214.9\text{ cm}^2$$

The hydraulic radius is

$$r_h = \frac{\text{area in flow}}{\text{wetted perimeter}} = \frac{214.9\text{ cm}^2}{39.27\text{ cm}} = \boxed{5.47\text{ cm}}$$

The answer is (A).

2. *method 1:* Use App. 7.A for a circular segment.

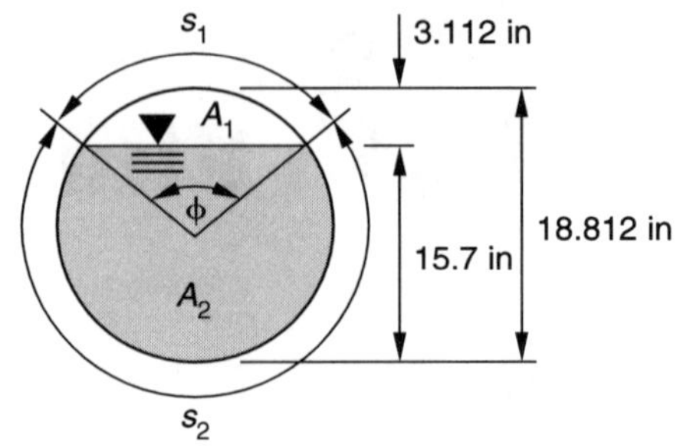

$$r = \frac{d}{2} = \frac{18.812\text{ in}}{2} = 9.406\text{ in}$$

$$\phi = (2)\left(\arccos\frac{r-d}{r}\right) = (2)\left(\arccos\frac{9.406\text{ in} - 3.112\text{ in}}{9.406\text{ in}}\right) = 1.675\text{ rad}$$

$$\sin\phi = 0.9946$$

$$A_1 = \tfrac{1}{2}r^2(\phi - \sin\phi) = \left(\tfrac{1}{2}\right)(9.406\text{ in})^2(1.675 - 0.9946) = 30.1\text{ in}^2$$

$$A_{\text{total}} = A_1 + A_2 = \frac{\pi}{4}D^2 = \left(\frac{\pi}{4}\right)(18.812\text{ in})^2 = 277.95\text{ in}^2$$

$$A_2 = A_{\text{total}} - A_1 = 277.95\text{ in}^2 - 30.1\text{ in}^2 = 247.85\text{ in}^2$$

$$s_1 = r\phi = (9.406\text{ in})(1.675) = 15.76\text{ in}$$

$$s_{\text{total}} = s_1 + s_2 = \pi D = \pi(18.812\text{ in}) = 59.1\text{ in}$$

$$s_2 = s_{\text{total}} - s_1 = 59.1\text{ in} - 15.76\text{ in} = 43.34\text{ in}$$

$$r_h = \frac{A_2}{s_2} = \frac{247.85\text{ in}^2}{43.34\text{ in}} = \boxed{5.719\text{ in}}$$

method 2: Use App. 16.A.

$$\frac{d}{D} = \frac{15.7\text{ in}}{18.812\text{ in}} = 0.83$$

From App. 16.A, $r_h/D = 0.3041$.

$$r_h = (0.3041)(18.812\text{ in}) = \boxed{5.72\text{ in}}$$

The answer is (C).

17 Fluid Dynamics

PRACTICE PROBLEMS

(Use g = 32.2 ft/sec^2 (9.81 m/s^2) and 60°F (16°C) water unless told to do otherwise.)

Conservation of Energy

1. 5.0 ft^3/sec (130 L/s) of water flows through a schedule-40 pipe that changes gradually in diameter from 6 in (154 mm) at point A to 18 in (429 mm) at point B. Point B is 15 ft (4.6 m) higher than point A. The respective pressures at points A and B are 10 psia (70 kPa) and 7 psia (48.3 kPa). All minor losses are insignificant. What are the direction of flow and velocity at point A?

(A) 3.2 ft/sec (1 m/s); from A to B
(B) 25 ft/sec (7.0 m/s); from A to B
(C) 3.2 ft/sec (1 m/s); from B to A
(D) 25 ft/sec (7.5 m/s); from B to A

2. Points A and B are separated by 3000 ft of new 6 in schedule-40 steel pipe. 750 gal/min of 60°F water flow from point A to point B. Point B is 60 ft above point A. What must be the pressure at point A if the pressure at B must be 50 psig?

(A) 87 psig
(B) 103 psig
(C) 125 psig
(D) 167 psig

3. *(Time limit: one hour)* A pipe network connects junctions A, B, C, and D as shown. All pipe sections have a C-value of 150. Water can be added and removed at any of the junctions to achieve the flows listed. Water flows from point A to point D. No flows are backward. The minimum allowable pressure anywhere in the system is 20 psi. All minor losses are insignificant. For simplification, use the nominal pipe sizes.

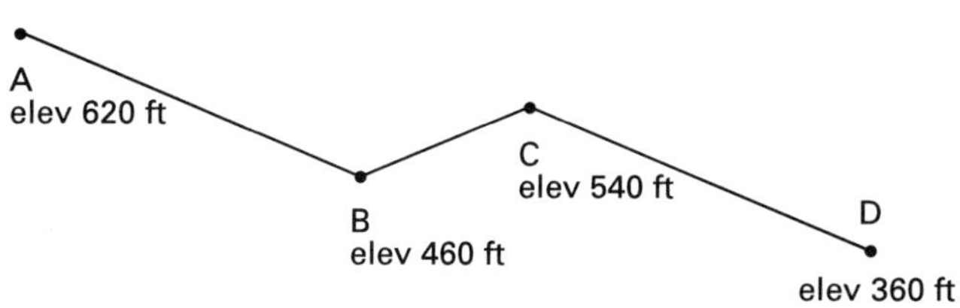

pipe section	length	diameter (in)	flow
A to B	20,000 ft	6	120 gal/min
B to C	10,000 ft	6	160 gal/min
C to D	30,000 ft	4	120 gal/min

(a) What is the pressure at point A?

(A) 14 psig
(B) 23 psig
(C) 31 psig
(D) 47 psig

(b) What is the elevation of the hydraulic grade line at point A referenced to point D?

(A) 330 ft
(B) 470 ft
(C) 610 ft
(D) 750 ft

Friction Loss

4. 1.5 ft^3/sec (40 L/s) of 70°F (20°C) water flows through 1200 ft (355 m) of 6 in (nominal) diameter new schedule-40 pipe. What is the friction loss?

(A) 4 ft (1.2 m)
(B) 18 ft (5.2 m)
(C) 36 ft (9.5 m)
(D) 70 ft (2.1 m)

5. 500 gal/min (30 L/s) of 100°F (40°C) water flows through 300 ft (90 m) of 6 in schedule-40 pipe. The pipe contains two 6 in flanged steel elbows, two full-open gate valves, a full-open 90° angle valve, and a swing check valve. The discharge is located 20 ft (6 m) higher than the entrance. What is the pressure difference between the two ends of the pipe?

(A) 12 psi (78 kPa)
(B) 21 psi (140 kPa)
(C) 45 psi (310 kPa)
(D) 87 psi (600 kPa)

6. 70°F (20°C) air is flowing at 60 ft/sec (18 m/s) through 300 ft (90 m) of 6 in schedule-40 pipe. The pipe contains two 6 in flanged steel elbows, two full-open gate valves, a full-open 90° angle valve, and a swing check valve. The discharge is located 20 ft (6 m) higher than the entrance. What is the pressure difference between the two ends of the pipe?

(A) 0.26 psi (1.8 kPa)
(B) 0.49 psi (3.2 kPa)
(C) 1.5 psi (10 kPa)
(D) 13 psi (90 kPa)

Reservoirs

7. Three reservoirs (A, B, and C) are interconnected with a common junction (point D) at elevation 25 ft above an arbitrary reference point. The water levels for reservoirs A, B, and C are at elevations of 50, 40, and 22 ft, respectively. The pipe from reservoir A to the junction is 800 ft of 3 in (nominal) steel pipe. The pipe from reservoir B to the junction is 500 ft of 10 in (nominal) steel pipe. The pipe from reservoir C to the junction is 1000 ft of 4 in (nominal) steel pipe. All pipes are schedule 40 with a friction factor of 0.02. All minor losses and velocity heads can be neglected. What are the direction of flow and pressure at point D?

(A) out of reservoir B; 500 psf
(B) out of reservoir B; 930 psf
(C) into reservoir B; 1100 psf
(D) into reservoir B; 1260 psf

Water Hammer

8. *(Time limit: one hour)* A cast-iron pipe has an inside diameter of 24 in (600 mm) and a wall thickness of 0.75 in (20 mm). The pipe's modulus of elasticity is 20×10^6 psi (140 GPa). The pipeline is 500 ft (150 m) long. 70°F (20°C) water is flowing at 6 ft/sec (2 m/s).

(a) If a valve is closed instantaneously, what will be the maximum pressure experienced in the pipe?

(A) 48 psi (330 kPa)
(B) 140 psi (970 kPa)
(C) 320 psi (2.5 MPa)
(D) 470 psi (3.2 MPa)

(b) If the pipe is 500 ft (150 m) long, over what length of time must the valve be closed to create a pressure equivalent to instantaneous closure?

(A) 0.25 sec
(B) 0.68 sec
(C) 1.6 sec
(D) 2.1 sec

Parallel Pipe Systems

9. 8 MGD (millions of gallons per day) (350 L/s) of 70°F (20°C) water flows into the smooth schedule-40 steel pipe network shown. Minor losses are insignificant.

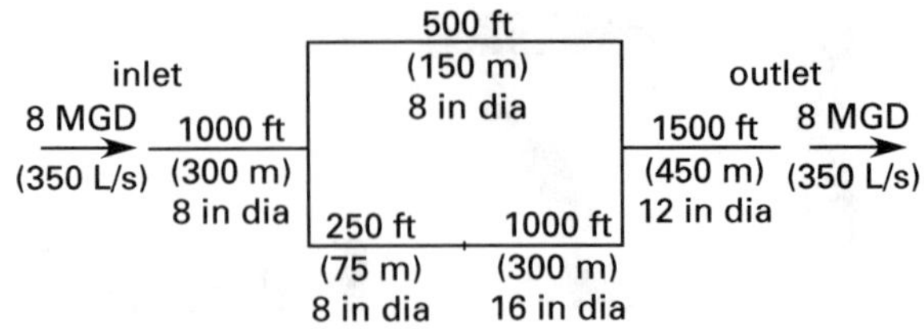

(a) What quantity of water is flowing in the upper branch?

(A) 1.2 ft^3/sec (0.034 m^3/s)
(B) 2.9 ft^3/sec (0.081 m^3/s)
(C) 4.1 ft^3/sec (0.11 m^3/s)
(D) 5.3 ft^3/sec (0.15 m^3/s)

(b) What is the energy loss per unit mass between the inlet and the outlet?

(A) 120 ft-lbf/lbm (0.37 kJ/kg)
(B) 300 ft-lbf/lbm (0.90 kJ/kg)
(C) 480 ft-lbf/lbm (1.4 kJ/kg)
(D) 570 ft-lbf/lbm (1.7 kJ/kg)

Pipe Networks

10. A single-loop pipe network is shown. The distance between each junction is 1000 ft. All junctions are on the same elevation. All pipes have a *C*-value of 100. The volumetric flow rates are to be determined to within 2 gal/min.

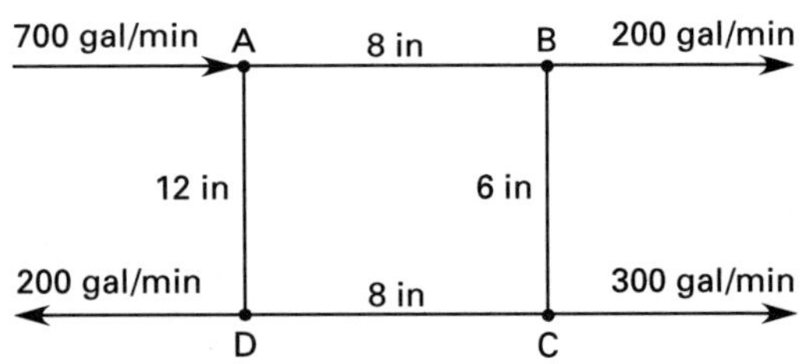

What is the flow rate between junctions B and C?

(A) 36 gal/min from B to C
(B) 58 gal/min from B to C
(C) 84 gal/min from C to B
(D) 112 gal/min from C to B

11. *(Time limit: one hour)* A double-loop pipe network is shown. The distance between each junction is 1000 ft. The water temperature is 60°F. Elevations and some pressure are known for the junctions. All pipes have a *C*-value of 100.

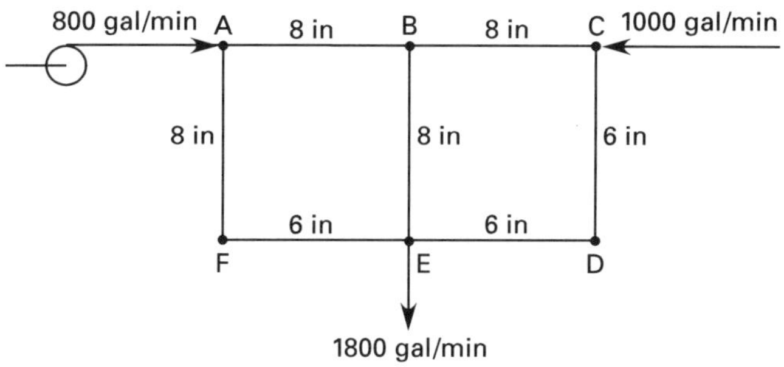

point	pressure	elevation
A		200 ft
B		150 ft
C	40 psig	300 ft
D		150 ft
E		200 ft
F		150 ft

(a) What is the flow rate between junctions B and E?

(A) 540 gal/min
(B) 620 gal/min
(C) 810 gal/min
(D) 980 gal/min

(b) What is the pressure at point D?

(A) 51 psig
(B) 73 psig
(C) 96 psig
(D) 120 psig

(c) If the pump receives water at 20 psig (140 kPa), what hydraulic power is required?

(A) 14 hp
(B) 28 hp
(C) 35 hp
(D) 67 hp

12. *(Time limit: one hour)* The water distribution network shown consists of class A cast-iron pipe installed 15 years ago. 1.5 MGD of 50°F water enters at junction A and leaves at junction B. The minimum acceptable pressure at point D is 40 psig (280 kPa).

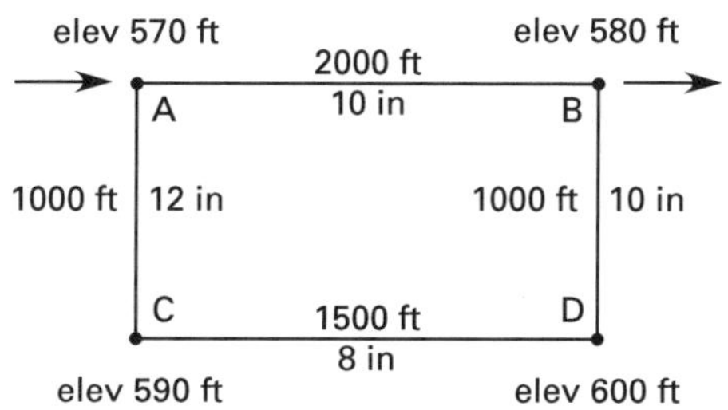

(a) What is the flow rate between junctions A and B?

(A) 320 gal/min
(B) 480 gal/min
(C) 590 gal/min
(D) 660 gal/min

(b) What is the pressure at point B?

(A) 32 psig
(B) 48 psig
(C) 57 psig
(D) 88 psig

Tank Discharge

13. The velocity of discharge from a fire hose is 50 ft/sec (15 m/s). The hose is oriented 45° from the horizontal. Disregarding air friction, what is the maximum range of the discharge?

(A) 45 ft (14 m)
(B) 78 ft (23 m)
(C) 91 ft (24 m)
(D) 110 ft (33 m)

14. A full cylindrical tank 40 ft (12 m) high has a constant diameter of 20 ft. The tank has a 4 in (100 mm) diameter hole in its bottom. The coefficient of discharge for the hole is 0.98. How long will it take for the water level to drop from 40 ft to 20 ft (12 m to 6 m)?

(A) 950 sec
(B) 1200 sec
(C) 1450 sec
(D) 1700 sec

Venturi Meters

15. A venturi meter with an 8 in diameter throat is installed in a 12 in diameter water line. The venturi is perfectly smooth, so that the discharge coefficient is 1.00. An attached mercury manometer registers a 4 in differential. What is the volumetric flow rate?

(A) 1.7 ft^3/sec
(B) 5.2 ft^3/sec
(C) 6.4 ft^3/sec
(D) 18 ft^3/sec

16. 60°F (15°C) benzene (specific gravity at 60°F (15°C) of 0.885) flows through an 8 in/3½ in (200 mm/90 mm) venturi meter whose coefficient of discharge is 0.99. A mercury manometer indicates a 4 in difference in the heights of the mercury columns. What is the volumetric flow rate of the benzene?

(A) 1.2 ft^3/sec (34 L/s)
(B) 9.1 ft^3/sec (250 L/s)
(C) 13 ft^3/sec (360 L/s)
(D) 27 ft^3/sec (760 L/s)

Orifice Meters

17. A sharp-edged orifice meter with a 0.2 ft diameter opening is installed in a 1.0 ft diameter pipe. 70°F water approaches the orifice at 2 ft/sec. What is the indicated pressure drop across the orifice meter?

(A) 5.9 psi
(B) 13 psi
(C) 22 psi
(D) 47 psi

18. A mercury manometer is used to measure a pressure difference across an orifice meter in a water line. The difference in mercury levels is 7 in (17.8 cm). What is the pressure differential?

(A) 1.7 psi (12 kPa)
(B) 3.2 psi (22 kPa)
(C) 7.9 psi (55 kPa)
(D) 23 psi (160 kPa)

19. A sharp-edged ISA orifice is used in a 12 in (300 mm) schedule-40 water line. (Figure 17.28 is applicable.) The water temperature is 70°F (20°C), and the flow rate is 10 ft^3/sec (250 L/s). The head loss across the orifice must not exceed 25 ft (7.5 m). What is the smallest orifice that can be used?

(A) 5.5 in (14 cm)
(B) 7.3 in (19 cm)
(C) 8.1 in (20 cm)
(D) 8.9 in (23 cm)

Impulse-Momentum

20. A pipe necks down from 24 in at point A to 12 in at point B. 8 ft^3/sec of 60°F water flow from point A to point B. The pressure head at point A is 20 ft. Friction is insignificant over the distance between points A and B. What are the magnitude and direction of the resultant force on the water?

(A) 2900 lbf; toward A
(B) 3500 lbf; toward A
(C) 2900 lbf; toward B
(D) 3500 lbf; toward B

21. A 2 in (50 mm) diameter horizontal water jet has an absolute velocity (with respect to a stationary point) of 40 ft/sec (12 m/s) as it strikes a curved blade. The blade is moving horizontally away with an absolute velocity of 15 ft/sec (4.5 m/s). Water is deflected 60° from the horizontal. What is the force on the blade?

(A) 18 lbf (80 N)
(B) 26 lbf (110 N)
(C) 35 lbf (160 N)
(D) 47 lbf (210 N)

Pumps

22. 2000 gal/min (125 L/s) of brine with a specific gravity of 1.2 pass through an 85% efficient pump. The centerlines of the pump's 12 in (300 mm) inlet and 8 in (200 mm) outlet are at the same elevation. The inlet suction gauge indicates 6 in (150 mm) of mercury below atmospheric. The discharge pressure gauge is located 4 ft (1.2 m) above the centerline of the pump's outlet and indicates 20 psig (138 kPa). What is the input power to the pump?

(A) 12 hp (8.9 kW)
(B) 36 hp (26 kW)
(C) 52 hp (39 kW)
(D) 87 hp (65 kW)

Turbines

23. 100 ft^3/sec (2.6 kL/s) of water pass through a horizontal turbine. The water's pressure is reduced from 30 psig (210 kPa) to 5 psi (35 kPa) vacuum. Disregarding friction, velocity, and other factors, what power is generated?

(A) 110 hp (82 kW)
(B) 380 hp (280 kW)
(C) 730 hp (540 kW)
(D) 920 hp (640 kW)

Drag

24. *(Time limit: one hour)* A car traveling through 70°F (20°C) air has the following characteristics.

frontal area	28 ft^2 (2.6 m^2)
mass	3300 lbm (1500 kg)
drag coefficient	0.42
rolling resistance	1% of weight
engine thermal efficiency	28%
fuel heating value	115,000 Btu/gal (460 MJ/L)

(a) Considering only the drag and rolling resistance, what is the fuel consumption when the car is traveling at 55 mi/hr (90 km/h)?

(A) 0.026 gal/mi (0.0043 L/km)
(B) 0.038 gal/mi (0.0065 L/km)
(C) 0.051 gal/mi (0.0097 L/km)
(D) 0.13 gal/mi (0.022 L/km)

(b) What is the percentage increase in fuel consumption at 65 mi/hr (105 km/h) compared to at 55 mi/hr (90 km/h)?

(A) 10%
(B) 20%
(C) 30%
(D) 40%

SOLUTIONS

1. *Customary U.S. Solution*

Assume schedule-40 pipe.

$$D_A = 0.5054 \text{ ft}$$
$$D_B = 1.4063 \text{ ft}$$

Let point A be at zero elevation.

The total energy at point A from Bernoulli's equation is

$$E_{tA} = E_p + E_v + E_z = \frac{p_A}{\rho} + \frac{v_A^2}{2g_c} + \frac{z_A g}{g_c}$$

At point A, the diameter is 6 in. The velocity at point A is

$$\dot{V} = v_A A_A = v_A\left(\frac{\pi}{4}\right)D_A^2$$

$$v_A = \left(\frac{4}{\pi}\right)\left(\frac{\dot{V}}{D_A^2}\right) = \left(\frac{4}{\pi}\right)\left(\frac{5.0\ \frac{\text{ft}^3}{\text{sec}}}{(0.5054\text{ ft})^2}\right) = \boxed{24.9 \text{ ft/sec}}$$

$$p_A = \left(10\ \frac{\text{lbf}}{\text{in}^2}\right)\left(\frac{144 \text{ in}^2}{1 \text{ ft}^2}\right) = 1440 \text{ lbf/ft}^2$$

$$z_A = 0$$

For water, $\rho \approx 62.4 \text{ lbm/ft}^3$.

$$E_{tA} = \frac{p_A}{\rho} + \frac{v_A^2}{2g_c} + \frac{z_A g}{g_c} = \frac{1440\ \frac{\text{lbf}}{\text{ft}^2}}{62.4\ \frac{\text{lbm}}{\text{ft}^3}} + \frac{\left(24.9\ \frac{\text{ft}}{\text{sec}}\right)^2}{(2)\left(32.2\ \frac{\text{lbm-ft}}{\text{lbf-sec}^2}\right)} + 0 = 32.7 \text{ ft-lbf/lbm}$$

Similarly, the total energy at point B is

$$v_B = \left(\frac{4}{\pi}\right)\left(\frac{\dot{V}}{D_B^2}\right) = \left(\frac{4}{\pi}\right)\left(\frac{5.0\ \frac{\text{ft}^3}{\text{sec}}}{(1.4063\text{ ft})^2}\right) = 3.22 \text{ ft/sec}$$

$$p_B = \left(7\ \frac{\text{lbf}}{\text{in}^2}\right)\left(\frac{144 \text{ in}^2}{1 \text{ ft}^2}\right) = 1008 \text{ lbf/ft}^2$$

$$z_B = 15 \text{ ft}$$

$$E_{tB} = \frac{p_B}{\rho} + \frac{v_B^2}{2g_c} + \frac{z_B g}{g_c} = \frac{1008\ \frac{\text{lbf}}{\text{ft}^2}}{62.4\ \frac{\text{lbm}}{\text{ft}^3}} + \frac{\left(3.22\ \frac{\text{ft}}{\text{sec}}\right)^2}{(2)\left(32.2\ \frac{\text{lbm-ft}}{\text{lbf-sec}^2}\right)} + \frac{(15\text{ ft})\left(32.2\ \frac{\text{ft}}{\text{sec}^2}\right)}{32.2\ \frac{\text{lbm-ft}}{\text{lbf-sec}^2}} = 31.3 \text{ ft-lbf/lbm}$$

Since $E_{tA} > E_{tB}$, $\boxed{\text{the flow is from point A to point B.}}$

The answer is (B).

SI Solution

Let point A be at zero elevation.

The total energy at point A from Bernoulli's equation is

$$E_{tA} = E_p + E_v + E_z = \frac{p_A}{\rho} + \frac{v_A^2}{2} + z_A g$$

At point A, the diameter is 154 mm (0.154 m). The velocity at point A is

$$\dot{V} = v_A A_A = v_A\left(\frac{\pi}{4}\right)D_A^2$$

$$v_A = \left(\frac{4}{\pi}\right)\left(\frac{\dot{V}}{D_A^2}\right) = \left(\frac{4}{\pi}\right)\left(\frac{\left(130\ \frac{\text{L}}{\text{s}}\right)\left(\frac{1 \text{ m}^3}{1000 \text{ L}}\right)}{(0.154 \text{ m})^2}\right) = \boxed{6.98 \text{ m/s}}$$

$$p_A = 69 \text{ kPa } (69\,000 \text{ Pa})$$

$$z_A = 0$$

For water, $\rho = 1000 \text{ kg/m}^3$.

$$E_{tA} = \frac{p_A}{\rho} + \frac{v_A^2}{2} + z_A g = \frac{70\,000 \text{ Pa}}{1000\ \frac{\text{kg}}{\text{m}^3}} + \frac{\left(6.98\ \frac{\text{m}}{\text{s}}\right)^2}{2} + 0 = 94.36 \text{ J/kg}$$

Similarly, the total energy at point B is

$$\begin{aligned} v_B &= \left(\frac{4}{\pi}\right)\left(\frac{\dot{V}}{D_B^2}\right) \\ &= \left(\frac{4}{\pi}\right)\left(\frac{\left(130\ \frac{\text{L}}{\text{s}}\right)\left(\frac{1\ \text{m}^3}{1000\ \text{L}}\right)}{(0.429\ \text{m})^2}\right) \\ &= 0.90\ \text{m/s} \\ p_B &= 48.3\ \text{kPa}\ (48\,300\ \text{Pa}) \\ z_B &= 4.6\ \text{m} \end{aligned}$$

$$\begin{aligned} E_{tB} &= \frac{p_B}{\rho} + \frac{v_B^2}{2} + z_B g \\ &= \frac{48\,300\ \text{Pa}}{1000\ \frac{\text{kg}}{\text{m}^3}} + \frac{\left(0.90\ \frac{\text{m}}{\text{s}}\right)^2}{2} \\ &\quad + (4.6\ \text{m})\left(9.81\ \frac{\text{m}}{\text{s}^2}\right) \\ &= 93.8\ \text{J/kg} \end{aligned}$$

Since $E_{tA} > E_{tB}$, the flow is from point A to point B.

The answer is (B).

2. $$\begin{aligned} \dot{V} &= \left(750\ \frac{\text{gal}}{\text{min}}\right)\left(0.002228\ \frac{\text{ft}^3\text{-min}}{\text{sec-gal}}\right) \\ &= 1.671\ \text{ft}^3/\text{sec} \end{aligned}$$

$D = 0.5054$ ft and $A = 0.2006\ \text{ft}^2$.

$$v = \frac{\dot{V}}{A} = \frac{1.671\ \frac{\text{ft}^3}{\text{sec}}}{0.2006\ \text{ft}^2} = 8.33\ \text{ft/sec}$$

For 60°F water,

$$\nu = 1.217 \times 10^{-5}\ \text{ft}^2/\text{sec} \quad \text{[App. 14.A]}$$

$$\text{Re} = \frac{vD}{\nu} = \frac{\left(8.33\ \frac{\text{ft}}{\text{sec}}\right)(0.5054\ \text{ft})}{1.217 \times 10^{-5}\ \frac{\text{ft}^2}{\text{sec}}} = 3.46 \times 10^5$$

For steel,

$$\begin{aligned} \epsilon &= 0.0002 \quad \text{[App. 17.A]} \\ \frac{\epsilon}{D} &= \frac{0.0002\ \text{ft}}{0.5054\ \text{ft}} \approx 0.0004 \\ f &= 0.0175 \quad \text{[App. 17.B]} \end{aligned}$$

From Eq. 17.22,

$$h_f = \frac{(0.0175)(3000\ \text{ft})\left(8.33\ \frac{\text{ft}}{\text{sec}}\right)^2}{(2)(0.5054\ \text{ft})\left(32.2\ \frac{\text{ft}}{\text{sec}^2}\right)} = 111.9\ \text{ft}$$

Use the Bernoulli equation. Since velocity is the same at points A and B, it may be omitted.

$$\frac{\left(144\ \frac{\text{in}^2}{\text{ft}^2}\right)p_1}{62.37\ \frac{\text{lbf}}{\text{ft}^3}} = \frac{\left(50\ \frac{\text{lbf}}{\text{in}^2}\right)\left(144\ \frac{\text{in}^2}{\text{ft}^2}\right)}{62.37\ \frac{\text{lbf}}{\text{ft}^3}} + 60\ \text{ft} + 111.9\ \text{ft}$$

$$p_1 = \boxed{124.5\ \text{lbf/in}^2\ \text{(psig)}}$$

The answer is (C).

3. From Eq. 17.31,

$$\begin{aligned} h_{f,\text{A-B}} &= \frac{(10.44)(20{,}000\ \text{ft})\left(120\ \frac{\text{gal}}{\text{min}}\right)^{1.85}}{(150)^{1.85}(6\ \text{in})^{4.8655}} \\ &= 22.6\ \text{ft} \end{aligned}$$

Investigate velocity head.

$$\begin{aligned} v &= \frac{\dot{V}}{A} \\ &= \frac{\left(120\ \frac{\text{gal}}{\text{min}}\right)\left(0.002228\ \frac{\text{ft}^3\text{-min}}{\text{sec-gal}}\right)}{\left(\frac{\pi}{4}\right)\left(\frac{6\ \text{in}}{12\ \frac{\text{in}}{\text{ft}}}\right)^2} = 1.36\ \text{ft/sec} \end{aligned}$$

$$\begin{aligned} h_v &= \frac{v^2}{2g} \\ &= \frac{\left(1.36\ \frac{\text{ft}}{\text{sec}}\right)^2}{(2)\left(32.2\ \frac{\text{ft}}{\text{sec}^2}\right)} = 0.029\ \text{ft} \end{aligned}$$

Velocity heads are low and can be disregarded.

$$\begin{aligned} h_{f,\text{B-C}} &= \frac{(10.44)(10{,}000\ \text{ft})\left(160\ \frac{\text{gal}}{\text{min}}\right)^{1.85}}{(150)^{1.85}(6\ \text{in})^{4.8655}} \\ &= 19.25\ \text{ft} \end{aligned}$$

$$\begin{aligned} h_{f,\text{C-D}} &= \frac{(10.44)(30{,}000\ \text{ft})\left(120\ \frac{\text{gal}}{\text{min}}\right)^{1.85}}{(150)^{1.85}(4\ \text{in})^{4.8655}} \\ &= 243.9\ \text{ft} \end{aligned}$$

Assume a pressure of 20 psig at point A.

$$h_{p,\text{A}} = \frac{\left(20\ \frac{\text{lbf}}{\text{in}^2}\right)\left(144\ \frac{\text{in}^2}{\text{ft}^2}\right)}{62.4\ \frac{\text{lbf}}{\text{ft}^3}} = 46.2\ \text{ft}$$

From the Bernoulli equation, ignoring velocity head,

$$\begin{aligned} h_{p,\text{A}} + z_\text{A} &= h_{p,\text{B}} + z_\text{B} + h_{f,\text{A-B}} \\ 46.2\ \text{ft} + 620\ \text{ft} &= h_{p,\text{B}} + 460\ \text{ft} + 22.6\ \text{ft} \\ h_{p,\text{B}} &= 183.6\ \text{ft} \end{aligned}$$

$$\begin{aligned} p_\text{B} = \gamma h &= \frac{\left(62.4\ \frac{\text{lbf}}{\text{ft}^3}\right)(183.6\ \text{ft})}{144\ \frac{\text{in}^2}{\text{ft}^2}} \\ &= 79.6\ \text{lbf/in}^2\ \text{(psig)} \end{aligned}$$

For B to C,

$$\begin{aligned} 183.6\ \text{ft} + 460\ \text{ft} &= h_{p,\text{C}} + 540\ \text{ft} + 19.25\ \text{ft} \\ h_{p,\text{C}} &= 84.35\ \text{ft} \end{aligned}$$

$$\begin{aligned} p_\text{C} &= \frac{(84.35\ \text{ft})\left(62.4\ \frac{\text{lbf}}{\text{ft}^3}\right)}{144\ \frac{\text{in}^2}{\text{ft}^2}} \\ &= 36.6\ \text{lbf/in}^2\ \text{(psig)} \end{aligned}$$

For C to D,

$$\begin{aligned} 84.35\ \text{ft} + 540\ \text{ft} &= h_{p,\text{D}} + 360\ \text{ft} + 243.9\ \text{ft} \\ h_{p,\text{D}} &= 20.45\ \text{ft} \end{aligned}$$

$$\begin{aligned} p_\text{D} &= \frac{(20.45\ \text{ft})\left(62.4\ \frac{\text{lbf}}{\text{ft}^3}\right)}{144\ \frac{\text{in}^2}{\text{ft}^2}} \\ &= 8.9\ \text{lbf/in}^2\ \text{(psig)}\quad [\text{too low}] \end{aligned}$$

Since p_D is too low, add $20 - 8.9 = 11.1\ \text{lbf/in}^2$ (psig) to each point.

$$\text{(a)}\quad p_\text{A} = 20.0\ \frac{\text{lbf}}{\text{in}^2} + 11.1\ \frac{\text{lbf}}{\text{in}^2} = \boxed{31.1\ \text{lbf/in}^2\ \text{(psig)}}$$

$$p_\text{B} = 79.6\ \frac{\text{lbf}}{\text{in}^2} + 11.1\ \frac{\text{lbf}}{\text{in}^2} = 90.7\ \text{lbf/in}^2\ \text{(psig)}$$

$$p_\text{C} = 36.6\ \frac{\text{lbf}}{\text{in}^2} + 11.1\ \frac{\text{lbf}}{\text{in}^2} = 47.7\ \text{lbf/in}^2\ \text{(psig)}$$

$$p_\text{D} = 8.9\ \frac{\text{lbf}}{\text{in}^2} + 11.1\ \frac{\text{lbf}}{\text{in}^2} = 20.0\ \text{lbf/in}^2\ \text{(psig)}$$

The answer is (C).

(b) The elevation of the hydraulic grade line above point D is the sum of the potential and static heads.

$$\begin{aligned} \Delta h_\text{A-D} &= z_\text{A} - z_\text{D} + \frac{p_\text{A}}{\gamma} \\ &= 620\ \text{ft} - 360\ \text{ft} \\ &\quad + \frac{\left(31.1\ \frac{\text{lbf}}{\text{in}^2}\right)\left(144\ \frac{\text{in}^2}{\text{ft}^2}\right)}{62.4\ \frac{\text{lbf}}{\text{ft}^3}} \\ &= \boxed{331.8\ \text{ft}} \end{aligned}$$

If the hydraulic grade line was with respect to point A, only the static head would be included.

The answer is (A).

4. *Customary U.S. Solution*

For 6 in schedule-40 pipe, the internal diameter, D_i, is 0.5054 ft. The internal area is 0.2006 ft^2.

The velocity, v, is calculated from the volumetric flow, $\dot{V}$, and the flow area, A_i, by

$$\text{v} = \frac{\dot{V}}{A_i} = \frac{1.5\ \frac{\text{ft}^3}{\text{sec}}}{0.2006\ \text{ft}^2} = 7.48\ \text{ft/sec}$$

For water at 70°F, the kinematic viscosity, ν, is $1.059 \times 10^{-5}\ \text{ft}^2/\text{sec}$ [App. 14.A].

Calculate the Reynolds number.

$$\begin{aligned} \text{Re} &= \frac{D_i \text{v}}{\nu} \\ &= \frac{(0.5054\ \text{ft})\left(7.48\ \frac{\text{ft}}{\text{sec}}\right)}{1.059 \times 10^{-5}\ \frac{\text{ft}^2}{\text{sec}}} \\ &= 3.57 \times 10^5 \end{aligned}$$

Since Re > 2100, the flow is turbulent. The friction loss coefficient can be determined from the Moody diagram.

For new steel pipe, the specific roughness, ϵ, is 0.0002 ft.

The relative roughness is

$$\frac{\epsilon}{D_i} = \frac{0.0002\ \text{ft}}{0.5054\ \text{ft}} = 0.0004$$

From the Moody diagram with $\text{Re} = 3.57 \times 10^5$ and $e/D_i = 0.0004$, the friction factor, f, can be determined as 0.0174.

Use Darcy's equation to compute the frictional loss.

$$h_f = \frac{fLv^2}{2D_i g} = \frac{(0.0174)(1200 \text{ ft})\left(7.48 \ \frac{\text{ft}}{\text{sec}}\right)^2}{(2)(0.5054 \text{ ft})\left(32.2 \ \frac{\text{ft}}{\text{sec}^2}\right)} = \boxed{35.9 \text{ ft}}$$

The answer is (C).

SI Solution

For 6 in pipe, the internal diameter is 154.1 mm and the internal area is 186.5×10^{-4} m^2.

The velocity, v, is calculated from the volumetric flow, $\dot{V}$, and the flow area, A_i, by

$$\text{v} = \frac{\dot{V}}{A_i} = \frac{\left(40 \ \frac{\text{L}}{\text{s}}\right)\left(0.001 \ \frac{\text{m}^3}{\text{L}}\right)}{186.5 \times 10^{-4} \text{ m}^2} = 2.145 \text{ m/s}$$

For water at 20°C, the kinematic viscosity is

$$\nu = \frac{\mu}{\rho} = \frac{1.0050 \times 10^{-3} \text{ Pa·s}}{998.23 \ \frac{\text{kg}}{\text{m}^3}} = 1.007 \times 10^{-6} \text{ m}^2/\text{s}$$

Calculate the Reynolds number.

$$\text{Re} = \frac{D_i \text{v}}{\nu} = \frac{(154.1 \text{ mm})\left(0.001 \ \frac{\text{m}}{\text{mm}}\right)\left(2.145 \ \frac{\text{m}}{\text{s}}\right)}{1.007 \times 10^{-6} \ \frac{\text{m}^2}{\text{s}}} = 3.282 \times 10^5$$

Since Re > 2100, the flow is turbulent. The friction loss coefficient can be determined from the Moody diagram.

For new steel pipe, the specific roughness, ϵ, is 6.0×10^{-5} m.

The relative roughness is

$$\frac{\epsilon}{D_i} = \frac{6.0 \times 10^{-5} \text{ m}}{0.1541 \text{ m}} = 0.0004$$

From the Moody diagram with Re $= 3.28 \times 10^5$ and $e/D_i = 0.0004$, the friction factor, f, can be determined as 0.0175.

Use Darcy's equation to compute the frictional loss.

$$h_f = \frac{fLv^2}{2D_i g} = \frac{(0.0175)(355 \text{ m})\left(2.145 \ \frac{\text{m}}{\text{s}}\right)^2}{(2)(0.1541 \text{ m})\left(9.81 \ \frac{\text{m}}{\text{s}^2}\right)} = \boxed{9.45 \text{ m}}$$

The answer is (C).

5. *Customary U.S. Solution*

For 6 in schedule-40 pipe, the internal diameter, D_i, is 0.5054 ft. The internal area is 0.2006 ft^2.

The velocity, v, is calculated from the volumetric flow, $\dot{V}$, and the flow area, A_i.

Convert the volumetric flow rate from gal/min to ft^3/sec.

$$\dot{V} = \left(500 \ \frac{\text{gal}}{\text{min}}\right)\left(\frac{1 \text{ ft}^3}{7.48 \text{ gal}}\right)\left(\frac{1 \text{ min}}{60 \text{ sec}}\right) = 1.114 \text{ ft}^3/\text{sec}$$

The velocity is

$$\text{v} = \frac{\dot{V}}{A_i} = \frac{1.114 \ \frac{\text{ft}^3}{\text{sec}}}{0.2006 \text{ ft}^2} = 5.55 \text{ ft/sec}$$

For water at 100°F, the kinematic viscosity, ν, is 0.739×10^{-5} ft^2/sec.

Calculate the Reynolds number.

$$\text{Re} = \frac{D_i \text{v}}{\nu} = \frac{(0.5054 \text{ ft})\left(5.55 \ \frac{\text{ft}}{\text{sec}}\right)}{0.739 \times 10^{-5} \ \frac{\text{ft}^2}{\text{sec}}} = 3.80 \times 10^5$$

Since Re > 2100, the flow is turbulent. The friction loss coefficient can be determined from the Moody diagram.

For new steel pipe, the specific roughness, ϵ, is 0.0002 ft.

The relative roughness is

$$\frac{\epsilon}{D_i} = \frac{0.0002 \text{ ft}}{0.5054 \text{ ft}} = 0.0004$$

From the Moody diagram with Re $= 3.80 \times 10^5$ and $e/D_i = 0.0004$, the friction factor, f, can be determined as 0.0173.

Use App. 17.D. The equivalent lengths of the valves and fittings are

standard radius elbow	2×8.9 ft	=	17.8 ft
gate valve (fully open)	2×3.2 ft	=	6.4 ft
90° angle valve (fully open)	1×63.0 ft	=	63.0 ft
swing check valve	1×63.0 ft	=	63.0 ft
			150.2 ft

The equivalent pipe length is the sum of the straight run of pipe and the equivalent length of pipe for the valves and fittings.

$$\begin{aligned} L_e &= L + L_{\text{fittings}} = 300 \text{ ft} + 150.2 \text{ ft} \\ &= 450.2 \text{ ft} \end{aligned}$$

Use Darcy's equation to compute the frictional loss.

$$\begin{aligned} h_f &= \frac{fL_e \text{v}^2}{2D_i g} \\ &= \frac{(0.0173)(450.2 \text{ ft})\left(5.55 \ \frac{\text{ft}}{\text{sec}}\right)^2}{(2)(0.5054 \text{ ft})\left(32.2 \ \frac{\text{ft}}{\text{sec}^2}\right)} \\ &= 7.37 \text{ ft} \end{aligned}$$

The total difference in head is the sum of the head loss through the pipe, valves, and fittings and the change in elevation.

$$\begin{aligned} \Delta h &= h_f + \Delta z = 7.37 \text{ ft} + 20 \text{ ft} \\ &= 27.37 \text{ ft} \end{aligned}$$

The pressure difference between the entrance and discharge can be determined from

$$\begin{aligned} \Delta p &= \gamma \Delta h = \rho h \times \left(\frac{g}{g_c}\right) \\ &= \frac{\left(62.0 \ \frac{\text{lbm}}{\text{ft}^3}\right)(27.37 \text{ ft})\left(32.2 \ \frac{\text{ft}}{\text{sec}^2}\right)\left(\frac{1 \text{ ft}^2}{144 \text{ in}^2}\right)}{32.2 \ \frac{\text{lbm-ft}}{\text{lbf-sec}^2}} \\ &= \boxed{11.8 \text{ lbf/in}^2 \ \ (11.8 \text{ psi})} \end{aligned}$$

The answer is (A).

SI Solution

For 6 in pipe, the internal diameter is 154.1 mm (0.1541 m).

The internal area is 186.5×10^{-4} m^2.

The velocity, v, is calculated from the volumetric flow, $\dot{V}$, and the flow area, A_i.

$$\begin{aligned} \text{v} &= \frac{\dot{V}}{A_i} \\ &= \frac{\left(30 \ \frac{\text{L}}{\text{s}}\right)\left(0.001 \ \frac{\text{m}^3}{\text{L}}\right)}{186.5 \times 10^{-4} \text{ m}^2} \\ &= 1.61 \text{ m/s} \end{aligned}$$

Use App. 14.B. For water at 40°C, the kinematic viscosity is

$$\nu = 6.611 \times 10^{-7} \text{ m}^2/\text{s}$$

$$\rho = 992.25 \text{ kg/m}^3$$

Calculate the Reynolds number.

$$\begin{aligned} \text{Re} &= \frac{D_i \text{v}}{\nu} \\ &= \frac{(0.1541 \text{ m})\left(1.61 \ \frac{\text{m}}{\text{s}}\right)}{6.611 \times 10^{-7} \ \frac{\text{m}^2}{\text{s}}} \\ &= 3.75 \times 10^5 \end{aligned}$$

Since Re > 2100, the flow is turbulent. The friction loss coefficient can be determined from the Moody diagram.

For new steel pipe, the specific roughness is 6.0×10^{-5} m.

The relative roughness is

$$\begin{aligned} \frac{\epsilon}{D_i} &= \frac{6.0 \times 10^{-5} \text{ m}}{0.1541 \text{ m}} \\ &= 0.0004 \end{aligned}$$

From the Moody diagram with Re $= 3.75 \times 10^5$ and $e/D_i = 0.0004$, the friction factor, f, can be determined as 0.0173.

Use App. 17.D. The equivalent lengths of the valves and fittings are

standard radius elbow	2×2.7 m	=	5.4 m
gate valve (fully open)	2×1.0 m	=	2.0 m
90° angle valve (fully open)	1×18.9 m	=	18.9 m
swing check valve	1×18.9 m	=	18.9 m
			45.2 m

The equivalent pipe length is the sum of the straight run of pipe and the equivalent length of pipe for the valves and fittings.

$$\begin{aligned} L_e &= L + L_{\text{fittings}} \\ &= 90 \text{ m} + 45.2 \text{ m} \\ &= 135.2 \text{ m} \end{aligned}$$

Use Darcy's equation to compute the frictional loss.

$$\begin{aligned} h_f &= \frac{fL\text{v}^2}{2D_i g} \\ &= \frac{(0.0173)(135.2\text{ m})\left(1.61\ \frac{\text{m}}{\text{s}^2}\right)^2}{(2)(0.1541\text{ m})\left(9.81\ \frac{\text{m}}{\text{s}^2}\right)} \\ &= 2.01\text{ m} \end{aligned}$$

The total difference in head is the sum of the head loss through the pipe, valves, and fittings and the change in elevation.

$$\begin{aligned} \Delta h &= h_f + \Delta z \\ &= 2.01\text{ m} + 6\text{ m} \\ &= 8.01\text{ m} \end{aligned}$$

The pressure difference between the entrance and discharge can be determined from

$$\begin{aligned} \Delta p &= \rho h g \\ &= \left(992.25\ \frac{\text{kg}}{\text{m}^3}\right)(8.01\text{ m})\left(9.81\ \frac{\text{m}}{\text{s}^2}\right) \\ &= \boxed{77\,969\text{ Pa}\ \ (78\text{ kPa})} \end{aligned}$$

The answer is (A).

6. *Customary U.S. Solution*

For 6 in schedule-40 pipe, the internal diameter, D_i, is 0.5054 ft. The internal area is 0.2006 ft^2.

For air at 70°F, the kinematic viscosity is 16.15×10^{-5} ft^2/sec.

Calculate the Reynolds number.

$$\begin{aligned} \text{Re} &= \frac{D_i \text{v}}{\nu} \\ &= \frac{(0.5054\text{ ft})\left(60\ \frac{\text{ft}}{\text{sec}}\right)}{16.15 \times 10^{-5}\ \frac{\text{ft}^2}{\text{sec}}} \\ &= 1.88 \times 10^5 \end{aligned}$$

Since Re > 2100, the flow is turbulent. The friction loss coefficient can be determined from the Moody diagram.

For new steel pipe, the specific roughness, ϵ, is 0.0002 ft.

The relative roughness is

$$\begin{aligned} \frac{\epsilon}{D_i} &= \frac{0.0002\text{ ft}}{0.5054\text{ ft}} \\ &= 0.0004 \end{aligned}$$

From the Moody diagram with $\text{Re} = 1.88 \times 10^5$ and $e/D_i = 0.0004$, the friction factor, f, can be determined as 0.0184.

From App. 17.D, the equivalent length of the valves and fittings is

standard radius elbow	2 × 8.9 ft	=	17.8 ft
gate valve (fully open)	2 × 3.2 ft	=	6.4 ft
90° angle valve (fully open)	1 × 63.0 ft	=	63.0 ft
swing check valve	1 × 63.0 ft	=	63.0 ft
			150.2 ft

The equivalent pipe length is the sum of the straight run of pipe and the equivalent length of pipe for the valves and fittings.

$$\begin{aligned} L_e &= L + L_{\text{fittings}} \\ &= 300\text{ ft} + 150.2\text{ ft} \\ &= 450.2\text{ ft} \end{aligned}$$

Use Darcy's equation to compute the frictional loss.

$$\begin{aligned} h_f &= \frac{fL_e\text{v}^2}{2D_i g} \\ &= \frac{(0.0184)(450.2\text{ ft})\left(60\ \frac{\text{ft}}{\text{sec}}\right)^2}{(2)(0.5054\text{ ft})\left(32.2\ \frac{\text{ft}}{\text{sec}^2}\right)} \\ &= 916.2\text{ ft} \end{aligned}$$

The difference in head is the sum of the head loss through the pipe, valves, and fittings and the change in elevation.

$$\begin{aligned} h &= h_f + \Delta z \\ &= 916.2\text{ ft} + 20\text{ ft} \\ &= 936.2\text{ ft} \end{aligned}$$

Assume the density of the air, ρ, is approximately 0.075 lbm/ft^3.

The pressure difference between the entrance and discharge can be determined from

$$\begin{aligned} \Delta p &= \gamma h = \rho h \times \left(\frac{g}{g_c}\right) \\ &= \left(0.075\ \frac{\text{lbm}}{\text{ft}^3}\right)(936.2\text{ ft}) \\ &\quad \times \left(\frac{32.2\ \frac{\text{ft}}{\text{sec}^2}}{32.2\ \frac{\text{lbm-ft}}{\text{lbf-sec}^2}}\right)\left(\frac{1\text{ ft}^2}{144\text{ in}^2}\right) \\ &= \boxed{0.49\text{ lbf/in}^2\ \ (0.49\text{ psi})} \end{aligned}$$

The answer is (B).

SI Solution

For 6 in pipe, the internal diameter, D_i, is 154.1 mm (0.1541 m) and the internal area is 186.5×10^{-4} m^2 [App. 16.C].

For air at 20°C, the kinematic viscosity, ν, is 1.51×10^{-5} m^2/s [App. 14.E].

Calculate the Reynolds number.

$$\begin{aligned}\text{Re} &= \frac{D_i \text{v}}{\nu} \\ &= \frac{(0.1541 \text{ m})\left(18 \ \frac{\text{m}}{\text{s}}\right)}{1.51 \times 10^{-5} \ \frac{\text{m}^2}{\text{s}}} \\ &= 1.84 \times 10^5\end{aligned}$$

Since Re > 2100, the flow is turbulent. The friction loss coefficient can be determined from the Moody diagram.

For new steel pipe, the specific roughness, ϵ, is 6.0×10^{-5} m.

The relative roughness is

$$\begin{aligned}\frac{\epsilon}{D_i} &= \frac{6.0 \times 10^{-5}}{0.1541 \text{ m}} \\ &= 0.0004\end{aligned}$$

From the Moody diagram with Re $= 1.84 \times 10^5$ and $e/D_i = 0.0004$, the friction factor, f, can be determined as 0.0185.

Compute the equivalent lengths of the valves and fittings. (Convert from App. 17.D.)

standard radius elbow	2×2.7 m	=	5.4 m
gate valve (fully open)	2×1.0 m	=	2.0 m
90° angle valve (fully open)	1×18.9 m	=	18.9 m
swing check valve	1×18.9 m	=	18.9 m
			45.2 m

The equivalent pipe length is the sum of the straight run of pipe and the equivalent length of pipe for the valves and fittings.

$$\begin{aligned}L_e &= L + L_{\text{fittings}} \\ &= 90 \text{ m} + 45.2 \text{ m} \\ &= 135.2 \text{ m}\end{aligned}$$

Use Darcy's equation to compute the frictional loss.

$$\begin{aligned}h_f &= \frac{f L \text{v}^2}{2 D_i g} \\ &= \frac{(0.0185)(135.2 \text{ m})\left(18 \ \frac{\text{m}}{\text{s}}\right)^2}{(2)(0.1541 \text{ m})\left(9.81 \ \frac{\text{m}}{\text{s}^2}\right)} \\ &= 268.0 \text{ m}\end{aligned}$$

The difference in head is the sum of the head loss through the pipe, valves, and fittings and the change in elevation.

$$\begin{aligned}\Delta h &= h_f + \Delta z \\ &= 268.0 \text{ m} + 6 \text{ m} \\ &= 274.0 \text{ m}\end{aligned}$$

Assume the density of the air, ρ, is approximately 1.20 kg/m^3.

The pressure difference between the entrance and discharge can be determined from

$$\begin{aligned}\Delta p &= \rho h g \\ &= \left(1.20 \ \frac{\text{kg}}{\text{m}^3}\right)(274.0 \text{ m})\left(9.81 \ \frac{\text{m}}{\text{s}^2}\right) \\ &= \boxed{3226 \text{ Pa} \ \ (3.23 \text{ kPa})}\end{aligned}$$

The answer is (B).

7. Assume that flows from reservoirs A and B are toward D and then toward C. Then,

$$\dot{V}_{\text{A-D}} + \dot{V}_{\text{B-D}} = \dot{V}_{\text{D-C}}$$

or

$$A_{\text{A}} \text{v}_{\text{A-D}} + A_{\text{B}} \text{v}_{\text{B-D}} - A_{\text{C}} \text{v}_{\text{D-C}} = 0$$

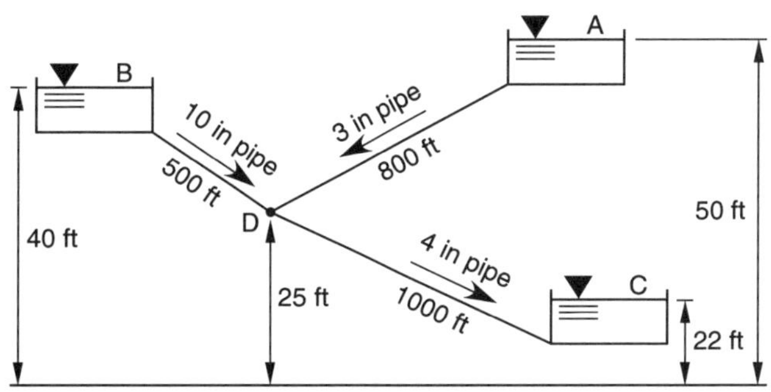

From App. 16.B, for schedule-40 pipe,

$$\begin{aligned}A_{\text{A}} &= 0.05134 \text{ ft}^2 & D_{\text{A}} &= 0.2557 \text{ ft} \\ A_{\text{B}} &= 0.5476 \text{ ft}^2 & D_{\text{B}} &= 0.8350 \text{ ft} \\ A_{\text{C}} &= 0.08841 \text{ ft}^2 & D_{\text{C}} &= 0.3355 \text{ ft}\end{aligned}$$

$$0.05134 \text{v}_{\text{A-D}} + 0.5476 \text{v}_{\text{B-D}} - 0.08841 \text{v}_{\text{D-C}} = 0 \quad \text{[Eq. 1]}$$

Ignoring the velocity heads, the conservation of energy equation between A and D is

$$z_{\text{A}} = \frac{p_{\text{D}}}{\gamma} + z_{\text{D}} + h_{f,\text{A-D}}$$

$$50 \text{ ft} = \frac{p_{\text{D}}}{62.4 \ \frac{\text{lbf}}{\text{ft}^3}} + 25 \text{ ft} + \frac{(0.02)(800 \text{ ft})(\text{v}_{\text{A-D}})^2}{(2)(0.2557 \text{ ft})\left(32.2 \ \frac{\text{ft}}{\text{sec}^2}\right)}$$

or

$$\text{v}_{\text{A-D}} = \sqrt{25.73 - 0.0165 p_\text{D}} \quad \text{[Eq. 2]}$$

Similarly, for B–D,

$$40 \text{ ft} = \frac{p_\text{D}}{62.4 \ \frac{\text{lbf}}{\text{ft}^3}} + 25 \text{ ft} + \frac{(0.02)(500 \text{ ft})(\text{v}_{\text{B-D}})^2}{(2)(0.8350 \text{ ft})\left(32.2 \ \frac{\text{ft}}{\text{sec}^2}\right)}$$

or

$$\text{v}_{\text{B-D}} = \sqrt{80.66 - 0.0862 p_\text{D}} \quad \text{[Eq. 3]}$$

For D–C,

$$22 \text{ ft} = \frac{p_\text{D}}{62.4 \ \frac{\text{lbf}}{\text{ft}^3}} + 25 \text{ ft} - \frac{(0.02)(1000 \text{ ft})(\text{v}_{\text{D-C}})^2}{(2)(0.3355 \text{ ft})\left(32.2 \ \frac{\text{ft}}{\text{sec}^2}\right)}$$

or

$$\text{v}_{\text{D-C}} = \sqrt{3.24 + 0.0173 p_\text{D}} \quad \text{[Eq. 4]}$$

Equations 1, 2, 3, and 4 must be solved simultaneously. To do this, assume a value for p_D. This value then determines all three velocities in Eqs. 2, 3, and 4. These velocities are substituted into Eq. 1. A trial and error solution yields

$$\text{v}_{\text{A-D}} = 3.21 \text{ ft/sec}$$
$$\text{v}_{\text{B-D}} = 0.408 \text{ ft/sec}$$
$$\text{v}_{\text{D-C}} = 4.40 \text{ ft/sec}$$

$$\boxed{p_\text{D} = 933.8 \text{ lbf/ft}^2 \text{ (psf)}}$$

$$\boxed{\text{Flow is from B to D.}}$$

The answer is (B).

8. *Customary U.S. Solution*

The composite modulus of elasticity of the pipe and water is given by Eq. 17.205.

For water at 70°F, $E_{\text{water}} = 320 \times 10^3 \text{ lbf/in}^2$.

For cast-iron pipe, $E_{\text{pipe}} = 20 \times 10^6 \text{ lbf/in}^2$.

$$E = \frac{E_{\text{water}} t_{\text{pipe}} E_{\text{pipe}}}{t_{\text{pipe}} E_{\text{pipe}} + d_{\text{pipe}} E_{\text{water}}}$$
$$= \frac{\left(320 \times 10^3 \ \frac{\text{lbf}}{\text{in}^2}\right)(0.75 \text{ in})\left(20 \times 10^6 \ \frac{\text{lbf}}{\text{in}^2}\right)}{(0.75 \text{ in})\left(20 \times 10^6 \ \frac{\text{lbf}}{\text{in}^2}\right) + (24 \text{ in})\left(320 \times 10^3 \ \frac{\text{lbf}}{\text{in}^2}\right)}$$
$$= 2.12 \times 10^5 \text{ lbf/in}^2$$

The speed of sound in the pipe is

$$a = \sqrt{\frac{E g_c}{\rho}}$$
$$= \sqrt{\frac{\left(2.12 \times 10^5 \ \frac{\text{lbf}}{\text{in}^2}\right)\left(\frac{144 \text{ in}^2}{1 \text{ ft}^2}\right)\left(32.2 \ \frac{\text{lbm-ft}}{\text{lbf-sec}^2}\right)}{62.3 \ \frac{\text{lbm}}{\text{ft}^3}}}$$
$$= 3972 \text{ ft/sec}$$

(a) The maximum pressure is given by Eq. 17.204.

$$\Delta p = \frac{\rho a \Delta \text{v}}{g_c}$$
$$= \left(\frac{\left(62.3 \ \frac{\text{lbm}}{\text{ft}^3}\right)\left(3972 \ \frac{\text{ft}}{\text{sec}}\right)\left(6 \ \frac{\text{ft}}{\text{sec}}\right)}{32.2 \ \frac{\text{lbm-ft}}{\text{lbf-sec}^2}}\right) \times \left(\frac{1 \text{ ft}^2}{144 \text{ in}^2}\right)$$
$$= \boxed{320.2 \text{ lbf/in}^2 \text{ (320.2 psi)}}$$

The answer is (C).

(b) The length of time the pressure is constant at the valve is given by

$$t = \frac{2L}{a} = \frac{(2)(500 \text{ ft})}{3972 \ \frac{\text{ft}}{\text{sec}}}$$
$$= \boxed{0.252 \text{ sec}}$$

The answer is (A).

SI Solution

The composite modulus of elasticity of the pipe and water is given by Eq. 17.205.

For water at 20°C, $E_{\text{water}} = 2.2 \times 10^9$ Pa.

For cast-iron pipe, $E_{\text{pipe}} = 1.4 \times 10^{11}$ Pa.

$$E = \frac{E_{\text{water}} t_{\text{pipe}} E_{\text{pipe}}}{t_{\text{pipe}} E_{\text{pipe}} + d_{\text{pipe}} E_{\text{water}}}$$
$$= \frac{(2.2 \times 10^9 \text{ Pa})(0.02 \text{ m})(1.4 \times 10^{11} \text{ Pa})}{(0.02 \text{ m})(1.4 \times 10^{11} \text{ Pa}) + (0.6 \text{ m})(2.2 \times 10^9 \text{ Pa})}$$
$$= 1.50 \times 10^9 \text{ Pa}$$

The speed of sound in the pipe is

$$a = \sqrt{\frac{E}{\rho}} = \sqrt{\frac{1.50 \times 10^9 \text{ Pa}}{1000 \ \frac{\text{kg}}{\text{m}^3}}} = 1225 \text{ m/s}$$

(a) The maximum pressure is given by

$$\Delta p = \rho a \Delta \text{v} = \left(1000 \ \frac{\text{kg}}{\text{m}^3}\right)\left(1225 \ \frac{\text{m}}{\text{s}}\right)\left(2 \ \frac{\text{m}}{\text{s}}\right) = \boxed{2.45 \times 10^6 \text{ Pa} \ \ (2450 \text{ kPa})}$$

The answer is (C).

(b) The length of time the pressure is constant at the valve is given by

$$t = \frac{2L}{a} = \frac{(2)(150 \text{ m})}{1225 \ \frac{\text{m}}{\text{s}}} = \boxed{0.245 \text{ s}}$$

The answer is (A).

9. *Customary U.S. Solution*

First it is necessary to collect data on schedule-40 pipe and water. The fluid viscosity, pipe dimensions, and other parameters can be found in various appendices in Chs. 14 and 16. At 70°F water, $\nu = 1.059 \times 10^{-5} \text{ ft}^2/\text{sec}$.

From Table 17.2, $\epsilon = 0.0002$ ft.

8 in pipe	$D = 0.6651$ ft $A = 0.3474 \text{ ft}^2$
12 in pipe	$D = 0.9948$ ft $A = 0.7773 \text{ ft}^2$
16 in pipe	$D = 1.25$ ft $A = 1.2272 \text{ ft}^2$

The flow quantity is converted from gallons per minute to cubic feet per second.

$$\dot{V} = \frac{(8 \text{ MGD})\left(10^6 \ \frac{\frac{\text{gal}}{\text{day}}}{\text{MGD}}\right)\left(0.002228 \ \frac{\frac{\text{ft}^3}{\text{sec}}}{\frac{\text{gal}}{\text{min}}}\right)}{\left(24 \ \frac{\text{hr}}{\text{day}}\right)\left(60 \ \frac{\text{min}}{\text{hr}}\right)} = 12.378 \text{ ft}^3/\text{sec}$$

For the inlet pipe, the velocity is

$$\text{v} = \frac{\dot{V}}{A} = \frac{12.378 \ \frac{\text{ft}^3}{\text{sec}}}{0.3474 \text{ ft}^2} = 35.63 \text{ ft/sec}$$

The Reynolds number is

$$\text{Re} = \frac{D\text{v}}{\nu} = \frac{(0.6651 \text{ ft})\left(35.63 \ \frac{\text{ft}}{\text{sec}}\right)}{1.059 \times 10^{-5} \ \frac{\text{ft}^2}{\text{sec}}} = 2.24 \times 10^6$$

The relative roughness is

$$\frac{\epsilon}{D} = \frac{0.0002 \text{ ft}}{0.6651 \text{ ft}} = 0.0003$$

From the Moody diagram, $f = 0.015$.

Equation 17.23(b) is used to calculate the frictional energy loss.

$$E_{f,1} = h_f \times \left(\frac{g}{g_c}\right) = \frac{fL\text{v}^2}{2Dg_c} = \frac{(0.015)(1000 \text{ ft})\left(35.63 \ \frac{\text{ft}}{\text{sec}}\right)^2}{(2)(0.6651 \text{ ft})\left(32.2 \ \frac{\text{lbm-ft}}{\text{lbf-sec}^2}\right)} = 444.6 \text{ ft-lbf/lbm}$$

For the outlet pipe, the velocity is

$$\text{v} = \frac{\dot{V}}{A} = \frac{12.378 \ \frac{\text{ft}^3}{\text{sec}}}{0.7773 \text{ ft}^2} = 15.92 \text{ ft/sec}$$

The Reynolds number is

$$\text{Re} = \frac{D\text{v}}{\nu} = \frac{(0.9948 \text{ ft})\left(15.92 \ \frac{\text{ft}}{\text{sec}}\right)}{1.059 \times 10^{-5} \ \frac{\text{ft}^2}{\text{sec}}} = 1.5 \times 10^6$$

The relative roughness is

$$\frac{\epsilon}{D} = \frac{0.0002 \text{ ft}}{0.9948 \text{ ft}} = 0.0002$$

From the Moody diagram, $f = 0.014$.

Equation 17.23(b) is used to calculate the frictional energy loss.

$$E_{f,2} = h_f \times \left(\frac{g}{g_c}\right) = \frac{fL\mathrm{v}^2}{2Dg_c}$$

$$= \frac{(0.014)(1500\ \text{ft})\left(15.92\ \frac{\text{ft}}{\text{sec}}\right)^2}{(2)(0.9948\ \text{ft})\left(32.2\ \frac{\text{lbm-ft}}{\text{lbf-sec}^2}\right)}$$

$$= 83.1\ \text{ft-lbf/lbm}$$

Assume a 50% split through the two branches. In the upper branch, the velocity is

$$\mathrm{v} = \frac{\dot{V}}{A} = \frac{\left(\frac{1}{2}\right)\left(12.378\ \frac{\text{ft}^3}{\text{sec}}\right)}{0.3474\ \text{ft}^2} = 17.81\ \text{ft/sec}$$

The Reynolds number is

$$\text{Re} = \frac{D\mathrm{v}}{\nu} = \frac{(0.6651\ \text{ft})\left(17.81\ \frac{\text{ft}}{\text{sec}}\right)}{1.059 \times 10^{-5}\ \frac{\text{ft}^2}{\text{sec}}}$$

$$= 1.1 \times 10^6$$

The relative roughness is

$$\frac{\epsilon}{D} = \frac{0.0002\ \text{ft}}{0.6651\ \text{ft}} = 0.0003$$

From the Moody diagram, $f = 0.015$.

For the 16 in pipe in the lower branch, the velocity is

$$\mathrm{v} = \frac{\dot{V}}{A} = \frac{\left(\frac{1}{2}\right)\left(12.378\ \frac{\text{ft}^3}{\text{sec}}\right)}{1.2272\ \text{ft}^2} = 5.04\ \text{ft/sec}$$

The Reynolds number is

$$\text{Re} = \frac{D\mathrm{v}}{\nu} = \frac{(1.25\ \text{ft})\left(5.04\ \frac{\text{ft}}{\text{sec}}\right)}{1.059 \times 10^{-5}\ \frac{\text{ft}^2}{\text{sec}}}$$

$$= 5.95 \times 10^5$$

The relative roughness is

$$\frac{\epsilon}{D} = \frac{0.0002\ \text{ft}}{1.25\ \text{ft}} = 0.00016$$

From the Moody diagram, $f = 0.015$.

These values of f for the two branches are fairly insensitive to changes in $\dot{V}$, so they will be used for the rest of the problem in the upper branch.

Eq. 17.23(b) is used to calculate the frictional energy loss in the upper branch.

$$E_{f,\text{upper}} = h_f \times \left(\frac{g}{g_c}\right) = \frac{fL\mathrm{v}^2}{2Dg_c}$$

$$= \frac{(0.015)(500\ \text{ft})\left(17.81\ \frac{\text{ft}}{\text{sec}}\right)^2}{(2)(0.6651\ \text{ft})\left(32.2\ \frac{\text{lbm-ft}}{\text{lbf-sec}^2}\right)}$$

$$= 55.5\ \text{ft-lbf/lbm}$$

To calculate a loss for any other flow in the upper branch,

$$E_{f,\text{upper }2} = E_{f,\text{upper}}\left(\frac{\dot{V}}{\left(\frac{1}{2}\right)\left(12.378\ \frac{\text{ft}^3}{\text{sec}}\right)}\right)^2$$

$$= \left(55.5\ \frac{\text{ft-lbf}}{\text{lbm}}\right)\left(\frac{\dot{V}}{6.189\ \frac{\text{ft}^3}{\text{sec}}}\right)^2$$

$$= 1.45\dot{V}^2$$

Similarly, for the lower branch, in the 8 in section,

$$E_{f,\text{lower, 8 in}} = \frac{(0.015)(250\ \text{ft})\left(17.81\ \frac{\text{ft}}{\text{sec}}\right)^2}{(2)(0.6651\ \text{ft})\left(32.2\ \frac{\text{lbm-ft}}{\text{lbf-sec}^2}\right)}$$

$$= 27.8\ \text{ft-lbf/lbm}$$

For the lower branch, in the 16 in section,

$$E_{f,\text{lower,16 in}} = \frac{(0.015)(1000\ \text{ft})\left(5.04\ \frac{\text{ft}}{\text{sec}}\right)^2}{(2)(1.25\ \text{ft})\left(32.2\ \frac{\text{lbm-ft}}{\text{lbf-sec}^2}\right)}$$

$$= 4.7\ \text{ft-lbf/lbm}$$

The total loss in the lower branch is

$$E_{f,\text{lower}} = E_{f,\text{lower,8 in}} + E_{f,\text{lower,16 in}}$$

$$= 27.8\ \frac{\text{ft-lbf}}{\text{lbm}} + 4.7\ \frac{\text{ft-lbf}}{\text{lbm}}$$

$$= 32.5\ \text{ft-lbf/lbm}$$

To calculate a loss for any other flow in the lower branch,

$$E_{f,\text{lower 2}} = E_{f,\text{lower}}\left(\frac{\dot{V}}{\left(\frac{1}{2}\right)\left(12.378\cdot\frac{\text{ft}^3}{\text{sec}}\right)}\right)^2$$

$$= \left(32.5\ \frac{\text{ft-lbf}}{\text{lbm}}\right)\left(\frac{\dot{V}}{6.189\ \frac{\text{ft}^3}{\text{sec}}}\right)^2$$

$$= 0.85\dot{V}^2$$

Let x be the fraction flowing in the upper branch. Then, because the friction losses are equal,

$$E_{f,\text{upper 2}} = E_{f,\text{lower 2}}$$
$$1.45x^2 = (0.85)(1-x)^2$$
$$x = 0.432$$

(a) $$\dot{V}_{\text{upper}} = (0.432)\left(12.378\ \frac{\text{ft}^3}{\text{sec}}\right) = \boxed{5.347\ \text{ft}^3/\text{sec}}$$

The answer is (D).

(b) $$\dot{V}_{\text{lower}} = (1-0.432)\left(12.378\ \frac{\text{ft}^3}{\text{sec}}\right) = 7.03\ \text{ft}^3/\text{sec}$$

$$E_{f,\text{total}} = E_{f,1} + E_{f,\text{lower 2}} + E_{f,2}$$
$$E_{f,\text{lower 2}} = 0.85\dot{V}^2_{\text{lower}} = (0.85)\left(7.03\ \frac{\text{ft}^3}{\text{sec}}\right)^2 = 42.0\ \text{ft}$$
$$E_{f,\text{total}} = 444.6\ \frac{\text{ft-lbf}}{\text{lbm}} + 42.0\ \frac{\text{ft-lbf}}{\text{lbm}} + 83.1\ \frac{\text{ft-lbf}}{\text{lbm}} = \boxed{569.7\ \text{ft-lbf/lbm}}$$

The answer is (D).

SI Solution

First it is necessary to collect data on schedule-40 pipe and water. The fluid viscosity, pipe dimensions, and other parameters can be found in various appendices in Chs. 14 and 16. At 20°C water, $\nu = 1.007\times10^{-6}\ \text{m}^2/\text{s}$.

From Table 17.2, $\epsilon = 6\times10^{-5}$ m.

8 in pipe	$D = 202.7$ mm $A = 322.7\times10^{-4}\ \text{m}^2$
12 in pipe	$D = 303.2$ mm $A = 721.9\times10^{-4}\ \text{m}^2$
16 in pipe	$D = 381$ mm $A = 1104\times10^{-4}\ \text{m}^2$

For the inlet pipe, the velocity is

$$\text{v} = \frac{\dot{V}}{A} = \frac{\left(350\ \frac{\text{L}}{\text{s}}\right)\left(\frac{1\ \text{m}^3}{1000\ \text{L}}\right)}{322.7\times10^{-4}\ \text{m}^2} = 10.85\ \text{m/s}$$

The Reynolds number is

$$\text{Re} = \frac{D\text{v}}{\nu} = \frac{(0.2027\ \text{m})\left(10.85\ \frac{\text{m}}{\text{s}}\right)}{1.007\times10^{-6}\ \frac{\text{m}^2}{\text{s}}} = 2.18\times10^6$$

The relative roughness is

$$\frac{\epsilon}{D} = \frac{6\times10^{-5}\ \text{m}}{0.2027\ \text{m}} = 0.0003$$

From the Moody diagram, $f = 0.015$.

Equation 17.23(a) is used to calculate the frictional energy loss.

$$E_{f,1} = h_f g = \frac{fL\text{v}^2}{2D} = \frac{(0.015)(300\ \text{m})\left(10.85\ \frac{\text{m}}{\text{s}}\right)^2}{(2)(0.2027\ \text{m})} = 1307\ \text{J/kg}$$

For the outlet pipe, the velocity is

$$\text{v} = \frac{\dot{V}}{A} = \frac{\left(350\ \frac{\text{L}}{\text{s}}\right)\left(\frac{1\ \text{m}^3}{1000\ \text{L}}\right)}{721.9\times10^{-4}\ \text{m}^2} = 4.848\ \text{m/s}$$

The Reynolds number is

$$\text{Re} = \frac{D\text{v}}{\nu} = \frac{(0.3032\ \text{m})\left(4.848\ \frac{\text{m}}{\text{s}}\right)}{1.007\times10^{-6}\ \frac{\text{m}^2}{\text{s}}} = 1.46\times10^6$$

The relative roughness is

$$\frac{\epsilon}{D} = \frac{6\times10^{-5}\ \text{m}}{0.3032\ \text{m}} = 0.0002$$

Water Resources
Flow of Fluids

From the Moody diagram, $f = 0.014$.

Equation 17.23(a) is used to calculate the frictional energy loss.

$$E_{f,2} = h_f g = \frac{fL\text{v}^2}{2D} = \frac{(0.014)(450 \text{ m})\left(4.848 \frac{\text{m}}{\text{s}}\right)^2}{(2)(0.3032 \text{ m})} = 244.2 \text{ J/kg}$$

Assume a 50% split through the two branches. In the upper branch, the velocity is

$$\text{v} = \frac{\dot{V}}{A} = \frac{\left(\frac{1}{2}\right)\left(350 \frac{\text{L}}{\text{s}}\right)\left(\frac{1 \text{ m}^3}{1000 \text{ L}}\right)}{322.7 \times 10^{-4} \text{ m}^2} = 5.423 \text{ m/s}$$

The Reynolds number is

$$\text{Re} = \frac{D\text{v}}{\nu} = \frac{(0.2027 \text{ m})\left(5.423 \frac{\text{m}}{\text{s}}\right)}{1.007 \times 10^{-6} \frac{\text{m}^2}{\text{s}}} = 1.1 \times 10^6$$

The relative roughness is

$$\frac{\epsilon}{D} = \frac{6 \times 10^{-5} \text{ m}}{0.2027 \text{ m}} = 0.0003$$

From the Moody diagram, $f = 0.015$.

For the 16 in pipe in the lower branch, the velocity is

$$\text{v} = \frac{\dot{V}}{A} = \frac{\left(\frac{1}{2}\right)\left(350 \frac{\text{L}}{\text{s}}\right)\left(\frac{1 \text{ m}^3}{1000 \text{ L}}\right)}{1104 \times 10^{-4} \text{ m}^2} = 1.585 \text{ m/s}$$

The Reynolds number is

$$\text{Re} = \frac{D\text{v}}{\nu} = \frac{(0.381 \text{ m})\left(1.585 \frac{\text{m}}{\text{s}}\right)}{1.007 \times 10^{-6} \frac{\text{m}^2}{\text{s}}} = 6.00 \times 10^5$$

The relative roughness is

$$\frac{\epsilon}{D} = \frac{6 \times 10^{-5} \text{ m}}{0.381 \text{ m}} = 0.00016$$

From the Moody diagram, $f = 0.015$.

These values of f for the two branches are fairly insensitive to changes in $\dot{V}$, so they will be used for the rest of the problem in the upper branch.

Eq. 17.23(a) is used to calculate the frictional energy loss in the upper branch.

$$E_{f,\text{upper}} = h_f g = \frac{fL\text{v}^2}{2D} = \frac{(0.015)(150 \text{ m})\left(5.423 \frac{\text{m}}{\text{s}}\right)^2}{(2)(0.2027 \text{ m})} = 163.2 \text{ J/kg}$$

To calculate a loss for any other flow in the upper branch,

$$E_{f,\text{upper 2}} = E_{f,\text{upper}}\left(\frac{\dot{V}}{\left(\frac{1}{2}\right)\left(0.350 \frac{\text{m}^3}{\text{s}}\right)}\right)^2 = \left(163.2 \frac{\text{J}}{\text{kg}}\right)\left(\frac{\dot{V}}{0.175 \frac{\text{m}^3}{\text{s}}}\right)^2 = 5329\dot{V}^2$$

Similarly, for the lower branch, in the 8 in section,

$$E_{f,\text{lower, 8 in}} = \frac{(0.015)(75 \text{ m})\left(5.423 \frac{\text{m}}{\text{s}}\right)^2}{(2)(0.2027 \text{ m})} = 81.61 \text{ J/kg}$$

For the lower branch, in the 16 in section,

$$E_{f,\text{lower,16 in}} = \frac{(0.015)(300 \text{ m})\left(1.585 \frac{\text{m}}{\text{s}}\right)^2}{(2)(0.381 \text{ m})} = 14.84 \text{ J/kg}$$

The total loss in the lower branch is

$$E_{f,\text{lower}} = E_{f,\text{lower,8 in}} + E_{f,\text{lower,16 in}} = 81.61 \frac{\text{J}}{\text{kg}} + 14.84 \frac{\text{J}}{\text{kg}} = 96.45 \text{ J/kg}$$

To calculate a loss for any other flow in the lower branch,

$$E_{f,\text{lower 2}} = E_{f,\text{lower}}\left(\frac{\dot{V}}{\left(\frac{1}{2}\right)\left(0.350 \frac{\text{m}^3}{\text{s}}\right)}\right)^2 = \left(96.45 \frac{\text{J}}{\text{kg}}\right)\left(\frac{\dot{V}}{0.175 \frac{\text{m}^3}{\text{s}}}\right)^2 = 3149\dot{V}^2$$

Let x be the fraction flowing in the upper branch. Then, because the friction losses are equal,

$$E_{f,\text{upper }2} = E_{f,\text{lower }2}$$
$$5329x^2 = (3149)(1-x)^2$$
$$x = 0.435$$

(a) $$\dot{V}_{\text{upper}} = (0.435)\left(0.350\ \frac{\text{m}^3}{\text{s}}\right) = \boxed{0.152\ \text{m}^3/\text{s}}$$

The answer is (D).

(b) $$\dot{V}_{\text{lower}} = (1-0.435)\left(0.350\ \frac{\text{m}^3}{\text{s}}\right) = 0.198\ \text{m}^3/\text{s}$$

$$E_{f,\text{total}} = E_{f,1} + E_{f,\text{lower }2} + E_{f,2}$$

$$E_{f,\text{lower }2} = 3149\dot{V}^2_{\text{lower}} = (3149)\left(0.198\ \frac{\text{m}^3}{\text{s}}\right)^2 = 123.5\ \text{J/kg}$$

$$E_{f,\text{total}} = 1307\ \frac{\text{J}}{\text{kg}} + 123.5\ \frac{\text{J}}{\text{kg}} + 244.2\ \frac{\text{J}}{\text{kg}} = \boxed{1675\ \text{J/kg}\ (1.7\ \text{kJ/kg})}$$

The answer is (D).

10. *steps 1, 2, and 3:*

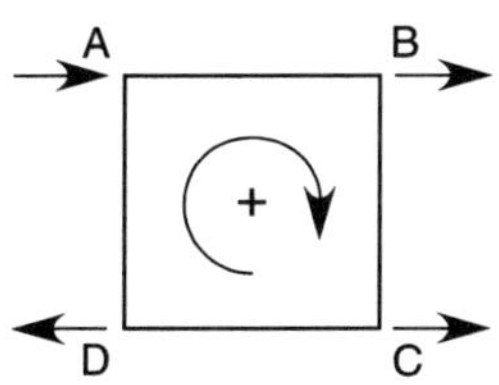

step 4: There is only one loop: ABCD.

step 5: pipe AB: $$K' = \frac{(10.44)(1000\ \text{ft})}{(100)^{1.85}(8\ \text{in})^{4.87}} = 8.33\times 10^{-5}$$

pipe BC: $$K' = \frac{(10.44)(1000\ \text{ft})}{(100)^{1.85}(6\ \text{in})^{4.87}} = 3.38\times 10^{-4}$$

pipe CD: $$K' = 8.33\times 10^{-5}\quad [\text{same as AB}]$$

pipe DA: $$K' = \frac{(10.44)(1000\ \text{ft})}{(100)^{1.85}(12\ \text{in})^{4.87}} = 1.16\times 10^{-5}$$

step 6: Assume the flows are as shown in the figure.

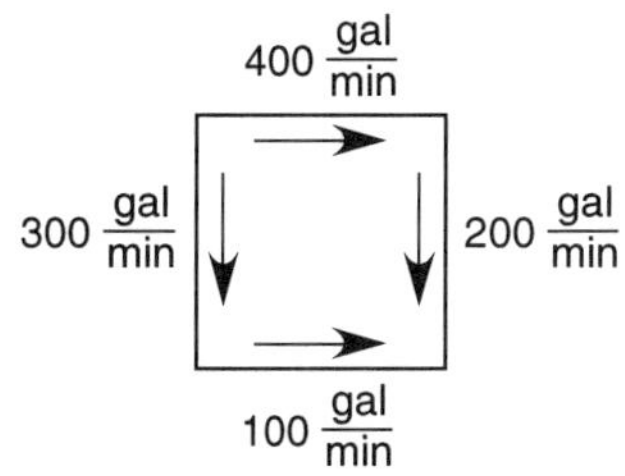

step 7:

$$\delta = \left(\frac{-1}{1.85}\right) \times \left[\frac{(8.33\times10^{-5})(400)^{1.85} + (3.38\times10^{-4})(200)^{1.85} - (8.33\times10^{-5})(100)^{1.85} - (1.16\times10^{-5})(300)^{1.85}}{(8.33\times10^{-5})(400)^{0.85} + (3.38\times10^{-4})(200)^{0.85} + (8.33\times10^{-5})(100)^{0.85} + (1.16\times10^{-5})(300)^{0.85}}\right]$$

$$= \left(\frac{-1}{1.85}\right)\left(\frac{10.67}{4.98\times10^{-2}}\right) = -116\ \text{gal/min}$$

step 8: The adjusted flows are shown.

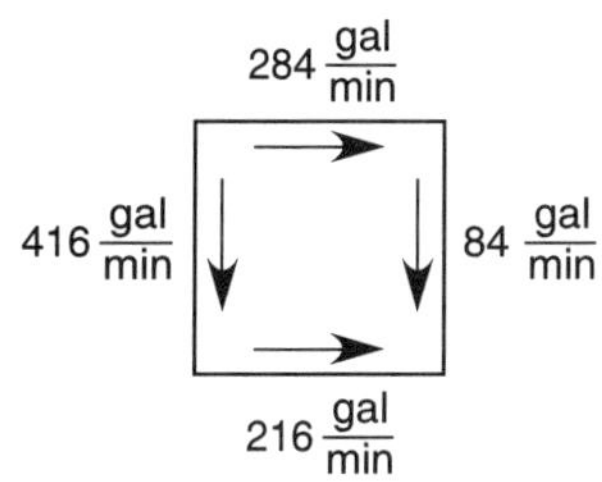

step 7: $\delta = -24$ gal/min
step 8: The adjusted flows are shown.

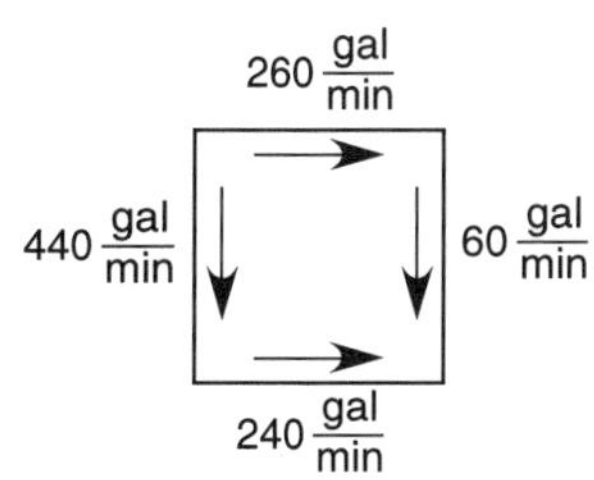

step 7: $\delta = -2$ gal/min [small enough]
step 8: The final adjusted flows are shown.

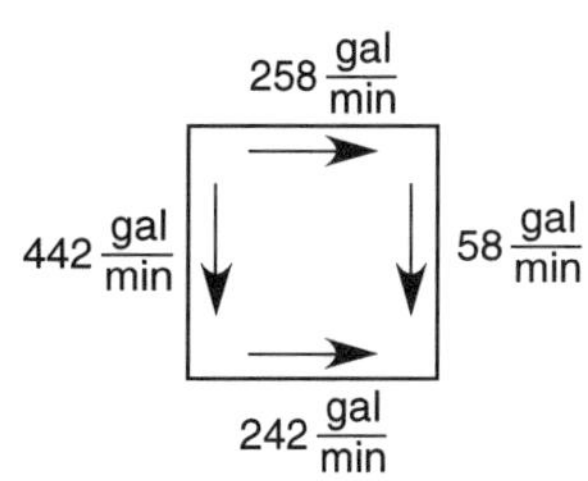

The answer is (B).

11. (a) This is a Hardy Cross problem. The pressure at point C does not change the solution procedure.

step 1: The Hazen-Williams roughness coefficient is given.

step 2: Choose clockwise as positive.

step 3: Nodes are already numbered.

step 4: Choose the loops as shown.

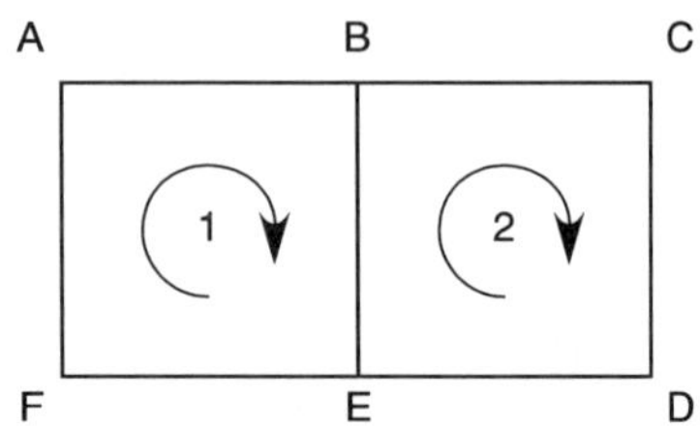

step 5: $\dot{V}$ is in gal/min.
d is in inches.
L is in feet.

Each pipe has the same length.

$$K'_{8\text{ in}} = \frac{(10.44)(1000\text{ ft})}{(100)^{1.85}(8\text{ in})^{4.87}} = 8.33\times10^{-5}$$

$$K'_{6\text{ in}} = \frac{(10.44)(1000\text{ ft})}{(100)^{1.85}(6\text{ in})^{4.87}} = 3.38\times10^{-4}$$

$$K'_{\text{CE}} = 2K'_{6\text{ in}} = 6.76\times10^{-4}$$

$$K'_{\text{EA}} = K'_{6\text{ in}} + K'_{8\text{ in}} = 4.21\times10^{-4}$$

step 6: Assume the flows shown.

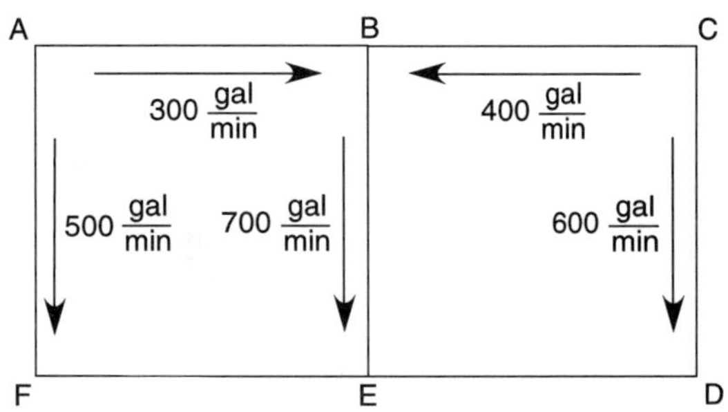

AB = 300 gal/min

BE = 700 gal/min

AE = 500 gal/min

CE = 600 gal/min

CB = 400 gal/min

step 7: If the elevations are included as part of the head loss,

$$\Sigma h = \Sigma K'\dot{V}_a^n + \delta\Sigma nK'\dot{V}_a^{n-1} + z_2 - z_1 = 0$$

However, since the loop closes on itself, $z_2 = z_1$, and the elevations can be omitted.

- First iteration:

Loop 1:

$$\delta_1 = \frac{-\left[(8.33\times10^{-5})(300)^{1.85} + (8.33\times10^{-5})(700)^{1.85} - (4.21\times10^{-4})(500)^{1.85}\right]}{(1.85)\left[(8.33\times10^{-5})(300)^{0.85} + (8.33\times10^{-5})(700)^{0.85} + (4.21\times10^{-4})(500)^{0.85}\right]}$$

$$= \frac{-(-22.97)}{0.213} = +108\text{ gal/min}$$

Loop 2:

$$\delta_2 = \frac{-\left[(6.76\times10^{-4})(600)^{1.85} - (8.33\times10^{-5})(700)^{1.85} - (8.33\times10^{-5})(400)^{1.85}\right]}{(1.85)\left[(6.76\times10^{-4})(600)^{0.85} + (8.33\times10^{-5})(700)^{0.85} + 8.33\times10^{-5})(400)^{0.85}\right]}$$

$$= \frac{-(72.5)}{0.353} = -205\text{ gal/min}$$

- Second iteration:

AB:	300 + 108	=	408 gal/min
BE:	700 + 108 − (−205)	=	1013 gal/min
AE:	500 − 108	=	392 gal/min
CE:	600 + (−205)	=	395 gal/min
CB:	400 − (−205)	=	605 gal/min

Loop 1:

$$\delta_1 = \frac{-(9.48)}{0.205} = -46\text{ gal/min}$$

Loop 2:

$$\delta_2 = \frac{-(1.08)}{0.292} = -3.7\text{ gal/min}\quad[\text{round to }-4]$$

- Third iteration:

AB:	408 + (−46)	=	362 gal/min
BE:	1013 + (−46) −(−4)	=	971 gal/min
AE:	392 − (−46)	=	438 gal/min
CE:	395 + (−4)	=	391 gal/min
CB:	605 − (−4)	=	609 gal/min

Loop 1:

$$\delta_1 = \frac{-(0.066)}{0.213} = -0.31\text{ gal/min}\quad[\text{round to }0]$$

Loop 2:

$$\delta_2 = \frac{-2.4}{0.29} = -8.3 \text{ gal/min} \quad [\text{round to } -8]$$

Use the following flows.

$$\begin{aligned} \text{AB: } & 362 + 0 = 362 \text{ gal/min} \\ \text{BE: } & 971 + 0 - (-8) = \boxed{979 \text{ gal/min}} \\ \text{AE: } & 438 - 0 = 438 \text{ gal/min} \\ \text{CE: } & 391 + (-8) = 383 \text{ gal/min} \\ \text{CB: } & 609 - (-8) = 617 \text{ gal/min} \end{aligned}$$

The answer is (D).

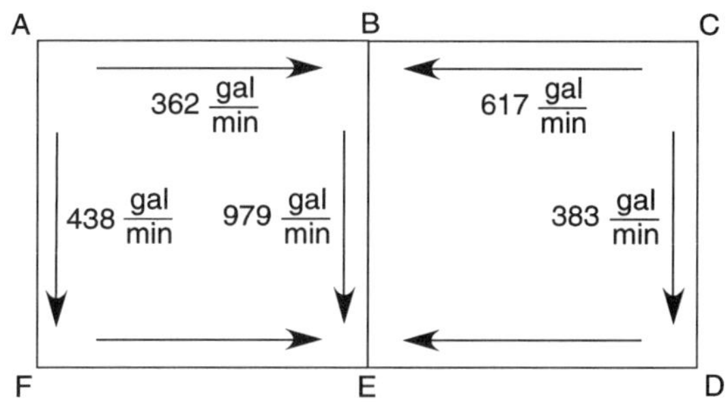

(b) The friction loss in each section is

$$\begin{aligned} h_{f,\text{AB}} &= (8.33 \times 10^{-5})(362)^{1.85} = 4.5 \text{ ft} \\ h_{f,\text{BE}} &= (8.33 \times 10^{-5})(979)^{1.85} = 28.4 \text{ ft} \\ h_{f,\text{AF}} &= (8.33 \times 10^{-5})(438)^{1.85} = 6.4 \text{ ft} \\ h_{f,\text{FE}} &= (3.38 \times 10^{-4})(438)^{1.85} = 26.0 \text{ ft} \\ h_{f,\text{CD}} &= h_{f,\text{DE}} = (3.38 \times 10^{-4})(383)^{1.85} = 20.3 \text{ ft} \\ h_{f,\text{CB}} &= (8.33 \times 10^{-5})(617)^{1.85} = 12.1 \text{ ft} \end{aligned}$$

$$\gamma = 62.37 \text{ lbf/ft}^3 \quad [\text{at } 60^\circ\text{F}]$$

The pressure at C is 40 psig.

$$h_\text{C} = \frac{\left(40 \frac{\text{lbf}}{\text{in}^2}\right)\left(144 \frac{\text{in}^2}{\text{ft}^2}\right)}{62.37 \frac{\text{lbf}}{\text{ft}^3}} = 92.4 \text{ ft}$$

Next, use the energy continuity equation between adjacent points.

$$h = h_\text{C} + z_\text{C} - z - h_f$$

$$\begin{aligned} h_\text{D} &= 92.4 \text{ ft} + 300 \text{ ft} - 150 \text{ ft} - 20.3 \text{ ft} = 222.1 \text{ ft} \\ h_\text{E} &= 222.1 \text{ ft} + 150 \text{ ft} - 200 \text{ ft} - 20.3 \text{ ft} = 151.8 \text{ ft} \\ h_\text{F} &= 151.8 \text{ ft} + 200 \text{ ft} - 150 \text{ ft} + 26 \text{ ft} = 227.8 \text{ ft} \\ h_\text{B} &= 92.4 \text{ ft} + 300 \text{ ft} - 150 \text{ ft} - 12.1 \text{ ft} = 230.3 \text{ ft} \\ h_\text{A} &= 230.3 \text{ ft} + 150 \text{ ft} - 200 \text{ ft} + 4.5 \text{ ft} = 184.8 \text{ ft} \end{aligned}$$

Using $p = \gamma h$,

$$p_\text{A} = \frac{\left(62.37 \frac{\text{lbf}}{\text{ft}^3}\right)(184.8 \text{ ft})}{144 \frac{\text{in}^2}{\text{ft}^2}} = 80.0 \text{ lbf/in}^2 \text{ (psig)}$$

$$p_\text{B} = \frac{\left(62.37 \frac{\text{lbf}}{\text{ft}^3}\right)(230.3 \text{ ft})}{144 \frac{\text{in}^2}{\text{ft}^2}} = 99.7 \text{ psig}$$

$$p_\text{C} = 40 \text{ psig} \quad [\text{given}]$$

$$p_\text{D} = \frac{\left(62.37 \frac{\text{lbf}}{\text{ft}^3}\right)(222.1 \text{ ft})}{144 \frac{\text{in}^2}{\text{ft}^2}} = \boxed{96.2 \text{ psig}}$$

$$p_\text{E} = \frac{\left(62.37 \frac{\text{lbf}}{\text{ft}^3}\right)(151.8 \text{ ft})}{144 \frac{\text{in}^2}{\text{ft}^2}} = 65.7 \text{ psig}$$

$$p_\text{F} = \frac{\left(62.37 \frac{\text{lbf}}{\text{ft}^3}\right)(227.8 \text{ ft})}{144 \frac{\text{in}^2}{\text{ft}^2}} = 98.7 \text{ psig}$$

The answer is (C).

(c) The pressure increase across the pump is

$$\left(80 \frac{\text{lbf}}{\text{in}^2} - 20 \frac{\text{lbf}}{\text{in}^2}\right)\left(144 \frac{\text{in}^2}{\text{ft}^2}\right) = 8640 \text{ lbf/ft}^2$$

The hydraulic horsepower is

$$P = \frac{\left(8640 \frac{\text{lbf}}{\text{ft}^2}\right)\left(800 \frac{\text{gal}}{\text{min}}\right)}{2.468 \times 10^5 \frac{\text{lbf-gal}}{\text{min-ft}^2\text{-hp}}} = \boxed{28 \text{ hp}}$$

The answer is (B).

12. This could be solved as a parallel pipe problem or as a pipe network problem.

(a) • Pipe network solution:

step 1: Use the Darcy equation since Hazen-Williams coefficients are not given (or assume C-values based on the age of the pipe).

step 2: Clockwise is positive.

step 3: All nodes are lettered.

step 4: There is only one loop.

step 5: $\epsilon \approx 0.0008$ ft. Assume full turbulence. For class A cast-iron pipe, $D_i = 10.1$ in.

$$\frac{\epsilon}{D} = \frac{0.0008 \text{ ft}}{\frac{10.1 \text{ in}}{12 \frac{\text{in}}{\text{ft}}}} \approx 0.001$$

$f \approx 0.020$ for full turbulence.

$$K'_{AB} = (1.251 \times 10^{-7}) \left[\frac{(0.02)(2000 \text{ ft})}{\left(\frac{10.1 \text{ in}}{12 \frac{\text{in}}{\text{ft}}} \right)^5} \right] = 11.8 \times 10^{-6}$$

$$K'_{BD} = (1.251 \times 10^{-7}) \left[\frac{(0.02)(1000 \text{ ft})}{\left(\frac{10.1 \text{ in}}{12 \frac{\text{in}}{\text{ft}}} \right)^5} \right] = 5.92 \times 10^{-6}$$

$$K'_{DC} = (1.251 \times 10^{-7}) \left[\frac{(0.02)(1500 \text{ ft})}{\left(\frac{8.13 \text{ in}}{12 \frac{\text{in}}{\text{ft}}} \right)^5} \right] = 26.3 \times 10^{-6}$$

$$K'_{CA} = (1.251 \times 10^{-7}) \left[\frac{(0.02)(1000 \text{ ft})}{\left(\frac{12.12 \text{ in}}{12 \frac{\text{in}}{\text{ft}}} \right)^5} \right] = 2.38 \times 10^{-6}$$

step 6: Assume $\dot{V}_{AB} = 1$ MGD.

$$\dot{V}_{ACDB} = 0.5 \text{ MGD}$$

Convert $\dot{V}$ to gal/min.

$$\dot{V}_{AB} = \frac{(1 \text{ MGD})(1 \times 10^6)}{\left(24 \frac{\text{hr}}{\text{day}}\right)\left(60 \frac{\text{min}}{\text{hr}}\right)} = 694 \text{ gal/min}$$

$$\dot{V}_{ACDB} = \frac{(0.5 \text{ MGD})(1 \times 10^6)}{\left(24 \frac{\text{hr}}{\text{day}}\right)\left(60 \frac{\text{min}}{\text{hr}}\right)} = 347 \text{ gal/min}$$

step 7: There is only one loop.

$$694 \frac{\text{gal}}{\text{min}} = (2)\left(347 \frac{\text{gal}}{\text{min}}\right)$$

$$\left(694 \frac{\text{gal}}{\text{min}}\right)^2 = (4)\left(347 \frac{\text{gal}}{\text{min}}\right)^2$$

$$\delta = \frac{(-1)(1 \times 10^{-6})(347)^2[(11.8)(4) - 5.92 - 26.3 - 2.38]}{(2)(1 \times 10^{-6})(347)[(11.8)(2) + 5.92 + 26.3 + 2.38]}$$

$$= -37.6 \text{ gal/min} \quad [\text{use } -38 \text{ gal/min}]$$

step 8:

$$\dot{V}_{AB} = 694 \frac{\text{gal}}{\text{min}} + \left(-38 \frac{\text{gal}}{\text{min}}\right) = 656 \text{ gal/min}$$

$$\dot{V}_{ACDB} = 347 \frac{\text{gal}}{\text{min}} - \left(-38 \frac{\text{gal}}{\text{min}}\right) = 385 \text{ gal/min}$$

Repeat step 7.

$$656 \frac{\text{gal}}{\text{min}} = (1.70)\left(385 \frac{\text{gal}}{\text{min}}\right)$$

$$\left(656 \frac{\text{gal}}{\text{min}}\right)^2 = (2.90)\left(385 \frac{\text{gal}}{\text{min}}\right)^2$$

$$\delta = \frac{(-1)(1 \times 10^{-6})(385)^2 \times [(11.8)(2.90) - 5.92 - 26.3 - 2.38]}{(2)(1 \times 10^{-6})(385) \times [(11.8)(1.70) + 5.92 + 26.3 + 2.38]} \approx 0$$

$$\dot{V}_{AB} = \boxed{656 \text{ gal/min}}$$

$$\dot{V}_{ACDB} = 385 \text{ gal/min}$$

Check the Reynolds number in leg AB to verify that $f = 0.02$ was a good choice.

$$A_{10 \text{ in pipe}} = \left(\frac{\pi}{4}\right)\left(\frac{10.1 \text{ in}}{12 \frac{\text{in}}{\text{ft}}}\right)^2 = 0.5564 \text{ [cast-iron pipe]}$$

The flow rate is

$$\left(656 \frac{\text{gal}}{\text{min}}\right)\left(0.002228 \frac{\text{ft}^3\text{-min}}{\text{sec-gal}}\right) = 1.46 \text{ ft}^3/\text{sec}$$

$$\text{v} = \frac{\dot{V}}{A} = \frac{1.46 \frac{\text{ft}^3}{\text{sec}}}{0.5564 \text{ ft}^2} = 2.62 \text{ ft/sec} \quad \text{[reasonable]}$$

Assume 50°F water.

$$\nu = 1.410 \times 10^{-5} \text{ ft}^2/\text{sec}$$

$$\text{Re} = \frac{\left(\frac{10.1 \text{ in}}{12 \frac{\text{in}}{\text{ft}}}\right)\left(2.62 \frac{\text{ft}}{\text{sec}}\right)}{1.410 \times 10^{-5} \frac{\text{ft}^2}{\text{sec}}} = 1.56 \times 10^5 \text{ [turbulent]}$$

The answer is (D).

- Alternative closed-form solution:

Use the Darcy equation. Assume $\epsilon = 0.0008$. The relative roughness is

$$\frac{\epsilon}{D} = \frac{0.0008 \text{ ft}}{\frac{10.1 \text{ in}}{12 \frac{\text{in}}{\text{ft}}}} = 0.00095 \quad \text{[use 0.001]}$$

Assume $v_{max} = 5$ ft/sec and 50°F temperature. The Reynolds number is

$$\text{Re} = \frac{vD}{\nu} = \frac{\left(5\ \frac{\text{ft}}{\text{sec}}\right)\left(\frac{10.1\ \text{in}}{12\ \frac{\text{in}}{\text{ft}}}\right)}{1.41 \times 10^{-5}\ \frac{\text{ft}^2}{\text{sec}}} = 2.98 \times 10^5$$

From the Moody diagram, $f \approx 0.0205$.

$$\begin{aligned} h_{f,\text{AB}} &= \frac{fLv^2}{2Dg} \\ &= \frac{(0.0205)(2000\ \text{ft})\dot{V}_{\text{AB}}^2}{(2)\left(\frac{10.1\ \text{in}}{12\ \frac{\text{in}}{\text{ft}}}\right)(0.556\ \text{ft}^2)^2\left(32.2\ \frac{\text{ft}^2}{\text{sec}}\right)} \\ &= 2.446\dot{V}_{\text{AB}}^2 \end{aligned}$$

$$\begin{aligned} h_{f,\text{ACDB}} &= \frac{(0.0205)(1000)\dot{V}_{\text{ACDB}}^2}{(2)\left(\frac{12.12}{12}\right)(0.801)^2(32.2)} \\ &\quad + \frac{(0.0205)(1500)\dot{V}_{\text{ACDB}}^2}{(2)\left(\frac{8.13}{12}\right)(0.360)^2(32.2)} \\ &\quad + \frac{(0.0205)(1000)\dot{V}_{\text{ACDB}}^2}{(2)\left(\frac{10.1}{12}\right)(0.556\ \text{ft}^2)^2(32.2)} \\ &= 0.4912\dot{V}_{\text{ACDB}}^2 + 5.438\dot{V}_{\text{ACDB}}^2 + 1.223\dot{V}_{\text{ACDB}}^2 \\ &= 7.152\dot{V}_{\text{ACDB}}^2 \end{aligned}$$

$$h_{f,\text{AB}} = h_{f,\text{ACDB}}$$

$$2.446\dot{V}_{\text{AB}}^2 = 7.152\dot{V}_{\text{ACDB}}^2$$

$$\dot{V}_{\text{AB}} = \sqrt{\frac{7.152}{2.446}}\dot{V}_{\text{ACDB}} = 1.71\dot{V}_{\text{ACDB}} \quad \text{[Eq. 1]}$$

The total flow rate is

$$\frac{1.5\ \text{MGD}}{\left(24\ \frac{\text{hr}}{\text{day}}\right)\left(60\ \frac{\text{min}}{\text{hr}}\right)} = 1041.7\ \text{gal/min}$$

$$\dot{V}_{\text{AB}} + \dot{V}_{\text{ACDB}} = 1041.7\ \text{gal/min} \quad \text{[Eq. 2]}$$

Solving Eqs. 1 and 2 simultaneously,

$$\dot{V}_{\text{AB}} = \boxed{657.3\ \text{gal/min}}$$

$$\dot{V}_{\text{ACDB}} = 384.4\ \text{gal/min}$$

The answer is (D).

This answer is insensitive to the $v_{max} = 5$ ft/sec assumption. A second iteration using actual velocities from these flow rates does not change the answer.

• The same technique can be used with the Hazen-Williams equation and an assumed value of C. If $C = 100$ is used, then

$$\dot{V}_{\text{AB}} = \boxed{725.4\ \text{gal/min}}$$

$$\dot{V}_{\text{ACDB}} = 316.3\ \text{gal/min}$$

(b) $f \approx 0.021$.

$$h_{f,\text{AB}} = \frac{(0.021)(2000\ \text{ft})\left(2.62\ \frac{\text{ft}}{\text{sec}}\right)^2}{(2)\left(\frac{10.1\ \text{in}}{12\ \frac{\text{in}}{\text{ft}}}\right)\left(32.2\ \frac{\text{ft}}{\text{sec}^2}\right)} = 5.32\ \text{ft}$$

For leg BD, use $f = 0.021$ (assumed).

$$\begin{aligned} v = \frac{\dot{V}}{A} &= \frac{\left(385\ \frac{\text{gal}}{\text{min}}\right)\left(0.002228\ \frac{\text{ft}^3\text{-min}}{\text{sec-gal}}\right)}{0.5564\ \text{ft}^2} \\ &= 1.54\ \text{ft/sec} \end{aligned}$$

$$h_{f,\text{DB}} = \frac{(0.021)(1000\ \text{ft})\left(1.54\ \frac{\text{ft}}{\text{sec}}\right)^2}{(2)\left(\frac{10\ \text{in}}{12\ \frac{\text{in}}{\text{ft}}}\right)\left(32.2\ \frac{\text{ft}}{\text{sec}^2}\right)} = 0.9\ \text{ft}$$

At 50°F, $\gamma = 62.4$ lbf/ft^3. From the Bernoulli equation (omitting the velocity term),

$$\frac{p_{\text{B}}}{\gamma} + z_{\text{B}} + h_{f,\text{DB}} = \frac{p_{\text{D}}}{\gamma} + z_{\text{D}}$$

$$\begin{aligned} p_{\text{B}} &= \left(\frac{62.4\ \frac{\text{lbf}}{\text{ft}^3}}{144\ \frac{\text{in}^2}{\text{ft}^2}}\right) \\ &\quad \times \left[\frac{\left(40\ \frac{\text{lbf}}{\text{in}^2}\right)\left(144\ \frac{\text{in}^2}{\text{ft}^2}\right)}{62.4\ \frac{\text{lbf}}{\text{ft}^3}} + 600\ \text{ft} - 0.9\ \text{ft} - 580\ \text{ft}\right] \\ &= \boxed{48.3\ \text{lbf/in}^2\ \text{(psi)}} \end{aligned}$$

The answer is (B).

$$\frac{p_A}{\gamma} + z_A = \frac{p_B}{\gamma} + z_B + h_{f,AB}$$

$$p_A = \left(\frac{62.4 \frac{\text{lbf}}{\text{ft}^3}}{144 \frac{\text{in}^2}{\text{ft}^2}}\right) \times \left[\frac{\left(48.3 \frac{\text{lbf}}{\text{in}^2}\right)\left(144 \frac{\text{in}^2}{\text{ft}^2}\right)}{62.4 \frac{\text{lbf}}{\text{ft}^3}} + 580 \text{ ft} + 5.32 \text{ ft} - 540 \text{ ft}\right]$$

$$= 67.9 \text{ lbf/in}^2 \text{ (psi)}$$

13. *Customary U.S. Solution*

Use projectile equations.

The maximum range of the discharge is given by

$$R = \text{v}_o^2\left(\frac{\sin 2\phi}{g}\right) = \left(50 \frac{\text{ft}}{\text{sec}}\right)^2\left(\frac{\sin\left((2)(45^\circ)\right)}{32.2 \frac{\text{ft}}{\text{sec}^2}}\right) = \boxed{77.64 \text{ ft}}$$

The answer is (B).

SI Solution

Use projectile equations.

The maximum range of the discharge is given by

$$R = \text{v}_o^2\left(\frac{\sin 2\phi}{g}\right) = \frac{\left(15 \frac{\text{m}}{\text{s}}\right)^2 \sin\left((2)(45^\circ)\right)}{9.81 \frac{\text{m}}{\text{s}^2}} = \boxed{22.94 \text{ m}}$$

The answer is (B).

14.

$$A_o = \left(\frac{\pi}{4}\right)\left(\frac{4 \text{ in}}{12 \frac{\text{in}}{\text{ft}}}\right)^2 = 0.08727 \text{ ft}^2$$

$$A_t = \left(\frac{\pi}{4}\right)(20 \text{ ft})^2 = 314.16 \text{ ft}^2$$

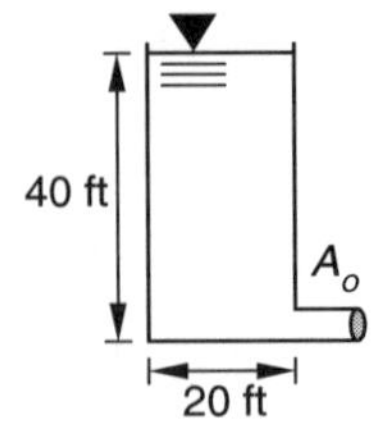

The time to drop from 40 ft to 20 ft is

$$t = \frac{(2)(314.16 \text{ ft}^2)\left(\sqrt{40 \text{ ft}} - \sqrt{20 \text{ ft}}\right)}{(0.98)(0.08727 \text{ ft}^2)\sqrt{(2)\left(32.2 \frac{\text{ft}}{\text{sec}^2}\right)}} = \boxed{1696 \text{ sec}}$$

The answer is (D).

15.

$$C_d = 1.00 \quad \text{[given]}$$

$$F_{va} = \frac{1}{\sqrt{1 - \left(\frac{D_2}{D_1}\right)^4}} = \frac{1}{\sqrt{1 - \left(\frac{8 \text{ in}}{12 \text{ in}}\right)^4}} = 1.116$$

$$A_2 = \left(\frac{\pi}{4}\right)\left(\frac{8 \text{ in}}{12 \frac{\text{in}}{\text{ft}}}\right)^2 = 0.3491 \text{ ft}^2$$

$$p_1 - p_2 = \left[\left(0.491 \frac{\text{lbf}}{\text{in}^3}\right)(4 \text{ in}) - \left(0.0361 \frac{\text{lbf}}{\text{in}^3}\right)(4 \text{ in})\right]\left(144 \frac{\text{in}^2}{\text{ft}^2}\right) = 262.0 \text{ lbf/ft}^2$$

$$Q = F_{va}C_dA_2\sqrt{\frac{2g(p_1 - p_2)}{\gamma}} = (1.116)(1)(0.3491 \text{ ft}^2) \times \sqrt{\frac{(2)\left(32.2 \frac{\text{ft}}{\text{sec}^2}\right)\left(262 \frac{\text{lbf}}{\text{ft}^2}\right)}{62.4 \frac{\text{lbf}}{\text{ft}^3}}} = \boxed{6.406 \text{ ft}^3\text{/sec (cfs)}}$$

The answer is (C).

16. *Customary U.S. Solution*

The volumetric flow rate of benzene through the venturi meter is given by

$$\dot{V} = C_f A_2\sqrt{\frac{2g(\rho_m - \rho)h}{\rho}}$$

The density of mercury, ρ_m, at 60°F is approximately 848 lbm/ft^3.

The density of the benzene at 60°F is

$$\begin{aligned}\rho &= (\text{SG})\rho_{\text{water}}\\ &= (0.885)\left(62.4\ \frac{\text{lbm}}{\text{ft}^3}\right)\\ &= 55.22\ \text{lbm/ft}^3\end{aligned}$$

The throat area is

$$\begin{aligned}A_2 &= \left(\frac{\pi}{4}\right)D_2^2\\ &= \left(\frac{\pi}{4}\right)(3.5\ \text{in})^2\left(\frac{1\ \text{ft}^2}{144\ \text{in}^2}\right) = 0.0668\ \text{ft}^2\end{aligned}$$

The flow coefficient is defined as

$$C_f = \frac{C_d}{\sqrt{1-\beta^4}}$$

β is the ratio of the throat to inlet diameters.

$$\beta = \frac{3.5\ \text{in}}{8\ \text{in}} = 0.4375$$

$$\begin{aligned}C_f &= \frac{C_d}{\sqrt{1-\beta^4}} = \frac{0.99}{\sqrt{1-(0.4375)^4}}\\ &= 1.00865\end{aligned}$$

Find the volumetric flow of benzene.

$$\begin{aligned}\dot{V} &= C_f A_2\sqrt{\frac{2g(\rho_m-\rho)h}{\rho}}\\ &= (1.00865)(0.0668\ \text{ft}^2)\\ &\quad\times\sqrt{\frac{(2)\left(32.2\ \frac{\text{ft}}{\text{sec}^2}\right)\left(848\ \frac{\text{lbm}}{\text{ft}^3} - 55.22\ \frac{\text{lbm}}{\text{ft}^3}\right)(4\ \text{in})\left(\frac{1\ \text{ft}}{12\ \text{in}}\right)}{55.22\ \frac{\text{lbm}}{\text{ft}^3}}}\\ &= \boxed{1.183\ \text{ft}^3/\text{sec}}\end{aligned}$$

The answer is (A).

SI Solution

The volumetric flow rate of benzene through the venturi meter is given by

$$\dot{V} = C_f A_2\sqrt{\frac{2g(\rho_m-\rho)h}{\rho}}$$

ρ_m is the density of mercury at 15°C; ρ_m is approximately 13 600 kg/m^3.

The density of the benzene at 15°C is

$$\begin{aligned}\rho &= (\text{SG})\rho_{\text{water}}\\ &= (0.885)\left(1000\ \frac{\text{kg}}{\text{m}^3}\right)\\ &= 885\ \text{kg/m}^3\end{aligned}$$

The throat area is

$$\begin{aligned}A_2 &= \left(\frac{\pi}{4}\right)D_2^2 = \left(\frac{\pi}{4}\right)(0.09\ \text{m})^2\\ &= 0.0064\ \text{m}^2\end{aligned}$$

The flow coefficient is defined as

$$C_f = \frac{C_d}{\sqrt{1-\beta^4}}$$

β is the ratio of the throat to inlet diameters.

$$\beta = \frac{9\ \text{cm}}{20\ \text{cm}} = 0.45$$

$$\begin{aligned}C_f &= \frac{C_d}{\sqrt{1-\beta^4}} = \frac{0.99}{\sqrt{1-(0.45)^4}}\\ &= 1.01094\end{aligned}$$

Find the volumetric flow of benzene.

$$\begin{aligned}\dot{V} &= C_f A_2\sqrt{\frac{2g(\rho_m-\rho)h}{\rho}}\\ &= (1.01094)(0.0064\ \text{m}^2)\\ &\quad\times\sqrt{\frac{(2)\left(9.81\ \frac{\text{m}}{\text{s}^2}\right)\times\left(13\,600\ \frac{\text{kg}}{\text{m}^3} - 885\ \frac{\text{kg}}{\text{m}^3}\right)(0.1\ \text{m})}{885\ \frac{\text{kg}}{\text{m}^3}}}\\ &= \boxed{0.0344\ \text{m}^3/\text{s}\ \ (34.4\ \text{L/s})}\end{aligned}$$

The answer is (A).

17. $D_o = 0.2\ \text{ft}$

$$\begin{aligned}\text{v}_o &= \text{v}\left(\frac{D}{D_o}\right)^2 = \left(2\ \frac{\text{ft}}{\text{sec}}\right)\left(\frac{1\ \text{ft}}{0.2\ \text{ft}}\right)^2\\ &= 50\ \text{ft/sec}\end{aligned}$$

For 70°F water,

$$\nu = 1.059 \times 10^{-5}\ \text{ft}^2/\text{sec}$$
$$\gamma = 62.3\ \text{lbf/ft}^3$$
$$\text{Re} = \frac{D_o \text{v}_o}{\nu} = \frac{(0.2\ \text{ft})\left(50\ \frac{\text{ft}}{\text{sec}}\right)}{1.059 \times 10^{-5}\ \frac{\text{ft}^2}{\text{sec}}} = 9.44 \times 10^5$$
$$A_o = \left(\frac{\pi}{4}\right)(0.2\ \text{ft})^2 = 0.0314\ \text{ft}^2$$
$$A_p = \left(\frac{\pi}{4}\right)(1\ \text{ft})^2 = 0.7854\ \text{ft}^2$$
$$\frac{A_o}{A_p} = \frac{0.0314\ \text{ft}^2}{0.7854\ \text{ft}^2} = 0.040$$

From Fig. 17.28,

$$C_f \approx 0.60$$
$$\dot{V} = A\text{v} = (0.7854\ \text{ft}^2)\left(2\ \frac{\text{ft}}{\text{sec}}\right) = 1.571\ \text{ft}^3/\text{sec}$$
$$\Delta p = \left(\frac{\gamma}{2g}\right)\left(\frac{\dot{V}}{C_f A_o}\right)^2$$
$$\Delta p = \left[\frac{62.3\ \frac{\text{lbf}}{\text{ft}^3}}{(2)\left(32.2\ \frac{\text{ft}}{\text{sec}^2}\right)}\right]\left[\frac{1.571\ \frac{\text{ft}^3}{\text{sec}}}{(0.6)(0.0314\ \text{ft}^2)}\right]^2 = 6727\ \text{lbf/ft}^2\ \text{(psf)}$$
$$\Delta p = \frac{6727\ \frac{\text{lbf}}{\text{ft}^2}}{144\ \frac{\text{in}^2}{\text{ft}^2}} = \boxed{46.7\ \text{lbf/in}^2\ \text{(psi)}}$$

The answer is (D).

18. *Customary U.S. Solution*

The pressure differential across the orifice meter is given by

$$\Delta p = p_1 - p_2 = (\rho_{\text{mercury}} - \rho_{\text{water}})h \times \left(\frac{g}{g_c}\right)$$

The densities of mercury and water are

$$\rho_{\text{mercury}} = 848\ \text{lbm/ft}^3$$
$$\rho_{\text{water}} = 62.4\ \text{lbm/ft}^3$$

Substituting gives

$$\Delta p = (\rho_{\text{mercury}} - \rho_{\text{water}})h \times \left(\frac{g}{g_c}\right)$$
$$= \frac{\left(848\ \frac{\text{lbm}}{\text{ft}^3} - 62.4\ \frac{\text{lbm}}{\text{ft}^3}\right)(7\ \text{in}) \times \left(\frac{1\ \text{ft}}{12\ \text{in}}\right)\left(32.2\ \frac{\text{ft}}{\text{sec}^2}\right)}{32.2\ \frac{\text{lbm-ft}}{\text{lbf-sec}^2}}$$
$$= \boxed{458.3\ \text{lbf/ft}^2 \quad (3.18\ \text{psi})}$$

The answer is (B).

SI Solution

The pressure differential across the orifice meter is given by

$$\Delta p = p_1 - p_2 = (\rho_{\text{mercury}} - \rho_{\text{water}})hg$$

The densities of mercury and water are

$$\rho_{\text{mercury}} = 13\,600\ \text{kg/m}^3$$
$$\rho_{\text{water}} = 1000\ \text{kg/m}^3$$

Substituting gives

$$\Delta p = (\rho_{\text{mercury}} - \rho_{\text{water}})gh$$
$$= \left(13\,600\ \frac{\text{kg}}{\text{m}^3} - 1000\ \frac{\text{kg}}{\text{m}^3}\right)(0.178\ \text{m})\left(9.81\ \frac{\text{m}}{\text{s}^2}\right)$$
$$= \boxed{22\,002\ \text{Pa} \quad (22.0\ \text{kPa})}$$

The answer is (B).

19. *Customary U.S. Solution*

For 12 in pipe (assuming schedule-40),

$$D_i = 0.99483\ \text{ft} \quad \text{[App. 16.B]}$$
$$A_i = 0.7773\ \text{ft}^2$$

The velocity is

$$\text{v} = \frac{\dot{V}}{A} = \frac{10\ \frac{\text{ft}^3}{\text{sec}}}{0.7773\ \text{ft}^2} = 12.87\ \text{ft/sec}$$

For water at 70°F, $\nu = 1.059 \times 10^{-5}\ \text{ft}^2/\text{sec}$.

The Reynolds number in the pipe is

$$\text{Re} = \frac{\text{v}D_i}{\nu} = \frac{\left(12.87\ \frac{\text{ft}}{\text{sec}}\right)(0.99483\ \text{ft})}{1.059 \times 10^{-5}\ \frac{\text{ft}^2}{\text{sec}}} = 1.21 \times 10^6 \quad \begin{bmatrix}\text{fully turbulent}\\ \text{in the pipe}\end{bmatrix}$$

Flow through the orifice will have a higher Reynolds number.

The volumetric flow rate through a sharp-edged orifice is given by

$$\dot{V} = C_f A_o \sqrt{\frac{2g(\rho_m - \rho)h}{\rho}}$$

In terms of pressure,

$$\dot{V} = C_f A_o \sqrt{\frac{2g_c(p_1 - p_2)}{\rho}}$$

Rearranging gives

$$C_f A_o = \frac{\dot{V}}{\sqrt{\frac{2g_c(p_1 - p_2)}{\rho}}}$$

p_1 is the upstream pressure, and p_2 is the downstream pressure.

The maximum head loss must not exceed 25 ft; therefore,

$$\frac{\left(\frac{g_c}{g}\right) \times (p_1 - p_2)}{\rho} = 25 \text{ ft}$$

$$\frac{g_c(p_1 - p_2)}{\rho} = (25 \text{ ft})g$$

Substituting gives

$$C_f A_0 = \frac{10 \ \frac{\text{ft}^3}{\text{sec}}}{\sqrt{(2)\left(32.2 \ \frac{\text{ft}}{\text{sec}^2}\right)(25 \text{ ft})}} = 0.249 \text{ ft}^2$$

Both C_f and A_o depend on the orifice diameter.

For a 7 in diameter orifice,

$$A_o = \frac{\pi D_o^2}{4} = \frac{\pi\left((7 \text{ in})\left(\frac{1 \text{ ft}}{12 \text{ in}}\right)\right)^2}{4} = 0.267 \text{ ft}^2$$

$$\frac{A_o}{A_1} = \frac{0.267 \text{ ft}^2}{0.7773 \text{ ft}^2} = 0.343$$

From a chart of flow coefficients (Fig. 17.28), for $A_o/A_1 = 0.343$ and fully turbulent flow,

$$C_f = 0.645$$

$$C_f A_o = (0.645)(0.267 \text{ ft}^2) = 0.172 \text{ ft}^2 < 0.249 \text{ ft}^2$$

Therefore, a 7 in diameter orifice is too small.

Try a 9 in diameter orifice.

$$A_o = \frac{\pi D_o^2}{4} = \frac{\pi\left((9 \text{ in})\left(\frac{1 \text{ ft}}{12 \text{ in}}\right)\right)^2}{4} = 0.442 \text{ ft}^2$$

$$\frac{A_o}{A_1} = \frac{0.442 \text{ ft}^2}{0.7773 \text{ ft}^2} = 0.569$$

From Fig. 17.28, for $A_o/A_1 = 0.569$ and fully turbulent flow,

$$C_f = 0.73$$

$$C_f A_o = (0.73)(0.442 \text{ ft}^2) = 0.323 \text{ ft}^2 > 0.249 \text{ ft}^2$$

Therefore, a 9 in orifice is too large.

Interpolating gives

$$D_o = 7 \text{ in} + \frac{(9 \text{ in} - 7 \text{ in})(0.249 \text{ ft}^2 - 0.172 \text{ ft}^2)}{0.323 \text{ ft}^2 - 0.172 \text{ ft}^2} = 8.0 \text{ in}$$

Further iterations yield

$$D_o = \boxed{8.1 \text{ in}}$$

$$C_f A_o = 0.243 \text{ ft}^2$$

The answer is (C).

SI Solution

For 300 mm pipe (assume the nominal diameter is the inner diameter), $D_i = 0.30$ m.

The velocity is

$$\text{v} = \frac{\dot{V}}{A} = \frac{\dot{V}}{\frac{\pi D_i^2}{4}} = \frac{\left(250 \ \frac{\text{L}}{\text{s}}\right)\left(\frac{1 \text{ m}^3}{1000 \text{ L}}\right)}{\frac{\pi(0.3 \text{ m})^2}{4}} = 3.54 \text{ m/s}$$

From App. 14.B, for water at 20°C,

$$\nu = 1.007 \times 10^{-6} \text{ m}^2/\text{s}$$

The Reynolds number in the pipe is

$$\begin{aligned}\text{Re} &= \frac{\text{v}D_i}{\nu} \\ &= \frac{\left(3.54\ \frac{\text{m}}{\text{s}}\right)(0.3\ \text{m})}{1.007\times 10^{-6}\ \frac{\text{m}^2}{\text{s}}} \\ &= 1.05\times 10^6 \quad \left[\begin{array}{c}\text{fully turbulent}\\ \text{in the pipe}\end{array}\right]\end{aligned}$$

Flow through the orifice will have a higher Reynolds number.

The volumetric flow rate through a sharp-edged orifice is given by

$$\dot{V} = C_f A_o \sqrt{\frac{2g(\rho_m - \rho)h}{\rho}}$$

In terms of pressure,

$$\dot{V} = C_f A_o \sqrt{\frac{2(p_1 - p_2)}{\rho}}$$

Rearranging gives

$$C_f A_o = \frac{\dot{V}}{\sqrt{\frac{2(p_1 - p_2)}{\rho}}}$$

p_1 is the upstream pressure, and p_2 is the downstream pressure.

The maximum head loss must not exceed 7.5 m; therefore,

$$\begin{aligned}\frac{p_1 - p_2}{g\rho} &= 7.5\ \text{m} \\ \frac{p_1 - p_2}{\rho} &= (7.5\ \text{m})g\end{aligned}$$

Substituting gives

$$\begin{aligned}C_f A_o &= \frac{0.25\ \frac{\text{m}^3}{\text{s}}}{\sqrt{(2)\left(9.81\ \frac{\text{m}}{\text{s}^2}\right)(7.5\ \text{m})}} \\ &= 0.021\ \text{m}^2\end{aligned}$$

Both C_f and A_o depend on the orifice diameter.

For an 18 cm diameter orifice,

$$\begin{aligned}A_o &= \frac{\pi D_o^2}{4} = \frac{\pi(0.18\ \text{m})^2}{4} = 0.0254\ \text{m}^2 \\ \frac{A_o}{A_1} &= \frac{0.0254\ \text{m}^2}{0.0707\ \text{m}^2} = 0.359\end{aligned}$$

From a chart of flow coefficients (Fig. 17.28), for $A_o/A_1 = 0.359$ and fully turbulent flow,

$$\begin{aligned}C_f &= 0.65 \\ C_f A_o &= (0.65)(0.0254\ \text{m}^2) = 0.0165\ \text{m}^2 < 0.021\ \text{m}^2\end{aligned}$$

Therefore, an 18 cm diameter orifice is too small.

Try a 23 cm diameter orifice.

$$\begin{aligned}A_o &= \frac{\pi D_o^2}{4} = \frac{\pi(0.23\ \text{m})^2}{4} = 0.0415\ \text{m}^2 \\ \frac{A_o}{A_1} &= \frac{0.0415\ \text{m}^2}{0.0707\ \text{m}^2} = 0.587\end{aligned}$$

From Fig. 17.28, for $A_o/A_1 = 0.587$ and fully turbulent flow,

$$\begin{aligned}C_f &= 0.73 \\ C_f A_o &= (0.73)(0.0415\ \text{m}^2) = 0.0303\ \text{m}^2 > 0.021\ \text{m}^2\end{aligned}$$

Therefore, a 23 cm orifice is too large.

Interpolating gives

$$\begin{aligned}D_o &= 18\ \text{cm} \\ &\quad + (23\ \text{cm} - 18\ \text{cm})\left(\frac{0.021\ \text{m}^2 - 0.0165\ \text{m}^2}{0.0303\ \text{m}^2 - 0.0165\ \text{m}^2}\right) \\ &= 19.6\ \text{cm}\end{aligned}$$

Further iteration yields

$$\begin{aligned}D_o &= \boxed{20.0\ \text{cm}} \\ C_f &= 0.675 \\ C_f A_o &= 0.021\ \text{m}^2\end{aligned}$$

The answer is (C).

20. $$A_\text{A} = \left(\frac{\pi}{4}\right)\left(\frac{24\ \text{in}}{12\ \frac{\text{in}}{\text{ft}}}\right)^2 = 3.142\ \text{ft}^2$$

$$A_\text{B} = \left(\frac{\pi}{4}\right)\left(\frac{12\ \text{in}}{12\ \frac{\text{in}}{\text{ft}}}\right)^2 = 0.7854\ \text{ft}^2$$

$$\text{v}_\text{A} = \frac{\dot{V}}{A} = \frac{8\ \frac{\text{ft}^3}{\text{sec}}}{3.142\ \text{ft}^2} = 2.546\ \text{ft/sec}$$

$$p_\text{A} = \gamma h = \left(62.4\ \frac{\text{lbf}}{\text{ft}^3}\right)(20\ \text{ft}) = 1248\ \text{lbf/ft}^2$$

Using the Bernoulli equation to solve for p_B,

$$\mathrm{v_B} = \frac{\dot{V}}{A} = \frac{8\ \frac{\text{ft}^3}{\text{sec}}}{0.7854\ \text{ft}^2} = 10.19\ \text{ft/sec}$$

$$p_B = 1248\ \frac{\text{lbf}}{\text{ft}^2} - \left[\frac{\left(10.19\ \frac{\text{ft}}{\text{sec}}\right)^2 - \left(2.546\ \frac{\text{ft}}{\text{sec}}\right)^2}{(2)\left(32.2\ \frac{\text{ft}}{\text{sec}^2}\right)}\right] \times \left(62.4\ \frac{\text{lbf}}{\text{ft}^3}\right) = 1153.67\ \text{lbf/ft}^2$$

With $\theta = 0$,

$$F_x = \left(1153.67\ \frac{\text{lbf}}{\text{ft}^2}\right)(0.7854\ \text{ft}^2) - \left(1248\ \frac{\text{lbf}}{\text{ft}^2}\right)(3.142\ \text{ft}^2) + \left[\frac{\left(8\ \frac{\text{ft}^3}{\text{sec}}\right)\left(62.4\ \frac{\text{lbf}}{\text{ft}^3}\right)}{32.2\ \frac{\text{ft}}{\text{sec}^2}}\right]\left(10.19\ \frac{\text{ft}}{\text{sec}} - 2.546\ \frac{\text{ft}}{\text{sec}}\right)$$

$$= \boxed{-2897\ \text{lbf on the fluid (toward A)}}$$

$$F_y = 0$$

The answer is (A).

21. *Customary U.S. Solution*

The mass flow rate of the water is

$$\dot{m} = \rho\dot{V} = \rho \mathrm{v} A = \frac{\rho \mathrm{v}\pi D^2}{4}$$

$$= \left(62.4\ \frac{\text{lbm}}{\text{ft}^3}\right)\left(40\ \frac{\text{ft}}{\text{sec}}\right)\left(\frac{\pi\left[(2\ \text{in})\left(\frac{1\ \text{ft}}{12\ \text{in}}\right)\right]^2}{4}\right)$$

$$= 54.45\ \text{lbm/sec}$$

The effective mass flow rate of the water is

$$\dot{m}_{\text{eff}} = \left(\frac{\mathrm{v} - \mathrm{v}_b}{\mathrm{v}}\right)\dot{m}$$

$$= \left(\frac{40\ \frac{\text{ft}}{\text{sec}} - 15\ \frac{\text{ft}}{\text{sec}}}{40\ \frac{\text{ft}}{\text{sec}}}\right)\left(54.45\ \frac{\text{lbm}}{\text{sec}}\right)$$

$$= 34.0\ \text{lbm/sec}$$

The force in the (horizontal) x-direction is given by

$$F_x = \frac{\dot{m}_{\text{eff}}(\mathrm{v} - \mathrm{v}_b)(\cos\theta - 1)}{g_c}$$

$$= \frac{\left(34.0\ \frac{\text{lbm}}{\text{sec}}\right)\left(40\ \frac{\text{ft}}{\text{sec}} - 15\ \frac{\text{ft}}{\text{sec}}\right)(\cos 60° - 1)}{32.2\ \frac{\text{lbm-ft}}{\text{lbf-sec}^2}}$$

$$= -13.2\ \text{lbf}\quad \text{[the force is acting to the left]}$$

The force in the (vertical) y-direction is given by

$$F_y = \frac{\dot{m}_{\text{eff}}(\mathrm{v} - \mathrm{v}_b)\sin\theta}{g_c}$$

$$= \frac{\left(34.0\ \frac{\text{lbm}}{\text{sec}}\right)\left(40\ \frac{\text{ft}}{\text{sec}} - 15\ \frac{\text{ft}}{\text{sec}}\right)(\sin 60°)}{32.2\ \frac{\text{lbm-ft}}{\text{lbf-sec}^2}}$$

$$= 22.9\ \text{lbf}\quad \text{[the force is acting upward]}$$

The net resultant force is

$$F = \sqrt{F_x^2 + F_y^2} = \sqrt{(-13.2\ \text{lbf})^2 + (22.9\ \text{lbf})^2} = \boxed{26.4\ \text{lbf}}$$

The answer is (B).

SI Solution

The mass flow rate of the water is

$$\dot{m} = \rho\dot{V} = \rho \mathrm{v} A = \frac{\rho \mathrm{v}\pi D^2}{4}$$

$$= \left(1000\ \frac{\text{kg}}{\text{m}^3}\right)\left(12\ \frac{\text{m}}{\text{s}}\right)\left(\frac{\pi(0.05\ \text{m})^2}{4}\right)$$

$$= 23.56\ \text{kg/s}$$

The effective mass flow rate of the water is

$$\dot{m}_{\text{eff}} = \left(\frac{\mathrm{v} - \mathrm{v}_b}{\mathrm{v}}\right)\dot{m}$$

$$= \left(\frac{12\ \frac{\text{m}}{\text{s}} - 4.5\ \frac{\text{m}}{\text{s}}}{12\ \frac{\text{m}}{\text{s}}}\right)\left(23.56\ \frac{\text{kg}}{\text{s}}\right)$$

$$= 14.73\ \text{kg/s}$$

The force in the (horizontal) x-direction is given by

$$F_x = \dot{m}_{\text{eff}}(\mathrm{v} - \mathrm{v}_b)(\cos\theta - 1)$$

$$= \left(14.73\ \frac{\text{kg}}{\text{s}}\right)\left(12\ \frac{\text{m}}{\text{s}} - 4.5\ \frac{\text{m}}{\text{s}}\right)(\cos 60° - 1)$$

$$= -55.2\ \text{N}\quad \text{[the force is acting to the left]}$$

The force in the (vertical) y-direction is given by

$$\begin{aligned}F_y &= \dot{m}_{\text{eff}}(\mathrm{v} - \mathrm{v}_b)\sin\theta \\ &= \left(14.73\ \frac{\text{kg}}{\text{s}}\right)\left(12\ \frac{\text{m}}{\text{s}} - 4.5\ \frac{\text{m}}{\text{s}}\right)(\sin 60^\circ) \\ &= 95.7\ \text{N} \quad [\text{the force is acting upward}]\end{aligned}$$

The net resultant force is

$$\begin{aligned}F &= \sqrt{F_x^2 + F_y^2} \\ &= \sqrt{(-55.2\ \text{N})^2 + (95.7\ \text{N})^2} \\ &= \boxed{110.5\ \text{N}}\end{aligned}$$

The answer is (B).

22. *Customary U.S. Solution*

The power that must be added to the pump is given by

$$P = \frac{\Delta h \dot{m} \times \left(\frac{g}{g_c}\right)}{\eta}$$

Assume schedule-40 pipe.

$$\begin{aligned}D_i &= 0.9948\ \text{ft} \quad [\text{App. 16.B}] \\ A_i &= 0.7773\ \text{ft}^2 \\ \mathrm{v} &= \frac{\dot{V}}{A_i} \\ &= \frac{\left(2000\ \frac{\text{gal}}{\text{min}}\right)\left(\frac{0.002228\ \frac{\text{ft}^3}{\text{sec}}}{1\ \frac{\text{gal}}{\text{min}}}\right)}{0.7773\ \text{ft}^2} \\ &= 5.73\ \text{ft/sec}\end{aligned}$$

(Note that the pressures are in terms of gage pressure, and the density of mercury is 0.491 lbm/in^3.)

$$\begin{aligned}p_i &= \left[14.7\ \frac{\text{lbf}}{\text{in}^2} - \frac{(6\ \text{in})\left(0.491\ \frac{\text{lbm}}{\text{in}^3}\right)\left(32.2\ \frac{\text{ft}}{\text{sec}^2}\right)}{32.2\ \frac{\text{lbm-ft}}{\text{lbf-sec}^2}}\right] \\ &\quad \times \left(\frac{144\ \text{in}^2}{1\ \text{ft}^2}\right) \\ &= 1692.6\ \text{lbf/ft}^2\end{aligned}$$

$$E_{ti} = \frac{p_i}{\rho} + \frac{\mathrm{v}_i^2}{2g_c} + \frac{z_i g}{g_c}$$

Since the pump inlet and outlet are at the same elevation, use $z = 0$ and $\rho = (\text{SG})\rho_{\text{water}}$.

$$\begin{aligned}E_{ti} &= \frac{p_i}{(\text{SG})\rho_{\text{water}}} + \frac{\mathrm{v}_i^2}{2g_c} + 0 \\ &= \frac{1692.6\ \frac{\text{lbf}}{\text{ft}^2}}{(1.2)\left(62.4\ \frac{\text{lbm}}{\text{ft}^3}\right)} + \frac{\left(5.73\ \frac{\text{ft}}{\text{sec}}\right)^2}{(2)\left(32.2\ \frac{\text{lbm-ft}}{\text{lbf-sec}^2}\right)} \\ &= 23.11\ \text{ft-lbf/lbm}\end{aligned}$$

Calculate the total head at the inlet.

$$\begin{aligned}h_{ti} &= E_{ti} \times \left(\frac{g_c}{g}\right) \\ &= \frac{\left(23.11\ \frac{\text{ft-lbf}}{\text{lbm}}\right)\left(32.2\ \frac{\text{lbm-ft}}{\text{lbf-sec}^2}\right)}{32.2\ \frac{\text{ft}}{\text{sec}^2}} \\ &= 23.11\ \text{ft}\end{aligned}$$

At the outlet side of the pump,

$$\begin{aligned}D_o &= 0.6651\ \text{ft} \\ A_o &= 0.3474\ \text{ft}^2 \\ Q &= \mathrm{v}_o A_o \\ \mathrm{v}_o &= \frac{Q}{A_o} \\ &= \frac{\left(2000\ \frac{\text{gal}}{\text{min}}\right)\left(\frac{0.002228\ \frac{\text{ft}^3}{\text{sec}}}{1\ \frac{\text{gal}}{\text{min}}}\right)}{0.3474\ \text{ft}^2} \\ &= 12.83\ \text{ft/sec}\end{aligned}$$

(Note that the pressures are in terms of gage pressure and the gauge is located 4 ft above the pump outlet, which adds 4 ft of pressure head at the pump outlet.)

$$\begin{aligned}p_o &= \left(14.7\ \frac{\text{lbf}}{\text{in}^2} + 20\ \frac{\text{lbf}}{\text{in}^2}\right)\left(\frac{144\ \text{in}^2}{1\ \text{ft}^2}\right) \\ &\quad + 4\ \text{ft}\left[\frac{(1.2)\left(62.4\ \frac{\text{lbm}}{\text{ft}^3}\right)\left(32.2\ \frac{\text{ft}}{\text{sec}^2}\right)}{32.2\ \frac{\text{lbm-ft}}{\text{lbf-sec}^2}}\right] \\ &= 5296\ \text{lbf/ft}^2\end{aligned}$$

$$E_{to} = \frac{p_o}{\rho} + \frac{\mathrm{v}_o^2}{2g_c} + \frac{z_o g}{g_c}$$

Since the pump inlet and outlet are at the same elevation, use $z = 0$ and $\rho = (\text{SG})\rho_{\text{water}}$.

$$E_{to} = \frac{p_o}{(\text{SG})\rho_{\text{water}}} + \frac{\text{v}_o^2}{2g_c} + 0$$

$$= \frac{5296 \ \frac{\text{lbf}}{\text{ft}^2}}{(1.2)\left(62.4 \ \frac{\text{lbm}}{\text{ft}^3}\right)} + \frac{\left(12.83 \ \frac{\text{ft}}{\text{sec}}\right)^2}{(2)\left(32.2 \ \frac{\text{lbm-ft}}{\text{lbf-sec}^2}\right)}$$

$$= 73.28 \text{ ft-lbf/lbm}$$

Calculate the total head at the outlet.

$$h_{to} = E_{to} \times \left(\frac{g_c}{g}\right)$$

$$= \left(73.28 \ \frac{\text{ft-lbf}}{\text{lbm}}\right)\left(\frac{32.2 \ \frac{\text{lbm-ft}}{\text{lbf-sec}^2}}{32.2 \ \frac{\text{ft}}{\text{sec}^2}}\right)$$

$$= 73.28 \text{ ft}$$

Compute the total head required across the pump.

$$\Delta h = h_{to} - h_{ti} = 73.28 \text{ ft} - 23.11 \text{ ft} = 50.17 \text{ ft}$$

The mass flow rate is

$$\dot{m} = \rho\dot{V}$$

In terms of the specific gravity, the mass flow rate is

$$\dot{m} = (\text{SG})\rho_{\text{water}}\dot{V}$$

$$= (1.2)\left(62.4 \ \frac{\text{lbm}}{\text{ft}^3}\right)\left(2000 \ \frac{\text{gal}}{\text{min}}\right)\left(\frac{0.002228 \ \frac{\text{ft}^3}{\text{sec}}}{1 \ \frac{\text{gal}}{\text{min}}}\right)$$

$$= 333.7 \text{ lbm/sec}$$

The power that must be added to the pump is

$$P = \frac{\Delta h\dot{m} \times \left(\frac{g}{g_c}\right)}{\eta}$$

$$= \frac{(50.17 \text{ ft})\left(333.7 \ \frac{\text{lbm}}{\text{sec}}\right)\left(\frac{32.2 \ \frac{\text{ft}}{\text{sec}^2}}{32.2 \ \frac{\text{lbm-ft}}{\text{lbf-sec}^2}}\right)}{(0.85)\left(550 \ \frac{\text{ft-lbf}}{\text{hp-sec}}\right)}$$

$$= \boxed{35.8 \text{ hp}}$$

(Note that it is not necessary to use absolute pressures as has been done in this solution.)

The answer is (B).

SI Solution

The power that must be added to the pump is given by

$$P = \frac{\Delta h\dot{m}g}{\eta}$$

Assume the pipe nominal diameter is equal to the internal diameter.

$$D_i = 0.30 \text{ m}$$

$$A_i = \frac{\pi D_i^2}{4} = x\frac{\pi(0.30 \text{ m})^2}{4} = 0.0707 \text{ m}^2$$

$$\text{v} = \frac{\dot{V}}{A_i} = \frac{0.125 \ \frac{\text{m}^3}{\text{s}}}{0.0707 \text{ m}^2} = 1.77 \text{ m/s}$$

(Note that the pressures are in terms of gage pressure, and the density of mercury is 13 600 kg/m^3.)

$$p_i = 1.013 \times 10^5 \text{ Pa} - (0.15 \text{ m})\left(13\,600 \ \frac{\text{kg}}{\text{m}^3}\right)\left(9.81 \ \frac{\text{m}}{\text{s}^2}\right) = 8.13 \times 10^4 \text{ Pa}$$

$$E_{ti} = \frac{p}{\rho} + \frac{\text{v}_i^2}{2} + z_i g$$

Since the pump inlet and outlet are at the same elevation, use $z = 0$ and $\rho = (\text{SG})\rho_{\text{water}}$.

$$E_{ti} = \frac{p}{(\text{SG})\rho_{\text{water}}} + \frac{\text{v}_i^2}{2} + 0$$

$$= \frac{8.13 \times 10^4 \text{ Pa}}{(1.2)\left(1000 \ \frac{\text{kg}}{\text{m}^3}\right)} + \frac{\left(1.77 \ \frac{\text{m}}{\text{s}}\right)^2}{2}$$

$$= 69.3 \text{ J/kg}$$

The total head at the inlet is

$$h_{ti} = \frac{E_{ti}}{g} = \frac{69.3 \ \frac{\text{J}}{\text{kg}}}{9.81 \ \frac{\text{m}}{\text{s}^2}} = 7.06 \text{ m}$$

Assume the pipe nominal diameter is equal to the internal diameter. On the outlet side of the pump,

$$D_i = 0.20 \text{ m}$$

$$A_o = \frac{\pi D_o^2}{4} = \frac{\pi(0.20 \text{ m})^2}{4} = 0.0314 \text{ m}^2$$

$$\text{v}_o = \frac{\dot{V}}{A_o} = \frac{0.125 \ \frac{\text{m}^3}{\text{s}}}{0.0314 \text{ m}^2} = 3.98 \text{ m/s}$$

Water Resources
Flow of Fluids

(Note that the pressures are in terms of gage pressure and the gauge is located 1.2 m above the pump outlet, which adds 1.2 m of pressure head at the pump outlet.)

$$\begin{aligned} p_o &= 1.013 \times 10^5 \text{ Pa} + 138 \times 10^3 \text{ Pa} \\ &\quad + (1.2 \text{ m})\left[(1.2)\left(1000 \ \frac{\text{kg}}{\text{m}^3}\right)\left(9.81 \ \frac{\text{m}}{\text{s}^2}\right)\right] \\ &= 2.53 \times 10^5 \text{ Pa} \end{aligned}$$

$$E_{to} = \frac{p_o}{\rho} + \frac{\text{v}_o^2}{2} + z_o g$$

Since the pump inlet and outlet are at the same elevation, use $z = 0$ and $\rho = (\text{SG})\rho_{\text{water}}$.

$$\begin{aligned} E_{to} &= \frac{p_o}{(\text{SG})\rho_{\text{water}}} + \frac{\text{v}_o^2}{2} + 0 \\ &= \frac{2.53 \times 10^5 \text{ Pa}}{(1.2)\left(1000 \ \frac{\text{kg}}{\text{m}^3}\right)} + \frac{\left(3.98 \ \frac{\text{m}}{\text{s}}\right)^2}{2} \\ &= 218.8 \text{ J/kg} \end{aligned}$$

The total head at the outlet is

$$\begin{aligned} h_{to} &= \frac{E_{to}}{g} = \frac{218.8 \ \frac{\text{J}}{\text{kg}}}{22.30 \ \frac{\text{m}}{\text{s}^2}} \\ &= 11.76 \text{ m} \end{aligned}$$

The total head required across the pump is

$$\begin{aligned} \Delta h &= h_{to} - h_{ti} \\ &= 22.30 \text{ m} - 7.04 \text{ m} \\ &= 15.26 \text{ m} \end{aligned}$$

The mass flow rate is

$$\dot{m} = \rho \dot{V}$$

In terms of the specific gravity, the mass flow rate is

$$\begin{aligned} \dot{m} &= (\text{SG})\rho_{\text{water}} Q \\ &= (1.2)\left(1000 \ \frac{\text{kg}}{\text{m}^3}\right)\left(0.125 \ \frac{\text{m}^3}{\text{s}}\right) \\ &= 150 \text{ kg/s} \end{aligned}$$

The power that must be added to the pump is

$$\begin{aligned} P &= \frac{\Delta h \dot{m} g}{\eta} \\ &= \frac{(15.26 \text{ m})\left(150 \ \frac{\text{kg}}{\text{s}}\right)\left(9.81 \ \frac{\text{m}}{\text{s}^2}\right)}{0.85} \\ &= \boxed{26\,417 \text{ W} \ \ (26.4 \text{ kW})} \end{aligned}$$

(Note that it is not necessary to use absolute pressures as has been done in this solution.)

The answer is (B).

23. *Customary U.S. Solution*

The power developed by the horizontal turbine is given by

$$P = \dot{m} h_{\text{loss}} \times \left(\frac{g}{g_c}\right)$$

The mass flow rate is

$$\begin{aligned} \dot{m} &= \dot{V}\rho \\ &= \left(100 \ \frac{\text{ft}^3}{\text{sec}}\right)\left(62.4 \ \frac{\text{lbm}}{\text{ft}^3}\right) \\ &= 6240 \text{ lbm/sec} \end{aligned}$$

The head loss across the horizontal turbine is given by

$$\begin{aligned} h_{\text{loss}} &= \left(\frac{\Delta p}{\rho}\right) \times \left(\frac{g_c}{g}\right) \\ &= \left(\frac{\left(30 \ \frac{\text{lbf}}{\text{in}^2} - \left(-5 \ \frac{\text{lbf}}{\text{in}^2}\right)\right)\left(\frac{144 \text{ in}^2}{1 \text{ ft}^2}\right)}{62.4 \ \frac{\text{lbm}}{\text{ft}^3}}\right) \\ &\quad \times \left(\frac{32.2 \ \frac{\text{lbm-ft}}{\text{lbf-sec}^2}}{32.2 \ \frac{\text{ft}}{\text{sec}^2}}\right) \\ &= 80.77 \text{ ft} \end{aligned}$$

From Table 18.5, the power developed is

$$\begin{aligned} P &= \dot{m} h_{\text{loss}} \times \left(\frac{g}{g_c}\right) \\ &= \frac{\left(6240 \ \frac{\text{lbm}}{\text{sec}}\right)(80.77 \text{ ft})\left(32.2 \ \frac{\text{ft}}{\text{sec}^2}\right)}{\left(32.2 \ \frac{\text{lbm-ft}}{\text{lbf-sec}^2}\right)\left(550 \ \frac{\text{ft-lbf}}{\text{hp-sec}}\right)} \\ &= \boxed{916 \text{ hp}} \end{aligned}$$

The answer is (D).

SI Solution

The power developed by the horizontal turbine is given by

$$P = \dot{m} h_{\text{loss}} g$$

The mass flow rate is

$$\begin{aligned} \dot{m} &= \dot{V}\rho = \left(2.6 \ \frac{\text{m}^3}{\text{s}}\right)\left(1000 \ \frac{\text{kg}}{\text{m}^3}\right) \\ &= 2600 \text{ kg/s} \end{aligned}$$

The head loss across the horizontal turbine is given by

$$h_{loss} = \frac{\Delta p}{\rho g}$$

$$= \frac{(210\ \text{kPa} - (-35\ \text{kPa}))\left(1000\ \frac{\text{Pa}}{\text{kPa}}\right)}{\left(1000\ \frac{\text{kg}}{\text{m}^3}\right)\left(9.81\ \frac{\text{m}}{\text{s}^2}\right)}$$

$$= 25.0\ \text{m}$$

From Table 18.5, the power developed is

$$P = \dot{m} h_{loss} g$$

$$= \left(2600\ \frac{\text{kg}}{\text{s}}\right)(25.0\ \text{m})\left(9.81\ \frac{\text{m}}{\text{s}^2}\right)$$

$$= \boxed{637\,650\ \text{W}\quad (638\ \text{kW})}$$

The answer is (D).

24. *Customary U.S. Solution*

(a) The drag on the car is given by

$$F_D = \frac{C_D A \rho \text{v}^2}{2g_c}$$

For air at 70°F,

$$\rho = \frac{p}{RT} = \frac{\left(14.7\ \frac{\text{lbf}}{\text{in}^2}\right)\left(144\ \frac{\text{in}^2}{\text{ft}^2}\right)}{\left(53.35\ \frac{\text{ft-lbf}}{\text{lbm-°R}}\right)(70°\text{F} + 460)}$$

$$= 0.0749\ \text{lbm/ft}^3$$

$$\text{v} = \left(55\ \frac{\text{mi}}{\text{hr}}\right)\left(5280\ \frac{\text{ft}}{\text{mi}}\right)\left(\frac{1\ \text{hr}}{3600\ \text{sec}}\right)$$

$$= 80.67\ \text{ft/sec}$$

Substituting gives

$$F_D = \frac{C_D A \rho \text{v}^2}{2g_c}$$

$$= \frac{(0.42)\left(28\ \text{ft}^2\right)\left(0.0749\ \frac{\text{lbm}}{\text{ft}^3}\right)\left(80.67\ \frac{\text{ft}}{\text{sec}}\right)^2}{(2)\left(32.2\ \frac{\text{lbm-ft}}{\text{lbf-sec}^2}\right)}$$

$$= 89.0\ \text{lbf}$$

The total resisting force is

$$F = F_D + \text{rolling resistance}$$

$$= 89.0\ \text{lbf} + (0.01)(3300\ \text{lbm}) \times \left(\frac{g}{g_c}\right)$$

$$= 89.0\ \text{lbf} + \frac{(0.01)(3300\ \text{lbm})\left(32.2\ \frac{\text{ft}}{\text{sec}^2}\right)}{32.2\ \frac{\text{lbm-ft}}{\text{lbf-sec}^2}}$$

$$= 122.0\ \text{lbf}$$

The power consumed is

$$P = F\text{v}$$

$$= \frac{(122.0\ \text{lbf})\left(80.67\ \frac{\text{ft}}{\text{sec}}\right)}{778\ \frac{\text{ft-lbf}}{\text{Btu}}}$$

$$= 12.65\ \text{Btu/sec}$$

The power available from the fuel is

$$P_A = (\text{engine thermal efficiency})(\text{fuel heating value})$$

$$= (0.28)\left(115{,}000\ \frac{\text{Btu}}{\text{gal}}\right)$$

$$= 32{,}200\ \text{Btu/gal}$$

The fuel consumption at 55 mi/hr is

$$\frac{P}{P_A} = \frac{12.65\ \frac{\text{Btu}}{\text{sec}}}{32{,}200\ \frac{\text{Btu}}{\text{gal}}}$$

$$= 3.93 \times 10^{-4}\ \text{gal/sec}$$

The fuel consumption is

$$\frac{3.93 \times 10^{-4}\ \frac{\text{gal}}{\text{sec}}}{\left(55\ \frac{\text{mi}}{\text{hr}}\right)\left(\frac{1\ \text{hr}}{3600\ \text{sec}}\right)} = \boxed{0.0257\ \text{gal/mi}}$$

The answer is (A).

(b) Similarly, the fuel consumption at 65 mi/hr is

$$\text{v} = \left(65\ \frac{\text{mi}}{\text{hr}}\right)\left(5280\ \frac{\text{ft}}{\text{mi}}\right)\left(\frac{1\ \text{hr}}{3600\ \text{sec}}\right)$$

$$= 95.33\ \text{ft/sec}$$

$$F_D = \frac{C_D A \rho \text{v}^2}{2g_c}$$

$$= \frac{(0.42)\left(28\ \text{ft}^2\right)\left(0.0749\ \frac{\text{lbm}}{\text{ft}^3}\right)\left(95.33\ \frac{\text{ft}}{\text{sec}}\right)^2}{(2)\left(32.2\ \frac{\text{lbm-ft}}{\text{lbf-sec}^2}\right)}$$

$$= 124.3\ \text{lbf}$$

The total resisting force is

$$F = F_D + \text{rolling resistance}$$

$$= 124.3\ \text{lbf} + (0.01)(3300\ \text{lbm}) \times \left(\frac{g}{g_c}\right)$$

$$= 124.3\ \text{lbf} + \frac{(0.01)(3300\ \text{lbm})\left(32.2\ \frac{\text{ft}}{\text{sec}^2}\right)}{32.2\ \frac{\text{lbm-ft}}{\text{lbf-sec}^2}}$$

$$= 157.3\ \text{lbf}$$

The power consumed is

$$P = F\text{v} = \frac{(157.3\ \text{lbf})\left(95.33\ \frac{\text{ft}}{\text{sec}}\right)}{778\ \frac{\text{ft-lbf}}{\text{Btu}}} = 19.27\ \text{Btu/sec}$$

The fuel consumption at 65 mi/hr is

$$\frac{P}{P_A} = \frac{19.27\ \frac{\text{Btu}}{\text{sec}}}{32{,}200\ \frac{\text{Btu}}{\text{gal}}} = 5.98 \times 10^{-4}\ \text{gal/sec}$$

The fuel consumption is

$$\frac{5.98 \times 10^{-4}\ \frac{\text{gal}}{\text{sec}}}{\left(65\ \frac{\text{mi}}{\text{hr}}\right)\left(\frac{1\ \text{hr}}{3600\ \text{sec}}\right)} = 0.0331\ \text{gal/mi}$$

The relative difference between the fuel consumption at 55 mi/hr and 65 mi/hr is

$$\frac{0.0331\ \frac{\text{gal}}{\text{mi}} - 0.0257\ \frac{\text{gal}}{\text{mi}}}{0.0257\ \frac{\text{gal}}{\text{mi}}} = \boxed{0.288\ \ (28.8\%)}$$

The answer is (C).

SI Solution

(a) The drag on the car is given by

$$F_D = \frac{C_D A \rho \text{v}^2}{2}$$

For air at 20°C,

$$\rho = \frac{p}{RT} = \frac{1.013 \times 10^5\ \text{Pa}}{\left(287\ \frac{\text{J}}{\text{kg·K}}\right)(20°\text{C} + 273)} = 1.205\ \text{kg/m}^3$$

$$\text{v} = \left(90\ \frac{\text{km}}{\text{h}}\right)\left(1000\ \frac{\text{m}}{\text{km}}\right)\left(\frac{1\ \text{h}}{3600\ \text{s}}\right) = 25.0\ \text{m/s}$$

Substituting gives

$$F_D = \frac{C_D A \rho \text{v}^2}{2} = \left(\tfrac{1}{2}\right)(0.42)(2.6\ \text{m}^2)\left(1.205\ \frac{\text{kg}}{\text{m}^3}\right)\left(25.0\ \frac{\text{m}}{\text{s}}\right)^2 = 411.2\ \text{N}$$

The total resisting force is

$$\begin{aligned} F &= F_D + \text{rolling resistance} \\ &= 411.2\ \text{N} + (0.01)(1500\ \text{kg})g \\ &= 411.2\ \text{N} + (0.01)(1500\ \text{kg})\left(9.81\ \frac{\text{m}}{\text{s}^2}\right) \\ &= 558.4\ \text{N} \end{aligned}$$

The power consumed is

$$P = F\text{v} = (558.4\ \text{N})\left(25\ \frac{\text{m}}{\text{s}}\right) = 13\,960\ \text{W}$$

The power available from the fuel is

$$\begin{aligned} P_A &= (\text{engine thermal efficiency})(\text{fuel heating value}) \\ &= (0.28)\left(4.6 \times 10^8\ \frac{\text{J}}{\text{L}}\right) \\ &= 1.288 \times 10^8\ \text{J/L} \end{aligned}$$

The fuel consumption at 90 km/h is

$$\frac{P}{P_A} = \frac{13\,960\ \text{W}}{1.288 \times 10^8\ \frac{\text{J}}{\text{L}}} = 1.08 \times 10^{-4}\ \text{L/s}$$

The fuel consumption is

$$\frac{1.08 \times 10^{-4}\ \frac{\text{L}}{\text{s}}}{\left(90\ \frac{\text{km}}{\text{h}}\right)\left(\frac{1\ \text{h}}{3600\ \text{s}}\right)} = \boxed{0.00434\ \text{L/km}}$$

The answer is (A).

(b) Similarly, the fuel consumption at 105 km/h is

$$\text{v} = \left(105\ \frac{\text{km}}{\text{h}}\right)\left(1000\ \frac{\text{m}}{\text{km}}\right)\left(\frac{1\ \text{h}}{3600\ \text{s}}\right) = 29.2\ \text{m/s}$$

$$D = \frac{C_D A \rho \text{v}^2}{2} = \left(\tfrac{1}{2}\right)(0.42)(2.6\ \text{m}^2)\left(1.205\ \frac{\text{kg}}{\text{m}^3}\right)\left(29.2\ \frac{\text{m}}{\text{s}}\right)^2 = 561.0\ \text{N}$$

The total resisting force is

$$\begin{aligned} F &= F_D + \text{rolling resistance} \\ &= 561.0\ \text{N} + (0.01)(1500\ \text{kg})g \\ &= 561.0\ \text{N} + (0.01)(1500\ \text{kg})\left(9.81\ \frac{\text{m}}{\text{s}^2}\right) \\ &= 708.2\ \text{N} \end{aligned}$$

The power consumed is

$$P = F\text{v} = (708.2\ \text{N})\left(29.2\ \frac{\text{m}}{\text{s}}\right) = 20\,679\ \text{W}$$

The fuel consumption at 105 km/h is

$$\frac{P}{P_A} = \frac{20\,679\ \text{W}}{1.288 \times 10^8\ \frac{\text{J}}{\text{L}}} = 1.61 \times 10^{-4}\ \text{L/s}$$

The fuel consumption is

$$\frac{1.61 \times 10^{-4}\ \frac{\text{L}}{\text{s}}}{\left(105\ \frac{\text{km}}{\text{h}}\right)\left(\frac{1\ \text{h}}{3600\ \text{s}}\right)} = 0.00552\ \text{L/km}$$

The relative difference between the fuel consumption at 90 km/h and 105 km/h is

$$\frac{0.00552\ \frac{\text{L}}{\text{km}} - 0.00434\ \frac{\text{L}}{\text{km}}}{0.00434\ \frac{\text{L}}{\text{km}}} = \boxed{0.272\ \ (27.2\%)}$$

The answer is (C).

18 Hydraulic Machines and Equipment

PRACTICE PROBLEMS

Pumping Power

1. 2000 gal/min of 60°F thickened sludge with a specific gravity of 1.2 flow through a pump with an inlet diameter of 12 in and an outlet of 8 in. The centerlines of the inlet and outlet are at the same elevation. The inlet pressure is 8 in of mercury (vacuum). A discharge pressure gauge located 4 ft above the pump discharge centerline reads 20 psig. The pump efficiency is 85%. All pipes are schedule-40. What is the input power of the pump?

(A) 26 hp
(B) 31 hp
(C) 37 hp
(D) 53 hp

2. 1.25 ft^3/sec (35 L/s) of 70°F (21°C) water are pumped from the bottom of a tank through 700 ft (230 m) of 4 in (10.2 cm) schedule-40 steel pipe. The line includes a 50 ft (15 m) rise in elevation, two right-angle elbows, a wide-open gate valve, and a swing check valve. All fittings and valves are regular screwed. The inlet pressure is 50 psig (345 kPa), and a working pressure of 20 psig (140 kPa) is needed at the end of the pipe. What is the hydraulic power for this pumping application?

(A) 16 hp (13 kW)
(B) 23 hp (17 kW)
(C) 49 hp (37 kW)
(D) 66 hp (50 kW)

3. 80 gal/min (5 L/s) of 80°F (27°C) water are lifted 12 ft (4 m) vertically by a pump through 50 ft (15 m) of a 2 in (5.1 cm) diameter rubber hose. The discharge end of the hose is submerged in 8 ft (2.5 m) of water as shown. What head is added by the pump?

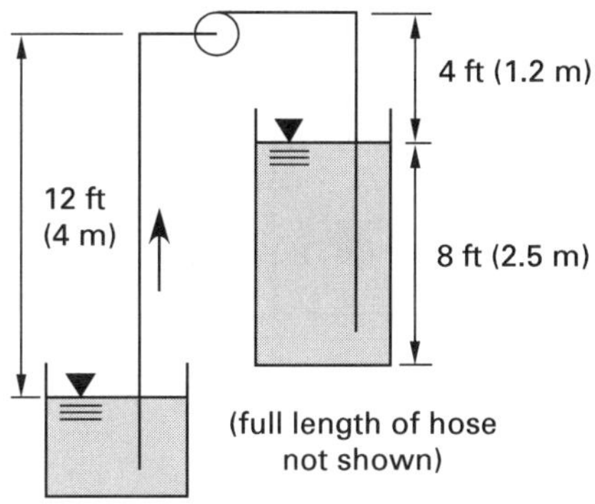

(A) 10 ft (3.0 m)
(B) 13 ft (4.3 m)
(C) 22 ft (6.6 m)
(D) 31 ft (9.3 m)

4. A 20 hp motor drives a centrifugal pump. The pump discharges 60°F (16°C) water at 12 ft/sec (4 m/s) into a 6 in (15.2 cm) steel schedule-40 line. The inlet is 8 in (20.3 cm) schedule-40 steel pipe. The pump suction is 5 psig (35 kPa) below standard atmospheric pressure. The friction head loss in the system is 10 ft (3.3 m). The pump efficiency is 70%. The suction and discharge lines are at the same elevation. What is the maximum height above the pump inlet that water is available at standard atmospheric pressure?

(A) 28 ft (6.9 m)
(B) 37 ft (11 m)
(C) 49 ft (15 m)
(D) 81 ft (25 m)

5. (*Time limit: one hour*) A pump station is used to fill a tank on a hill above from a lake below. The flow rate is 10,000 gal/hr (10.5 L/s) of 60°F (16°C) water. The atmospheric pressure is 14.7 psia (101 kPa). The pump is 12 ft (4 m) above the lake, and the tank surface level is 350 ft (115 m) above the pump. The discharge line is 4 in (10.2 cm) diameter schedule-40 steel pipe. The equivalent length of the inlet line between the lake and the pump is 300 ft (100 m). The total equivalent length between the lake and the tank is 7000 ft (2300 m), including all fittings, bends, screens, and valves. The cost of electricity is $0.04 per kW-hr. The overall efficiency of the pump and motor set is 70%.

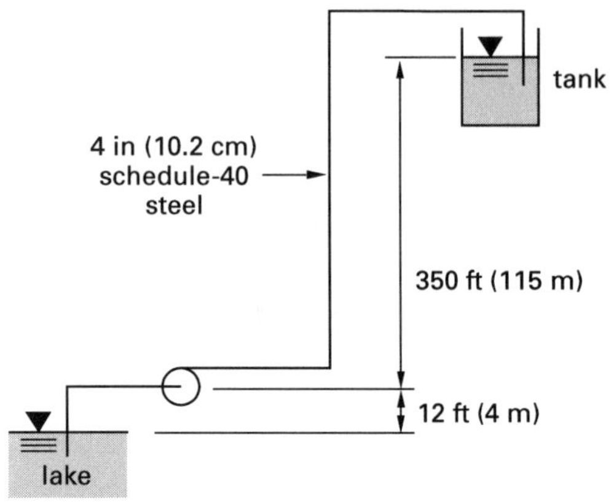

(a) What does it cost to operate the pump for 1 hr?

(A) $0.1
(B) $1
(C) $3
(D) $6

(b) What motor power is required?

(A) 10 hp (7.5 kW)
(B) 30 hp (25 kW)
(C) 50 hp (40 kW)
(D) 75 hp (60 kW)

(c) What is the NPSHA for this application?

(A) 4 ft (1.2 m)
(B) 8 ft (2.4 m)
(C) 12 ft (3.6 m)
(D) 16 ft (4.5 m)

6. (*Time limit: one hour*) A town with a stable, constant population of 10,000 produces sewage at the average rate of 100 gallons per capita day (gpcd), with peak flows of 250 gpcd. The pipe to the pumping station is 5000 ft in length and has a *C*-value of 130. The elevation drop along the length is 48 ft. Minor losses in infiltration are insignificant. The pump's maximum suction lift is 10 ft.

(a) If all diameters are available and the pipe flows 100% full under gravity flow, what minimum pipe diameter is required?

(A) 8 in
(B) 12 in
(C) 14 in
(D) 18 in

(b) If constant-speed pumps are used, what is the minimum number of pumps (disregarding spares and backups) that should be used?

(A) 2
(B) 3
(C) 4
(D) 5

(c) If variable-speed pumps are used, what is the minimum number of pumps (disregarding spares and backups) that should be used?

(A) 1
(B) 2
(C) 3
(D) 4

(d) If three constant-speed pumps are used with a fourth pump as backup, what motor power is required?

(A) 2 hp
(B) 5 hp
(C) 8 hp
(D) 12 hp

(e) If two variable-speed pumps are used with a third pump as backup, what motor power is required?

(A) 3 hp
(B) 8 hp
(C) 12 hp
(D) 18 hp

(f) Which of the following are ways of controlling sump pump on-off cycles?

I. detecting sump levels
II. detecting pressure in the sump
III. detecting incoming flow rates
IV. using fixed run times
V. detecting outgoing flow rates
VI. operating manually

(A) I, II, and III
(B) I, II, IV, and V
(C) I, III, and V
(D) I, II, IV, V, and VI

(g) With intermittent fan operation, approximately how many air changes should the wet well receive per hour?

(A) 6
(B) 12
(C) 20
(D) 30

(h) With continuous fan operation, approximately how many air changes should the dry well receive per hour?

(A) 6
(B) 12
(C) 20
(D) 30

7. (*Time limit: one hour*) A pump transfers 3.5 MGD of filtered water from the clear well of a 10 ft × 20 ft (plan) rapid sand filter to a higher elevation. The pump efficiency is 85%, and the motor driving pump has an efficiency of 90%. Minor losses are insignificant. Refer to the following illustration for additional information.

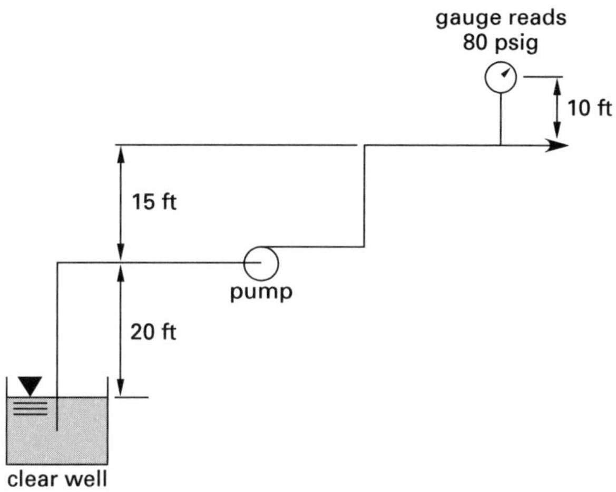

(a) What is the static suction lift?

(A) 15 ft
(B) 20 ft
(C) 35 ft
(D) 40 ft

(b) What is the static discharge head?

(A) 15 ft
(B) 20 ft
(C) 35 ft
(D) 40 ft

(c) Based on the information given, what is the approximate total dynamic head?

(A) 45 ft
(B) 185 ft
(C) 210 ft
(D) 230 ft

(d) What motor power is required?

(A) 50 hp
(B) 100 hp
(C) 150 hp
(D) 200 hp

(e) If the filter output is increased by 25%, how much surplus will accumulate in 8 hr?

(A) 60,000 gal
(B) 180,000 gal
(C) 250,000 gal
(D) 370,000 gal

Pumping Other Fluids

8. (*Time limit: one hour*) Gasoline with a specific gravity of 0.7 and viscosity of 6×10^{-6} ft^2/sec (5.6×10^{-7} m^2/s) is transferred from a tanker to a storage tank. The interior of the storage tank is maintained at atmospheric pressure by a vapor-recovery system. The free surface in the storage tank is 60 ft (20 m) above the tanker's free surface. The pipe consists of 500 ft (170 m) of 3 in (7.62 cm) schedule-40 steel pipe with six flanged elbows and two wide-open gate valves. The pump and motor both have individual efficiencies of 88%. Electricity costs $0.045 per kW-hr. The pump's performance data (based on cold, clear water) are known.

flow rate gpm (L/s)	head ft (m)
0 (0)	127 (42)
100 (6.3)	124 (41)
200 (12)	117 (39)
300 (18)	108 (36)
400 (24)	96 (32)
500 (30)	80 (27)
600 (36)	55 (18)

(a) What is the transfer rate?

(A) 150 gal/min (9.2 L/s)
(B) 180 gal/min (11 L/s)
(C) 200 gal/min (12 L/s)
(D) 230 gal/min (14 L/s)

(b) What is the total cost of operating the pump for 1 hr?

(A) $0.20
(B) $0.80
(C) $1.30
(D) $2.70

Specific Speed

9. A double-suction water pump moving 300 gal/sec (1.1 kL/s) turns at 900 rpm. The pump adds 20 ft (7 m) of head to the water. What is the specific speed?

(A) 3000 rpm (52 rpm)
(B) 6000 rpm (100 rpm)
(C) 9000 rpm (160 rpm)
(D) 12,000 rpm (210 rpm)

Cavitation

10. 100 gal/min (6.3 L/s) of pressurized hot water at 281°F and 80 psia (138°C and 550 kPa) discharges through 30 ft (10 m) of 1.5 in (3.81 cm) schedule-40 steel pipe into a 2 psig (14 kPa) tank. The inlet and outlet are both 20 ft (6 m) below the surface of the water when the tank is full. The inlet line contains two

Water Resources
Flow of Fluids

wide-open gate valves and two long-radius elbows. All components are regular screwed. The pump's NPSHR is 10 ft (3 m) for this application. The kinematic viscosity of 281°F (138°C) water is 0.239×10^{-5} ft^2/sec (0.222×10^{-6} m^2/s) and the vapor pressure is 50.02 psia (3.431 bar). Will the pump cavitate?

(A) yes; NPSHA = 4 ft (1.2 m)
(B) yes; NPSHA = 9 ft (2.7 m)
(C) no; NPSHA = 24 ft (7.2 m)
(D) no; NPSHA = 68 ft (21 m)

11. The velocity of the tip of a marine propeller is 4.2 times the boat velocity. The propeller is located 8 ft (3 m) below the surface. The temperature of the seawater is 68°F (20°C). The density is approximately 64.0 lbm/ft^3 (1024 kg/m^3), and the salt content is 2.5% by weight. What is the practical maximum boat velocity, as limited strictly by cavitation?

(A) 9.1 ft/sec (2.7 m/s)
(B) 12 ft/sec (3.8 m/s)
(C) 15 ft/sec (4.5 m/s)
(D) 22 ft/sec (6.6 m/s)

Pump and System Curves

12. The inlet of a centrifugal water pump is 7 ft (2.3 m) above the free surface from which it draws. The inlet consists of 12 ft (4 m) of 2 in (5.08 cm) schedule-40 steel pipe and contains one long-radius elbow and one check valve. The discharge line is 2 in (5.08 cm) schedule-40 steel pipe and includes two long-radius elbows and an 80 ft (27 m) run. The discharge is 20 ft (6.3 m) above the free surface. All components are regular screwed. The water temperature is 70°F (21°C). The following pump curve data are applicable.

flow rate gpm (L/s)	head ft (m)
0 (0)	110 (37)
10 (0.6)	108 (36)
20 (1.2)	105 (35)
30 (1.8)	102 (34)
40 (2.4)	98 (33)
50 (3.2)	93 (31)
60 (3.6)	87 (29)
70 (4.4)	79 (26)
80 (4.8)	66 (22)
90 (5.7)	50 (17)

(a) What is the flow rate?

(A) 44 gal/min (2.9 L/s)
(B) 69 gal/min (4.5 L/s)
(C) 82 gal/min (5.5 L/s)
(D) 95 gal/min (6.2 L/s)

(b) What can be said about the use of this pump in this installation?

(A) A different pump should be used.
(B) The pump is operating near its most efficient point.
(C) Pressure fluctuations could result from surging.
(D) Overloading will not be a problem.

Affinity Laws

13. A pump was intended to run at 1750 rpm when driven by a 0.5 hp (0.37 kW) motor. What is the required power rating of a motor that will turn the pump at 2000 rpm?

(A) 0.25 hp (0.19 kW)
(B) 0.45 hp (0.34 kW)
(C) 0.65 hp (0.49 kW)
(D) 0.75 hp (0.55 kW)

14. (*Time limit: one hour*) A centrifugal pump running at 1400 rpm has the curve shown. The pump will be installed in an existing pipeline with known head requirements given by the formula $H = 30 + 2Q^2$. H is the system head in feet of water. Q is the flow rate in ft^3/sec.

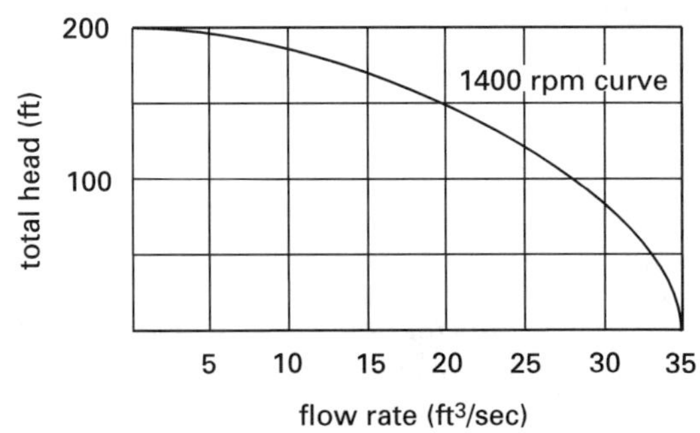

(a) What is the flow rate if the pump is turned at 1400 rpm?

(A) 2000 gal/min
(B) 3500 gal/min
(C) 4000 gal/min
(D) 4500 gal/min

(b) What power is required to drive the pump?

(A) 190 hp
(B) 210 hp
(C) 230 hp
(D) 260 hp

(c) What is the flow rate if the pump is turned at 1200 rpm?

(A) 2000 gal/min
(B) 3500 gal/min
(C) 4000 gal/min
(D) 4500 gal/min

Turbines

15. A horizontal turbine reduces 100 ft^3/sec of water from 30 psia to 5 psia. Friction is negligible. What power is developed?

(A) 350 hp
(B) 500 hp
(C) 650 hp
(D) 800 hp

16. 1000 ft^3/sec of 60°F water flow from a high reservoir through a hydroelectric turbine installation, exiting 625 ft lower. The head loss due to friction is 58 ft. The turbine efficiency is 89%. What power is developed in the turbines?

(A) 40 kW
(B) 18 MW
(C) 43 MW
(D) 71 MW

17. Water at 500 psig and 60°F (3.5 MPa and 16°C) drives a 250 hp (185 kW) turbine at 1750 rpm against a back pressure of 30 psig (210 kPa). The water discharges through a 4 in (100 mm) diameter nozzle at 35 ft/sec (10.5 m/s). The water is deflected 80° by a single blade moving directly away at 10 ft/sec (3 m/s).

(a) What type of turbine is appropriate for this application?

(A) impulse turbine
(B) low-head turbine
(C) reaction turbine
(D) radial-flow turbine

(b) What is the total force acting on the blade?

(A) 100 lbf (450 N)
(B) 140 lbf (570 N)
(C) 160 lbf (720 N)
(D) 280 lbf (1300 N)

18. (*Time limit: one hour*) A Francis-design hydraulic reaction turbine with 22 in (560 mm) diameter blades runs at 610 rpm. The turbine develops 250 hp (185 kW) when 25 ft^3/sec (700 L/s) of water flow through it. The pressure head at the turbine entrance is 92.5 ft (30.8 m). The elevation of the turbine above the tailwater level is 5.26 ft (1.75 m). The inlet and outlet velocity are both 12 ft/sec (3.6 m/s).

(a) What is the effective head?

(A) 90 ft (30 m)
(B) 95 ft (31 m)
(C) 100 ft (33 m)
(D) 105 ft (35 m)

(b) What is the overall turbine efficiency?

(A) 81%
(B) 88%
(C) 93%
(D) 96%

(c) What will be the turbine speed if the effective head is 225 ft (75 m)?

(A) 600 rpm
(B) 920 rpm
(C) 1100 rpm
(D) 1400 rpm

(d) What horsepower is developed if the effective head is 225 ft (75 m)?

(A) 560 hp (420 kW)
(B) 630 hp (470 kW)
(C) 750 hp (560 kW)
(D) 840 hp (630 kW)

(e) What is the flow rate if the effective head is 225 ft (75 m)?

(A) 25 ft^3/sec (700 L/s)
(B) 38 ft^3/sec (1100 L/s)
(C) 56 ft^3/sec (1600 L/s)
(D) 64 ft^3/sec (1800 L/s)

SOLUTIONS

1. $$\left(2000\ \frac{\text{gal}}{\text{min}}\right)\left(0.002228\ \frac{\frac{\text{ft}^3}{\text{sec}}}{\frac{\text{gal}}{\text{min}}}\right) = 4.456\ \text{ft}^3/\text{sec}$$

Assume schedule-40 steel pipe. From App. 16.B,

$$12'':\quad D_1 = 0.9948\ \text{ft} \qquad A_1 = 0.7773\ \text{ft}^2$$
$$8'':\quad D_2 = 0.6651\ \text{ft} \qquad A_2 = 0.3473\ \text{ft}^2$$

$$\begin{aligned} p_1 &= \left[14.7\ \frac{\text{lbf}}{\text{in}^2} - (8\ \text{in})\left(0.491\ \frac{\text{lbf}}{\text{in}^3}\right)\right]\left(144\ \frac{\text{in}^2}{\text{ft}^2}\right) \\ &= 1551.2\ \text{lbf/ft}^2 \end{aligned}$$

$$\begin{aligned} p_2 &= \left(14.7\ \frac{\text{lbf}}{\text{in}^2} + 20\ \frac{\text{lbf}}{\text{in}^2}\right)\left(144\ \frac{\text{in}^2}{\text{ft}^2}\right) \\ &\quad + (4\ \text{ft})(1.2)\left(62.4\ \frac{\text{lbf}}{\text{ft}^3}\right) \\ &= 5296.3\ \text{lbf/ft}^2 \end{aligned}$$

$$\text{v}_1 = \frac{4.456\ \frac{\text{ft}^3}{\text{sec}}}{0.7773\ \text{ft}^2} = 5.73\ \text{ft/sec}$$

$$\text{v}_2 = \frac{4.456\ \frac{\text{ft}^3}{\text{sec}}}{0.3473\ \text{ft}^2} = 12.83\ \text{ft/sec}$$

From Eq. 18.9, the total heads at points 1 and 2 are

$$h_{t,1} = \frac{1551.2\ \frac{\text{lbf}}{\text{ft}^2}}{\left(62.4\ \frac{\text{lbf}}{\text{ft}^3}\right)(1.2)} + \frac{\left(5.73\ \frac{\text{ft}}{\text{sec}}\right)^2}{(2)\left(32.2\ \frac{\text{ft}}{\text{sec}^2}\right)} = 21.23\ \text{ft}$$

$$h_{t,2} = \frac{5296.3\ \frac{\text{lbf}}{\text{ft}^2}}{\left(62.4\ \frac{\text{lbf}}{\text{ft}^3}\right)(1.2)} + \frac{\left(12.83\ \frac{\text{ft}}{\text{sec}}\right)^2}{(2)\left(32.2\ \frac{\text{ft}}{\text{sec}^2}\right)} = 73.29\ \text{ft}$$

The pump must add 73.29 ft − 21.23 ft = 52.06 ft of head.

The power required is given in Table 18.5.

$$\begin{aligned} P_{\text{ideal}} &= \frac{(52.06\ \text{ft})(1.2)\left(4.456\ \frac{\text{ft}^3}{\text{sec}}\right)\left(62.4\ \frac{\text{lbf}}{\text{ft}^3}\right)}{550\ \frac{\text{ft-lbf}}{\text{hp-sec}}} \\ &= 31.58\ \text{hp} \end{aligned}$$

The input horsepower is

$$P_{\text{in}} = \frac{P_{\text{ideal}}}{\eta} = \frac{31.58\ \text{hp}}{0.85} = \boxed{37.15\ \text{hp}}$$

The answer is (C).

2. *Customary U.S. Solution*

From App. 16.B, data for 4 in schedule-40 steel pipe are

$$D_i = 0.3355\ \text{ft}$$
$$A_i = 0.08841\ \text{ft}^2$$

The velocity in the pipe is

$$\text{v} = \frac{\dot{V}}{A} = \frac{1.25\ \frac{\text{ft}^3}{\text{sec}}}{0.08841\ \text{ft}^2} = 14.139\ \text{ft/sec}$$

From Table 17.3, typical equivalent lengths for schedule-40, screwed steel fittings for 4 in pipes are

90° elbow: 13 ft

gate valve: 2.5 ft

check valve: 38.0 ft

The total equivalent length is

$$(2)(13\ \text{ft}) + (1)(2.5\ \text{ft}) + (1)(38\ \text{ft}) = 66.5\ \text{ft}$$

At 70°F, from App. 14.A, the density of water is 62.3 lbm/ft^3 and the kinematic viscosity of water, ν, is $1.059 \times 10^{-5}\ \text{ft}^2/\text{sec}$. The Reynolds number is

$$\begin{aligned} \text{Re} &= \frac{D\text{v}}{\nu} = \frac{(0.3355\ \text{ft})\left(14.139\ \frac{\text{ft}}{\text{sec}}\right)}{1.059 \times 10^{-5}\ \frac{\text{ft}^2}{\text{sec}}} \\ &= 4.479 \times 10^5 \end{aligned}$$

From App. 17.A, for steel, $\epsilon = 0.0002$ ft.

So,

$$\frac{\epsilon}{D} = \frac{0.0002\ \text{ft}}{0.3355\ \text{ft}} \approx 0.0006$$

From App. 17.B, the friction factor is $f = 0.01835$.

The friction head is given by Eq. 18.6.

$$\begin{aligned} h_f &= \frac{fL\text{v}^2}{2Dg} \\ &= \frac{(0.01835)(700\ \text{ft} + 66.5\ \text{ft})\left(14.139\ \frac{\text{ft}}{\text{sec}}\right)^2}{(2)(0.3355\ \text{ft})\left(32.2\ \frac{\text{ft}}{\text{sec}^2}\right)} \\ &= 130.1\ \text{ft} \end{aligned}$$

The total dynamic head is given by Eq. 18.9. Point s is taken as the bottom of the supply tank.

$$h = \frac{(p_d - p_s)g_c}{\rho g} + \frac{\text{v}_d^2 - \text{v}_s^2}{2g} + z_d - z_s$$

$$\text{v}_s \approx 0$$

$$z_d - z_s = 50\ \text{ft} \quad \text{[given as rise in elevation]}$$

The discharge and suction pressures are

$$p_d = 20 \text{ psig}$$
$$p_s = 50 \text{ psig}$$

$$h = \frac{\left(20\ \frac{\text{lbf}}{\text{in}^2} - 50\ \frac{\text{lbf}}{\text{in}^2}\right) \times \left(144\ \frac{\text{in}^2}{\text{ft}^2}\right)\left(32.2\ \frac{\text{ft-lbm}}{\text{lbf-sec}^2}\right)}{\left(62.3\ \frac{\text{lbm}}{\text{ft}^3}\right)\left(32.2\ \frac{\text{ft}}{\text{sec}^2}\right)} + \frac{\left(14.139\ \frac{\text{ft}}{\text{sec}}\right)^2}{(2)\left(32.2\ \frac{\text{ft}}{\text{sec}^2}\right)} + 50 \text{ ft}$$
$$= -16.2 \text{ ft}$$

The head added is

$$h_A = h + h_f = -16.2 \text{ ft} + 130.1 \text{ ft}$$
$$= 113.9 \text{ ft}$$

The mass flow rate is

$$\dot{m} = \rho\dot{V}$$
$$= \left(62.3\ \frac{\text{lbm}}{\text{ft}^3}\right)\left(1.25\ \frac{\text{ft}^3}{\text{sec}}\right)$$
$$= 77.875 \text{ lbm/sec}$$

From Table 18.5, the hydraulic horsepower is

$$\text{WHP} = \left(\frac{h_A\dot{m}}{550}\right) \times \left(\frac{g}{g_c}\right)$$
$$= \left(\frac{(113.9 \text{ ft})\left(77.875\ \frac{\text{lbm}}{\text{sec}}\right)}{550\ \frac{\text{ft-lbf}}{\text{hp-sec}}}\right) \times \left(\frac{32.2\ \frac{\text{ft}}{\text{sec}^2}}{32.2\ \frac{\text{ft-lbm}}{\text{lbf-sec}^2}}\right)$$
$$= \boxed{16.13 \text{ hp}}$$

The answer is (A).

SI Solution

From App. 16.C, data for 4 in schedule-40 steel pipe are

$$D_i = 102.3 \text{ mm}$$
$$A_i = 82.19 \times 10^{-4} \text{ m}^2$$

The velocity in the pipe is

$$\text{v} = \frac{\dot{V}}{A} = \frac{\left(35\ \frac{\text{L}}{\text{s}}\right)\left(\frac{1 \text{ m}^3}{1000 \text{ L}}\right)}{82.19 \times 10^{-4} \text{ m}^2} = 4.26 \text{ m/s}$$

From Table 17.3, typical equivalent lengths for schedule-40, screwed steel fittings for 4 in pipes are

90° elbow: 13 ft

gate valve: 2.5 ft

check valve: 38.0 ft

The total equivalent length is

$$(2)(13 \text{ ft}) + (1)(2.5 \text{ ft}) + (1)(38 \text{ ft}) = 66.5 \text{ ft}$$
$$(66.5 \text{ ft})\left(0.3048\ \frac{\text{m}}{\text{ft}}\right) = 20.27 \text{ m}$$

At 21°C, from App. 14.B, the water properties are

$$\rho = 998 \text{ kg/m}^3$$
$$\mu = 0.9827 \times 10^{-3} \text{ Pa·s}$$
$$\nu = \frac{\mu}{\rho} = \frac{0.9827 \times 10^{-3} \text{ Pa·s}}{998\ \frac{\text{kg}}{\text{m}^3}}$$
$$= 9.85 \times 10^{-7} \text{ m}^2\text{/s}$$

The Reynolds number is

$$\text{Re} = \frac{D\text{v}}{\nu} = \frac{(102.3 \text{ mm})\left(\frac{1 \text{ m}}{1000 \text{ mm}}\right)\left(4.26\ \frac{\text{m}}{\text{s}}\right)}{9.85 \times 10^{-7}\ \frac{\text{m}^2}{\text{s}}}$$
$$= 4.424 \times 10^5$$

From Table 17.2, for steel, $\epsilon = 6 \times 10^{-5}$ m.

So,

$$\frac{\epsilon}{D} = \frac{6.0 \times 10^{-5} \text{ m}}{(102.3 \text{ mm})\left(\frac{1 \text{ m}}{1000 \text{ mm}}\right)} = 0.0006$$

From App. 17.B, the friction factor is $f = 0.01836$.

From Eq. 18.6, the friction head is

$$h_f = \frac{fL\text{v}^2}{2Dg}$$
$$= \frac{(0.01836)(230 \text{ m} + 20.27 \text{ m})\left(4.26\ \frac{\text{m}}{\text{s}}\right)^2}{(2)(102.3 \text{ mm})\left(\frac{1 \text{ m}}{1000 \text{ mm}}\right)\left(9.81\ \frac{\text{m}}{\text{s}^2}\right)}$$
$$= 41.5 \text{ m}$$

Water Resources
Flow of Fluids

The total dynamic head is given by Eq. 18.9. Point s is taken as the bottom of the supply tank.

$$h = \frac{p_d - p_s}{\rho g} + \frac{v_d^2 - v_s^2}{2g} + z_d - z_s$$

$$v_s \approx 0$$

$$z_d - z_s = 15 \text{ m} \quad \text{[given as rise in elevation]}$$

The difference between discharge and suction pressure is

$$p_d - p_s = 140 \text{ kPa} - 345 \text{ kPa} = -205 \text{ kPa}$$

$$h = \frac{(-205 \text{ kPa})\left(1000 \ \frac{\text{Pa}}{\text{kPa}}\right)}{\left(998 \ \frac{\text{kg}}{\text{m}^3}\right)\left(9.81 \ \frac{\text{m}}{\text{s}^2}\right)} + \frac{\left(4.26 \ \frac{\text{m}}{\text{s}}\right)^2}{(2)\left(9.81 \ \frac{\text{m}}{\text{s}^2}\right)} + 15 \text{ m}$$

$$= -5.0 \text{ m}$$

The head added by the pump is

$$h_A = h + h_f = -5.0 \text{ m} + 41.5 \text{ m} = 36.5 \text{ m}$$

The mass flow rate is

$$\dot{m} = \rho \dot{V} = \left(998 \ \frac{\text{kg}}{\text{m}^3}\right)\left(35 \ \frac{\text{L}}{\text{s}}\right)\left(\frac{1 \text{ m}^3}{1000 \text{ L}}\right) = 34.93 \text{ kg/s}$$

From Table 18.6, the hydraulic power is

$$W\text{kW} = \frac{(9.81)h_A\dot{m}}{1000} = \frac{\left(9.81 \ \frac{\text{m}}{\text{s}^2}\right)(36.5 \text{ m})\left(34.93 \ \frac{\text{kg}}{\text{s}}\right)}{1000 \ \frac{\text{W}}{\text{kW}}} = \boxed{12.51 \text{ kW}}$$

The answer is (A).

3. *Customary U.S. Solution*

The area of the rubber hose is

$$A = \left(\frac{\pi}{4}\right)D^2 = \left(\frac{\pi}{4}\right)\left(\frac{2 \text{ in}}{12 \ \frac{\text{in}}{\text{ft}}}\right)^2 = 0.0218 \text{ ft}^2$$

The velocity of water in the hose is

$$v = \frac{\dot{V}}{A} = \frac{\left(80 \ \frac{\text{gal}}{\text{min}}\right)\left(0.002228 \ \frac{\frac{\text{ft}^3}{\text{sec}}}{\frac{\text{gal}}{\text{min}}}\right)}{0.0218 \text{ ft}^2} = 8.176 \text{ ft/sec}$$

At 80°F from App. 14.A, the kinematic viscosity of water is $\nu = 0.93 \times 10^{-5}$ ft^2/sec.

The Reynolds number is

$$\text{Re} = \frac{vD}{\nu} = \frac{\left(8.176 \ \frac{\text{ft}}{\text{sec}}\right)(2 \text{ in})\left(\frac{1 \text{ ft}}{12 \text{ in}}\right)}{0.93 \times 10^{-5} \ \frac{\text{ft}^2}{\text{sec}}} = 1.47 \times 10^5$$

Assume that the rubber hose is smooth. From App. 17.B, the friction factor is $f = 0.0166$.

From Eq. 18.6, the friction head is

$$h_f = \frac{fLv^2}{2Dg} = \frac{(0.0166)(50 \text{ ft})\left(8.176 \ \frac{\text{ft}}{\text{sec}}\right)^2}{(2)(2 \text{ in})\left(\frac{1 \text{ ft}}{12 \text{ in}}\right)\left(32.2 \ \frac{\text{ft}}{\text{sec}^2}\right)} = 5.17 \text{ ft}$$

Neglecting entrance and exit losses, the head added by the pump is

$$h_A = h_f + h_z = 5.17 \text{ ft} + 12 \text{ ft} - 4 \text{ ft} = \boxed{13.17 \text{ ft}}$$

The answer is (B).

SI Solution

The area of the rubber hose is

$$A = \left(\frac{\pi}{4}\right)D^2 = \left(\frac{\pi}{4}\right)(5.1 \text{ cm})^2\left(\frac{1 \text{ m}}{100 \text{ cm}}\right)^2 = 0.00204 \text{ m}^2$$

The velocity of water in the hose is

$$v = \frac{\dot{V}}{A} = \frac{\left(5 \ \frac{\text{L}}{\text{s}}\right)\left(\frac{1 \text{ m}^3}{1000 \text{ L}}\right)}{0.00204 \text{ m}^2} = 2.45 \text{ m/s}$$

From App. 14.B, at 27°C the water data are

$$\rho = 996.5\ \text{kg/m}^3$$
$$\mu = 0.8565 \times 10^{-3}\ \text{Pa·s}$$
$$\nu = \frac{\mu}{\rho} = \frac{0.8565 \times 10^{-3}\ \text{Pa·s}}{996.5\ \frac{\text{kg}}{\text{m}^3}} = 8.60 \times 10^{-7}\ \text{m}^2/\text{s}$$

The Reynolds number is

$$\text{Re} = \frac{vD}{\nu} = \frac{\left(2.45\ \frac{\text{m}}{\text{s}}\right)(5.1\ \text{cm})\left(\frac{1\ \text{m}}{100\ \text{cm}}\right)}{8.60 \times 10^{-7}\ \frac{\text{m}^2}{\text{s}}} = 1.45 \times 10^5$$

Assume that the rubber hose is smooth. From App. 17.B, the friction factor is $f \approx 0.0166$.

From Eq. 18.6, the friction head is

$$h_f = \frac{fLv^2}{2Dg} = \frac{(0.0166)(15\ \text{m})\left(2.45\ \frac{\text{m}}{\text{s}}\right)^2}{(2)(5.1\ \text{cm})\left(\frac{1\ \text{m}}{100\ \text{cm}}\right)\left(9.81\ \frac{\text{m}}{\text{s}^2}\right)} = 1.49\ \text{m}$$

Neglecting entrance and exit losses, the head added by the pump is

$$h_A = h_f + h_z = 1.49\ \text{m} + 4\ \text{m} - 1.2\ \text{m} = \boxed{4.29\ \text{m}}$$

The answer is (B).

4. *Customary U.S. Solution*

From App. 16.B, the diameters (inside) for 8 in and 6 in schedule-40 steel pipe are

$$D_1 = 7.981\ \text{in}$$
$$D_2 = 6.065\ \text{in}$$

From App. 14.A, at 60°F the density of water is 62.37 lbm/ft^3.

The mass flow rate through 6 in pipe is

$$\dot{m} = A_2 v_2 \rho = \left(\frac{\pi}{4}\right)(6.065\ \text{in})^2\left(\frac{1\ \text{ft}^2}{144\ \text{in}^2}\right)\left(12\ \frac{\text{ft}}{\text{sec}}\right)\left(62.37\ \frac{\text{lbm}}{\text{ft}^3}\right) = 150.2\ \text{lbm/sec}$$

The inlet (suction) pressure is

$$14.7\ \text{psia} - 5\ \text{psig} = 9.7\ \text{psia} = \left(9.7\ \frac{\text{lbf}}{\text{in}^2}\right)\left(144\ \frac{\text{in}^2}{\text{ft}^2}\right) = 1397\ \text{lbf/ft}^2$$

From Table 18.5, the head added by the pump is

$$h_A = \left(\frac{(550)(\text{BHP})\eta}{\dot{m}}\right) \times \left(\frac{g_c}{g}\right) = \left(\frac{\left(550\ \frac{\text{ft-lbf}}{\text{hp-sec}}\right)(20\ \text{hp})(0.70)}{150.2\ \frac{\text{lbm}}{\text{sec}}}\right)\left(\frac{32.2\ \frac{\text{ft-lbm}}{\text{lbf-sec}^2}}{32.2\ \frac{\text{ft}}{\text{sec}^2}}\right) = 51.26\ \text{ft}$$

At 1:

$$p_1 = 1397\ \text{lbf/ft}^2$$
$$z_1 = 0$$
$$v_1 = \frac{v_2 A_2}{A_1} = v_2\left(\frac{D_2}{D_1}\right)^2 = \left(12\ \frac{\text{ft}}{\text{sec}}\right)\left(\frac{6.065\ \text{in}}{7.981\ \text{in}}\right)^2 = 6.93\ \text{ft/sec}$$

At 2:

$$p_2 = \left(14.7\ \frac{\text{lbf}}{\text{in}^2}\right)\left(144\ \frac{\text{in}^2}{\text{ft}^2}\right) = 2117\ \text{lbf/ft}^2$$
$$v_2 = 12\ \text{ft/sec}\quad [\text{given}]$$

Let z_3 be the additional head above atmospheric.

Water Resources Flow of Fluids

From Eq. 18.9(b), the head added by the pump is

$$h_A = \frac{(p_2 - p_1)g_c}{\rho g} + \frac{v_2^2 - v_1^2}{2g} + z_2 - z_1 + h_f + z_3$$

$$51.26 \text{ ft} = \left(\frac{2117 \frac{\text{lbf}}{\text{ft}^2} - 1397 \frac{\text{lbf}}{\text{ft}^2}}{62.37 \frac{\text{lbm}}{\text{ft}^3}}\right) \times \left(\frac{32.2 \frac{\text{ft-lbm}}{\text{lbf-sec}^2}}{32.2 \frac{\text{ft}}{\text{sec}^2}}\right) + \frac{\left(12 \frac{\text{ft}}{\text{sec}}\right)^2 - \left(6.93 \frac{\text{ft}}{\text{sec}}\right)^2}{(2)\left(32.2 \frac{\text{ft}}{\text{sec}^2}\right)} + 0 - 0 + 10 \text{ ft} + z_3$$

$$z_3 = \boxed{28.2 \text{ ft}}$$

The answer is (A).

SI Solution

From App. 16.C, the inside diameters for 8 in and 6 in steel schedule-40 pipe are

$$D_1 = 202.7 \text{ mm}$$
$$D_2 = 154.1 \text{ mm}$$

From App. 14.B, at 16°C the density of water is 998.83 kg/m^3.

The mass flow rate through the 6 in pipe is

$$\begin{aligned}\dot{m} &= A_2 v_2 \rho \\ &= \left(\frac{\pi}{4}\right)(154.1 \text{ mm})^2 \left(\frac{1 \text{ m}}{1000 \text{ mm}}\right)^2 \left(4 \frac{\text{m}}{\text{s}}\right) \times \left(998.83 \frac{\text{kg}}{\text{m}^3}\right) \\ &= 74.5 \text{ kg/s}\end{aligned}$$

The inlet (suction) pressure is

$$101.3 \text{ kPa} - 35 \text{ kPa} = 66.3 \text{ kPa}$$

From Table 18.6, the head added by the pump is

$$\begin{aligned}h_A &= \frac{(1000)(\text{BkW})\eta}{(9.81)\dot{m}} \\ &= \frac{\left(1000 \frac{\text{W}}{\text{kW}}\right)(20 \text{ hp})\left(\frac{0.7457 \text{ kW}}{\text{hp}}\right)(0.70)}{\left(9.81 \frac{\text{m}}{\text{s}^2}\right)\left(74.5 \frac{\text{kg}}{\text{s}}\right)} \\ &= 14.28 \text{ m}\end{aligned}$$

At 1:

$$\begin{aligned}p_1 &= 66.3 \text{ kPa} \\ z_1 &= 0 \\ v_1 &= v_2 \left(\frac{A_2}{A_1}\right) \\ &= v_2 \left(\frac{D_2}{D_1}\right)^2 \\ &= \left(4 \frac{\text{m}}{\text{s}}\right)\left(\frac{154.1 \text{ mm}}{202.7 \text{ mm}}\right)^2 \\ &= 2.31 \text{ m/s}\end{aligned}$$

At 2:

$$\begin{aligned}p_2 &= 101.3 \text{ kPa} \\ v_2 &= 4 \text{ m/s} \quad \text{[given]}\end{aligned}$$

From Eq. 18.9(a), the head added by the pump is

$$h_A = \frac{p_2 - p_1}{\rho g} + \frac{v_2^2 - v_1^2}{2g} + z_2 - z_1 + h_f + z_3$$

$$14.28 \text{ m} = \frac{(101.3 \text{ kPa} - 66.3 \text{ kPa})\left(1000 \frac{\text{Pa}}{\text{kPa}}\right)}{\left(998.83 \frac{\text{kg}}{\text{m}^3}\right)\left(9.81 \frac{\text{m}}{\text{s}^2}\right)} + \frac{\left(4 \frac{\text{m}}{\text{s}}\right)^2 - \left(2.31 \frac{\text{m}}{\text{s}}\right)^2}{(2)\left(9.81 \frac{\text{m}}{\text{s}^2}\right)} + 0 - 0 + 3.3 \text{ m} + z_3$$

$$z_3 = \boxed{6.86 \text{ m}}$$

The answer is (A).

5. *Customary U.S. Solution*

The flow rate is

$$\dot{V} = \left(10{,}000 \frac{\text{gal}}{\text{hr}}\right)\left(0.1337 \frac{\text{ft}^3}{\text{gal}}\right) = 1337 \text{ ft}^3/\text{hr}$$

From App. 16.B, data for 4 in schedule-40 steel pipe are

$$D_i = 0.3355 \text{ ft}$$
$$A_i = 0.08841 \text{ ft}^2$$

The velocity in the pipe is

$$\begin{aligned}v = \frac{\dot{V}}{A} &= \frac{\left(1337 \frac{\text{ft}^3}{\text{hr}}\right)\left(\frac{1 \text{ hr}}{3600 \text{ sec}}\right)}{0.08841 \text{ ft}^2} \\ &= 4.20 \text{ ft/sec}\end{aligned}$$

From App. 14.A, at 60°F the kinematic viscosity of water is

$$\nu = 1.217 \times 10^{-5}\ \text{ft}^2/\text{sec}$$
$$\rho = 62.37\ \text{lbm/ft}^3$$

The Reynolds number is

$$\text{Re} = \frac{Dv}{\nu} = \frac{(0.3355\ \text{ft})\left(4.20\ \frac{\text{ft}}{\text{sec}}\right)}{1.217 \times 10^{-5}\ \frac{\text{ft}^2}{\text{sec}}} = 1.16 \times 10^5$$

From App. 17.A, for welded and seamless steel, $\epsilon = 0.0002$ ft.

$$\frac{\epsilon}{D} = \frac{0.0002\ \text{ft}}{0.3355\ \text{ft}} \approx 0.0006$$

From App. 17.B, the friction factor, f, is 0.0205. The friction head is

$$h_f = \frac{fLv^2}{2Dg} = \frac{(0.0205)(7000\ \text{ft})\left(4.2\ \frac{\text{ft}}{\text{sec}}\right)^2}{(2)(0.3355\ \text{ft})\left(32.2\ \frac{\text{ft}}{\text{sec}^2}\right)} = 117.2\ \text{ft}$$

The head added by the pump is

$$\begin{aligned} h_A &= h_f + h_z \\ &= 117.2\ \text{ft} + 12\ \text{ft} + 350\ \text{ft} \\ &= 479.5\ \text{ft} \end{aligned}$$

From Table 18.5, the hydraulic horsepower is

$$\text{WHP} = \frac{h_A Q(\text{SG})}{3956} = \frac{(479.2\ \text{ft})\left(10{,}000\ \frac{\text{gal}}{\text{hr}}\right)\left(\frac{1\ \text{hr}}{60\ \text{min}}\right)(1)}{3956\ \frac{\text{ft-gal}}{\text{hp-min}}} = 20.2\ \text{hp}$$

From Eq. 18.16, the overall efficiency of the pump is

$$\eta = \frac{\text{WHP}}{\text{EHP}}$$
$$\text{EHP} = \frac{20.2\ \text{hp}}{0.7} = 28.9\ \text{hp}$$

(a) At \$0.04/kW-hr, power costs for 1 hr are

$$(28.9\ \text{hp})\left(\frac{0.7457\ \text{kW}}{\text{hp}}\right)(1\ \text{hr})\left(\frac{\$0.04}{\text{kW-hr}}\right) = \boxed{\$0.86\ \text{per hour}}$$

The answer is (B).

(b) The motor horsepower, EHP, is 28.9 hp. Select a $\boxed{\text{30 hp motor.}}$

The answer is (B).

(c) From Eq. 18.5(b),

$$h_{\text{atm}} = \left(\frac{p_{\text{atm}}}{\rho}\right)\left(\frac{g_c}{g}\right) = \left(\frac{\left(14.7\ \frac{\text{lbf}}{\text{in}^2}\right)\left(144\ \frac{\text{in}^2}{\text{ft}^2}\right)}{62.37\ \frac{\text{lbm}}{\text{ft}^3}}\right)\left(\frac{32.2\ \frac{\text{ft}}{\text{sec}^2}}{32.2\ \frac{\text{ft-lbm}}{\text{lbf-sec}^2}}\right) = 33.94\ \text{ft}$$

The friction losses due to 300 ft is

$$h_{f(s)} = \left(\frac{300\ \text{ft}}{7000\ \text{ft}}\right)h_f = \left(\frac{300\ \text{ft}}{7000\ \text{ft}}\right)(117.2\ \text{ft}) = 5.0\ \text{ft}$$

From App. 14.A, the vapor pressure head at 60°F is 0.59 ft psia.

The NPSHA from Eq. 18.30(a) is

$$\begin{aligned} \text{NPSHA} &= h_{\text{atm}} + h_{z(s)} - h_{f(s)} - h_{\text{vp}} \\ &= 33.94\ \text{ft} - 12\ \text{ft} - 5.0\ \text{ft} - 0.59\ \text{ft} \\ &= \boxed{16.35\ \text{ft}} \end{aligned}$$

The answer is (D).

SI Solution

From App. 16.C, data for 4 in schedule-40 steel pipe are

$$D_i = 102.3\ \text{mm}$$
$$A_i = 82.19 \times 10^{-4}\ \text{m}^2$$

The velocity in the pipe is

$$\text{v} = \frac{\dot{V}}{A} = \frac{\left(10.5\ \frac{\text{L}}{\text{s}}\right)\left(\frac{1\ \text{m}^3}{1000\ \text{L}}\right)}{82.19 \times 10^{-4}\ \text{m}^2} = 1.28\ \text{m/s}$$

From App. 14.B, at 16°C the water data are

$$\rho = 998.83 \text{ kg/m}^3$$
$$\mu = 1.1261 \times 10^{-3} \text{ Pa·s}$$

The Reynolds number is

$$\text{Re} = \frac{\rho \text{v} D}{\mu} = \frac{\left(998.83 \ \frac{\text{kg}}{\text{m}^3}\right)\left(1.28 \ \frac{\text{m}}{\text{s}}\right) \times (102.3 \text{ mm})\left(\frac{1 \text{ m}}{1000 \text{ mm}}\right)}{1.1261 \times 10^{-3} \text{ Pa·s}} = 1.16 \times 10^5$$

From Table 17.2, for welded and seamless steel, $\epsilon = 6.0 \times 10^{-5}$ m.

$$\frac{\epsilon}{D} = \frac{6.0 \times 10^{-5} \text{ m}}{(102.3 \text{ mm})\left(\frac{1 \text{ m}}{1000 \text{ mm}}\right)} = 0.0006$$

From App. 17.B, the friction factor is $f = 0.0205$.

From Eq. 18.6, the friction head is

$$h_f = \frac{fL\text{v}^2}{2Dg} = \frac{(0.0205)(2300 \text{ m})\left(1.28 \ \frac{\text{m}}{\text{s}}\right)^2}{(2)(102.3 \text{ mm})\left(\frac{1 \text{ m}}{1000 \text{ mm}}\right)\left(9.81 \ \frac{\text{m}}{\text{s}^2}\right)} = 38.5 \text{ m}$$

The head added by the pump is

$$h_A = h_f + h_z = 38.5 \text{ m} + 4 \text{ m} + 115 \text{ m} = 157.5 \text{ m}$$

From Table 18.6, the hydraulic power is

$$\text{WkW} = \frac{(9.81)h_A Q(\text{SG})}{1000} = \frac{\left(9.81 \ \frac{\text{m}}{\text{s}^2}\right)(157.5 \text{ m})\left(10.5 \ \frac{\text{L}}{\text{s}}\right)(1)}{1000 \ \frac{\text{W·L}}{\text{kW·kg}}} = 16.22 \text{ kW}$$

From Eq. 18.16,

$$\text{EHP} = \frac{\text{WHP}}{\eta_{\text{overall}}} = \frac{16.22 \text{ kW}}{0.7} = 23.2 \text{ kW}$$

(a) At \$0.04/kW·h, power costs for 1 h are

$$(23.2 \text{ kW})(1 \text{ h})\left(\frac{\$0.04}{\text{kW·h}}\right) = \boxed{\$0.93 \text{ per hour}}$$

The answer is (B).

(b) The required motor power is 23.2 kW. Select the next higher standard motor size.

The answer is (B).

(c) From Eq. 18.5(a),

$$h_{\text{atm}} = \frac{p}{\rho g} = \frac{(101 \text{ kPa})(1000 \text{ Pa})}{\left(998.83 \ \frac{\text{kg}}{\text{m}^3}\right)\left(9.81 \ \frac{\text{m}}{\text{s}^2}\right)} = 10.31 \text{ m}$$

The friction loss due to 100 m is

$$h_{f(s)} = \left(\frac{100 \text{ m}}{2300 \text{ m}}\right)h_f = \left(\frac{100 \text{ m}}{2300 \text{ m}}\right)(38.5 \text{ m}) = 1.67 \text{ m}$$

The vapor pressure at 16°C is 0.01818 bar.

From Eq. 18.5(a),

$$h_{\text{vp}} = \frac{p_{\text{vp}}}{g\rho} = \frac{(0.01818 \text{ bar})\left(1 \times 10^5 \ \frac{\text{Pa}}{\text{bar}}\right)}{\left(9.81 \ \frac{\text{m}}{\text{s}^2}\right)\left(998.83 \ \frac{\text{kg}}{\text{m}^3}\right)} = 0.19 \text{ m}$$

The NPSHA from Eq. 18.30(a) is

$$\text{NPSHA} = h_{\text{atm}} + h_{z(s)} - h_{f(s)} - h_{\text{vp}} = 10.31 \text{ m} - 4 \text{ m} - 1.67 \text{ m} - 0.19 \text{ m} = \boxed{4.45 \text{ m}}$$

The answer is (D).

6. (a) $h_f = \Delta z = 48$ ft since $\Delta p = 0$ and $\Delta\text{v} = 0$ for open channel flow.

$$Q = \frac{\left(250 \ \frac{\text{gal}}{\text{person·day}}\right)(10{,}000 \text{ people})}{\left(24 \ \frac{\text{hr}}{\text{day}}\right)\left(60 \ \frac{\text{min}}{\text{hr}}\right)} = 1736 \text{ gal/min}$$

Given $C = 130$, solving for d from Eq. 17.31,

$$d_{\text{in}}^{4.8655} = \frac{(10.44)(5000 \text{ ft})\left(1736 \ \frac{\text{gal}}{\text{min}}\right)^{1.85}}{(130)^{1.85}(48 \text{ ft})} = 131{,}462$$

$$d = \boxed{11.27 \text{ in} \quad [\text{round to 12 in minimum}]}$$

The answer is (B).

(b) Without having a specific pump curve, the number of pumps can only be specified based on general rules. Use the *Ten States' Standards*, which states:

- No station will have less than two identical pumps.
- Capacity must not be reduced with one pump out of service.

$$\boxed{\text{Two pumps are required, plus one backup.}}$$

The answer is (A).

(c) With a variable-speed pump, it will be possible to adjust to the wide variations in flow (100 to 250 gpcd). It may be possible to operate with one pump and one backup.

The answer is (B).

(d) With three constant-speed pumps,

$$Q = \frac{1736 \ \frac{\text{gal}}{\text{min}}}{3} = 579 \text{ gal/min at maximum capacity}$$

From Table 18.5, assuming specific gravity ≈ 1.00 and using $\eta_{\text{pump}} = 0.80$,

$$\text{rated motor power} = \frac{(10 \text{ ft})\left(579 \ \frac{\text{gal}}{\text{min}}\right)}{\left(3956 \ \frac{\text{gal-ft}}{\text{min-hp}}\right)(0.80)} = \boxed{1.83 \text{ hp} \quad [\text{use 2.0 hp}]}$$

The answer is (A).

(e) With two variable-speed pumps,

$$Q = \frac{1736 \ \frac{\text{gal}}{\text{min}}}{2} = 868 \text{ gal/min}$$

$$\text{rated motor power} = \frac{(10 \text{ ft})\left(868 \ \frac{\text{gal}}{\text{min}}\right)}{\left(3956 \ \frac{\text{gal-ft}}{\text{min-hp}}\right)(0.80)} = \boxed{2.74 \text{ hp} \quad [\text{use 3.0 hp}]}$$

The answer is (A).

(f) Assume the question refers to demand control rather than to motor speed control. Methods of control are

- variable run times
- pressure change in sump
- flow rate incoming
- sump level
- flow rate outgoing
- manual on/off

The answer is (B).

(g) In the wet well, the pump (and perhaps motor) is submerged. Forced ventilation air will prevent a concentration of explosive methane. From the *Ten States' Standards*,

$$\boxed{\begin{array}{l}\text{12 air changes per hour if continuous;}\\ \text{30 per hour if intermittent.}\end{array}}$$

The answer is (D).

(h) In the dry well,

$$\boxed{\begin{array}{l}\text{6 air changes per hour if continuous;}\\ \text{30 per hour if intermittent.}\end{array}}$$

The answer is (A).

7. (a) $h_{p(s)} = \boxed{20 \text{ ft}}$

The answer is (B).

(b) $h_{p(d)} = \boxed{15 \text{ ft}}$

The answer is (A).

(c) There is no pipe size specified, so h_{v} cannot be calculated. Even so, v is typically in the 5 to 10 ft/sec range, and $h_{\text{v}} \approx 0$.

Since pipe lengths are not given, assume $h_f \approx 0$.

$$20 \text{ ft} + 15 \text{ ft} + \frac{\left(80 \ \frac{\text{lbf}}{\text{in}^2}\right)\left(144 \ \frac{\text{in}^2}{\text{ft}^2}\right)}{62.4 \ \frac{\text{lbf}}{\text{ft}^3}} + 10 \text{ ft} = \boxed{229.6 \text{ ft of water}}$$

The answer is (D).

(d) The flow rate is

$$\frac{(3.5 \text{ MGD})\left(62.4 \ \frac{\text{lbf}}{\text{ft}^3}\right)}{0.64632 \ \frac{\text{MGD}}{\frac{\text{ft}^3}{\text{sec}}}} = 337.9 \text{ lbf/sec}$$

From Table 18.5, the rated motor output power does not depend on the motor efficiency.

$$P = \frac{(229.6 \text{ ft})\left(337.9 \frac{\text{lbf}}{\text{sec}}\right)}{\left(550 \frac{\text{ft-lbf}}{\text{hp-sec}}\right)(0.85)} = 166.0 \text{ hp}$$

Use a 200 hp motor.

The answer is (D).

(e) The chosen motor has more power than is necessary.

$$\begin{aligned}\dot{m} &= \left(\frac{(200 \text{ hp})\left(550 \frac{\text{ft-lbf}}{\text{hp-sec}}\right)(0.85)}{229.6 \text{ ft}}\right) \\ &\quad \times \left(\frac{0.64632 \frac{\text{MGD}}{\frac{\text{ft}^3}{\text{sec}}}}{62.4 \frac{\text{lbf}}{\text{ft}^3}}\right) \\ &= 4.20 \text{ MGD}\end{aligned}$$

The excess required storage is

$$\begin{aligned}&\left(\frac{8 \text{ hr}}{24 \frac{\text{hr}}{\text{day}}}\right)\big((1.25)(3.5 \text{ MGD}) - 4.20 \text{ MGD}\big) \\ &\quad = \boxed{0.058 \text{ MG } (60{,}000 \text{ gal})}\end{aligned}$$

The answer is (A).

8. *Customary U.S. Solution*

From App. 16.B, the pipe data for 3 in schedule-40 steel pipe are

$$D_i = 0.2557 \text{ ft}$$
$$A_i = 0.05134 \text{ ft}^2$$

From App. 17.D, the equivalent length for various fittings is

$$\text{flanged elbow, } L_e = 4.4 \text{ ft}$$
$$\text{wide-open gate valve, } L_e = 2.8 \text{ ft}$$

The total equivalent length of pipe and fittings is

$$\begin{aligned}L_e &= 500 \text{ ft} + (6)(4.4 \text{ ft}) + (2)(2.8 \text{ ft}) \\ &= 532 \text{ ft}\end{aligned}$$

As a first estimate, assume the flow rate is 100 gal/min.

The velocity in the pipe is

$$\begin{aligned}\text{v} = \frac{\dot{V}}{A} &= \frac{\left(100 \frac{\text{gal}}{\text{min}}\right)\left(0.002228 \frac{\frac{\text{ft}^3}{\text{sec}}}{\frac{\text{gal}}{\text{min}}}\right)}{0.05134 \text{ ft}^2} \\ &= 4.34 \text{ ft/sec}\end{aligned}$$

The Reynolds number is

$$\begin{aligned}\text{Re} = \frac{\text{v}D}{\nu} &= \frac{\left(4.34 \frac{\text{ft}}{\text{sec}}\right)(0.2557 \text{ ft})}{6 \times 10^{-6} \frac{\text{ft}^2}{\text{sec}}} \\ &= 1.85 \times 10^5\end{aligned}$$

From App. 17.A, $\epsilon = 0.0002$ ft.

So,

$$\frac{\epsilon}{D} = \frac{0.0002 \text{ ft}}{0.2557 \text{ ft}} \approx 0.0008$$

From the friction factor table, $f \approx 0.0204$.

For higher flow rates, f approaches 0.0186. Since the chosen flow rate was almost the lowest, $f = 0.0186$ should be used.

From Eq. 18.6, the friction head loss is

$$\begin{aligned}h_f &= \frac{fL\text{v}^2}{2Dg} \\ &= \frac{(0.0186)(532 \text{ ft})\left(4.34 \frac{\text{ft}}{\text{sec}}\right)^2}{(2)(0.2557 \text{ ft})\left(32.2 \frac{\text{ft}}{\text{sec}^2}\right)} \\ &= 11.3 \text{ ft of gasoline}\end{aligned}$$

This neglects the small velocity head. The other system points can be found using

$$\frac{h_{f_1}}{h_{f_2}} = \left(\frac{Q_1}{Q_2}\right)^2$$

$$\begin{aligned}h_{f_2} &= h_{f_1}\left(\frac{Q_2}{100 \frac{\text{gal}}{\text{min}}}\right)^2 \\ &= (11.3 \text{ ft})\left(\frac{Q_2}{100 \frac{\text{gal}}{\text{min}}}\right)^2 \\ &= 0.00113Q_2^2\end{aligned}$$

Q (gal/min)	h_f (ft)	$h_f + 60$ (ft)
100	11.3	71.3
200	45.2	105.2
300	101.7	161.7
400	180.8	240.8
500	282.5	342.5
600	406.8	466.8

(a) Plot the system and pump curves.

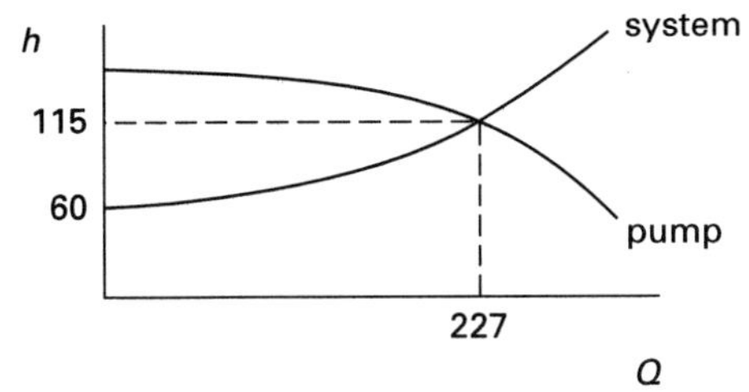

$$h = 115 \text{ ft}$$

$$\boxed{Q = 227 \text{ gal/min}}$$

The answer is (D).

(b) From Table 18.5, the hydraulic horsepower is

$$\text{WHP} = \frac{h_A Q(\text{SG})}{3956} = \frac{(115 \text{ ft})\left(227 \ \frac{\text{gal}}{\text{min}}\right)(0.7)}{3956 \ \frac{\text{ft-gal}}{\text{hp-min}}}$$

$$= 4.62 \text{ hp}$$

$$\frac{(4.62 \text{ hp})\left(0.7457 \ \frac{\text{kW}}{\text{hp}}\right) \times (1 \text{ hr})\left(0.045 \ \frac{\$}{\text{kW-hr}}\right)}{(0.88)(0.88)} = \boxed{\$0.20}$$

The answer is (A).

SI Solution

From App. 16.C, the pipe data for 3 in schedule-40 pipe are

$$D_i = 77.92 \text{ mm}$$
$$A_i = 47.69 \times 10^{-4} \text{ m}^2$$

From App. 17.D, the equivalent length for various fittings is

$$\text{flanged elbow, } L_e = 4.4 \text{ ft}$$
$$\text{wide-open gate valve, } L_e = 2.8 \text{ ft}$$

The total equivalent length of pipe and fittings is

$$L_e = 170 \text{ m} + (6)(4.4 \text{ ft})\left(0.3048 \ \frac{\text{m}}{\text{ft}}\right) + (2)(2.8 \text{ ft})\left(0.3048 \ \frac{\text{m}}{\text{ft}}\right)$$
$$= 180 \text{ m}$$

As a first estimate, assume the flow rate is 6.3 L/s. The velocity in the pipe is

$$\text{v} = \frac{\dot{V}}{A} = \frac{\left(6.3 \ \frac{\text{L}}{\text{s}}\right)\left(\frac{1 \text{ m}^3}{1000 \text{ L}}\right)}{47.69 \times 10^{-4} \text{ m}^2} = 1.32 \text{ m/s}$$

The Reynolds number is

$$\text{Re} = \frac{\text{v}D}{\nu} = \frac{\left(1.32 \ \frac{\text{m}}{\text{s}}\right)(77.92 \text{ mm})\left(\frac{1 \text{ m}}{1000 \text{ mm}}\right)}{5.6 \times 10^{-7} \ \frac{\text{m}^2}{\text{s}}}$$
$$= 1.75 \times 10^5$$

From Table 17.2, $\epsilon = 6.0 \times 10^{-5}$ m.

$$\frac{\epsilon}{D} = \frac{6.0 \times 10^{-5} \text{ m}}{(77.92 \text{ mm})\left(\frac{1 \text{ m}}{1000 \text{ mm}}\right)}$$
$$\approx 0.0008$$

From the friction factor table (App. 17.B), $f = 0.0205$.

For higher flow rates, f approaches 0.0186. Since the chosen flow rate was almost the lowest, $f = 0.0186$ should be used.

From Eq. 17.22, the friction head loss is

$$h_f = \frac{fL\text{v}^2}{2Dg} = \frac{(0.0186)(180 \text{ m})\left(1.32 \ \frac{\text{m}}{\text{s}}\right)^2}{(2)(77.92 \text{ mm})\left(\frac{1 \text{ m}}{1000 \text{ mm}}\right)\left(9.81 \ \frac{\text{m}}{\text{s}^2}\right)}$$
$$= 3.82 \text{ m of gasoline}$$

This neglects the small velocity head. The other system points can be found using

$$\frac{h_{f_1}}{h_{f_2}} = \left(\frac{Q_1}{Q_2}\right)^2$$

$$h_{f_2} = h_{f_1}\left(\frac{Q_2}{Q_1}\right)^2 = (3.82 \text{ m})\left(\frac{Q_2}{6.3 \ \frac{\text{L}}{\text{s}}}\right)^2$$
$$= 0.0962 Q_2^2$$

Q (L/s)	h_f (m)	$h_f + 20$ (m)
6.3	3.82	23.82
12	13.85	33.85
18	31.2	51.2
24	55.4	75.4
30	86.6	106.6
36	124.7	144.7

(a) Plot the system and pump curves.

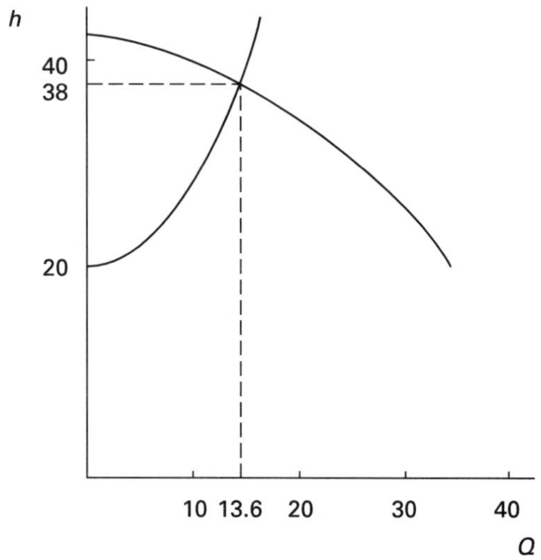

$$h = 38.0 \text{ m}$$

$$\boxed{Q = 13.6 \text{ L/s}}$$

The answer is (D).

(b) From Table 18.6, the hydraulic power is

$$\begin{aligned} \text{WkW} &= \frac{9.81 h_A Q(\text{SG})}{1000} \\ &= \frac{\left(9.81 \ \frac{\text{m}}{\text{s}^2}\right)(38.0 \text{ m})\left(13.6 \ \frac{\text{L}}{\text{s}}\right)(0.7)}{1000 \ \frac{\text{W}}{\text{kW}}} \\ &= 3.55 \text{ kW} \end{aligned}$$

The cost per hour is

$$\begin{aligned} &= \left(\frac{3.55 \text{ kW}}{(0.88)(0.88)}\right)(1 \text{ h})\left(0.045 \ \frac{\$}{\text{kW·h}}\right) \\ &= \boxed{\$0.21} \end{aligned}$$

The answer is (A).

9. *Customary U.S. Solution*

From Eq. 18.28(b), the specific speed is

$$n_s = \frac{n\sqrt{Q}}{h_A^{0.75}}$$

For a double-suction pump, Q in the preceding equation is half of the full flow rate.

$$\begin{aligned} n_s &= \frac{(900 \text{ rpm})\sqrt{\left(300 \ \frac{\text{gal}}{\text{sec}}\right)\left(60 \ \frac{\text{sec}}{\text{min}}\right)\left(\frac{1}{2}\right)}}{(20 \text{ ft})^{0.75}} \\ &= \boxed{9028 \text{ rpm}} \end{aligned}$$

The answer is (C).

SI Solution

From Eq. 18.28(a), the specific speed is

$$n_s = \frac{n\sqrt{\dot{V}}}{h_A^{0.75}}$$

For a double-suction pump, $\dot{V}$ in the preceding equation is half of the full flow rate.

$$\begin{aligned} n_s &= \frac{(900 \text{ rpm})\sqrt{\left(1.1 \ \frac{\text{kL}}{\text{s}}\right)\left(\frac{1 \text{ m}^3}{1 \text{ kL}}\right)\left(\frac{1}{2}\right)}}{(7 \text{ m})^{0.75}} \\ &= \boxed{155.1 \text{ rpm}} \end{aligned}$$

The answer is (C).

10.

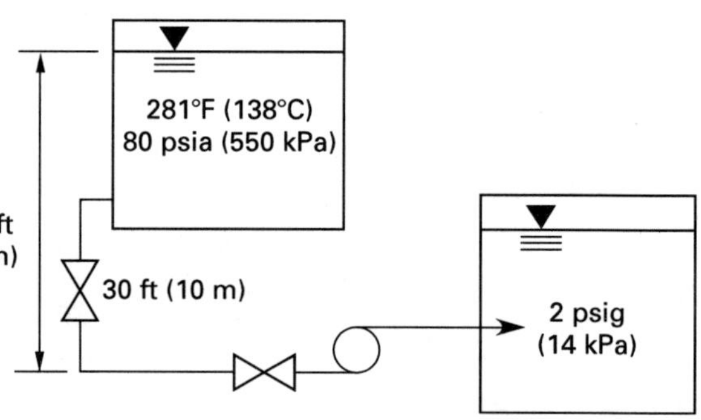

Customary U.S. Solution

From App. 16.B, data for 1.5 in schedule-40 steel pipe are

$$D_i = 0.1342 \text{ ft}$$

$$A_i = 0.01414 \text{ ft}^2$$

The velocity in the pipe is

$$\begin{aligned} \text{v} = \frac{\dot{V}}{A} &= \frac{\left(100 \ \frac{\text{gal}}{\text{min}}\right)\left(0.002228 \ \frac{\text{ft}^3\text{-min}}{\text{sec-gal}}\right)}{0.01414 \text{ ft}^2} \\ &= 15.76 \text{ ft/sec} \end{aligned}$$

From App. 17.D, for screwed steel fittings, the approximate equivalent lengths for fittings are

inlet (square mouth): $L_e = 3.1$ ft

long radius 90° ell: $L_e = 3.4$ ft

wide-open gate valves: $L_e = 1.2$ ft

The total equivalent length is

$$30 \text{ ft} + 3.1 \text{ ft} + (2)(3.4 \text{ ft}) + (2)(1.2 \text{ ft}) = 42.3 \text{ ft}$$

From App. 17.A, for steel, $\epsilon = 0.0002$ ft, so

$$\frac{\epsilon}{D} = \frac{0.0002 \text{ ft}}{0.1342 \text{ ft}} = 0.0015$$

At 281°F, $\nu = 0.239 \times 10^{-5}$ ft^2/sec. The Reynolds number is

$$\text{Re} = \frac{D\text{v}}{\nu} = \frac{(0.1342 \text{ ft})\left(15.76 \frac{\text{ft}}{\text{sec}}\right)}{0.239 \times 10^{-5} \frac{\text{ft}^2}{\text{sec}}} = 8.85 \times 10^5$$

From App. 17.B, the friction factor is $f = 0.022$.

From Eq. 18.6, the friction head is

$$h_f = \frac{fL\text{v}^2}{2Dg} = \frac{(0.022)(42.3 \text{ ft})\left(15.76 \frac{\text{ft}}{\text{sec}}\right)^2}{(2)(0.1342 \text{ ft})\left(32.2 \frac{\text{ft}}{\text{sec}^2}\right)} = 26.74 \text{ ft}$$

At 281°F, the saturated vapor pressure of pressurized water is

$$p_{\text{vapor}} = 50.02 \text{ psia} \quad \text{[from steam tables]}$$

$$\rho = \frac{1}{\upsilon_f} = \frac{1}{0.01727 \frac{\text{ft}^3}{\text{lbm}}} = 57.9 \text{ lbm/ft}^3$$

From Eq. 18.5(b),

$$h_{\text{vp}} = \left(\frac{p_{\text{vapor}}}{\rho}\right) \times \left(\frac{g_c}{g}\right) = \left(\frac{\left(50.02 \frac{\text{lbf}}{\text{in}^2}\right)\left(144 \frac{\text{in}^2}{\text{ft}^2}\right)}{57.9 \frac{\text{lbm}}{\text{ft}^3}}\right) \times \left(\frac{32.2 \frac{\text{ft-lbm}}{\text{lbf-sec}^2}}{32.2 \frac{\text{ft}}{\text{sec}^2}}\right) = 124.4 \text{ ft}$$

From Eq. 18.5(b), the pressure head is

$$h_p = \left(\frac{p}{\rho}\right) \times \left(\frac{g_c}{g}\right) = \left(\frac{\left(80 \frac{\text{lbf}}{\text{in}^2}\right)\left(144 \frac{\text{in}^2}{\text{ft}^2}\right)}{57.9 \frac{\text{lbm}}{\text{ft}^3}}\right) \times \left(\frac{32.2 \frac{\text{ft-lbm}}{\text{lbf-sec}^2}}{32.2 \frac{\text{ft}}{\text{sec}^2}}\right) = 199.0 \text{ ft}$$

From Eq. 18.30(a), the NPSHA is

$$\begin{aligned} \text{NPSHA} &= h_p + h_{z(s)} - h_{f(s)} - h_{\text{vp}} \\ &= 199.0 \text{ ft} + 20 \text{ ft} - 26.74 \text{ ft} - 124.4 \text{ ft} \\ &= 67.9 \text{ ft} \end{aligned}$$

Since NPSHR = 10 ft, **the pump will not cavitate.**

(Note that a pump may not actually be needed in this configuration.)

The answer is (D).

SI Solution

From App. 16.C, data for 1.5 in schedule-40 steel pipe are

$$D_i = 40.89 \text{ mm}$$
$$A_i = 13.13 \times 10^{-4} \text{ m}^2$$

The velocity in the pipe is

$$\text{v} = \frac{\dot{V}}{A} = \frac{\left(6.3 \frac{\text{L}}{\text{s}}\right)\left(\frac{1 \text{ m}^3}{1000 \text{ L}}\right)}{13.13 \times 10^{-4} \text{ m}^2} = 4.80 \text{ m/s}$$

From App. 17.D, for screwed steel fittings, the approximate equivalent lengths for fittings are

inlet (square mouth): $L_e = 3.1$ ft

long radius 90° ell: $L_e = 3.4$ ft

wide-open gate valves: $L_e = 1.2$ ft

The total equivalent length is

$$30 \text{ ft} + 3.1 \text{ ft} + (2)(3.4 \text{ ft}) + (2)(1.2 \text{ ft}) = 42.3 \text{ ft}$$

$$(42.3 \text{ ft})\left(0.3048 \frac{\text{m}}{\text{ft}}\right) = 12.89 \text{ m}$$

From Table 17.2, for steel, $\epsilon = 6.0 \times 10^{-5}$ m.

$$\frac{\epsilon}{D} = \frac{6.0 \times 10^{-5} \text{ m}}{(40.89 \text{ mm})\left(\frac{1 \text{ m}}{1000 \text{ mm}}\right)} \approx 0.0015$$

At 138°C, $\nu = 0.222 \times 10^{-6}$ m²/s. The Reynolds number is

$$\begin{aligned}\text{Re} &= \frac{D\text{v}}{\nu} \\ &= \frac{(40.89\text{ mm})\left(\frac{1\text{ m}}{1000\text{ mm}}\right)\left(4.80\ \frac{\text{m}}{\text{s}}\right)}{0.222\times 10^{-6}\ \frac{\text{m}^2}{\text{s}}} \\ &= 8.84\times 10^5\end{aligned}$$

From App. 17.B, the friction factor is $f = 0.022$.

From Eq. 18.6, the friction head is

$$\begin{aligned}h_f &= \frac{fL\text{v}^2}{2Dg} \\ &= \frac{(0.022)(12.89\text{ m})\left(4.8\ \frac{\text{m}}{\text{s}}\right)^2}{(2)(40.89\text{ mm})\left(\frac{1\text{ m}}{1000\text{ mm}}\right)\left(9.81\ \frac{\text{m}}{\text{s}^2}\right)} \\ &= 8.14\text{ m}\end{aligned}$$

From Eq. 18.5(a),

$$h_{\text{vp}} = \frac{p_{\text{vapor}}}{\rho g}$$

At 138°C, the saturated vapor pressure of pressurized water is

$$\begin{aligned}p_{\text{vapor}} &= 3.431\text{ bar}\quad\text{[from steam tables]} \\ \rho &= \frac{1}{\upsilon_f} \\ &= \frac{1}{\left(1.0777\ \frac{\text{cm}^3}{\text{g}}\right)\left(1000\ \frac{\text{g}}{\text{kg}}\right)\left(\frac{1\text{ m}^3}{(100\text{ cm})^3}\right)} \\ &= 927.9\text{ kg/m}^3 \\ h_{\text{vp}} &= \frac{(3.431\text{ bar})\left(10^5\ \frac{\text{Pa}}{\text{bar}}\right)}{\left(927.9\ \frac{\text{kg}}{\text{m}^3}\right)\left(9.81\ \frac{\text{m}}{\text{s}^2}\right)} \\ &= 37.69\text{ m}\end{aligned}$$

From Eq. 18.5(a), the pressure head is

$$\begin{aligned}h_p &= \frac{p}{\rho g} \\ &= \frac{(550\text{ kPa})\left(1000\ \frac{\text{Pa}}{\text{kPa}}\right)}{\left(927.9\ \frac{\text{kg}}{\text{m}^3}\right)\left(9.81\ \frac{\text{m}}{\text{s}^2}\right)} \\ &= 60.42\text{ m}\end{aligned}$$

From Eq. 18.30(a), the NPSHA is

$$\begin{aligned}\text{NPSHA} &= h_p + h_{z(s)} - h_{f(s)} - h_{\text{vp}} \\ &= 60.42\text{ m} + 6\text{ m} - 8.14\text{ m} - 37.69\text{ m} \\ &= 20.6\text{ m}\end{aligned}$$

Since NPSHR is 3 m, **the pump will not cavitate.**

(Note that a pump may not actually be needed in this configuration.)

The answer is (D).

11. The solvent is the water (fresh), and the solution is the seawater. Since seawater contains approximately 2½% salt by weight, 100 lbm of seawater will yield 2.5 lbm salt and 97.5 lbm water. The molecular weight of salt is 23.0 + 35.5 = 58.5 lbm/lbmol. The number of moles of salt in 100 lbm of seawater is

$$n_{\text{salt}} = \frac{2.5\text{ lbm}}{58.5\ \frac{\text{lbm}}{\text{lbmol}}} = 0.043\text{ lbmol}$$

Similarly, the molecular weight of water is 18.016 lbm/lbmol. The number of moles of water is

$$n_{\text{water}} = \frac{97.5\text{ lbm}}{18.016\ \frac{\text{lbm}}{\text{lbmol}}} = 5.412\text{ lbmol}$$

The mole fraction of water is

$$\frac{5.412\text{ lbmol}}{5.412\text{ lbmol} + 0.043\text{ lbmol}} = 0.992$$

Customary U.S. Solution

Cavitation will occur when

$$h_{\text{atm}} - h_{\text{v}} < h_{\text{vp}}$$

The density of seawater is 64.0 lbm/ft³.

From Eq. 18.5(b), the atmospheric head is

$$\begin{aligned}h_{\text{atm}} &= \left(\frac{p}{\rho}\right)\times\left(\frac{g_c}{g}\right) \\ &= \left(\frac{\left(14.7\ \frac{\text{lbf}}{\text{in}^2}\right)\left(\frac{144\text{ in}^2}{\text{ft}^2}\right)}{64.0\ \frac{\text{lbm}}{\text{ft}^3}}\right)\times\left(\frac{32.2\ \frac{\text{ft-lbm}}{\text{lbf-sec}^2}}{32.2\ \frac{\text{ft}}{\text{sec}^2}}\right) \\ &= 33.075\text{ ft}\quad\text{[ft of seawater]} \\ h_{\text{depth}} &= 8\text{ ft}\quad\text{[given]}\end{aligned}$$

From Eq. 18.7, the velocity head is

$$h_v = \frac{v^2_{propeller}}{2g} = \frac{(4.2v_{boat})^2}{(2)\left(32.2\ \frac{ft}{sec^2}\right)} = 0.2739v^2_{boat}$$

The vapor pressure of 68°F freshwater is $p_{vp} = 0.3391$ psia.

From App. 14.A, the density of water at 68°F is 62.32 lbm/ft^3. Raoult's law predicts the actual vapor pressure of the solution.

$$p_{vapor,solution} = (p_{vapor,solvent})\begin{pmatrix}\text{mole fraction}\\ \text{of the solvent}\end{pmatrix}$$

$$p_{vapor,seawater} = (0.992)(0.3391\ \text{psia}) = 0.3364\ \text{psia}$$

From Eq. 18.5(b), the vapor pressure head is

$$h_{vapor,seawater} = \left(\frac{p}{\rho}\right)\times\left(\frac{g_c}{g}\right) = \left(\frac{\left(0.3364\ \frac{lbf}{in^2}\right)\left(144\ \frac{in^2}{ft^2}\right)}{64.0\ \frac{lbm}{ft^3}}\right) \times \left(\frac{32.2\ \frac{ft\text{-}lbm}{lbf\text{-}sec^2}}{32.2\ \frac{ft}{sec^2}}\right) = 0.7569\ \text{ft}$$

Then,

$$8\ \text{ft} + 33.075\ \text{ft} - 0.2739v^2_{boat} = 0.7569\ \text{ft}$$

$$v_{boat} = \boxed{12.13\ \text{ft/sec}}$$

The answer is (B).

SI Solution

Cavitation will occur when

$$h_{atm} - h_v < h_{vp}$$

The density of seawater is 1024 kg/m^3.

From Eq. 18.5(a), the atmospheric head is

$$h_{atm} = \frac{p}{\rho g} = \frac{(101.3\ \text{kPa})\left(1000\ \frac{Pa}{kPa}\right)}{\left(1024\ \frac{kg}{m^3}\right)\left(9.81\ \frac{m}{s^2}\right)} = 10.08\ \text{m}$$

$$h_{depth} = 3\ \text{m}\quad [\text{given}]$$

From Eq. 18.7, the velocity head is

$$h_v = \frac{v^2_{propeller}}{2g} = \frac{(4.2v_{boat})^2}{(2)\left(9.81\ \frac{m}{s^2}\right)} = 0.899v^2_{boat}$$

The vapor pressure of 20°C freshwater is

$$p_{vp} = (0.02339\ \text{bar})\left(100\ \frac{kPa}{bar}\right) = 2.339\ \text{kPa}$$

From App. 14.B, the density of water at 20°C is 998.23 kg/m^3. Raoult's law predicts the actual vapor pressure of the solution.

$$p_{vapor,solution} = (p_{vapor,solvent})\begin{pmatrix}\text{mole fraction}\\ \text{of the solvent}\end{pmatrix}$$

The solvent is the freshwater and the solution is the seawater.

The mole fraction of water is 0.992.

$$p_{vapor,seawater} = (0.992)(2.339\ \text{kPa}) = 2.320\ \text{kPa}$$

From Eq. 18.5(a), the vapor pressure head is

$$h_{vapor,seawater} = \frac{(2.320\ \text{kPa})\left(1000\ \frac{Pa}{kPa}\right)}{\left(9.81\ \frac{m}{s^2}\right)\left(1024\ \frac{kg}{m^3}\right)} = 0.231\ \text{m}$$

Then,

$$3\ \text{m} + 10.08\ \text{m} - 0.899v^2_{boat} = 0.231\ \text{m}$$

$$v_{boat} = \boxed{3.78\ \text{m/s}}$$

The answer is (B).

12.

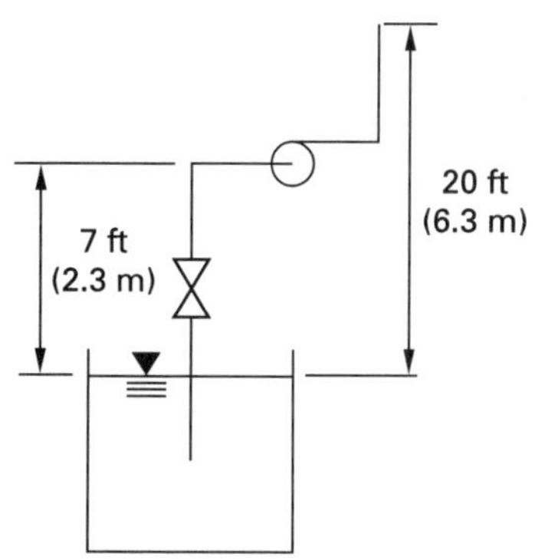

Customary U.S. Solution

From App. 17.D, the approximate equivalent lengths of various screwed steel fittings are

inlet: $L_e = 8.5$ ft

check valve: $L_e = 19$ ft

long radius elbows: $L_e = 3.6$ ft

The total equivalent length of the 2 in line is

$$\begin{aligned} L_e &= 12 \text{ ft} + 8.5 \text{ ft} + 19.0 \text{ ft} + (3)(3.6 \text{ ft}) + 80 \text{ ft} \\ &= 130.3 \text{ ft} \end{aligned}$$

From App. 16.B, for schedule-40 2 in pipe, the pipe data are

$$D_i = 0.1723 \text{ ft}$$
$$A_i = 0.0233 \text{ ft}^2$$

Since the flow rate is unknown, it must be assumed in order to find velocity. Assume 90 gal/min.

$$\dot{V} = \left(90 \ \frac{\text{gal}}{\text{min}}\right)\left(0.002228 \ \frac{\frac{\text{ft}^3}{\text{sec}}}{\frac{\text{gal}}{\text{min}}}\right) = 0.2005 \text{ ft}^3/\text{sec}$$

The velocity is

$$\text{v} = \frac{\dot{V}}{A_i} = \frac{0.2005 \ \frac{\text{ft}^3}{\text{sec}}}{0.0233 \text{ ft}^2} = 8.605 \text{ ft/sec}$$

From App. 14.A, the kinematic viscosity of water at 70°F is $\nu = 1.059 \times 10^{-5}$ ft^2/sec.

The Reynolds number is

$$\begin{aligned} \text{Re} &= \frac{D\text{v}}{\nu} = \frac{(0.1723 \text{ ft})\left(8.605 \ \frac{\text{ft}}{\text{sec}}\right)}{1.059 \times 10^{-5} \ \frac{\text{ft}^2}{\text{sec}}} \\ &= 1.4 \times 10^5 \end{aligned}$$

From App. 17.A, the specific roughness of steel pipe is

$$\epsilon = 0.0002 \text{ ft}$$
$$\frac{\epsilon}{D} = \frac{0.0002 \text{ ft}}{0.1723 \text{ ft}} = 0.0012$$

From App. 17.B, $f = 0.022$. At 90 gal/min, the friction loss in the line from Eq. 18.6 is

$$\begin{aligned} h_f &= \frac{fL\text{v}^2}{2Dg} \\ &= \frac{(0.022)(130.3 \text{ ft})\left(8.605 \ \frac{\text{ft}}{\text{sec}}\right)^2}{(2)(0.1723 \text{ ft})\left(32.2 \ \frac{\text{ft}}{\text{sec}^2}\right)} = 19.1 \text{ ft} \end{aligned}$$

From Eq. 18.7, the velocity head at 90 gal/min is

$$h_\text{v} = \frac{\text{v}^2}{2g} = \frac{\left(8.605 \ \frac{\text{ft}}{\text{sec}}\right)^2}{(2)\left(32.2 \ \frac{\text{ft}}{\text{sec}^2}\right)} = 1.1 \text{ ft}$$

In general, the friction head and velocity head are proportional to v^2 and Q^2.

$$h_f = (19.1 \text{ ft})\left(\frac{Q_2}{90 \ \frac{\text{gal}}{\text{min}}}\right)^2$$

$$h_\text{v} = (1.1 \text{ ft})\left(\frac{Q_2}{90 \ \frac{\text{gal}}{\text{min}}}\right)^2$$

(Note that the 7 ft dimension is included in the 20 ft dimension.)

The total system head is

$$\begin{aligned} h &= h_z + h_\text{v} + h_f \\ &= 20 \text{ ft} + (1.1 + 19.1 \text{ ft})\left(\frac{Q_2}{90 \ \frac{\text{gal}}{\text{min}}}\right)^2 \end{aligned}$$

From this equation, the following table for system head can be generated.

Q_2 (gal/min)	system head, h (ft)
0	20.0
10	20.2
20	21.0
30	22.2
40	24.0
50	26.2
60	29.0
70	32.2
80	36.0
90	40.2
100	44.9
110	50.2

(a) The intersection point of the system curve and the pump curve defines the operating flow rate. The flow rate is **95 gal/min.**

The answer is (D).

(b) The intersection point is not in an efficient range for the pump because it is so far down on the system curve that the pumping efficiency will be low.

A different pump should be used.

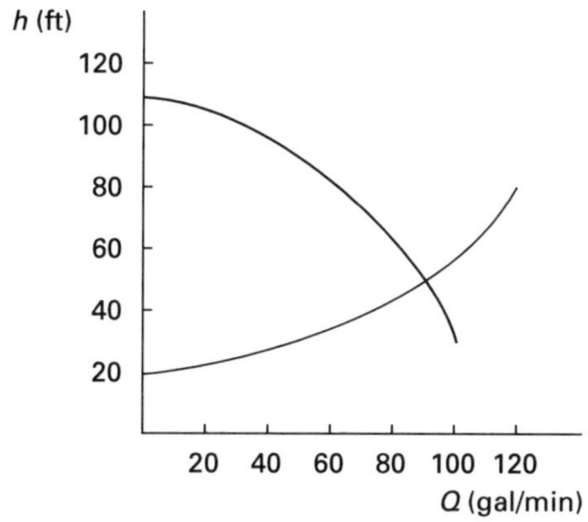

The answer is (A).

SI Solution

Use the approximate quivalent lengths of various screwed steel fittings from the customary U.S. solution. The total equivalent length of 5.08 cm schedule-40 pipe is

$$L_e = 4\text{ m} + \big(8.5\text{ ft} + 19.0\text{ ft} + (3)(3.6\text{ ft})\big)\left(0.3048\ \frac{\text{m}}{\text{ft}}\right) + 27\text{ m} = 42.67\text{ m}$$

From App. 16.C, for 2 in schedule-40 pipe, the pipe data are

$$D_i = 52.50\text{ mm}$$
$$A_i = 21.65 \times 10^{-4}\text{ m}^2$$

Since the flow rate is unknown, it must be assumed in order to find velocity. Assume 6 L/s.

$$\dot{V} = \left(6\ \frac{\text{L}}{\text{s}}\right)\left(\frac{1\text{ m}^3}{1000\text{ L}}\right) = 6 \times 10^{-3}\text{ m}^3/\text{s}$$

The velocity is

$$\text{v} = \frac{\dot{V}}{A_i} = \frac{6 \times 10^{-3}\ \frac{\text{m}^3}{\text{s}}}{21.65 \times 10^{-4}\text{ m}^2} = 2.77\text{ m/s}$$

From App. 14.B, the absolute viscosity of water at 21°C is $\mu = 0.9827 \times 10^{-3}$ Pa·s.

The density of water is $\rho = 998\text{ kg/m}^3$.

The Reynolds number is

$$\text{Re} = \frac{\rho \text{v} D_i}{\mu} = \frac{\left(998\ \frac{\text{kg}}{\text{m}^3}\right)\left(2.77\ \frac{\text{m}}{\text{s}}\right) \times (52.50\text{ mm})\left(\frac{1\text{ m}}{1000\text{ mm}}\right)}{0.9827 \times 10^{-3}\text{ Pa·s}} = 1.5 \times 10^5$$

From Table 17.2, the specific roughness of steel pipe is

$$\epsilon = 6.0 \times 10^{-5}\text{ m}$$

$$\frac{\epsilon}{D} = \frac{6.0 \times 10^{-5}\text{ m}}{(52.50\text{ mm})\left(\frac{1\text{ m}}{1000\text{ mm}}\right)} \approx 0.0012$$

From App. 17.B, $f = 0.022$. At 6 L/s, the friction loss in the line from Eq. 18.6 is

$$h_f = \frac{fL\text{v}^2}{2Dg} = \frac{(0.022)(42.67\text{ m})\left(2.77\ \frac{\text{m}}{\text{s}}\right)^2}{(2)(52.50\text{ mm})\left(\frac{1\text{ m}}{1000\text{ mm}}\right)\left(9.81\ \frac{\text{m}}{\text{s}^2}\right)} = 6.99\text{ m}$$

At 6 L/s, the velocity head from Eq. 18.7 is

$$h_\text{v} = \frac{\text{v}^2}{2g} = \frac{\left(2.77\ \frac{\text{m}}{\text{s}}\right)^2}{(2)\left(9.81\ \frac{\text{m}}{\text{s}^2}\right)} = 0.39\text{ m}$$

In general, the friction head and velocity head are proportional to v^2 and Q^2.

$$h_f = (6.99\text{ m})\left(\frac{Q_2}{6\ \frac{\text{L}}{\text{s}}}\right)^2$$

$$h_\text{v} = (0.39\text{ m})\left(\frac{Q_2}{6\ \frac{\text{L}}{\text{s}}}\right)^2$$

(Note that the 2.3 m dimension is included in the 6.3 m dimension.)

The total system head is

$$h = h_z + h_\text{v} + h_f = 6.3\text{ m} + (0.39\text{ m} + 6.99\text{ m})\left(\frac{Q_2}{6\ \frac{\text{L}}{\text{s}}}\right)^2$$

From this equation the following table for the system head can be generated.

Q_2 (L/s)	h (m)
0	6.3
0.6	6.37
1.2	6.60
1.8	6.96
2.4	7.48
3.2	8.40
3.6	8.96
4.4	10.27
4.8	11.02
5.7	12.96
6.5	14.96
7.0	16.35
7.5	17.83

The intersection point of the system curve and the pump curve will define the operating flow rate.

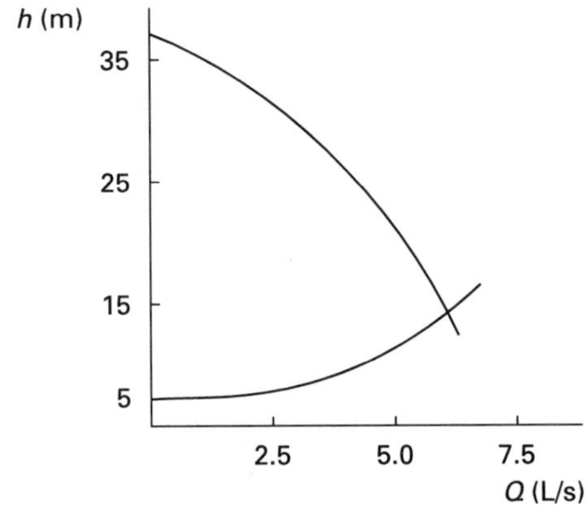

(a) The flow rate is $\boxed{6.2 \text{ L/s.}}$

The answer is (D).

(b) The intersection point is not in an efficient range of the pump because it is so far down on the system curve that the pumping efficiency will be low.

$\boxed{\text{A different pump should be used.}}$

The answer is (A).

13. From Eq. 18.52,

$$P_2 = P_1\left(\frac{\rho_2 n_2^3 D_2^5}{\rho_1 n_1^3 D_1^5}\right)$$
$$= P_1\left(\frac{n_2}{n_1}\right)^3 \quad [\rho_2 = \rho_1 \text{ and } D_2 = D_1]$$

Customary U.S. Solution

$$P_2 = (0.5 \text{ hp})\left(\frac{2000 \text{ rpm}}{1750 \text{ rpm}}\right)^3 = \boxed{0.746 \text{ hp}}$$

The answer is (D).

SI Solution

$$P_2 = P_1\left(\frac{n_2}{n_1}\right)^3$$
$$= (0.37 \text{ kW})\left(\frac{2000 \text{ rpm}}{1750 \text{ rpm}}\right)^3$$
$$= \boxed{0.55 \text{ kW}}$$

The answer is (D).

14. (a) Random values of Q are chosen, and the corresponding values of H are determined by the formula $H = 30 + 2Q^2$.

Q (ft^3/sec)	H (ft)
0	30
2.5	42.5
5	80
7.5	142.5
10	230
15	480
20	830
25	1280
30	1830

The intersection of the system curve and the 1400 rpm pump curve defines the operating point at that rpm.

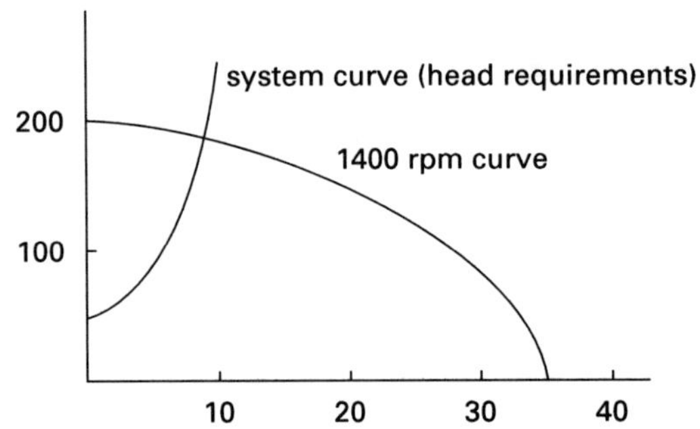

From the intersection of the graphs, at 1400 rpm the flow rate is approximately 9 ft^3/sec and the corresponding head is $30 + (2)(9)^2 \approx 192$ ft.

$$Q = \left(9 \ \frac{\text{ft}^3}{\text{sec}}\right)\left(448.8 \ \frac{\frac{\text{gal}}{\text{min}}}{\frac{\text{ft}^3}{\text{sec}}}\right) = \boxed{4039 \text{ gal/min}}$$

The answer is (C).

(b) From Table 18.5, the hydraulic horsepower is

$$\text{WHP} = \frac{h_A \dot{V}(\text{SG})}{8.814}$$
$$= \frac{(192 \text{ ft})\left(9 \ \frac{\text{ft}^3}{\text{sec}}\right)(1)}{8.814 \ \frac{\text{ft}^4}{\text{hp-sec}}}$$
$$= 196 \text{ hp}$$

From Eq. 18.28(b), the specific speed is

$$n_s = \frac{n\sqrt{Q}}{h_A^{0.75}} = \frac{(1400 \text{ rpm})\sqrt{4039 \ \frac{\text{gal}}{\text{min}}}}{(192 \text{ ft})^{0.75}}$$
$$= 1725$$

From Fig. 18.8 with curve E, $\eta \approx 87\%$.

The minimum pump motor power should be

$$\frac{196 \text{ hp}}{0.87} = \boxed{225 \text{ hp}}$$

The answer is (C).

(c) From Eq. 18.41,

$$Q_2 = Q_1\left(\frac{n_2}{n_1}\right) = \left(4039 \ \frac{\text{gal}}{\text{min}}\right)\left(\frac{1200 \text{ rpm}}{1400 \text{ rpm}}\right)$$
$$= \boxed{3462 \text{ gal/min}}$$

The answer is (B).

15. Since turbines are essentially pumps running backward, use Table 18.5.

$$\Delta p = \left(30 \ \frac{\text{lbf}}{\text{in}^2} - 5 \ \frac{\text{lbf}}{\text{in}^2}\right)\left(144 \ \frac{\text{in}^2}{\text{ft}^2}\right) = 3600 \text{ lbf/ft}^2$$

$$P = \frac{\left(3600 \ \frac{\text{lbf}}{\text{ft}^2}\right)\left(100 \ \frac{\text{ft}^3}{\text{sec}}\right)}{550 \ \frac{\text{ft-lbf}}{\text{hp-sec}}} = \boxed{654.5 \text{ hp}}$$

The answer is (C).

16. The flow rate is

$$\gamma\dot{V} = \left(62.4 \ \frac{\text{lbf}}{\text{ft}^3}\right)\left(1000 \ \frac{\text{ft}^3}{\text{sec}}\right) = 6.24 \times 10^4 \text{ lbf/sec}$$

The head available for work is

$$\Delta h = 625 \text{ ft} - 58 \text{ ft} = 567 \text{ ft}$$

Use Table 18.5. The power is

$$P = (0.89)\left(6.24 \times 10^4 \ \frac{\text{lbf}}{\text{sec}}\right)(567 \text{ ft})$$
$$= 3.149 \times 10^7 \text{ ft-lbf/sec}$$

Convert from hp to kW.

$$\frac{\left(3.149 \times 10^7 \ \frac{\text{ft-lbf}}{\text{sec}}\right)\left(0.7457 \ \frac{\text{kW}}{\text{hp}}\right)}{550 \ \frac{\text{ft-lbf}}{\text{hp-sec}}}$$
$$= \boxed{4.27 \times 10^4 \text{ kW (43 MW)}}$$

The answer is (C).

17. *Customary U.S. Solution*

(a) From App. 14.A, the density of water at 60°F is 62.37 lbm/ft^3. From Eq. 18.5(b), the head dropped is

$$h = \left(\frac{\Delta p}{\rho}\right) \times \left(\frac{g_c}{g}\right)$$
$$= \left(\frac{\left(500 \ \frac{\text{lbf}}{\text{in}^2} - 30 \ \frac{\text{lbf}}{\text{in}^2}\right)\left(144 \ \frac{\text{in}^2}{\text{ft}^2}\right)}{62.37 \ \frac{\text{lbm}}{\text{ft}^3}}\right) \times \left(\frac{32.2 \ \frac{\text{ft-lbm}}{\text{lbf-sec}^2}}{32.2 \ \frac{\text{ft}}{\text{sec}^2}}\right)$$
$$= 1085 \text{ ft}$$

From Eq. 18.55(b), the specific speed of a turbine is

$$n_s = \frac{n\sqrt{P}}{(h_t)^{1.25}}$$
$$= \frac{(1750 \text{ rpm})\sqrt{250 \text{ hp}}}{(1085 \text{ ft})^{1.25}}$$
$$= 4.443$$

Since the lowest suggested value of n_s for a reaction turbine is 10,

$$\boxed{\text{recommend an impulse (Pelton) turbine.}}$$

The answer is (A).

(b) The flow rate, $\dot{V}$, is

$$\dot{V} = A\text{v} = \left(\frac{\pi}{4}\right)\left(\frac{4 \text{ in}}{12 \ \frac{\text{in}}{\text{ft}}}\right)^2\left(35 \ \frac{\text{ft}}{\text{sec}}\right)$$
$$= 3.054 \text{ ft}^3/\text{sec}$$

The flow rate, considering the blade's moving away at 10 ft/sec, is

$$\dot{V}' = \frac{\left(35 \ \frac{\text{ft}}{\text{sec}} - 10 \ \frac{\text{ft}}{\text{sec}}\right)\left(3.054 \ \frac{\text{ft}^3}{\text{sec}}\right)}{35 \ \frac{\text{ft}}{\text{sec}}}$$
$$= 2.181 \text{ ft}^3/\text{sec}$$

The forces in the x-direction and the y-direction are

$$F_x = \left(\frac{\dot{V}'\rho}{g_c}\right)(\text{v}_j - \text{v}_b)(\cos\theta - 1)$$

$$= \left(\frac{\left(2.181\ \frac{\text{ft}^3}{\text{sec}}\right)\left(62.37\ \frac{\text{lbm}}{\text{ft}^3}\right)}{32.2\ \frac{\text{lbm-ft}}{\text{lbf-sec}^2}}\right) \times \left(35\ \frac{\text{ft}}{\text{sec}} - 10\ \frac{\text{ft}}{\text{sec}}\right)(\cos 80^\circ - 1)$$

$$= -87.27\ \text{lbf}$$

$$F_y = \left(\frac{\dot{V}'\rho}{g_c}\right)(\text{v}_j - \text{v}_b)\sin\theta$$

$$= \left(\frac{\left(2.181\ \frac{\text{ft}^3}{\text{sec}}\right)\left(62.37\ \frac{\text{lbm}}{\text{ft}^3}\right)}{32.2\ \frac{\text{lbm-ft}}{\text{lbf-sec}^2}}\right) \times \left(35\ \frac{\text{ft}}{\text{sec}} - 10\ \frac{\text{ft}}{\text{sec}}\right)\sin 80^\circ$$

$$= 104.0\ \text{lbf}$$

$$R = \sqrt{F_x^2 + F_y^2}$$

$$= \sqrt{(-87.27\ \text{lbf})^2 + (104.0\ \text{lbf})^2} = \boxed{135.8\ \text{lbf}}$$

The answer is (B).

SI Solution

(a) From App. 14.B, the density of water at 16°C is 998.83 kg/m^3. From Eq. 18.5(a), the head dropped is

$$h = \frac{\Delta p}{\rho g}$$

$$= \frac{(3.5\ \text{MPa})\left(10^6\ \frac{\text{Pa}}{\text{MPa}}\right) - (210\ \text{kPa})\left(1000\ \frac{\text{Pa}}{\text{kPa}}\right)}{\left(998.83\ \frac{\text{kg}}{\text{m}^3}\right)\left(9.81\ \frac{\text{m}}{\text{s}^2}\right)}$$

$$= 335.8\ \text{m}$$

From Eq. 18.55(a), the specific speed of a turbine is

$$n_s = \frac{n\sqrt{P}}{(h_t)^{1.25}}$$

$$= \frac{(1750\ \text{rpm})\sqrt{185\ \text{kW}}}{(335.8\ \text{m})^{1.25}}$$

$$= 16.56$$

Since the lowest suggested value of n_s for a reaction turbine is 38,

recommend an impulse (Pelton) turbine.

The answer is (A).

(b) The flow rate, $\dot{V}$, is

$$\dot{V} = A\text{v} = \left(\frac{\pi}{4}\right)\left((100\ \text{mm})\left(\frac{1\ \text{m}}{1000\ \text{mm}}\right)\right)^2\left(10.5\ \frac{\text{m}}{\text{s}}\right)$$

$$= 0.08247\ \text{m}^3/\text{s}$$

The flow rate, considering the blade's moving away at 3 m/s, is

$$\dot{V}' = \frac{\left(10.5\ \frac{\text{m}}{\text{s}} - 3\ \frac{\text{m}}{\text{s}}\right)\left(0.08247\ \frac{\text{m}^3}{\text{s}}\right)}{10.5\ \frac{\text{m}}{\text{s}}}$$

$$= 0.05891\ \text{m}^3/\text{s}$$

The forces in the x-direction and the y-direction are

$$F_x = \dot{V}'\rho(\text{v}_j - \text{v}_b)(\cos\theta - 1)$$

$$= \left(0.05891\ \frac{\text{m}^3}{\text{s}}\right)\left(998.83\ \frac{\text{kg}}{\text{m}^3}\right) \times \left(10.5\ \frac{\text{m}}{\text{s}} - 3\ \frac{\text{m}}{\text{s}}\right)(\cos 80^\circ - 1)$$

$$= -364.7\ \text{N}$$

$$F_y = \dot{V}'\rho(\text{v}_j - \text{v}_b)\sin\theta$$

$$= \left(0.05891\ \frac{\text{m}^3}{\text{s}}\right)\left(998.83\ \frac{\text{kg}}{\text{m}^3}\right) \times \left(10.5\ \frac{\text{m}}{\text{s}} - 3\ \frac{\text{m}}{\text{s}}\right)\sin 80^\circ$$

$$= 434.6\ \text{N}$$

$$R = \sqrt{F_x^2 + F_y^2}$$

$$= \sqrt{(-364.7\ \text{N})^2 + (434.6\ \text{N})^2} = \boxed{567.3\ \text{N}}$$

The answer is (B).

18. *Customary U.S. Solution*

(a) The total effective head is due to the pressure head, velocity head, and tailwater head.

$$h_{\text{eff}} = h_p + h_{\text{v}} - h_{z(\text{tailwater})}$$

$$= 92.5\ \text{ft} + \frac{\left(12\ \frac{\text{ft}}{\text{sec}}\right)^2}{(2)\left(32.2\ \frac{\text{ft}}{\text{sec}^2}\right)} - (-5.26\ \text{ft})$$

$$= \boxed{100\ \text{ft}}$$

The answer is (C).

(b) From Table 18.5, the theoretical hydraulic horsepower is

$$P_{\text{th}} = \frac{h_A\dot{V}(\text{SG})}{8.814} = \frac{(100\ \text{ft})\left(25\ \frac{\text{ft}^3}{\text{sec}}\right)(1)}{8.814\ \frac{\text{ft}^4}{\text{hp-sec}}}$$

$$= 283.6\ \text{hp}$$

The overall turbine efficiency is

$$\eta = \frac{P_{\text{brake}}}{P_{\text{th}}} = \frac{250 \text{ hp}}{283.6 \text{ hp}} = \boxed{0.882 \quad (88.2\%)}$$

The answer is (B).

(c) From Eq. 18.42 (the affinity laws),

$$\begin{aligned} n_2 &= n_1\sqrt{\frac{h_2}{h_1}} = (610 \text{ rpm})\sqrt{\frac{225 \text{ ft}}{100 \text{ ft}}} \\ &= \boxed{915 \text{ rpm}} \end{aligned}$$

The answer is (B).

(d) Combine Eqs. 18.42 and 18.43.

$$\begin{aligned} P_2 &= P_1\left(\frac{h_2}{h_1}\right)^{1.5} \\ &= (250 \text{ hp})\left(\frac{225 \text{ ft}}{100 \text{ ft}}\right)^{1.5} \\ &= \boxed{843.8 \text{ hp}} \end{aligned}$$

The answer is (D).

(e) Combine Eqs. 18.41 and 18.42.

$$\begin{aligned} Q_2 &= Q_1\sqrt{\frac{h_2}{h_1}} \\ &= \left(25 \ \frac{\text{ft}^3}{\text{sec}}\right)\sqrt{\frac{225 \text{ ft}}{100 \text{ ft}}} \\ &= \boxed{37.5 \text{ ft}^3/\text{sec}} \end{aligned}$$

The answer is (B).

SI Solution

(a) The total effective head is due to the pressure head, velocity head, and tailwater head.

$$\begin{aligned} h_{\text{eff}} &= h_p + h_{\text{v}} - h_{z(\text{tailwater})} \\ &= 30.8 \text{ m} + \frac{\left(3.6 \ \frac{\text{m}}{\text{s}}\right)^2}{(2)\left(9.81 \ \frac{\text{m}}{\text{s}^2}\right)} - (-1.75 \text{ m}) \\ &= \boxed{33.21 \text{ m}} \end{aligned}$$

The answer is (C).

(b) From Table 18.6, the theoretical hydraulic kilowatt is

$$\begin{aligned} P_{\text{th}} &= \frac{9.81 h_A Q(\text{SG})}{1000} \\ &= \frac{\left(9.81 \ \frac{\text{m}}{\text{s}^2}\right)(33.21 \text{ m})\left(700 \ \frac{\text{L}}{\text{s}}\right)(1)}{1000 \ \frac{\text{W}}{\text{kW}}} \\ &= 228.1 \text{ kW} \end{aligned}$$

The overall turbine efficiency is

$$\begin{aligned} \eta &= \frac{P_{\text{brake}}}{P_{\text{th}}} = \frac{185 \text{ kW}}{228.1 \text{ kW}} \\ &= \boxed{0.811 \quad (81.1\%)} \end{aligned}$$

The answer is (A).

(c) From Eq. 18.42 (the affinity laws),

$$\begin{aligned} h_2 &= n_1\sqrt{\frac{h_2}{h_1}} = (610 \text{ rpm})\sqrt{\frac{75 \text{ m}}{33.21 \text{ m}}} \\ &= \boxed{917 \text{ rpm}} \end{aligned}$$

The answer is (B).

(d) Combine Eqs. 18.42 and 18.43.

$$\begin{aligned} P_2 &= P_1\left(\frac{\text{h}_2}{h_1}\right)^{1.5} = (185 \text{ kW})\left(\frac{75 \text{ m}}{33.21 \text{ m}}\right)^{1.5} \\ &= \boxed{627.3 \text{ kW}} \end{aligned}$$

The answer is (D).

(e) Combine Eqs. 18.41 and 18.42.

$$\begin{aligned} Q_2 &= Q_1\sqrt{\frac{h_2}{h_1}} = \left(700 \ \frac{\text{L}}{\text{s}}\right)\sqrt{\frac{75 \text{ m}}{33.21 \text{ m}}} \\ &= \boxed{1051.9 \text{ L/s}} \end{aligned}$$

The answer is (B).

19 Open Channel Flow

PRACTICE PROBLEMS

Rectangular Channels

1. A wooden flume ($n = 0.012$) with a rectangular cross section is 2 ft wide. The flume carries 3 ft^3/sec of water down a 1% slope. What is the depth of flow?

(A) 0.3 ft
(B) 0.4 ft
(C) 0.5 ft
(D) 0.6 ft

2. A rectangular open channel is to be constructed with smooth concrete on a slope of 0.08. The design flow rate is 17 m^3/s. The channel design must be optimum. What are the channel dimensions?

(A) 0.6 m deep; 2.1 m wide
(B) 0.8 m deep; 1.6 m wide
(C) 1.0 m deep; 1.3 m wide
(D) 1.2 m deep; 1.1 m wide

3. (*Time limit: one hour*) A dam's outfall structure has three rectangular inlets sized 1 ft high by 2 ft wide as shown. The orifice coefficient for each inlet is 0.7. The vertical riser joins with a 100 ft long square box culvert ($n = 0.013$) placed on a 0.05 slope. The box culvert drains into a tailwater basin whose water level remains constant at the top of the culvert opening. During steady-state operation, the water level in the vertical barrel of the structure is 4 ft above the top of the box culvert. The orifice coefficient of the box culvert entrance is 2/3. What are the dimensions of the box culvert?

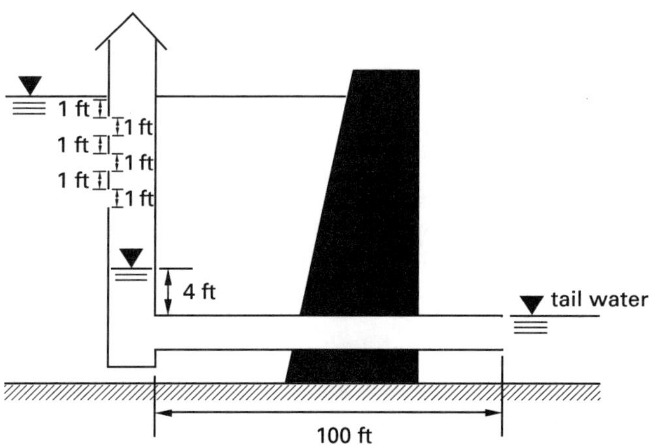

(A) 1.5 ft by 1.5 ft
(B) 2.0 ft by 2.0 ft
(C) 2.5 ft by 2.5 ft
(D) 4.0 ft by 4.0 ft

Circular Channels

4. A 24 in diameter pipe ($n = 0.013$) was installed 30 years ago on a 0.001 slope. Recent tests indicate that the full-flow capacity of the pipe is 6.0 ft^3/sec.

(a) What was the original full-flow capacity?

(A) 4.1 ft^3/sec
(B) 6.3 ft^3/sec
(C) 6.7 ft^3/sec
(D) 7.2 ft^3/sec

(b) What was the original full-flow velocity?

(A) 1.8 ft/sec
(B) 2.3 ft/sec
(C) 2.7 ft/sec
(D) 3.7 ft/sec

(c) What is the current full-flow velocity?

(A) 1.9 ft/sec
(B) 2.4 ft/sec
(C) 3.1 ft/sec
(D) 5.8 ft/sec

(d) What is the current Manning coefficient?

(A) 0.011
(B) 0.013
(C) 0.016
(D) 0.024

5. A circular sewer is to be installed on a 1% grade. Its Manning coefficient is $n = 0.013$. The maximum full-flow capacity is to be 3.5 ft^3/sec.

(a) What size pipe would you recommend?

(A) 8 in
(B) 12 in
(C) 16 in
(D) 18 in

(b) Assuming that the diameter chosen is 12 in, what is the full-flow capacity?

(A) 2.7 ft^3/sec
(B) 3.3 ft^3/sec
(C) 3.6 ft^3/sec
(D) 4.5 ft^3/sec

(c) What is the full-flow velocity?

(A) 2.9 ft/sec
(B) 3.3 ft/sec
(C) 4.1 ft/sec
(D) 4.6 ft/sec

(d) What is the depth of flow when the flow is 0.7 ft^3/sec?

(A) 4 in
(B) 7 in
(C) 9 in
(D) 12 in

(e) What minimum velocity will prevent solids from settling out in the sewer?

(A) 1 to 1.5 ft/sec
(B) 1.5 to 2.0 ft/sec
(C) 2.0 to 2.5 ft/sec
(D) 2.5 to 3.0 ft/sec

6. A 4 ft diameter concrete storm drain on a 0.02 slope carries water at a depth of 1.5 ft. Manning's roughness coefficient for a full storm drain is $n_{full} = 0.013$, but n varies with depth.

(a) What is the velocity of the water in the storm drain?

(A) 3 ft/sec
(B) 6 ft/sec
(C) 8 ft/sec
(D) 11 ft/sec

(b) What is the maximum velocity that can be achieved in the storm drain?

(A) 8 ft/sec
(B) 13 ft/sec
(C) 17 ft/sec
(D) 21 ft/sec

(c) What is the maximum capacity of the storm drain?

(A) 130 ft^3/sec
(B) 160 ft^3/sec
(C) 190 ft^3/sec
(D) 210 ft^3/sec

7. A circular storm sewer ($n = 0.012$) is being designed to carry a peak flow of 5 ft^3/sec. To allow for excess capacity, the depth at peak flow is to be 75% of the sewer diameter. The sewer is to be installed in a bed with a slope of 2%. What is the required sewer diameter?

(A) 12 in
(B) 16 in
(C) 20 in
(D) 24 in

Trapezoidal Channels

8. The trapezoidal channel shown has a Manning coefficient of $n = 0.013$ and is laid at a slope of 0.002. The depth of flow is 2.0 ft. What is the flow rate?

(A) 120 ft^3/sec
(B) 150 ft^3/sec
(C) 180 ft^3/sec
(D) 210 ft^3/sec

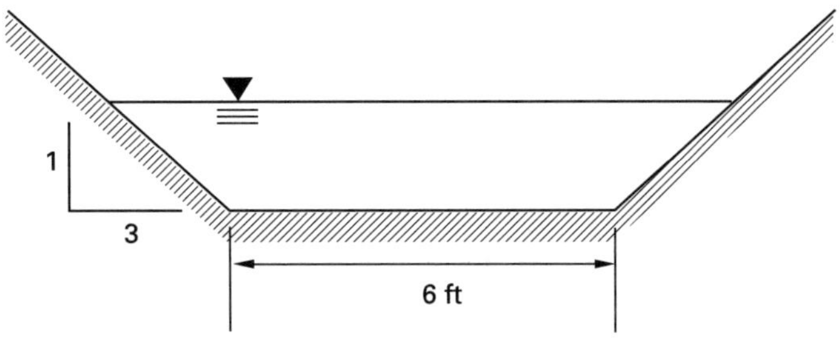

9. A trapezoidal open channel is to be constructed with smooth concrete on a slope of 0.08. The design flow rate is 17 m^3/s. The channel design must be optimum. What are the channel dimensions?

(A) depth 0.8 m; base 1.0 m
(B) depth 1.0 m; base 1.2 m
(C) depth 1.2 m; base 1.4 m
(D) depth 1.4 m; base 1.8 m

Weirs

10. A sharp-crested rectangular weir with two end contractions is 5 ft wide. The weir height is 6 ft. The head over the weir is 0.43 ft. What is the flow rate?

(A) 3.9 ft^3/sec
(B) 4.3 ft^3/sec
(C) 4.6 ft^3/sec
(D) 6.4 ft^3/sec

11. A weir is constructed with a trapezoidal opening. The base of the opening is 18 in wide, and the sides have a slope of 4:1 (vertical:horizontal). The depth of flow over the weir is 9 in. What is the rate of discharge?

(A) 1.8 ft^3/sec
(B) 3.3 ft^3/sec
(C) 5.1 ft^3/sec
(D) 8.9 ft^3/sec

12. (*Time limit: one hour*) A 17 ft by 20 ft rectangular tank is fed from the bottom and drains through a double-constricted rectangular weir into a trough. The weir opening is to be designed using the following parameters.

discharge rate:	2 MGD minimum
	4 MGD average
	8 MGD maximum
minimum tank freeboard:	4 in
minimum head over weir:	10 in
maximum weir elevation:	590 ft
tank bottom elevation:	580 ft

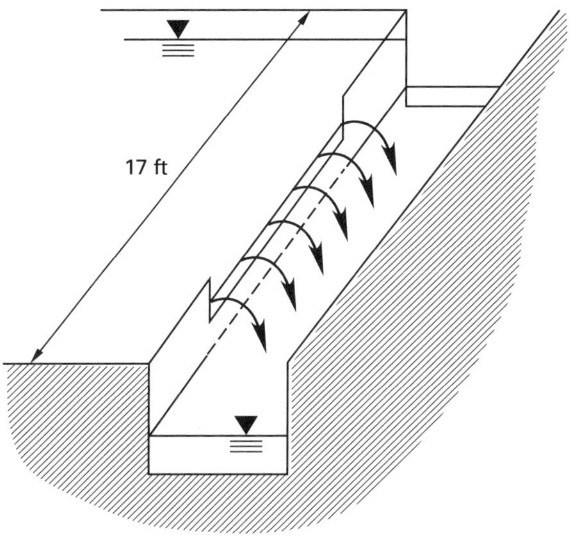

(a) What is the required weir opening length?

(A) 1.2 ft
(B) 1.4 ft
(C) 2.1 ft
(D) 2.7 ft

(b) What is the surface elevation in the tank at maximum flow?

(A) 581.9 ft
(B) 589.8 ft
(C) 592.6 ft
(D) 593.1 ft

(c) What is the surface elevation in the tank at minimum flow?

(A) 590.8 ft
(B) 591.2 ft
(C) 591.7 ft
(D) 592.0 ft

Hydraulic Jumps and Drops

13. Water flowing in a rectangular channel 5 ft wide experiences a hydraulic jump. The depth of flow upstream from the jump is 1 ft. The depth of flow downstream of the jump is 2.4 ft. What quantity is flowing?

(A) 39 ft^3/sec
(B) 45 ft^3/sec
(C) 52 ft^3/sec
(D) 57 ft^3/sec

14. A hydraulic jump forms at the toe of a spillway. The water surface levels are 0.2 ft and 6 ft above the apron before and after the jump, respectively. The velocity before the jump is 54.7 ft/sec. What is the energy loss in the jump?

(A) 41 ft-lbf/lbm
(B) 58 ft-lbf/lbm
(C) 65 ft-lbf/lbm
(D) 92 ft-lbf/lbm

15. Water flowing in a rectangular channel 6 ft wide experiences a hydraulic jump. The depth of flow upstream from the jump is 1.2 ft. The depth of flow downstream of the jump is 3.8 ft.

(a) What quantity is flowing?

(A) 75 ft^3/sec
(B) 120 ft^3/sec
(C) 180 ft^3/sec
(D) 210 ft^3/sec

(b) What is the energy loss in the jump?

(A) 1 ft-lbf/lbm
(B) 3 ft-lbf/lbm
(C) 5 ft-lbf/lbm
(D) 12 ft-lbf/lbm

Spillways

16. A spillway operates with 2 ft of head. The toe of the spillway is 40 ft below the crest of the spillway. The spillway discharge coefficient is 3.5.

(a) What is the discharge per foot of crest?

(A) 9.9 ft^3/sec
(B) 11 ft^3/sec
(C) 15 ft^3/sec
(D) 27 ft^3/sec

(b) What is the depth of flow at the toe?

(A) 0.19 ft
(B) 0.35 ft
(C) 0.78 ft
(D) 1.4 ft

17. 10,000 ft^3/sec of water flow over a 100 ft wide spillway and continue down a constant-width discharge chute on a 5% grade. The discharge chute has a Manning coefficient of $n = 0.012$.

(a) What is the normal depth of the water flowing down the chute?

(A) 1.9 ft
(B) 2.2 ft
(C) 2.7 ft
(D) 3.5 ft

(b) What is the critical depth of the water flowing down the chute?

(A) 3.8 ft
(B) 5.1 ft
(C) 6.8 ft
(D) 7.7 ft

(c) What type of flow occurs on the chute?

(A) tranquil
(B) subcritical
(C) rapid
(D) nonsteady

(d) If a hydraulic jump forms at the bottom of the 5% slope where it joins with a horizontal apron, what is the depth after the jump?

(A) 9.8 ft
(B) 12 ft
(C) 16 ft
(D) 21 ft

18. A spillway has a crest length of 60 ft. The stilling basin below the crest has a width of 60 ft and a level bottom. The spillway discharge coefficient is 3.7. The head loss in the chute is 20% of the difference in level between the reservoir surface and the stilling basin bottom.

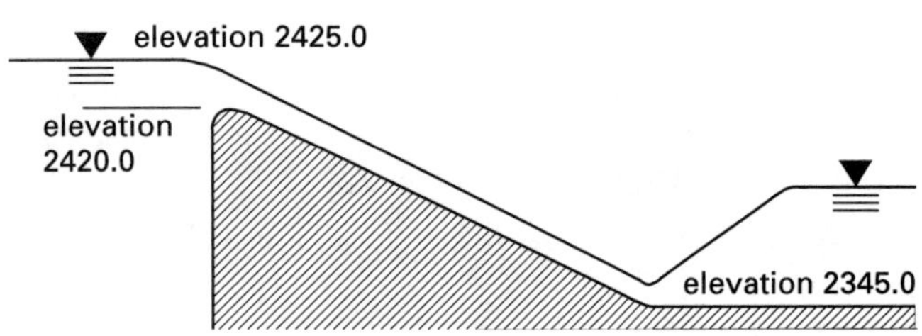

(a) What is the flow rate over the spillway?

(A) 1800 ft^3/sec
(B) 2100 ft^3/sec
(C) 2500 ft^3/sec
(D) 2900 ft^3/sec

(b) What is the depth of flow at the toe of the spillway?

(A) 0.65 ft
(B) 1.2 ft
(C) 1.6 ft
(D) 2.0 ft

(c) What tailwater depth is required to cause a hydraulic jump to form at the toe?

(A) 4 ft
(B) 7 ft
(C) 10 ft
(D) 13 ft

(d) What energy is lost in the hydraulic jump?

(A) 30 ft-lbf/lbm
(B) 50 ft-lbf/lbm
(C) 75 ft-lbf/lbm
(D) 100 ft-lbf/lbm

Parshall Flumes

19. A Parshall flume has a throat width of 6 ft. The upstream head measured from the throat floor is 18 in.

(a) What is the flow rate?

(A) 41 ft^3/sec
(B) 46 ft^3/sec
(C) 54 ft^3/sec
(D) 63 ft^3/sec

(b) Which of the following is not normally considered to be a feature of Parshall flumes?

(A) They are self-cleansing.
(B) They can be operated by unskilled personnel.
(C) Head loss is small.
(D) They can be placed in temporary installations.

Backwater Curves

20. A rectangular channel 8 ft wide carries a flow of 150 ft^3/sec. The channel slope is 0.0015, and the Manning coefficient is $n = 0.015$. A weir installed across the channel raises the depth at the weir to 6 ft.

(a) What is the normal depth of flow?

(A) 2.9 ft
(B) 3.3 ft
(C) 3.6 ft
(D) 4.5 ft

(b) What is the distance between the weir and the point where the depth is 5 ft?

(A) 610 ft
(B) 840 ft
(C) 1100 ft
(D) 1700 ft

(c) What is the distance between the points where the depth is 4 ft and 5 ft?

(A) 700 ft
(B) 800 ft
(C) 950 ft
(D) 1100 ft

Culverts

21. A 42 in diameter concrete culvert is 250 ft long and laid at a slope of 0.006. The culvert entrance is flush and square-edged. The outlet is submerged to just above the crown of the barrel, and the headwater is 5.0 ft above the crown of the culvert's inlet. What is the capacity?

(A) 60 ft^3/sec
(B) 80 ft^3/sec
(C) 100 ft^3/sec
(D) 120 ft^3/sec

22. (*Time limit: one hour*) Flow in a 6 ft diameter, newly formed concrete culvert is 150 ft^3/sec at a point where the depth of flow is 3 ft.

(a) What is the hydraulic radius?

(A) 1.1 ft
(B) 1.5 ft
(C) 2.3 ft
(D) 3.0 ft

(b) What is the slope of the pipe?

(A) 0.0043
(B) 0.0080
(C) 0.010
(D) 0.030

(c) What type of flow occurs?

(A) subcritical
(B) critical
(C) supercritical
(D) choked

(d) What is the hydraulic radius if the pipe flows full?

(A) 1.5 ft
(B) 3.0 ft
(C) 4.0 ft
(D) 6.0 ft

(e) What is the flow quantity if the culvert flows full?

(A) 100 ft^3/sec
(B) 120 ft^3/sec
(C) 200 ft^3/sec
(D) 300 ft^3/sec

(f) If the flow is 150 ft^3/sec, what is the critical depth?

(A) 1.1 ft
(B) 1.5 ft
(C) 3.3 ft
(D) 4.3 ft

(g) What is the critical slope?

(A) 0.002
(B) 0.003
(C) 0.004
(D) 0.05

(h) If the exit depth is less than critical, which culvert flow types are most probable?

(A) 1, 5
(B) 1, 3, 5
(C) 1, 2, 3
(D) 5 only

(i) What is the Froude number for a depth of flow of 3 ft?

(A) 0.86
(B) 0.99
(C) 1.22
(D) 2.84

(j) After the water exits the culvert, it flows at a velocity of 15 ft/sec and a depth of 2 ft. Assuming a hydraulic jump is possible, what will be the depth after the jump?

(A) 2.5 ft
(B) 3.0 ft
(C) 3.3 ft
(D) 4.4 ft

23. (*Time limit: one hour*) A dam spills 2000 m^3/s of water over its crest as shown. The crest height is 30 m. The width of the dam is 150 m. The depth at point B is 0.6 m. The stilling basin, constructed of rough concrete, is rectangular in shape, 150 m in width, and has a slope of 0.002.

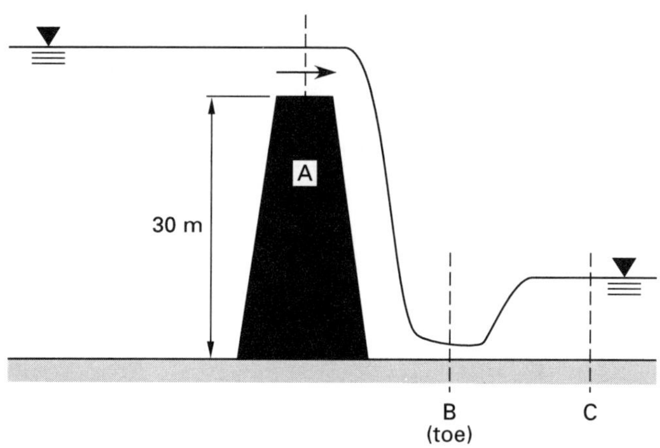

(a) What is the head over the crest?

(A) 0.8 m
(B) 1.2 m
(C) 2.2 m
(D) 3.4 m

(b) What is the total upstream specific energy head?

(A) 32.2 m
(B) 34.1 m
(C) 36.1 m
(D) 38.0 m

(c) What is the total specific energy head at point B?

(A) 25.8 m
(B) 32.2 m
(C) 34.1 m
(D) 43.3 m

(d) What is the friction head loss between points A and B?

(A) 2.2 m
(B) 4.2 m
(C) 8.3 m
(D) 9.8 m

(e) What is the normal downstream depth?

(A) 2.0 m
(B) 2.6 m
(C) 5.6 m
(D) 10.0 m

(f) What is the critical depth?

(A) 2.0 m
(B) 2.6 m
(C) 7.0 m
(D) 8.3 m

(g) At what point will a hydraulic jump occur?

(A) on the spillway slope
(B) at point B
(C) downstream of point B
(D) nowhere (a hydraulic jump will not occur)

(h) The stilling basin feeds into a natural channel composed of graded material from silt to cobble size. What is the maximum velocity in this channel to prevent erosion?

(A) 1.2 m/s
(B) 6.7 m/s
(C) 15.0 m/s
(D) 26.0 m/s

(i) If the natural channel has the same geometry as the stilling basin (i.e., rectangular with a width of 150 m) and the depth remains the same, what slope is required to maintain the maximum velocity and prevent erosion?

(A) 0.0001
(B) 0.0003
(C) 0.002
(D) 0.02

(j) Assuming that the slope in the natural channel remains at 0.002, what is the friction loss over a 1000 m reach?

(A) 0.1 m
(B) 0.6 m
(C) 1.5 m
(D) 2.0 m

SOLUTIONS

1.

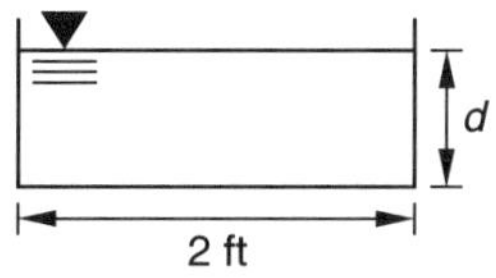

$$R = \frac{2d}{2+2d} = \frac{d}{1+d}$$

From Eq. 19.13,

$$Q = \left(\frac{1.49}{n}\right) AR^{\frac{2}{3}}\sqrt{S}$$

$$3\ \frac{\text{ft}^3}{\text{sec}} = \left(\frac{1.49}{0.012}\right)(2d)\left(\frac{d}{d+1}\right)^{\frac{2}{3}}\sqrt{0.01}$$

$$0.120805 = \left(\frac{d^{\frac{5}{2}}}{d+1}\right)^{\frac{2}{3}}$$

$$0.042 = \frac{d^{\frac{5}{2}}}{d+1}$$

By trial and error, $d = \boxed{0.314\ \text{ft.}}$

The answer is (A).

2. From App. 19.A, $n = 0.011$. $d = w/2$ for an optimum channel.

$$A = wd = \frac{w^2}{2}$$

$$P = w + 2\left(\frac{w}{2}\right) = 2w$$

$$R = \frac{A}{P} = \frac{\frac{w^2}{2}}{2w} = \frac{w}{4}$$

$$17\ \frac{\text{m}^3}{\text{s}} = \left(\frac{w^2}{2}\right)\left(\frac{1.0}{0.011}\right)\left(\frac{w}{4}\right)^{\frac{2}{3}}\sqrt{0.08}$$

$$d = \boxed{0.79\ \text{m}}$$

$$w = \boxed{1.57\ \text{m}}$$

The answer is (B).

3.

$$Q_o = C_d A_o \sqrt{2gh}$$

$$Q_1 = (0.7)(2\ \text{ft}^2)\sqrt{(2)\left(32.2\ \frac{\text{ft}}{\text{sec}^2}\right)(1.5\ \text{ft})}$$

$$= (11.235)\sqrt{1.5}\quad [\text{top inlet}]$$

$$Q_1 + Q_2 + Q_3 = (11.235)\left(\sqrt{1.5} + \sqrt{3.5} + \sqrt{5.5}\right)$$

$$= 61.13\ \text{ft}^3/\text{sec}$$

Use Eq. 19.100 since C_d is given. Culvert size (and hence R) is unknown. To get a trial value, initially disregard friction.

$$H = 4\ \text{ft} + (100\ \text{ft})(0.05) = 9\ \text{ft}$$

$$\text{v} = \sqrt{2gh} = \sqrt{(2)\left(32.2\ \frac{\text{ft}}{\text{sec}^2}\right)(9\ \text{ft})}$$

$$= 24.07\ \text{ft/sec}$$

$$A = \frac{Q}{C_d\text{v}} = \frac{61.13\ \frac{\text{ft}^3}{\text{sec}}}{\left(\frac{2}{3}\right)\left(24.07\ \frac{\text{ft}}{\text{sec}}\right)} = 3.81\ \text{ft}^2$$

$$\text{box size} = \sqrt{3.81\ \text{ft}^2} = 1.95\ \text{ft}$$

(Round to 2¼ ft^2 to account for friction.)

Use Eq. 19.100.

$$R = \frac{(2.25\ \text{ft})^2}{(4)(2.25\ \text{ft})} = 0.5625\ \text{ft}$$

$$\text{v} = \sqrt{\frac{9\ \text{ft}}{\frac{1+0}{(2)\left(32.2\ \frac{\text{ft}}{\text{sec}^2}\right)} + \frac{(0.013)^2(100\ \text{ft})}{(2.21)(0.5625\ \text{ft})^{\frac{4}{3}}}}}$$

$$= 16.77\ \text{ft/sec}$$

$$\text{box size} = \sqrt{\frac{61.13\ \frac{\text{ft}^3}{\text{sec}}}{\left(\frac{2}{3}\right)\left(16.77\ \frac{\text{ft}}{\text{sec}}\right)}}$$

$$= \boxed{2.34\ \text{ft}\quad [28\ \text{in or more}]}$$

The answer is (C).

4. From Eq. 19.13,

$$\text{(a)}\quad Q_o = \left(\frac{1.49}{0.013}\right)\left(\frac{\pi}{4}\right)\left(\frac{24\ \text{in}}{12\ \frac{\text{in}}{\text{ft}}}\right)^2\left[\frac{24\ \text{in}}{\left(12\ \frac{\text{in}}{\text{ft}}\right)(4)}\right]^{\frac{2}{3}}$$

$$\times\sqrt{0.001}$$

$$= \boxed{7.173\ \text{ft}^3/\text{sec}}$$

The answer is (D).

$$\text{(b)}\qquad \text{v}_o = \frac{Q_o}{A} = \frac{7.173\ \frac{\text{ft}^3}{\text{sec}}}{\left(\frac{\pi}{4}\right)(2\ \text{ft})^2}$$

$$= \boxed{2.28\ \text{ft/sec}}$$

The answer is (B).

(c)

$$\text{v}_{\text{current}} = \frac{6.00\ \frac{\text{ft}^3}{\text{sec}}}{\left(\frac{\pi}{4}\right)(2\ \text{ft})^2} = \boxed{1.91\ \text{ft/sec}}$$

The answer is (A).

(d) Since Q is inversely proportional to n,

$$n_{\text{current}} = \left(\frac{Q_o}{Q_{\text{current}}}\right) n_o = \left(\frac{7.173\ \frac{\text{ft}^3}{\text{sec}}}{6\ \frac{\text{ft}^3}{\text{sec}}}\right)(0.013) = \boxed{0.01554}$$

The answer is (C).

5. (a) From Eq. 19.16,

$$D = (1.33)\left[\frac{\left(3.5\ \frac{\text{ft}^3}{\text{sec}}\right)(0.013)}{\sqrt{0.01}}\right]^{\frac{3}{8}} = 0.99\ \text{ft}$$

$$\boxed{\text{Use 12 in pipe.}}$$

The answer is (B).

(b) From Eq. 19.13,

$$Q = \left(\frac{1.49}{0.013}\right)\left(\frac{\pi}{4}\right)(1\ \text{ft})^2\left(\frac{1\ \text{ft}}{4}\right)^{\frac{2}{3}}\sqrt{0.01} = \boxed{3.57\ \text{ft}^3/\text{sec}}$$

The answer is (C).

(c)

$$\text{v} = \frac{Q}{A} = \frac{3.57\ \frac{\text{ft}^3}{\text{sec}}}{\left(\frac{\pi}{4}\right)(1\ \text{ft})^2} = \boxed{4.55\ \text{ft/sec}}$$

The answer is (D).

(d)

$$\frac{Q}{Q_{\text{full}}} = \frac{0.7\ \frac{\text{ft}^3}{\text{sec}}}{3.57\ \frac{\text{ft}^3}{\text{sec}}} \approx 0.2$$

From App. 19.C, letting n vary with depth,

$$\frac{d}{D} \approx 0.35$$

$$d = (0.35)(12\ \text{in}) = \boxed{4.2\ \text{in}}$$

The answer is (A).

(e) To be self-cleansing, $\text{v} = \boxed{2\text{ to }2.5\ \text{ft/sec.}}$

The answer is (C).

6. *Full flowing:*

$$R = \frac{D}{4} = \frac{4\ \text{ft}}{4} = 1\ \text{ft}$$

From Eqs. 19.12 and 19.13,

$$\text{v} = \left(\frac{1.49}{0.013}\right)(1\ \text{ft})^{\frac{2}{3}}\sqrt{0.02} = 16.21\ \text{ft/sec}$$

$$Q = A\text{v} = \left(\frac{\pi}{4}\right)(4\ \text{ft})^2\left(16.21\ \frac{\text{ft}}{\text{sec}}\right) = 203.7\ \text{ft}^3/\text{sec}$$

Actual:

$$\frac{d}{D} = \frac{1.5\ \text{ft}}{4\ \text{ft}} = 0.375$$

From App. 19.C, assuming that n varies with depth,

$$\frac{\text{v}}{\text{v}_{\text{full}}} = 0.68$$

(a) $\text{v} = (0.68)\left(16.21\ \frac{\text{ft}}{\text{sec}}\right) = \boxed{11.0\ \text{ft/sec.}}$

The answer is (D).

(b) From App. 19.C, the maximum value of $\text{v}/\text{v}_{\text{full}}$ is 1.04 (at $d/D \approx 0.9$).

$$\text{v}_{\text{max}} = (1.04)\left(16.21\ \frac{\text{ft}}{\text{sec}}\right) = \boxed{16.86\ \text{ft/sec}}$$

The answer is (C).

(c) Similarly, $Q_{\text{max}}/Q_{\text{full}} = 1.02$ (at $d/D \approx 0.96$).

$$Q_{\text{max}} = (1.02)\left(203.7\ \frac{\text{ft}^3}{\text{sec}}\right) = \boxed{207.8\ \text{ft}^3/\text{sec}}$$

The answer is (D).

7. From App. 16.A,

$$A = 0.6318D^2$$

$$R = 0.3017D$$

From Eq. 19.13,

$$5\ \frac{\text{ft}^3}{\text{sec}} = (0.6318D^2)\left(\frac{1.49}{0.012}\right)(0.3017D)^{\frac{2}{3}}\sqrt{0.02}$$
$$1.0 = D^{\frac{8}{3}}$$
$$D = \boxed{1.0\ \text{ft}\ \ (12\ \text{in})}$$

The answer is (A).

8.

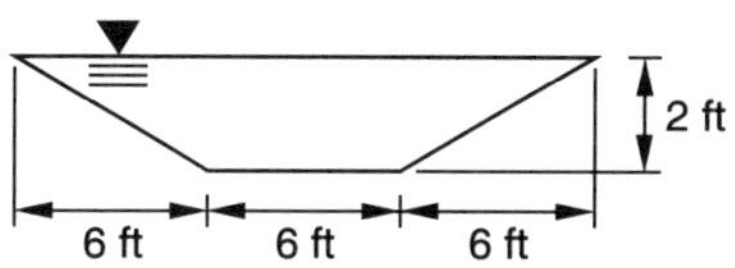

$$A = (6\ \text{ft})(2\ \text{ft}) + \frac{(2)(6\ \text{ft})(2\ \text{ft})}{2} = 24\ \text{ft}^2$$
$$P = 6\ \text{ft} + 2\sqrt{(6\ \text{ft})^2 + (2\ \text{ft})^2} = 18.65\ \text{ft}$$
$$R = \frac{24\ \text{ft}^2}{18.65\ \text{ft}} = 1.287\ \text{ft}$$
$$Q = (24\ \text{ft}^2)\left(\frac{1.49}{0.013}\right)(1.287\ \text{ft})^{\frac{2}{3}}\sqrt{0.002}$$
$$= \boxed{145.6\ \text{ft}^3/\text{sec}}$$

The answer is (B).

9. The optimum trapezoidal channel has side slopes of 60° and a depth that is twice the hydraulic radius.

From App. 19.A, $n = 0.011$ for smooth concrete.

Rearranging Eq. 19.13,

$$AR^{\frac{2}{3}} = \frac{Qn}{\sqrt{S}} = \frac{\left(17\ \frac{\text{m}^3}{\text{s}}\right)(0.011)}{\sqrt{0.08}} = 0.661$$

From Table 19.2,

$$A = \left(b + \frac{d}{\tan\theta}\right)d$$
$$R = \frac{bd\sin\theta + d^2\cos\theta}{b\sin\theta + 2d}$$

Since $R = d/2$ and $b = 2d/\sqrt{3}$,

$$AR^{\frac{2}{3}} = \left(\frac{2d}{\sqrt{3}} + \frac{d}{\tan\theta}\right)d\left(\frac{d}{2}\right)^{\frac{2}{3}} = 0.661$$

Simplifying, $d^{\frac{8}{3}} = 0.606$ with $\theta = 60°$, the optimal channel has a depth, d, of 0.828 m and a base, b, of 0.957 m.

The answer is (A).

10. From Eq. 19.52,

$$b = 5\ \text{ft} - (0.1)(2)(0.43\ \text{ft}) = 4.914\ \text{ft}$$

From Eq. 19.50,

$$C_1 = \left(0.6035 + (0.0813)\left(\frac{0.43\ \text{ft}}{6\ \text{ft}}\right) + \frac{0.000295}{6}\right) \times \left(1 + \frac{0.00361}{0.43\ \text{ft}}\right)^{\frac{3}{2}}$$
$$= (0.60938)(1.0126) = 0.617$$

From Eq. 19.49,

$$Q = \left(\tfrac{2}{3}\right)(0.617)(4.914\ \text{ft})\sqrt{(2)\left(32.2\ \frac{\text{ft}}{\text{sec}^2}\right)}(0.43\ \text{ft})^{\frac{3}{2}}$$
$$= \boxed{4.574\ \text{ft}^3/\text{sec}}$$

The answer is (C).

11. This is a Cipoletti weir. From Eq. 19.57,

$$Q = (3.367)\left(\frac{18\ \text{in}}{12\ \frac{\text{in}}{\text{ft}}}\right)\left(\frac{9\ \text{in}}{12\ \frac{\text{in}}{\text{ft}}}\right)^{\frac{3}{2}} = \boxed{3.28\ \text{ft}^3/\text{sec}}$$

The answer is (B).

12.

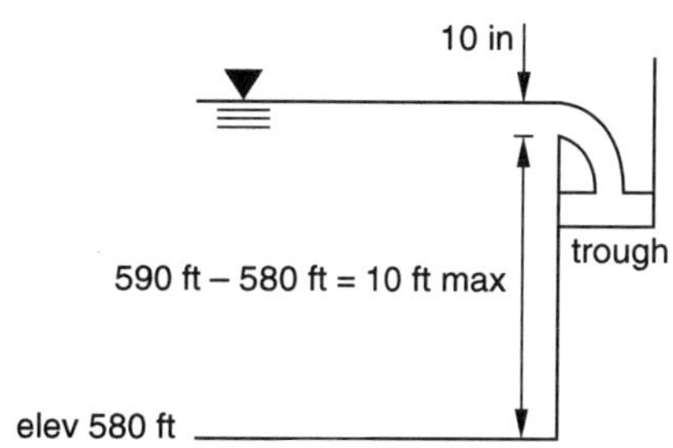

(a) Use Eq. 19.50.

$$C_1 \approx 0.602 + (0.083)\left(\frac{10\ \text{in}}{(590\ \text{ft} - 580\ \text{ft})\left(12\ \frac{\text{in}}{\text{ft}}\right)}\right)$$
$$= 0.609$$

Use Eq. 19.49 with $C_1 = 0.609$.

$$\begin{aligned} Q_{\text{MGD}} &= \left(0.6463\ \frac{\text{MGD-sec}}{\text{ft}^3}\right) Q_{\text{ft}^3/\text{sec}} \\ &= \left(0.6463\ \frac{\text{MGD-sec}}{\text{ft}^3}\right)\left(\tfrac{2}{3}\right)(0.609)bH^{\frac{3}{2}} \\ &\quad \times \sqrt{(2)\left(32.2\ \frac{\text{ft}}{\text{sec}^2}\right)} \\ &= 2.11bH^{\frac{3}{2}} \end{aligned}$$

To satisfy the minimum head requirement, the minimum flow must be used. The minimum head requirement will then be automatically satisfied at maximum flow.

At minimum flow, $Q = 2$ MGD.

$$H = \frac{10 \text{ in}}{12\ \frac{\text{in}}{\text{ft}}} = 0.833 \text{ ft}$$

$$b_{\text{effective}} = \frac{2 \text{ MGD}}{(2.11)(0.833 \text{ ft})^{\frac{3}{2}}} = 1.25 \text{ ft}$$

From Eq. 19.52,

$$b_{\text{actual}} = 1.25 \text{ ft} + (0.1)(2)(0.833 \text{ ft}) = \boxed{1.42 \text{ ft}}$$

The answer is (B).

(b) At maximum flow,

$$\begin{aligned} 8 \text{ MGD} &= (2.11)(1.25)H^{\frac{3}{2}} \\ H &\approx 2.1 \text{ ft} \end{aligned}$$

Recalculate C_1 from Eq. 19.50.

$$\begin{aligned} C_1 &\approx 0.602 + (0.083)\left(\frac{2.1 \text{ ft}}{590 \text{ ft} - 580 \text{ ft}}\right) \\ &= 0.619 \end{aligned}$$

From Eq. 19.49,

$$\begin{aligned} Q &= \left(0.6463\ \frac{\text{MGD-sec}}{\text{ft}^3}\right)\left(\tfrac{2}{3}\right)(0.619)bH^{\frac{3}{2}} \\ &\quad \times \sqrt{(2)\left(32.2\ \frac{\text{ft}}{\text{sec}^2}\right)} \\ &= 2.14bH^{\frac{3}{2}} \end{aligned}$$

Use Eq. 19.52 to calculate $b_{\text{effective}}$. Continue iterating until H stabilizes.

$$\begin{aligned} b_{\text{effective}} &= 1.42 \text{ ft} - (0.1)(2)(2.1 \text{ ft}) = 1.0 \text{ ft} \\ 8 \text{ MGD} &= (2.14)(1)H^{\frac{3}{2}} \\ H &= 2.41 \text{ ft} \\ b_{\text{effective}} &= 1.42 \text{ ft} - (0.1)(2)(2.41 \text{ ft}) = 0.94 \text{ ft} \\ 8 \text{ MGD} &= (2.14)(0.94)H^{\frac{3}{2}} \\ H &= 2.51 \text{ ft} \end{aligned}$$

$$\begin{aligned} b_{\text{effective}} &= 1.42 \text{ ft} - (0.1)(2)(2.51 \text{ ft}) = 0.918 \text{ ft} \\ 8 \text{ MGD} &= (2.14)(0.918)H^{\frac{3}{2}} \\ H &= 2.55 \text{ ft} \\ b_{\text{effective}} &= 1.42 \text{ ft} - (0.1)(2)(2.55 \text{ ft}) = 0.920 \text{ ft} \\ 8 \text{ MGD} &= (2.14)(0.920)H^{\frac{3}{2}} \\ H &= 2.55 \text{ ft} \quad \text{[acceptable]} \end{aligned}$$

Use a freeboard elevation of

$$580 \text{ ft} + 10 \text{ ft} + 2.55 \text{ ft} + \frac{4 \text{ in}}{12\ \frac{\text{in}}{\text{ft}}} = 592.88 \text{ ft}$$

The elevation at maximum flow is

$$\begin{aligned} 592.88 \text{ ft} - \text{freeboard height} &= 592.88 \text{ ft} - \frac{4 \text{ in}}{12\ \frac{\text{in}}{\text{ft}}} \\ &= \boxed{592.55 \text{ ft } (592.6 \text{ ft})} \end{aligned}$$

The answer is (C).

(c) The channel elevation is

$$580 \text{ ft} + 10 \text{ ft} = 590 \text{ ft}$$

The elevation at minimum flow is

$$590 \text{ ft} + 0.833 \text{ ft} = \boxed{590.833 \text{ ft}}$$

The answer is (A).

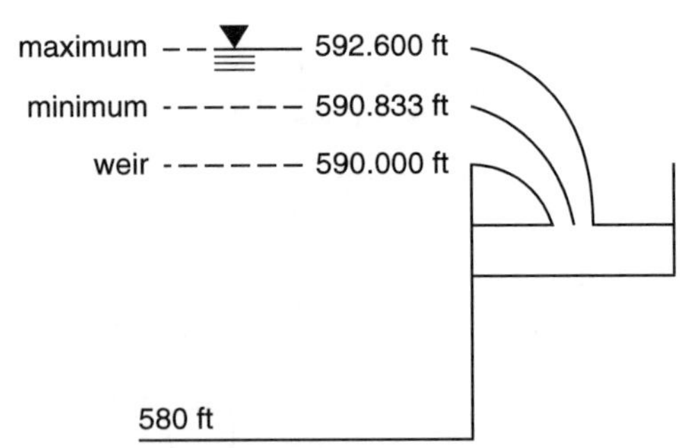

13. From Eq. 19.93,

$$\mathrm{v}_1 = \sqrt{\left(\frac{32.2\ \frac{\text{ft}}{\text{sec}^2}}{2}\right)\left(\frac{2.4\ \text{ft}}{1\ \text{ft}}\right)(1\ \text{ft} + 2.4\ \text{ft})}$$
$$= 11.46\ \text{ft/sec}$$

$$Q = A\mathrm{v} = (5\ \text{ft})(1\ \text{ft})\left(11.46\ \frac{\text{ft}}{\text{sec}}\right) = \boxed{57.30\ \text{ft}^3/\text{sec}}$$

The answer is (D).

14. From Eq. 19.8,

$$(0.2\ \text{ft})(\text{width})\left(54.7\ \frac{\text{ft}}{\text{sec}}\right) = (6\ \text{ft})(\text{width})\mathrm{v}_2$$
$$\mathrm{v}_2 = 1.82\ \text{ft/sec}$$

From Eq. 19.94,

$$\Delta E = 0.2\ \text{ft} + \frac{\left(54.7\ \frac{\text{ft}}{\text{sec}}\right)^2}{(2)\left(32.2\ \frac{\text{ft}}{\text{sec}^2}\right)} - 6\ \text{ft} - \frac{\left(1.82\ \frac{\text{ft}}{\text{sec}}\right)^2}{(2)\left(32.2\ \frac{\text{ft}}{\text{sec}^2}\right)}$$
$$= \boxed{40.61\ \text{ft (ft-lbf/lbm)}}$$

The answer is (A).

15. (a) From Eq. 19.93,

$$\mathrm{v}_1 = \sqrt{\left(\frac{\left(32.2\ \frac{\text{ft}}{\text{sec}^2}\right)(3.8\ \text{ft})}{(2)(1.2\ \text{ft})}\right)(1.2\ \text{ft} + 3.8\ \text{ft})}$$
$$= 15.97\ \text{ft/sec}$$

$$Q = A\mathrm{v} = (1.2\ \text{ft})(6\ \text{ft})\left(15.97\ \frac{\text{ft}}{\text{sec}}\right)$$
$$= \boxed{115\ \text{ft}^3/\text{sec}}$$

The answer is (B).

(b)
$$\mathrm{v}_2 = \frac{Q}{A} = \frac{115\ \frac{\text{ft}^3}{\text{sec}}}{(3.8\ \text{ft})(6\ \text{ft})} = 5.04\ \text{ft/sec}$$

From Eq. 19.94,

$$\Delta E = \left[1.2\ \text{ft} + \frac{\left(15.97\ \frac{\text{ft}}{\text{sec}}\right)^2}{(2)\left(32.2\ \frac{\text{ft}}{\text{sec}^2}\right)}\right] - \left[3.8\ \text{ft} + \frac{\left(5.04\ \frac{\text{ft}}{\text{sec}}\right)^2}{(2)\left(32.2\ \frac{\text{ft}}{\text{sec}^2}\right)}\right]$$
$$= \boxed{0.97\ \text{ft (ft-lbf/lbm)}}$$

The answer is (A).

16. Disregard the velocity of approach. Use $C_s = 3.5$ in Eq. 19.60.

(a)
$$Q = (3.5)(1\ \text{ft})(2\ \text{ft})^{\frac{3}{2}}$$
$$= \boxed{9.9\ \text{ft}^3/\text{sec per ft of width}}$$

The answer is (A).

(b) The upstream energy is

$$E_1 = d = 2\ \text{ft} \quad [\mathrm{v}_1 \approx 0]$$

At the toe, from Eq. 19.72,

$$E_2 = E_1 + z_1 - z_2$$
$$= 2\ \text{ft} + 40\ \text{ft} - 0$$
$$= 42\ \text{ft}$$

$$E_2 = d_2 + \frac{\mathrm{v}_2^2}{2g} = d_2 + \frac{Q^2}{2gA_2^2}$$

$$d_2 + \frac{\left(9.9\ \frac{\text{ft}^3}{\text{sec}}\right)^2}{(2)\left(32.2\ \frac{\text{ft}}{\text{sec}^2}\right)\left(1\ \frac{\text{ft}}{\text{sec}}\right)^2 d_2^2} = 42\ \text{ft}$$

By trial and error, $d_2 = \boxed{0.19\ \text{ft.}}$

The answer is (A).

17.

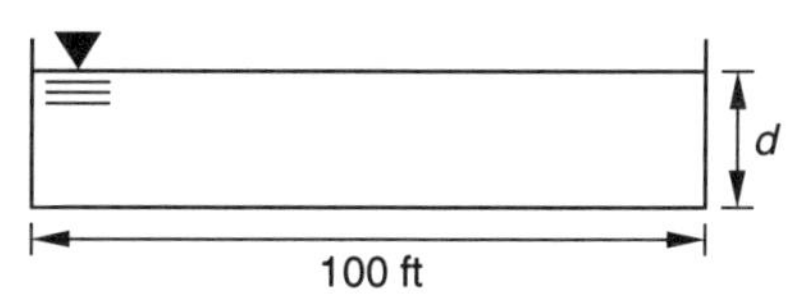

(a)
$$R = \frac{(100\ \text{ft})d}{100\ \text{ft} + 2d} = \frac{50d}{50 + d}$$

From Eq. 19.13,

$$10{,}000\ \frac{\text{ft}^3}{\text{sec}} = \left(\frac{1.49}{0.012}\right)(100d)\left(\frac{50d}{50 + d}\right)^{\frac{2}{3}}\sqrt{0.05}$$

$$0.26538 = \left(\frac{d^{\frac{5}{2}}}{50 + d}\right)^{\frac{2}{3}}$$

$$0.1367 = \frac{d^{\frac{5}{2}}}{50 + d}$$

By trial and error, $d = \boxed{2.2\ \text{ft.}}$

The answer is (B).

(b) From Eq. 19.74,

$$d_c = \sqrt[3]{\frac{\left(10{,}000\ \frac{\text{ft}^3}{\text{sec}}\right)^2}{\left(32.2\ \frac{\text{ft}}{\text{sec}^2}\right)(100\ \text{ft})^2}} = \boxed{6.77\ \text{ft}}$$

The answer is (C).

(c) Since $d < d_c$, $\boxed{\text{flow is rapid (shooting).}}$

The answer is (C).

(d)

$$\text{v} = \frac{Q}{A} = \frac{10{,}000\ \frac{\text{ft}^3}{\text{sec}}}{(100\ \text{ft})(2.2\ \text{ft})} = 45.45\ \text{ft/sec}$$

From Eq. 19.91,

$$\begin{aligned} d_2 &= -\left(\tfrac{1}{2}\right)(2.2\ \text{ft}) \\ &\quad + \sqrt{\frac{(2)\left(45.45\ \frac{\text{ft}}{\text{sec}}\right)^2(2.2\ \text{ft})}{32.2\ \frac{\text{ft}}{\text{sec}^2}} + \frac{(2.2\ \text{ft})^2}{4}} \\ &= \boxed{15.74\ \text{ft}} \end{aligned}$$

The answer is (C).

18. (a) From Eq. 19.60 using $C_s = 3.7$,

$$H = 2425\ \text{ft} - 2420\ \text{ft} = 5\ \text{ft}$$

$$Q = (3.7)(60\ \text{ft})(5\ \text{ft})^{\frac{3}{2}} = \boxed{2482\ \text{ft}^3/\text{sec}}$$

The answer is (C).

(b) Disregarding the velocity of approach, the initial energy is

$$E_1 = (\text{elev})_1 = 2425\ \text{ft}$$

At the toe (point 2),

$$E_1 = E_2 + h_f$$

$$\begin{aligned} 2425 &= (\text{elev})_2 + d_2 + \frac{Q^2}{2gw^2d_2^2} \\ &\quad + (0.20)(2425\ \text{ft} - 2345\ \text{ft}) \end{aligned}$$

$$(1 - 0.20) \times (2425\ \text{ft} - 2345\ \text{ft}) = d_2 + \frac{\left(2482\ \frac{\text{ft}^3}{\text{sec}}\right)^2}{(2)\left(32.2\ \frac{\text{ft}}{\text{sec}^2}\right)(60\ \text{ft})^2 d_2^2}$$

$$64\ \text{ft} = d_2 + \frac{26.57}{d_2^2}$$

$d_{\text{toe}} = d_1$ is implicitly a conjugate depth.

By trial and error, $d_2 = \boxed{0.647\ \text{ft.}}$

The answer is (A).

(c) If there were a hydraulic jump, it would be between the conjugate depths, d_1 to d_2.

$$d_1 = 0.647\ \text{ft}$$

$$\begin{aligned} \text{v}_1 &= \frac{Q}{A_1} = \frac{Q}{d_1 w} = \frac{2482\ \frac{\text{ft}^3}{\text{sec}}}{(0.647\ \text{ft})(60\ \text{ft})} \\ &= 63.94\ \text{ft/sec} \end{aligned}$$

From Eq. 19.91,

$$\begin{aligned} d_2 &= -\tfrac{1}{2}d_1 + \sqrt{\frac{2\text{v}_1^2 d_1}{g} + \frac{d_1^2}{4}} \\ &= \left(-\tfrac{1}{2}\right)(0.647\ \text{ft}) \\ &\quad + \sqrt{\frac{(2)\left(63.9\ \frac{\text{ft}}{\text{sec}}\right)^2(0.647\ \text{ft})}{32.2\ \frac{\text{ft}}{\text{sec}^2}} + \frac{(0.647\ \text{ft})^2}{4}} \\ &= 12.49\ \text{ft} \end{aligned}$$

The answer is (D).

(d)

$$\text{v}_2 = \frac{Q}{A_2} = \frac{Q}{d_2 w} = \frac{2482\ \frac{\text{ft}^3}{\text{sec}}}{(12.49\ \text{ft})(60\ \text{ft})} = 3.31\ \text{ft/sec}$$

Use d_2 and v_2 in Eq. 19.94.

$$\begin{aligned} \Delta E &= \left(d_1 + \frac{\text{v}_1^2}{2g}\right) - \left(d_2 + \frac{\text{v}_2^2}{2g}\right) \\ &= \left(0.647\ \text{ft} + \frac{\left(63.9\ \frac{\text{ft}}{\text{sec}}\right)^2}{(2)\left(32.2\ \frac{\text{ft}}{\text{sec}^2}\right)}\right) \\ &\quad - \left(12.49\ \text{ft} + \frac{\left(3.31\ \frac{\text{ft}}{\text{sec}}\right)^2}{(2)\left(32.2\ \frac{\text{ft}}{\text{sec}^2}\right)}\right) \\ &= \boxed{51.4\ \text{ft (ft-lbf/lbm)}} \end{aligned}$$

The answer is (B).

19. (a) From Eq. 19.64,

$$n = (1.522)(6\text{ ft})^{0.026} = 1.595$$

From Eq. 19.63,

$$Q = (4)(6\text{ ft})\left(\frac{18\text{ in}}{12\,\frac{\text{in}}{\text{ft}}}\right)^{1.595} = \boxed{45.8\text{ ft}^3/\text{sec}}$$

The answer is (B).

(b) The flume is self-cleansing, the head loss is small, and operation by unskilled personnel is possible. Parshall flumes are usually permanent structures.

The answer is (D).

20. (a) The normal depth is d.

$$R = \frac{8d}{8+2d}$$

From Eq. 19.13,

$$150\,\frac{\text{ft}^3}{\text{sec}} = (8d)\left(\frac{1.49}{0.015}\right)\left(\frac{8d}{8+2d}\right)^{\frac{2}{3}}\sqrt{0.0015}$$

$$4.875 = d\left(\frac{8d}{8+2d}\right)^{\frac{2}{3}}$$

By trial and error, $d = \boxed{3.28\text{ ft.}}$

The answer is (B).

(b) To get the backwater curve, choose depths and compute distances to those depths.

Preliminary parameters:

d	$A=8d$	$v=\frac{Q}{A}$	$\frac{v^2}{2g}$	$E=d+\frac{v^2}{2g}$	$P=2d+8$	R
6	48	3.12	0.151	6.151	20	2.40
5	40	3.75	0.218	5.218	18	2.22
4	32	4.69	0.342	4.342	16	2.00
3.28	26.24	5.72	0.502	3.782	14.56	1.80

The average velocities and average hydraulic radii are

d	v	v_{ave}	R	R_{ave}
6.00	3.12		2.40	
		3.435		2.31
5.00	3.75		2.22	
		4.22		2.11
4.00	4.69		2.00	
		5.21		1.90
3.28	5.72		1.80	

From Eq. 19.87, the average energy gradient is

$$S_{ave} = \left[\frac{n v_{ave}}{(1.486)(R_{ave})^{\frac{2}{3}}}\right]^2$$

d	S_{ave}	ΔE	$S_o - S_{ave}$
6.00			
	0.000394	0.933	0.001106
5.00			
	0.000671	0.876	0.000829
4.00			
	0.001175	0.552	0.000325
3.28			

From Eq. 19.89,

$$L_{6\text{-}5}:\quad \frac{\Delta E}{S_o - S_{ave}} = \frac{0.933\text{ ft}}{0.001106} = \boxed{844\text{ ft}}$$

The answer is (B).

(c) $$L_{5\text{-}4}:\quad \frac{0.876\text{ ft}}{0.000829} = \boxed{1057\text{ ft}}$$

The answer is (D).

21. For concrete, $n = 0.012$.

The pipe diameter is

$$D = \frac{42\text{ in}}{12\,\frac{\text{in}}{\text{ft}}} = 3.5\text{ ft}$$

The hydraulic radius is

$$R = \frac{D}{4} = \frac{3.5\text{ ft}}{4} = 0.875\text{ ft}$$

From a table of orifice coefficients, Table 17.5, $C_d = 0.82$.

$$h_1 = 5\text{ ft} + 3.5\text{ ft} + (250\text{ ft})(0.006) = 10\text{ ft}$$

Determine the culvert flow type.

$$\frac{h_1 - z}{D} = \frac{10\text{ ft} - (250\text{ ft})(0.006)}{3.5\text{ ft}} = 2.4$$

The tailwater submerges the culvert outlet, so this is type-4 flow.

$h_1 = 10$ ft

$h_4 = D = 3.5$ ft

$$Q = C_d A_0 \sqrt{2g\left(\frac{h_1 - h_4}{1 + \frac{29C_d^2 n^2 L}{R^{\frac{4}{3}}}}\right)}$$

$$= \left(\frac{(0.82\pi)(3.5\ \text{ft})^2}{4}\right) \times \sqrt{(2)\left(32.2\ \frac{\text{ft}}{\text{sec}^2}\right) \times \left(\frac{10\ \text{ft} - 3.5\ \text{ft}}{1 + \frac{(29)(0.82)^2(0.012)^2(250\ \text{ft})}{(0.875\ \text{ft})^{\frac{4}{3}}}}\right)}$$

$$= \boxed{119}$$

The answer is (D).

22. (a) For a circular channel flowing half-full, the hydraulic radius is

$$R = \frac{D}{4} = \frac{6\ \text{ft}}{4} = \boxed{1.5\ \text{ft}}$$

The answer is (B).

(b) From App. 19.A, $n = 0.012$. From Eq. 19.13,

$$Q = \left(\frac{1.49}{n}\right) AR^{\frac{2}{3}}\sqrt{S}$$

Rearranging, and recognizing that the culvert flows half full,

$$S = \left(\frac{Qn}{1.49AR^{\frac{2}{3}}}\right)^2$$

$$= \left(\frac{\left(150\ \frac{\text{ft}^3}{\text{sec}}\right)(0.012)}{(1.49)\left(\frac{\pi(6\ \text{ft})^2}{4}\right)(0.5)(1.5\ \text{ft})^{\frac{2}{3}}}\right)^2$$

$$= \boxed{0.0043}$$

The answer is (A).

(c) Equation 19.77 could be used, but App. 19.D is more convenient for use with circular channels. Entering with $Q = 150\ \text{ft}^3/\text{sec}$ and $D = 6$ ft, $d_c = 3.3$ ft. The actual flow depth is less than the critical depth; therefore, the flow is supercritical.

The answer is (C).

(d) For a circular channel flowing full, the hydraulic radius is the same as for half-full flow.

$$R = \frac{D}{4} = \frac{6\ \text{ft}}{4} = \boxed{1.5\ \text{ft}}$$

The answer is (A).

(e) From Eq. 19.13,

$$Q = \left(\frac{1.49}{n}\right) AR^{\frac{2}{3}}\sqrt{S}$$

$$= \left(\frac{1.49}{0.012}\right)\left(\frac{\pi(6\ \text{ft})^2}{4}\right)(1.5\ \text{ft})^{\frac{2}{3}}\sqrt{0.0043}$$

$$= \boxed{300\ \text{ft}^3/\text{sec}}$$

The answer is (D).

(f) This part was solved in part (c): $d_c = \boxed{3.3\ \text{ft.}}$

The answer is (C).

(g) From Eq. 19.13, use $d = d_c$.

$$Q = \left(\frac{1.49}{n}\right) AR^{\frac{2}{3}}\sqrt{S}$$

R is calculated using the formula in Table 19.2.

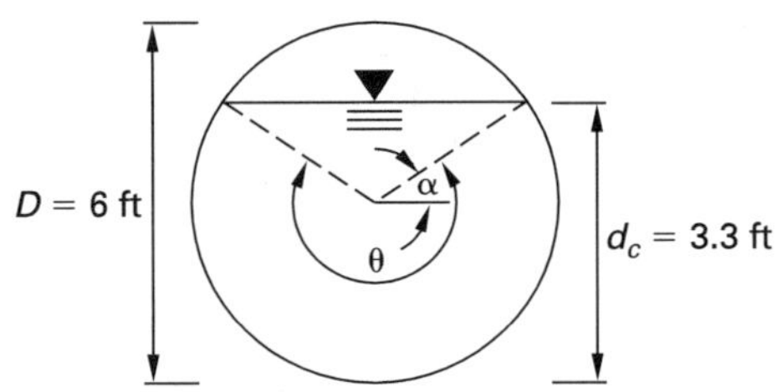

$$\sin\alpha = \frac{0.3}{3}$$

$$\alpha = 5.74°$$

$$\theta = 180° + 2\alpha = 180° + (2)(5.74°)$$

$$= 191.48°$$

Convert to radians.

$$\alpha = (191.48°)\left(\frac{2\pi\ \text{rad}}{360°}\right)$$

$$= 3.34\ \text{rad}$$

$$R = \tfrac{1}{4}\left(1 - \frac{\sin\theta}{\theta}\right) D$$

$$= \left(\tfrac{1}{4}\right)\left(1 - \frac{\sin(3.34\ \text{rad})}{3.34\ \text{rad}}\right)(6\ \text{ft})$$

$$= 1.59\ \text{ft}$$

$$A = \tfrac{1}{8}(\theta - \sin\theta)D^2$$

$$= \left(\tfrac{1}{8}\right)(3.34\ \text{rad} - \sin(3.34\ \text{rad}))(6\ \text{ft})^2$$

$$= 15.9\ \text{ft}^2$$

$$S = \left(\frac{Qn}{1.49AR^{\frac{2}{3}}}\right)^2$$

$$= \left(\frac{\left(150\ \frac{\text{ft}^3}{\text{sec}}\right)(0.012)}{(1.49)(15.9\ \text{ft}^2)(1.59\ \text{ft})^{\frac{2}{3}}}\right)^2$$

$$= \boxed{0.003}$$

The answer is (B).

(h) The pipe flows partially full, and the exit is not submerged. This eliminates types 4 and 6 and leaves types 1, 2, 3, and 5. Type 2 is eliminated because the flow is not at critical depth at the outlet. Type 3 is tranquil flow throughout, so it is also eliminated. Distinguishing between types 1 and 5 cannot be done without further information about the entrance conditions.

The answer is (A).

(i)

$$\text{v} = \left(\frac{1.49}{n}\right) R^{\frac{2}{3}}\sqrt{S}$$

$$= \left(\frac{1.49}{0.012}\right)(1.5\ \text{ft})^{\frac{2}{3}}\sqrt{0.0043}$$

$$= 10.67\ \text{ft/sec}$$

$$D_h = \frac{A}{w} = \frac{\left(\frac{1}{2}\right)\left(\frac{\pi}{4}\right)(6\ \text{ft})^2}{6\ \text{ft}}$$

$$= 2.36\ \text{ft}$$

From Eq. 19.78,

$$\text{Fr} = \frac{\text{v}}{\sqrt{gD_h}}$$

$$= \frac{10.67\ \frac{\text{ft}}{\text{sec}}}{\sqrt{\left(32.2\ \frac{\text{ft}}{\text{sec}^2}\right)(2.36\ \text{ft})}}$$

$$= \boxed{1.22}$$

Since Fr > 1, the flow is supercritical.

The answer is (C).

(j) From Eq. 19.91,

$$d_2 = -\tfrac{1}{2}d_1 + \sqrt{\frac{2\text{v}_1^2 d_1}{g} + \frac{d_1^2}{4}}$$

$$= -\left(\tfrac{1}{2}\right)(2\ \text{ft}) + \sqrt{\frac{(2)\left(15\ \frac{\text{ft}}{\text{sec}}\right)^2(2\ \text{ft})}{32.2\ \frac{\text{ft}}{\text{sec}^2}} + \frac{(2\ \text{ft})^2}{4}}$$

$$= \boxed{4.4\ \text{ft}}$$

The answer is (D).

23. (a) Rearrange Eq. 19.60.

$$H = \left(\frac{Q}{C_s b}\right)^{\frac{2}{3}} = \left(\frac{2000\ \frac{\text{m}^3}{\text{s}}}{(2.2)(150\ \text{m})}\right)^{\frac{2}{3}}$$

$$= \boxed{3.32\ \text{m}}$$

The answer is (D).

(b)

$$\text{v}_1 = \frac{Q}{A_1} = \frac{2000\ \frac{\text{m}^3}{\text{s}}}{(150\ \text{m})(3.32\ \text{m})}$$

$$= 4.02\ \text{m/s}$$

From Eq. 19.96,

$$E_1 = y_{\text{crest}} + H + \frac{\text{v}_1^2}{2g}$$

$$= 30\ \text{m} + 3.32\ \text{m} + \frac{\left(4.02\ \frac{\text{m}}{\text{s}}\right)^2}{(2)\left(9.81\ \frac{\text{m}}{\text{s}^2}\right)}$$

$$= \boxed{34.1\ \text{m}}$$

The answer is (B).

(c)

$$\text{v}_2 = \frac{Q}{A_2} = \frac{2000\ \frac{\text{m}^3}{\text{s}}}{(150\ \text{m})(0.6\ \text{m})}$$

$$= 22.2\ \text{m/s}$$

$$E_2 = d_2 + \frac{\text{v}_2^2}{2g}$$

$$= 0.6\ \text{m} + \frac{\left(22.2\ \frac{\text{m}}{\text{s}}\right)^2}{(2)\left(9.81\ \frac{\text{m}}{\text{s}^2}\right)}$$

$$= \boxed{25.8\ \text{m}}$$

The answer is (A).

(d) The friction loss is

$$E_1 - E_2 = 34.1\ \text{m} - 25.8\ \text{m}$$

$$= \boxed{8.3\ \text{m}}$$

The answer is (C).

(e) From Eq. 19.13,

$$Q = \left(\frac{1.00}{n}\right) AR^{\frac{2}{3}}\sqrt{S}$$

$$2000\ \frac{\text{m}^3}{\text{s}} = \left(\frac{1.00}{0.016}\right)(150\ \text{m})d_n\left(\frac{(150\ \text{m})d_n}{2d_n + 150\ \text{m}}\right)^{\frac{2}{3}} \times \sqrt{0.002}$$

Water Resources Flow of Fluids

By trial and error, $d_n = \boxed{2.6 \text{ m.}}$

The answer is (B).

(f) From Eq. 19.74,

$$\begin{aligned} d_c &= \left(\frac{Q^2}{gw^2}\right)^{\frac{1}{3}} \\ &= \left(\frac{\left(2000 \ \frac{\text{m}^3}{\text{s}}\right)^2}{\left(9.81 \ \frac{\text{m}}{\text{s}^2}\right)(150 \text{ m})^2}\right)^{\frac{1}{3}} \\ &= \boxed{2.6 \text{ m}} \end{aligned}$$

The answer is (B).

(g) The normal depth of flow coincides with the critical depth. A hydraulic jump is an abrupt rise from below the critical depth to above the critical depth. In this case the water will rise gradually and no jump occurs.

The answer is (D).

(h) From Table 19.7, the maximum velocity (assuming the water is clear) is

$$\text{v}_{\text{max}} = \left(4.0 \ \frac{\text{ft}}{\text{sec}}\right)\left(0.3 \ \frac{\text{m}}{\text{ft}}\right) = \boxed{1.2 \text{ m/s}}$$

The answer is (A).

(i) From App. 19.A, $n = 0.025$.

$$\begin{aligned} R &= \frac{bd}{2d + b} \\ &= \frac{(150 \text{ m})(2.6 \text{ m})}{(2)(2.6 \text{ m}) + 150 \text{ m}} \\ &= 2.5 \text{ m} \end{aligned}$$

Rearranging Eq. 19.12,

$$\begin{aligned} S &= \left(\frac{\text{v}n}{R^{\frac{2}{3}}}\right)^2 \\ &= \left(\frac{\left(1.2 \ \frac{\text{m}}{\text{s}}\right)(0.025)}{(2.5 \text{ m})^{\frac{2}{3}}}\right)^2 \\ &= \boxed{0.0003} \end{aligned}$$

The answer is (B).

(j) From Eq. 19.29,

$$h_f = LS = (1000 \text{ m})(0.002) = \boxed{2 \text{ m}}$$

The answer is (D).

20 Meteorology, Climatology, and Hydrology

PRACTICE PROBLEMS

Hydrographs

1. A 2 h storm over a 111 km^2 area produces a total runoff volume of 4×10^6 m^3 with a peak discharge of 260 m^3/s.

(a) What is the total excess precipitation?

(A) 1.4 cm
(B) 2.6 cm
(C) 3.6 cm
(D) 4.0 cm

(b) What is the unit hydrograph peak discharge?

(A) 72 $m^3/s \cdot cm$
(B) 120 $m^3/s \cdot cm$
(C) 210 $m^3/s \cdot cm$
(D) 260 $m^3/s \cdot cm$

(c) If a 2 h storm producing 6.5 cm of runoff is to be used to design a culvert, what is the design flood hydrograph volume?

(A) 4.0×10^6 m^3
(B) 7.2×10^6 m^3
(C) 2.6×10^7 m^3
(D) 3.6×10^7 m^3

(d) What is the design discharge?

(A) 89 m^3/s
(B) 130 m^3/s
(C) 260 m^3/s
(D) 470 m^3/s

(e) The recurrence interval of the 6.5 cm storm is 50 yr, and the culvert is to be designed for a 30 yr life. What is the probability that the capacity will be exceeded during the design life?

(A) 0%
(B) 33%
(C) 45%
(D) 92%

(f) The unit hydrograph represents water flowing into a stream from

I. base flow
II. evapotranspiration
III. overland flow
IV. surface flow
V. interflow

(A) I only
(B) I and III
(C) II, III, and IV
(D) III, IV, and V

Use the following information for parts (g) through (j).

Stream discharges recorded during a 3 h storm on a 40 km^2 watershed are as follows.

t (h)	$Q\left(\frac{m^3}{s}\right)$
0	17
1	16
2	48
3	90
4	108
5	85
6	58
7	38
8	26
9	20
10	18
11	17
12	16

(g) What is the unit hydrograph peak discharge from surface runoff for this 3 h storm?

(A) 5.2 $m^3/s \cdot cm$
(B) 29 $m^3/s \cdot cm$
(C) 65 $m^3/s \cdot cm$
(B) 120 $m^3/s \cdot cm$

(h) What is the peak discharge for a 6 h storm producing 5 cm of runoff in this watershed?

(A) 92 m^3/s
(B) 140 m^3/s
(C) 170 m^3/s
(D) 300 m^3/s

(i) What is the peak discharge for a 5 h storm producing 5 cm of runoff in this watershed?

(A) 63 $m^3/s \cdot cm$
(B) 90 $m^3/s \cdot cm$
(C) 110 $m^3/s \cdot cm$
(D) 260 $m^3/s \cdot cm$

(j) What is the design flood hydrograph volume for a 5 h storm producing 5 cm of runoff in this watershed?

(A) 1.3×10^4 m^3
(B) 4.0×10^5 m^3
(C) 8.3×10^5 m^3
(D) 2.0×10^6 m^3

2. (*Time limit: one hour*) A stream gauging station recorded the following discharges from a 1.2 mi^2 drainage area after a storm.

t (hr)	$Q \left(\frac{\text{ft}^3}{\text{sec}}\right)$
0	102
1	99
2	101
3	215
4	507
5	625
6	455
7	325
8	205
9	145
10	100
11	70
12	55
13	49
14	43
15	38

(a) Draw the actual hydrograph. (b) Separate the groundwater and surface water. (c) Draw the unit hydrograph.

(d) What is the length of the direct runoff recession limb?

(A) 5 hr
(B) 6 hr
(C) 10 hr
(D) 13 hr

Rational Equation

3. A 0.5 mi^2 drainage area has a suggested runoff coefficient of 0.6 and a time of concentration of 60 min. The drainage area is in Steel region #3, and a 10 yr storm is to be used for design purposes. What is the peak runoff?

(A) 310 ft^3/sec
(B) 390 ft^3/sec
(C) 460 ft^3/sec
(D) 730 ft^3/sec

4. Four contiguous 5 ac watersheds are served by an adjacent 1200 ft storm drain (n = 0.013 and slope = 0.005). Inlets to the storm drain are placed every 300 ft along the storm drain. The inlet time for each area served by an inlet is 15 min, and the area's runoff coefficient is 0.55. A storm to be used for design purposes has the following characteristics (I is in in/hr, t is in min).

$$I = \frac{100}{t_c + 10}$$

All flows are maximum, and all pipe sizes are available. n is constant. What is the diameter of the last section of storm drain?

(A) 28 in
(B) 32 in
(C) 36 in
(D) 42 in

5. (*Time limit: one hour*) A 75 ac urbanized section of land drains into a rectangular 5 ft by 7 ft channel that directs runoff through a round culvert under a roadway. The culvert is concrete and is 60 in in diameter. It is 36 ft long and placed on a 1% slope.

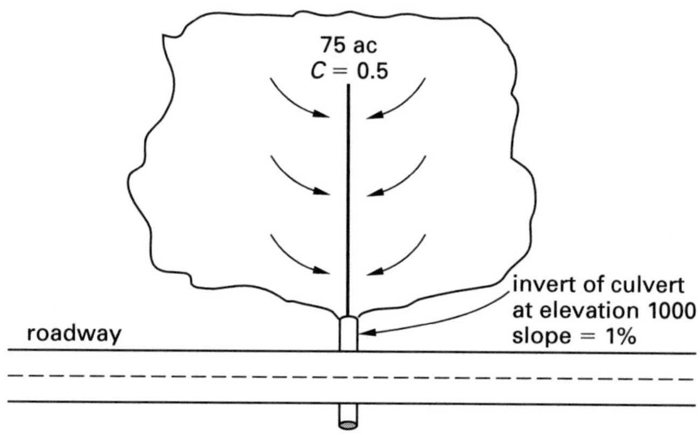

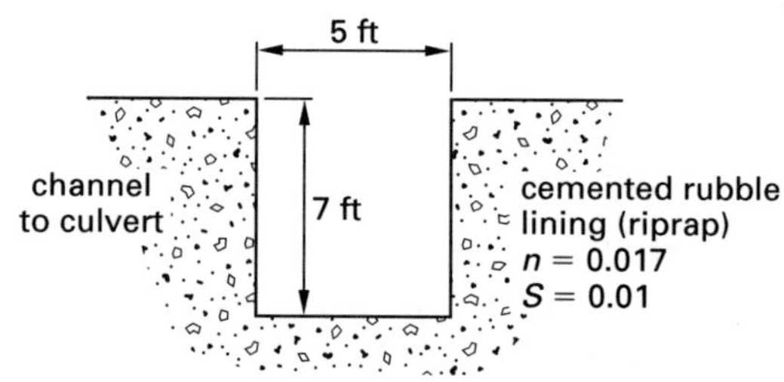

It is desired to evaluate the culvert design based on a 50 yr storm. It is known that the time of concentration to the entrance of the culvert is 30 min. The corresponding rainfall intensity is 5.3 in/hr.

(a) If the minimum road surface elevation is 1010 ft, will the road surface be flooded?

(A) yes; the culvert needs 10 ft^3/sec additional capacity
(B) yes; the culvert needs 70 ft^3/sec additional capacity
(C) no; the culvert has 20 ft^3/sec excess capacity
(D) no; the culvert has 70 ft^3/sec excess capacity

(b) What is the depth of flow elevation upstream at the culvert entrance after 30 min?

(A) 2.7 ft
(B) 3.1 ft
(C) 3.6 ft
(D) 4.1 ft

(c) Assuming n varies with depth of flow, what is the depth of flow at the culvert exit after 30 min?

(A) 2.7 ft
(B) 3.1 ft
(C) 3.6 ft
(D) 4.1 ft

Hydrograph Synthesis

6. (*Time limit: one hour*) A standard 4 hr storm produces 2 in net of runoff. A stream gauging report is produced from successive sampling every few hours. Two weeks later, the first 4 hr of an 8 hr storm over the same watershed produces 1 in of runoff. The second 4 hr produces 2 in of runoff. Neglecting ground water, draw a hydrograph of the 8 hr storm.

t (hr)	$Q\left(\frac{\text{ft}^3}{\text{sec}}\right)$
0	0
2	100
4	350
6	600
8	420
10	300
12	250
14	150
16	100
18	–
20	50
22	–
24	0

Reservoir Sizing

7. A class A evaporation pan located near a reservoir shows a 1 day evaporation loss of 0.8 in. If the pan coefficient is 0.7, what is the approximate evaporation loss in the reservoir?

(A) 0.56 in
(B) 0.63 in
(C) 0.98 in
(D) 1.1 in

8. A reservoir has a total capacity of 7 volume units. At the beginning of a study, the reservoir contains 5.5 units. The monthly demand on the reservoir from a nearby city is 0.7 units. The monthly inflow to the reservoir is normally distributed with a mean of 0.9 units and a standard deviation of 0.2 units. Simulate one year of reservoir operation with a 99.97% confidence level.

9. Repeat Prob. 8 assuming that the monthly demand on the reservoir is normally distributed with a mean of 0.7 units and a standard deviation of 0.2 units.

10. (*Time limit: one hour*) A reservoir is needed to provide 20 ac-ft of water each month. The inflow for each of 13 representative months is given. The reservoir starts full. Size the reservoir.

month	inflow (ac-ft)
February	30
March	60
April	20
May	10
June	5
July	10
August	5
September	10
October	20
November	90
December	85
January	75
February	50

(A) 40 ac-ft
(B) 60 ac-ft
(C) 90 ac-ft
(D) 120 ac-ft

11. (*Time limit: one hour*) A reservoir with a constant draft of 240 MG/mi^2-yr is being designed. The inflow distribution is known.

month	inflow (MG/mi^2)	inflow (m^3/m^2)
January	20	0.029
February	30	0.044
March	45	0.066
April	30	0.044
May	40	0.058
June	30	0.044
July	15	0.022
August	5	0.007
September	15	0.022
October	60	0.088
November	90	0.132
December	40	0.058

(a) What should be the minimum reservoir size?

(A) 25 MG/mi^2
(B) 35 MG/mi^2
(C) 45 MG/mi^2
(D) 65 MG/mi^2

(b) When would the reservoir start to spill?

(A) at the end of March

(B) at the end of June

(C) at the end of September

(D) at the end of March and the end of October

SOLUTIONS

1. (a) From Eq. 20.21,

$$P_{\text{ave}} = \frac{V}{A_d} = \frac{4 \times 10^6 \text{ m}^3}{(111 \text{ km}^2)\left(1000 \ \frac{\text{m}}{\text{km}}\right)^2} = \boxed{0.036 \text{ m } (3.6 \text{ cm})}$$

The answer is (C).

(b) The unit hydrograph discharge is the peak discharge divided by the average precipitation.

$$Q_{\text{hydrograph,unit}} = \frac{Q_p}{P} = \frac{260 \ \frac{\text{m}^3}{\text{s}}}{3.6 \text{ cm}} = \boxed{72.2 \text{ m}^3/\text{s·cm}}$$

The answer is (A).

(c) The design flood hydrograph volume for a 6.5 cm storm is determined by multiplying the unit hydrograph volume by 6.5. For the unit hydrograph,

$$V_{\text{hydrograph}} = \frac{V}{P} = \frac{4 \times 10^6 \text{ m}^3}{3.6 \text{ cm}} = 1.11 \times 10^6 \text{ m}^3/\text{cm}$$

For the 6.5 cm storm,

$$V = \left(1.11 \times 10^6 \ \frac{\text{m}^3}{\text{cm}}\right)(6.5 \text{ cm}) = \boxed{7.2 \times 10^6 \text{ m}^3}$$

The answer is (B).

(d) The design discharge is determined by multiplying the unit hydrograph discharge by 6.5 cm.

$$Q_p = (Q_{\text{hydrograph}})(6.5) = \left(72.2 \ \frac{\text{m}^3}{\text{s·cm}}\right)(6.5 \text{ cm}) = \boxed{469.3 \text{ m}^3/\text{s}}$$

The answer is (D).

(e) From Eq. 20.20,

$$p\{F \text{ event in } n \text{ years}\} = 1 - \left(1 - \frac{1}{F}\right)^n$$

$$p\{50 \text{ yr flood in } 30 \text{ yr}\} = 1 - \left(1 - \frac{1}{50}\right)^{30} = \boxed{0.45 \ \ (45\%)}$$

The answer is (C).

(f) The unit hydrograph represents all discharge into a stream except for groundwater or base flow, which are separated out. Evapotranspiration refers to water that is returned to the atmosphere and, therefore, is not measured in the stream discharge. The unit hydrograph includes overland flow, surface flow, and interflow.

The answer is (D).

(g) The actual runoff is plotted. To separate base flow from overland flow, use the straight line method. Draw a horizontal line from the start of the rising limb to the falling limb. In the table, subtract the base flow from the overland flow.

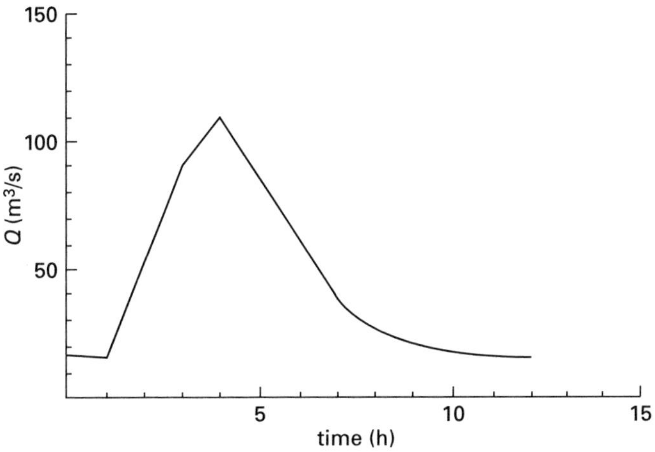

hour	runoff	ground water	surface water	surface water / 3.14
0	17	16	1	0.32
1	16	16	0	0
2	48	16	32	10.19
3	90	16	74	23.56
4	108	16	92	29.29
5	85	16	69	21.97
6	58	16	42	13.37
7	38	16	22	7.00
8	26	16	10	3.18
9	20	16	4	1.27
10	18	16	2	0.64
11	17	16	1	0.32
12	16	16	0	0
		total	349 m^3/s	

The average precipitation is given by Eq. 20.21.

$$P = \frac{V}{A_d}$$

The volume of runoff is the area under the separated hydrograph curve.

$$V = (349\ \text{m}^3/\text{s})(1\ \text{h})\left(3600\ \frac{\text{s}}{\text{h}}\right) = 1\,256\,400\ \text{m}^3$$

$$P = \frac{1\,256\,400\ \text{m}^3}{(40\ \text{km}^2)\left(1000\ \frac{\text{m}}{\text{km}}\right)^2} = 0.0314\ \text{m}\quad (3.14\ \text{cm})$$

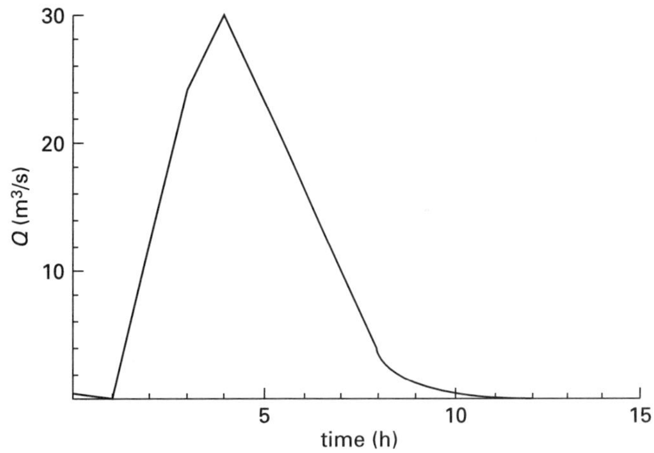

The unit hydrograph peak discharge is

$$\frac{108\ \frac{\text{m}^3}{\text{s}} - 16\ \frac{\text{m}^3}{\text{s}}}{3.14\ \text{cm}} = \boxed{29.3\ \text{m}^3/\text{s}\cdot\text{cm}}$$

The answer is (B).

(h) There are two methods for constructing hydrographs for a longer storm than that of the unit hydrograph. For storms that are whole multiples of the unit hydrograph duration, the lagging storm method is simplest.

In a table, add another unit hydrograph separated by $t_r = 3$ h. Then, add the ordinates for a hydrograph of $Nt_r = (2)(3\ \text{h}) = 6$ h duration. The ordinates must then be divided by $N = 2$ to get the unit hydrograph ordinates.

hour	unit hydrograph 3 h storm	second storm	total	unit hydrograph 6 h storm
0	0.32	0	0.32	0.16
1	0	0	0	0
2	10.19	0	10.19	5.09
3	23.56	0.32	23.88	11.94
4	29.29	0	29.29	14.65
5	21.97	10.19	32.16	16.08
6	13.37	23.56	36.93	18.47
7	7.00	29.29	36.29	18.15
8	3.18	21.97	25.15	12.58
9	1.27	13.37	14.65	7.32
10	0.64	7.00	7.64	3.82
11	0.32	3.18	3.50	1.75
12	0	1.27	1.27	0.64
13	0	0.64	0.64	0.32
14	0	0.32	0.32	0.16
15	0	0	0	0

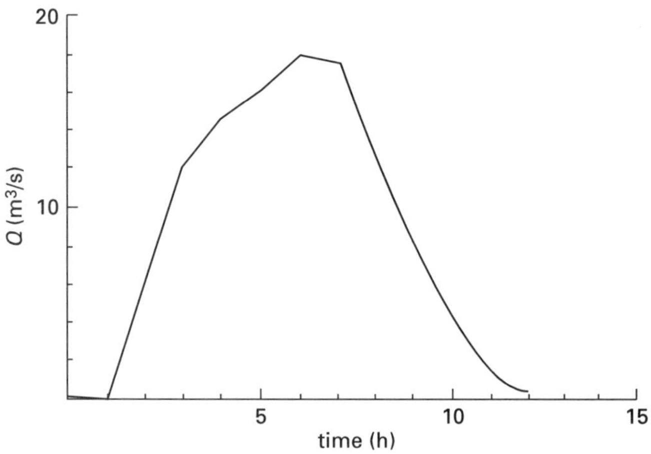

The peak discharge for a 5 cm storm is calculated from the peak unit hydrograph discharge.

$$Q_{p,\text{unit hydrograph}} = 18.47\ \text{m}^3/\text{s·cm}$$

$$Q_{p,5\ \text{cm storm}} = (5\ \text{cm})\left(18.47\ \frac{\text{m}^3}{\text{s·cm}}\right) = \boxed{92.35\ \text{m}^3/\text{s}}$$

The answer is (A).

(i) If a storm is not a multiple of the unit hydrograph duration, the S-curve method must be used to construct a hydrograph for that storm. In the table, add the ordinates of five 3 h unit hydrographs, each offset by 3 h, to produce an S-curve.

hour	3 h unit hydro- graph	second storm	third storm	fourth storm	fifth storm	total (S-curve)
0	0.32	0	0	0	0	0.32
1	0	0	0	0	0	0
2	10.19	0	0	0	0	10.19
3	23.56	0.32	0	0	0	23.88
4	29.29	0	0	0	0	29.29
5	21.97	10.19	0	0	0	32.16
6	13.37	23.56	0.32	0	0	37.25
7	7.00	29.29	0	0	0	36.29
8	3.18	21.97	10.19	0	0	35.33
9	1.27	13.37	23.56	0.32	0	38.52
10	0.64	7.00	29.29	0	0	36.93
11	0.32	3.18	21.97	10.19	0	35.66
12	0	1.27	13.37	23.56	0.32	38.52
13	0	0.64	7.00	29.29	0	36.93
14	0	0.32	3.18	21.97	10.19	35.66
15	0	0	1.27	13.37	23.56	38.52

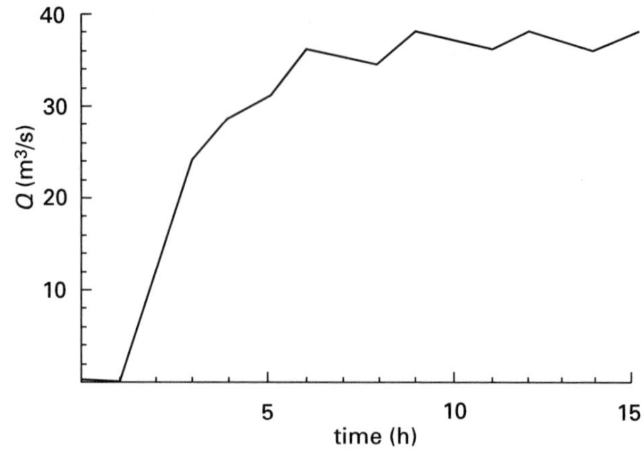

After approximately 5 h, the S-curve levels off to a steady value. (It actually alternates slightly between 35 and 39 m^3/s.) The 5 h unit hydrograph is calculated by tabulating another S-curve lagging the first by 5 h, finding the difference, and scaling to a ratio of 3:5. (The negative values obtained reflect the approximate nature of the method.)

hour	S-curve	lagging S-curve	5 h hydrograph
0	0.32	0	0.19
1	0	0	0
2	10.19	0	6.11
3	23.88	0	14.32
4	29.29	0	17.57
5	32.16	0.32	19.10
6	37.25	0	22.35
7	36.29	10.19	15.66
8	35.33	23.88	6.88
9	38.52	29.29	5.54
10	36.93	32.16	2.87
11	35.66	37.25	−0.96
12	38.52	36.29	1.34
13	36.93	35.33	0.96
14	35.66	38.52	−1.72
15	38.52	36.93	0.96
		total	111.17

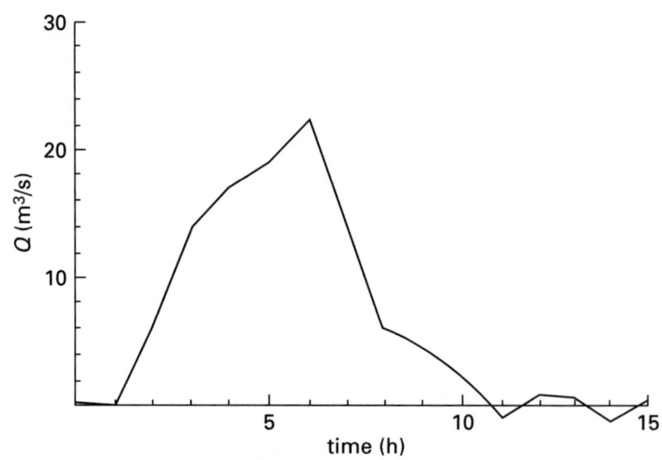

The peak discharge for a 5 cm storm is calculated from the peak unit hydrograph discharge.

$$Q_{p,\text{unit hydrograph}} = 22.35 \text{ m}^3/\text{s}$$

$$Q_{p,5 \text{ cm storm}} = (5 \text{ cm})\left(22.35 \, \frac{\text{m}^3}{\text{s·cm}}\right)$$

$$= \boxed{111.75 \text{ m}^3/\text{s·cm}}$$

The answer is (C).

(j) The design flood volume is derived from the 5 h unit hydrograph.

$$V_{\text{unit hydrograph}} = \left(111.17 \, \frac{\text{m}^3}{\text{s·cm}}\right)(1 \text{ h})(3600 \text{ s/h})$$

$$= 400\,212 \text{ m}^3/\text{cm}$$

For a 5 cm storm,

$$V_{5 \text{ cm storm}} = \left(400\,212 \, \frac{\text{m}^3}{\text{cm}}\right)(5 \text{ cm})$$

$$= \boxed{2.0 \times 10^6 \text{ m}^3}$$

The answer is (D).

2. (a) The actual hydrograph is a plot of time versus flow quantity, both of which are given in the problem statement.

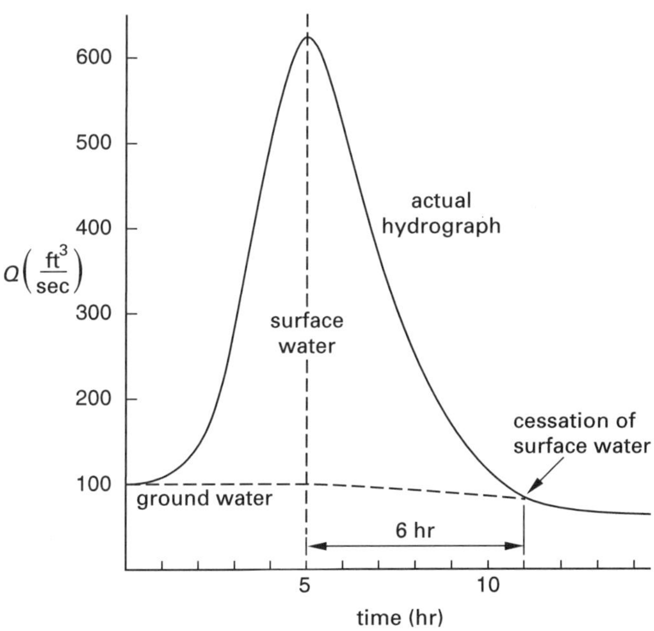

(b) Use the fixed-base method to separate the base flow from the overland flow. The base flow before the storm is projected to a point directly under the peak (in this case a more or less horizontal line), and then a straight line connects the projection to the falling limb. The connection point is 5 to 7 hr later, so use 6 hr.

(c) The ordinates of the unit hydrograph are found by separating the base flow and then dividing the ordinates of the actual hydrograph by the average precipitation. The average precipitation is determined by tabulating the surface water flow and adding to find the total volume.

hour	runoff (ft^3/sec)	ground water (ft^3/sec)	surface water (ft^3/sec)	surface water/2.43 (ft^3/sec)
0	102	100	0	0
1	99	100	0	0
2	101	100	0	0
3	215	100	115	47.3
4	507	100	407	167.5
5	625	100	525	215.0
6	455	93	362	149.0
7	325	87	238	97.9
8	205	80	125	51.4
9	145	74	71	29.2
10	100	67	33	13.6
11	70	60	10	4.1
12	55	55	0	0
13	49	49	0	0
14	43	43	0	0
15	38	38	0	0
		total	1886 ft^3/sec	

The total volume of surface water is the total flow multiplied by the time interval (1 hr).

$$V = \left(1886 \, \frac{\text{ft}^3}{\text{sec}}\right)(1 \text{ hr})\left(3600 \, \frac{\text{sec}}{\text{hr}}\right)$$

$$= 6.79 \times 10^6 \text{ ft}^3$$

The watershed area is

$$A_d = (1.2 \text{ mi}^2)\left(5280 \, \frac{\text{ft}}{\text{mi}}\right)^2$$

$$= 3.35 \times 10^7 \text{ ft}^2$$

The average precipitation is the total volume divided by the drainage area.

$$P = \frac{(6.79 \times 10^6 \text{ ft}^3)\left(12 \, \frac{\text{in}}{\text{ft}}\right)}{3.35 \times 10^7 \text{ ft}^2}$$

$$= 2.43 \text{ in}$$

The ordinates for the unit hydrograph are tabulated in the last column of the table.

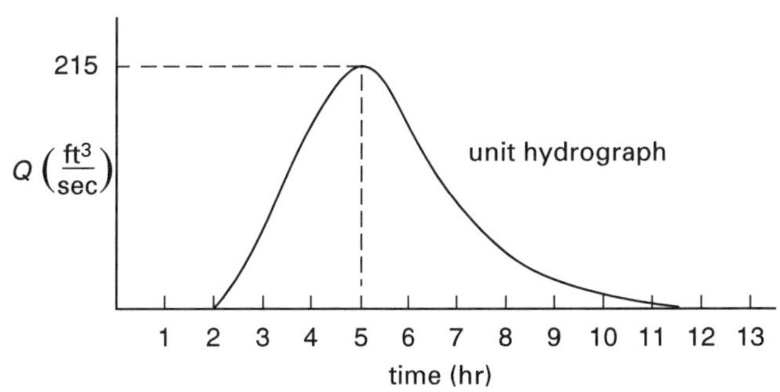

(d) The time base for direct runoff is about 11 hr − 2 hr = 9 hr. The time base includes all the time in which direct runoff is observed, that is, both the rising and falling limbs.

The actual hydrograph is as shown. From the graph,

$$t_b \approx \boxed{6 \text{ hr}}$$

The answer is (B).

3. Use Table 20.2.

$$K = 170$$
$$b = 23$$

From Eq. 20.14, the intensity is

$$I = \frac{170}{60 \text{ min} + 23} = 2.05 \text{ in/hr}$$
$$A_d = (0.5)(640 \text{ ac}) = 320 \text{ ac}$$

From Eq. 20.36,

$$Q_p = CIA_d = (0.6)\left(2.05 \ \frac{\text{in}}{\text{hr}}\right)(320 \text{ ac})$$
$$= \boxed{393.6 \text{ ft}^3/\text{sec}}$$

The answer is (B).

4.

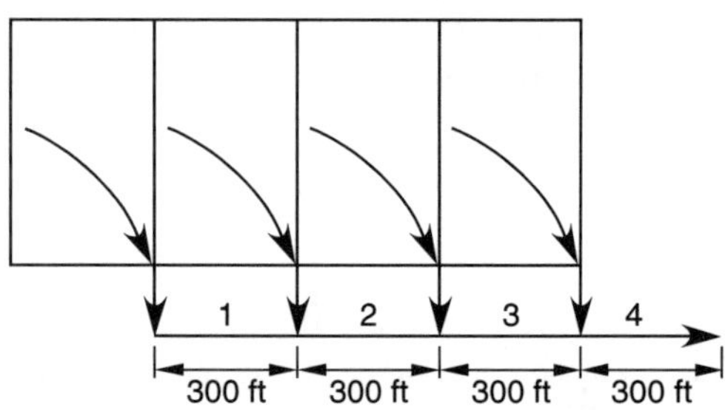

• First inlet:

Peak flow for the first inlet begins at $t = 15$ min.

$$I = \frac{100}{15 \text{ min} + 10} = 4 \text{ in/hr}$$

From Eq. 20.36,

$$Q_p = CIA_d = (0.55)\left(4 \ \frac{\text{in}}{\text{hr}}\right)(5 \text{ ac}) = 11 \text{ ft}^3/\text{sec}$$

Use Eq. 19.16.

$$D = (1.33)\left(\frac{nQ}{\sqrt{S}}\right)^{\frac{3}{8}} = (1.33)\left[\frac{(0.013)\left(11 \ \frac{\text{ft}^3}{\text{sec}}\right)}{\sqrt{0.005}}\right]^{\frac{3}{8}}$$
$$= 1.73 \text{ ft} \quad (21 \text{ in})$$

$$\text{v}_{\text{full}} = \frac{Q}{A} = \frac{11 \ \frac{\text{ft}^3}{\text{sec}}}{\left(\frac{\pi}{4}\right)\left(\frac{21 \text{ in}}{12 \ \frac{\text{in}}{\text{ft}}}\right)^2}$$
$$= 4.6 \text{ ft/sec}$$

(Note that an integer-inch pipe diameter must be used to calculate velocity, because 1.73 ft diameter pipes are not manufactured.)

• Second inlet:

The flow time from inlet 1 to inlet 2 is

$$\frac{300 \text{ ft}}{4.6 \ \frac{\text{ft}}{\text{sec}}} = 65.2 \text{ sec} \quad (1.09 \text{ min})$$

The intensity at $t = 15.0 \text{ min} + 1.09 \text{ min} = 16.09$ min is

$$I = \frac{100}{16.09 \text{ min} + 10} = 3.83 \text{ in/hr}$$

The runoff is

$$Q = (0.55)\left(3.83 \ \frac{\text{in}}{\text{hr}}\right)(10 \text{ ac}) = 21.07 \text{ ft}^3/\text{sec}$$

$$D = (1.33)\left[\frac{(0.013)\left(21.07 \ \frac{\text{ft}^3}{\text{sec}}\right)}{\sqrt{0.005}}\right]^{\frac{3}{8}}$$
$$= 2.21 \text{ ft} \quad (27 \text{ in})$$

$$\text{v}_{\text{full}} = \frac{21.07 \ \frac{\text{ft}^3}{\text{sec}}}{\left(\frac{\pi}{4}\right)\left(\frac{27 \text{ in}}{12 \ \frac{\text{in}}{\text{ft}}}\right)^2}$$
$$\approx 5.3 \text{ ft/sec}$$

• Third inlet:

The flow time from inlet 2 is

$$t = \frac{L}{\text{v}} = \frac{300 \text{ ft}}{\left(5.3 \ \frac{\text{ft}}{\text{sec}}\right)\left(60 \ \frac{\text{sec}}{\text{min}}\right)} = 0.94 \text{ min}$$
$$t_c = 16.09 \text{ min} + 0.94 \text{ min} = 17.03 \text{ min}$$
$$I = \frac{100}{17.03 \text{ min} + 10} = 3.70 \text{ in/hr}$$

$$Q = (0.55)\left(3.70\ \frac{\text{in}}{\text{hr}}\right)(15) = 30.53\ \text{ft}^3/\text{sec}$$

$$D = (1.33)\left[\frac{(0.013)\left(30.53\ \frac{\text{ft}^3}{\text{sec}}\right)}{\sqrt{0.005}}\right]^{\frac{3}{8}}$$

$$= 2.54\ \text{ft}\quad(31\ \text{in})$$

$$\text{v} = \frac{Q}{A} = \frac{30.53\ \frac{\text{ft}^3}{\text{sec}}}{\left(\frac{\pi}{4}\right)\left(\frac{31\ \text{in}}{12\ \frac{\text{in}}{\text{ft}}}\right)^2}$$

$$= 5.82\ \text{ft/sec}$$

- Fourth inlet:

The flow time from inlet 3 is

$$\frac{300\ \text{ft}}{\left(5.82\ \frac{\text{ft}}{\text{sec}}\right)\left(60\ \frac{\text{sec}}{\text{min}}\right)} = 0.86\ \text{min}$$

$$t_c = 17.03\ \text{min} + 0.86\ \text{min} = 17.89\ \text{min}$$

$$I = \frac{100}{17.89\ \text{min} + 10} = 3.59\ \text{in/hr}$$

$$Q = (0.55)\left(3.59\ \frac{\text{in}}{\text{hr}}\right)(20\ \text{ac}) = 39.49\ \text{ft}^3/\text{sec}$$

$$D = (1.33)\left[\frac{(0.013)\left(39.49\ \frac{\text{ft}^3}{\text{sec}}\right)}{\sqrt{0.005}}\right]^{\frac{3}{8}}$$

$$= 2.80\ \text{ft}\quad(33.6\ \text{in})$$

Use 36 in pipe.

The answer is (C).

5.

$$I = 5.3\ \text{in/hr}$$

$$Q = CIA = (0.5)\left(5.3\ \frac{\text{in}}{\text{hr}}\right)(75\ \text{ac}) = 198.8\ \text{ac-in/hr}\quad(200\ \text{ft}^3/\text{sec})$$

(a) Take a worst-case approach. See if the culvert has a capacity equal to or greater than 200 ft^3/sec when water is at the elevation of the roadway.

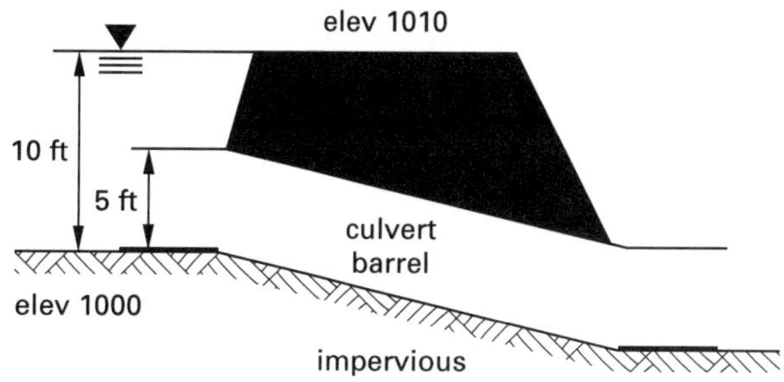

Disregard barrel friction. $K_e = 0.90$ for the projecting, square end. Assume Type 6 flow.

$$H = 10\ \text{ft} + (36\ \text{ft})(0.01) - 5\ \text{ft} = 5.36\ \text{ft}$$

From Eq. 19.100,

$$\text{v} = \sqrt{\frac{\frac{5.36\ \text{ft}}{1+0.9}}{(2)\left(32.2\ \frac{\text{ft}}{\text{sec}^2}\right)}} = 13.48\ \text{ft/sec}$$

$$Q = A\text{v} = \left(\frac{\pi}{4}\right)(5\ \text{ft})^2\left(13.48\ \frac{\text{ft}}{\text{sec}}\right) = 265\ \text{ft}^3/\text{sec}$$

The excess capacity is

$$265\ \frac{\text{ft}^3}{\text{sec}} - 200\ \frac{\text{ft}^3}{\text{sec}} = 65\ \text{ft}^3/\text{sec}$$

There will be no flooding, as the culvert has excess capacity.

The answer is (D).

- Alternate method:

$$C_d = 0.72\quad\text{[Table 17.5]}$$

$$H = 5.36\ \text{ft}$$

$$Q = C_d A\sqrt{2gh} = (0.72)\left(\frac{\pi}{4}\right)(5\ \text{ft})^2 \times \sqrt{(2)\left(32.2\ \frac{\text{ft}}{\text{sec}^2}\right)(5.36\ \text{ft})}$$

$$= 263\ \text{ft}^3/\text{sec}$$

(b) For the rubble channel,

$$n = 0.017$$

$$S = 0.01$$

$$R = \frac{5d}{2d+5}$$

$$C = \left(\frac{1.49}{0.017}\right)\left(\frac{5d}{2d+5}\right)^{\frac{1}{6}}$$

$$Q = A\text{v} = (5d)\left(\frac{1.49}{0.017}\right)\left(\frac{5d}{2d+5}\right)^{0.667}\sqrt{0.01}$$

$$200\ \frac{\text{ft}^3}{\text{sec}} = (43.8d)\left(\frac{5d}{2d+5}\right)^{0.667}$$

By trial and error,

$$d = \text{depth in channel} = \boxed{3.55 \text{ ft}}$$

The answer is (C).

(c) The full capacity of the culvert is calculated from the Chezy-Manning equation.

$$R = \frac{D}{4} = \frac{5 \text{ ft}}{4} = 1.25 \text{ ft}$$
$$A = \left(\frac{\pi}{4}\right) D^2 = \left(\frac{\pi}{4}\right)(5 \text{ ft})^2 = 19.63 \text{ ft}^2$$
$$n = 0.013 \quad \text{[assumed]}$$

From Eq. 19.13,

$$\begin{aligned} Q_{\text{full}} &= \left(\frac{1.49}{n}\right) A R^{\frac{2}{3}} \sqrt{S} \\ &= \left(\frac{1.49}{0.013}\right)(19.63 \text{ ft}^2)(1.25 \text{ ft})^{\frac{2}{3}} \sqrt{0.01} \\ &= 261 \text{ ft}^3/\text{sec} \end{aligned}$$

This is close to (but not the same as) the 265 ft^3/sec calculated for the pressure flow.

$$\frac{Q}{Q_{\text{full}}} = \frac{200 \ \frac{\text{ft}^3}{\text{sec}}}{261 \ \frac{\text{ft}^3}{\text{sec}}} = 0.77$$

Use the table of circular channel ratios. From App. 19.C,

$$\frac{d}{D} \approx 0.72$$

$$d_{\text{barrel}} = (0.72)(5 \text{ ft}) = \boxed{3.6 \text{ ft}}$$

The answer is (C).

6. Since the first storm produces 1 in net and the second storm produces 2 in, divide all runoffs by two. Offset the second storm by 4 hr.

hour	first storm	second storm	total
0	0		0
2	50		50
4	175	0	175
6	300	100	400
8	210	350	560
10	150	600	750
12	125	420	545
14	75	300	375
16	50	250	300
18	≈ 38	150	188
20	25	100	125
22	≈ 12	≈ 75	87
24	0	50	50
26		≈ 25	25
28		0	0

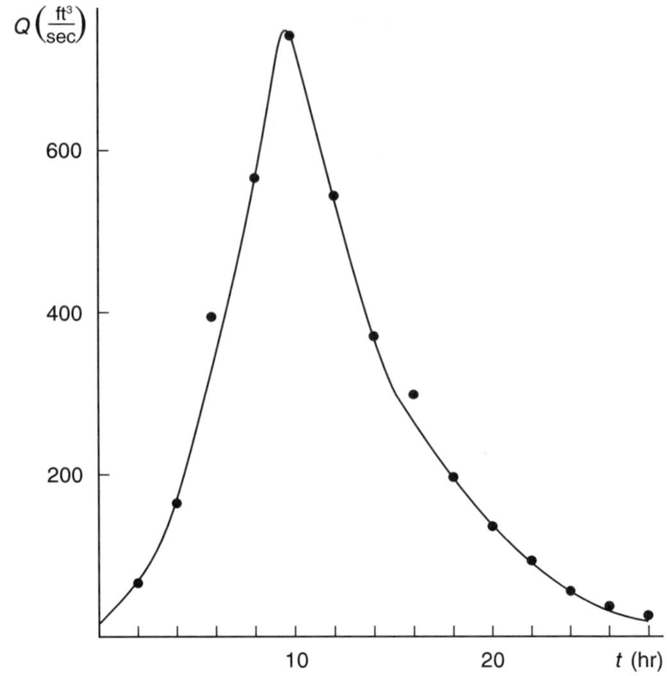

7. From Eq. 20.48,

$$E_{\text{reservoir}} = C_p E_p = (0.7)(0.8 \text{ in}) = \boxed{0.56 \text{ in}}$$

The answer is (A).

8. For 99.97% confidence, $z = \pm 3$. The inflow distribution is a normal distribution that extends $\pm 3\sigma$, or from 0.3 to 1.5. Take the cell width as $\frac{1}{2}\sigma$ or 0.1. Then, for the first cell,

actual endpoints: 0.3 to 0.4

midpoint: 0.35

z limits: -3.0 to -2.5

area under curve: $0.5 - 0.49 = 0.01$

The following inflow distribution is produced similarly.

endpoints	midpoint	z limits	area under curve	cum. × 100
0.3 to 0.4	0.35	−3.0 to −2.5	0.01	1
0.4 to 0.5	0.45	−2.5 to −2.0	0.02	3
0.5 to 0.6	0.55	−2.0 to −1.5	0.04	7
0.6 to 0.7	0.65	−1.5 to −1.0	0.09	16
0.7 to 0.8	0.75	−1.0 to −0.5	0.15	31
0.8 to 0.9	0.85	−0.5 to 0	0.19	50
0.9 to 1.0	0.95	0 to 0.5	0.19	69
1.0 to 1.1	1.05	0.5 to 1.0	0.15	84
1.1 to 1.2	1.15	1.0 to 1.5	0.09	93
1.2 to 1.3	1.25	1.5 to 2.0	0.04	97
1.3 to 1.4	1.35	2.0 to 2.5	0.02	99
1.4 to 1.5	1.45	2.5 to 3.0	0.01	100

Choose 12 random numbers less than 100 from App. 20.B. Use the third row, reading to the right.

- Inflow distribution:

month	random number	corresponding midpoint
1	06	0.55
2	40	0.85
3	18	0.75
4	73	1.05
5	97	1.25
6	72	1.05
7	89	1.15
8	83	1.05
9	24	0.75
10	41	0.85
11	88	1.15
12	86	1.15

Simulate the reservoir operation.

month	starting volume +	inflow −	constant use	= ending volume +	spill
1	5.5	0.55	0.7	5.35	
2	5.35	0.85	0.7	5.5	
3	5.5	0.75	0.7	5.55	
4	5.55	1.05	0.7	5.9	
5	5.9	1.25	0.7	6.45	
6	6.45	1.05	0.7	6.8	
7	6.8	1.15	0.7	7.0	0.25
8	7.0	1.05	0.7	7.0	0.35
9	7.0	0.75	0.7	7.0	0.05
10	7.0	0.85	0.7	7.0	0.15
11	7.0	1.15	0.7	7.0	0.45
12	7.0	1.15	0.7	7.0	0.45

9. Use the same simulation procedure for inflow as in Prob. 8. Proceed similarly.

- Demand distribution:

endpoints	midpoint	z limits	cum. × 100
0.1 to 0.2	0.15	−3.0 to −2.5	1
0.2 to 0.3	0.25	−2.5 to −2.0	3
0.3 to 0.4	0.35	−2.0 to −1.5	7
0.4 to 0.5	0.45	−1.5 to −1.0	16
0.5 to 0.6	0.55	−1.0 to −0.5	31
0.6 to 0.7	0.65	−0.5 to 0	50
0.7 to 0.8	0.75	0 to 0.5	69
0.8 to 0.9	0.85	0.5 to 1.0	84
0.9 to 1.0	0.95	1.0 to 1.5	93
1.0 to 1.1	1.05	1.5 to 2.0	97
1.1 to 1.2	1.15	2.0 to 2.5	99
1.2 to 1.3	1.25	2.5 to 3.0	100

Choose 12 random numbers. Use the fourth row, reading to the right.

- Demand distribution:

month	random number	use
1	04	0.35
2	75	0.85
3	41	0.65
4	44	0.65
5	89	0.95
6	39	0.65
7	42	0.65
8	09	0.45
9	42	0.65
10	11	0.45
11	58	0.75
12	04	0.35

Simulate the reservoir operation.

month	starting volume +	inflow −	monthly demand =	ending volume +	spill
1	5.5	0.55	0.35	5.7	
2	5.7	0.85	0.85	5.7	
3	5.7	0.75	0.65	5.8	
4	5.8	1.05	0.65	6.2	
5	6.2	1.25	0.95	6.5	
6	6.5	1.05	0.65	6.9	
7	6.9	1.15	0.65	7.0	0.4
8	7.0	1.05	0.45	7.0	0.6
9	7.0	0.75	0.65	7.0	0.1
10	7.0	0.85	0.45	7.0	0.4
11	7.0	1.15	0.75	7.0	0.4
12	7.0	1.15	0.35	7.0	0.8

10. Note that all dates correspond to the end of the month. Draw the mass diagram.

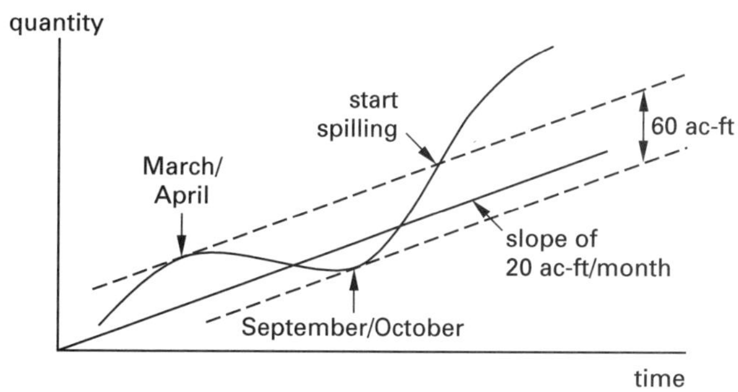

If the reservoir had been full in March or April, the maximum shortfall would have been 60 ac-ft, and the reservoir would be essentially empty in September and October.

60 ac-ft

The answer is (B).

11. (a) Assume the reservoir is initially empty. Draw the mass diagram.

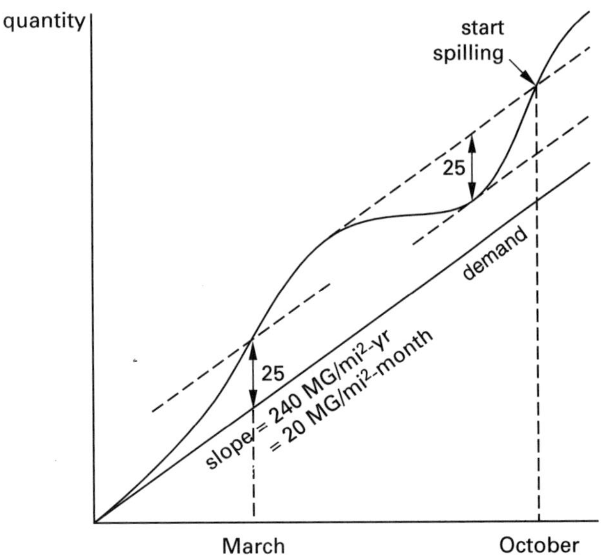

$$\text{capacity} = \boxed{25 \text{ MG/mi}^2}$$

The answer is (A).

(b) The reservoir would start spilling around the end of March. It would also begin spilling around the end of October.

The answer is (D).

21 Groundwater

PRACTICE PROBLEMS

Wells

1. A well extends from the ground surface at elevation 383 ft through a gravel bed to a layer of bedrock at elevation 289 ft. The screened well is 1500 ft from a river whose surface level is 363 ft. The well is pumped by a 10 in diameter schedule-40 steel pipe which draws 120,000 gal/day. The hydraulic conductivity of the well is 1600 gal/day-ft^2. The pump discharges into a piping network whose friction head is 100 ft. What net power is required for steady flow?

(A) 1.1 hp
(B) 2.6 hp
(C) 7.8 hp
(D) 15 hp

2. (*Time limit: one hour*) An aquifer consists of a homogeneous material 300 ft thick. The surface of the water table in this aquifer is 100 ft below ground surface. An 18 in diameter well extends through the top 100 ft and then 200 ft below the water table, for a total depth of 300 ft. The aquifer transmissivity is 10,000 gal/day-ft. The well's radius of influence is 900 ft with a 20 ft drawdown at the well.

(a) The geologic formation in which the well is installed is called

(A) an aquifuge.
(B) an artesian well.
(C) a connate aquifer.
(D) an unconfined aquifer.

(b) What is the hydraulic conductivity of the aquifer?

(A) 50 gal/day-ft^2
(B) 500 gal/day-ft^2
(C) 1300 gal/day-ft^2
(D) 10,000 gal/day-ft^2

(c) What steady discharge is possible?

(A) 0.26 ft^3/sec
(B) 0.52 ft^3/sec
(C) 1.2 ft^3/sec
(D) 5.2 ft^3/sec

(d) What is the drawdown 100 ft from the well?

(A) 1 ft
(B) 6 ft
(C) 13 ft
(D) 18 ft

(e) Assuming a reasonable pump efficiency, what horsepower motor should be selected to achieve the steady discharge?

(A) 3.6 hp
(B) 5.5 hp
(C) 7.5 hp
(D) 12.2 hp

The same well conditions are found at an adjacent site where the aquifer is overlain by a material of low permeability that is 200 ft thick. The piezometric surface is 100 ft below the ground surface.

(f) The geologic formation in which the well is installed is called

(A) an aquifuge.
(B) a confined aquifer.
(C) a spring.
(D) a vadose well.

(g) The 18 in diameter well extends 300 ft below the ground surface to the bottom of the aquifer. The aquifer transmissivity is 10,000 gal/day-ft. What is the hydraulic conductivity of the aquifer?

(A) 50 gal/day-ft^2
(B) 100 gal/day-ft^2
(C) 500 gal/day-ft^2
(D) 800 gal/day-ft^2

(h) What is the steady discharge in the well if the radius of influence is 900 ft with a 20 ft drawdown at the well?

(A) 0.27 ft^3/sec
(B) 0.62 ft^3/sec
(C) 1.2 ft^3/sec
(D) 3.5 ft^3/sec

(i) At what distance from the well is the drawdown equal to 10 ft?

(A) 3 ft
(B) 26 ft
(C) 130 ft
(D) 450 ft

(j) After pumping is stopped, groundwater flows in one direction through the aquifer. If the aquifer is only 100 ft wide and the porosity is 0.4, what is the area in clear flow?

(A) 2000 ft^2
(B) 4000 ft^2
(C) 8000 ft^2
(D) 20,000 ft^2

Flow Nets

3. (*Time limit: one hour*) An impervious concrete dam is shown. The dam reduces seepage by using two impervious sheets extending 5 m below the dam bottom.

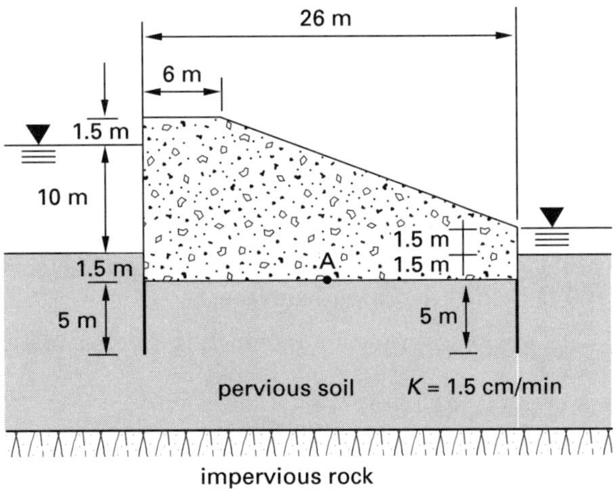

(a) Sketch the flow net.

(b) Determine the seepage per meter of width.

(A) 0.03 m^3/min
(B) 0.1 m^3/min
(C) 0.3 m^3/min
(D) 0.9 m^3/min

(c) What is the uplift pressure on the dam at point A, midway between the left and right edges?

(A) 17 kPa
(B) 33 kPa
(C) 56 kPa
(D) 91 kPa

4. (*Time limit: one hour*) A cofferdam in a river is shown. Sheet piles extend below the mud line, and the cofferdam floor is unlined. (The figure is drawn to scale.) Draw the flow net.

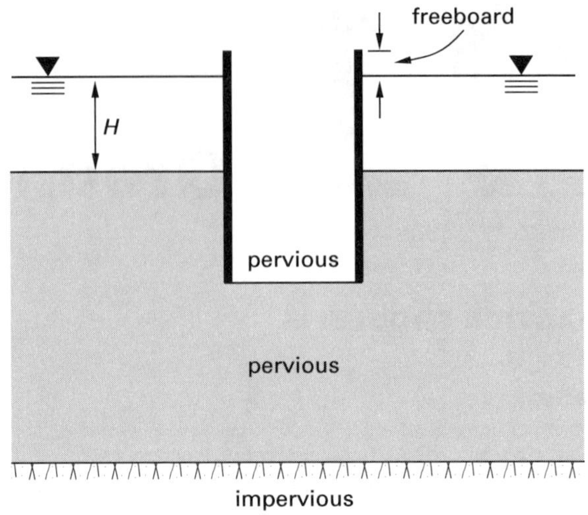

5. (*Time limit: one hour*) The concrete dam shown is drawn to scale. The silt has a hydraulic conductivity of 0.15 ft/hr.

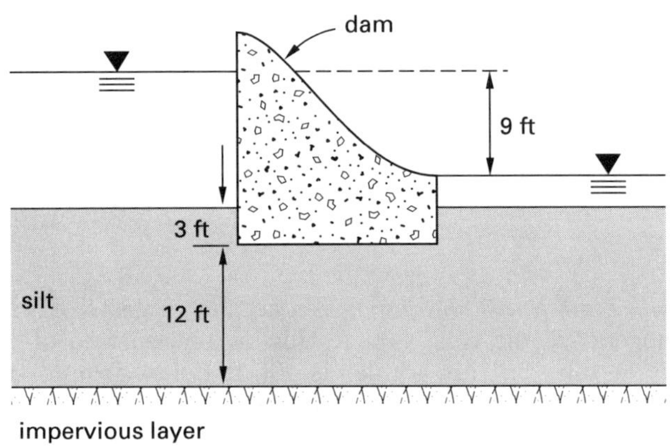

(a) Sketch the flow net for the dam.

(b) What is the approximate seepage rate per foot of width?

(A) 0.05 ft^3/hr-ft width
(B) 0.20 ft^3/hr-ft width
(C) 0.40 ft^3/hr-ft width
(D) 0.90 ft^3/hr-ft width

SOLUTIONS

1.

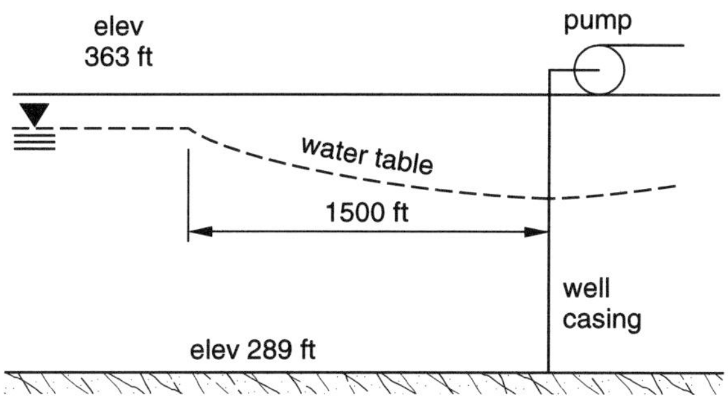

From App. 16.B, for schedule-40 pipe,

$$D_i = 0.835 \text{ ft}$$
$$A_i = 0.5476 \text{ ft}^2$$
$$Q = \left(120{,}000 \ \frac{\text{gal}}{\text{day}}\right)\left(1.547\times10^{-6} \ \frac{\text{ft}^3\text{-day}}{\text{sec-gal}}\right) = 0.1856 \text{ ft}^3/\text{sec}$$
$$K_p = \left(1600 \ \frac{\text{gal}}{\text{day-ft}^2}\right)\left(0.1337 \ \frac{\text{ft}^3}{\text{gal}}\right) = 213.9 \text{ ft}^3/\text{day-ft}^2$$
$$y_1 = 363 \text{ ft} - 289 \text{ ft} = 74 \text{ ft}$$
$$r_1 = 1500 \text{ ft}$$
$$r_2 = \frac{0.835 \text{ ft}}{2} = 0.4175 \text{ ft}$$

From Eq. 21.25,

$$0.1856 \ \frac{\text{ft}^3}{\text{sec}} = \frac{\pi\left(213.9 \ \frac{\text{ft}^3}{\text{day-ft}^2}\right)\left((74 \text{ ft})^2 - y_2^2\right)}{\left(86{,}400 \ \frac{\text{sec}}{\text{day}}\right)\ln\left(\frac{1500 \text{ ft}}{0.4175 \text{ ft}}\right)}$$
$$195.36 \text{ ft}^2 = (74 \text{ ft})^2 - y_2^2$$
$$y_2 = 72.67 \text{ ft}$$

The drawdown is

$$d = 74 \text{ ft} - 72.67 \text{ ft} = 1.33 \text{ ft}$$

d is small compared to y.

The velocity of the water in the pipe is

$$\text{v} = \frac{Q}{A} = \frac{0.1856 \ \frac{\text{ft}^3}{\text{sec}}}{0.5476 \text{ ft}^2} = 0.34 \text{ ft/sec}$$

This velocity is too small to include the velocity head.

The suction lift is

$$383 \text{ ft} - 363 \text{ ft} + 1.33 \text{ ft} = 21.33 \text{ ft}$$
$$H = 21.33 \text{ ft} + 100 \text{ ft} = 121.33 \text{ ft}$$

The mass flow of the water is

$$\dot{m} = \left(0.1856 \ \frac{\text{ft}^3}{\text{sec}}\right)\left(62.4 \ \frac{\text{lbm}}{\text{ft}^3}\right) = 11.58 \text{ lbm/sec}$$

From Table 18.5, the net water horsepower is

$$P = \frac{h_A\dot{m}}{550}\times\frac{g}{g_c} = \frac{(121.33 \text{ ft})\left(11.58 \ \frac{\text{lbm}}{\text{sec}}\right)}{550 \ \frac{\text{ft-lbf}}{\text{hp-sec}}}\times\frac{32.2 \ \frac{\text{ft}}{\text{sec}^2}}{32.2 \ \frac{\text{ft-lbm}}{\text{lbf-sec}^2}} = \boxed{2.55 \text{ hp}}$$

The answer is (B).

2. (a) The geologic formation is an unconfined aquifer. If the aquifer were overlain by an impermeable layer, it would be a confined, or artesian, aquifer.

The answer is (D).

(b) From Eq. 21.13,

$$K = \frac{T}{Y} = \frac{10{,}000 \ \frac{\text{gal}}{\text{day-ft}}}{200 \text{ ft}} = \boxed{50 \text{ gal/day-ft}^2}$$

The answer is (A).

(c)

$$y_1 = 200 \text{ ft at } r_1 = 900 \text{ ft}$$
$$y_2 = 200 \text{ ft} - 20 \text{ ft} = 180 \text{ ft at } r_2$$
$$r_2 = \frac{\frac{18 \text{ in}}{2}}{12 \ \frac{\text{in}}{\text{ft}}} = 0.75 \text{ ft}$$

From Eq. 21.25,

$$Q = \frac{\pi K(y_1^2 - y_2^2)}{\ln\left(\frac{r_1}{r_2}\right)} = \frac{\pi\left(50 \ \frac{\text{gal}}{\text{day-ft}^2}\right)\left(0.13368 \ \frac{\text{ft}^3}{\text{gal}}\right)\times\left((200 \text{ ft})^2 - (180 \text{ ft})^2\right)}{\ln\left(\frac{900 \text{ ft}}{0.75 \text{ ft}}\right)\left(86{,}400 \ \frac{\text{sec}}{\text{day}}\right)} = \boxed{0.261 \text{ ft}^3/\text{sec}}$$

The answer is (A).

(d)

$$y_1 = 200 \text{ ft at } r_1$$
$$r_1 = 900 \text{ ft}$$
$$r_2 = 100 \text{ ft}$$
$$Q = 0.261 \text{ ft}^3/\text{sec}$$

Rearranging Eq. 21.25,

$$y_2^2 = y_1^2 - \frac{Q \ln\left(\frac{r_1}{r_2}\right)}{\pi K}$$
$$= (200 \text{ ft})^2 - \frac{\left(0.261 \frac{\text{ft}^3}{\text{sec}}\right) \ln\left(\frac{900 \text{ ft}}{100 \text{ ft}}\right)\left(86{,}400 \frac{\text{sec}}{\text{day}}\right)}{\pi\left(50 \frac{\text{gal}}{\text{day-ft}^2}\right)\left(0.13368 \frac{\text{ft}^3}{\text{gal}}\right)}$$
$$y_2 = 194 \text{ ft}$$

The drawdown is

$$200 \text{ ft} - 194 \text{ ft} = \boxed{6 \text{ ft}}$$

The answer is (B).

(e) This is a low-capacity pump, so the efficiency will be lower. Assume a pump efficiency of 0.65. From Table 18.5, the hydraulic horsepower is

$$P = \frac{h_A \dot{V}(\text{SG})}{8.814}$$
$$= \frac{(100 \text{ ft} + 20 \text{ ft})\left(0.261 \frac{\text{ft}^3}{\text{sec}}\right)(1)}{\left(8.814 \frac{\text{sec}}{\text{ft}^4\text{-hp}}\right)(0.65)}$$
$$= \boxed{5.47 \text{ hp}}$$

Choose a standard motor size of 7.5 hp.

The answer is (C).

(f) The geologic formation is a confined aquifer. The well is an artesian well.

The answer is (B).

(g) The aquifer depth, Y, is 100 ft. Y is the thickness of the aquifer, not the height of the water table or piezometric surface.

From Eq. 21.13,

$$K = \frac{T}{Y} = \frac{10{,}000 \frac{\text{gal}}{\text{day-ft}}}{100 \text{ ft}}$$
$$= \boxed{100 \text{ gal/day-ft}^2}$$

The answer is (B).

(h)

$$r_1 = 900 \text{ ft}$$
$$y_1 = 200 \text{ ft at } r_1$$
$$r_2 = \frac{\frac{18 \text{ in}}{2}}{12 \frac{\text{in}}{\text{ft}}} = 0.75 \text{ ft}$$
$$y_2 = 200 \text{ ft} - 20 \text{ ft} = 180 \text{ ft at } r_2$$

From Eq. 21.27,

$$Q = \frac{2\pi K(y_1 - y_2)Y}{\ln\left(\frac{r_1}{r_2}\right)}$$
$$= \frac{2\pi\left(100 \frac{\text{gal}}{\text{day-ft}^2}\right)(200 \text{ ft} - 180 \text{ ft})(100 \text{ ft})}{\ln\left(\frac{900 \text{ ft}}{0.75 \text{ ft}}\right)}$$
$$= \frac{\left(1.772 \times 10^5 \frac{\text{gal}}{\text{day}}\right)\left(0.002228 \frac{\text{ft}^3\text{-min}}{\text{gal-sec}}\right)}{\left(24 \frac{\text{hr}}{\text{day}}\right)\left(60 \frac{\text{min}}{\text{hr}}\right)}$$
$$= \boxed{0.274 \text{ ft}^3/\text{sec}}$$

The answer is (A).

(i) Rearrange Eq. 21.27 to find the drawdown distance, r_2.

$$y_2 = 200 \text{ ft} - 10 \text{ ft} = 190 \text{ ft}$$
$$\ln\left(\frac{r_1}{r_2}\right) = 2\pi K(y_1 - y_2)\left(\frac{Y}{Q}\right)$$
$$= \frac{2\pi\left(100 \frac{\text{gal}}{\text{day-ft}^2}\right)(200 \text{ ft} - 190 \text{ ft})(100 \text{ ft})}{177{,}239 \frac{\text{gal}}{\text{day}}}$$
$$= 3.545$$
$$r_2 = \frac{r_1}{e^{3.545}} = \frac{900 \text{ ft}}{e^{3.545}}$$
$$= \boxed{26 \text{ ft}}$$

The answer is (B).

(j) Darcy's law assumes the total cross-sectional area. To obtain the cross-sectional area of the pores (voids), multiply by the porosity.

$$A_{\text{clear flow}} = nA = nbY$$
$$= (0.4)(100 \text{ ft})(100 \text{ ft})$$
$$= \boxed{4000 \text{ ft}^2}$$

The answer is (B).

3. (a) One possible flow net is shown.

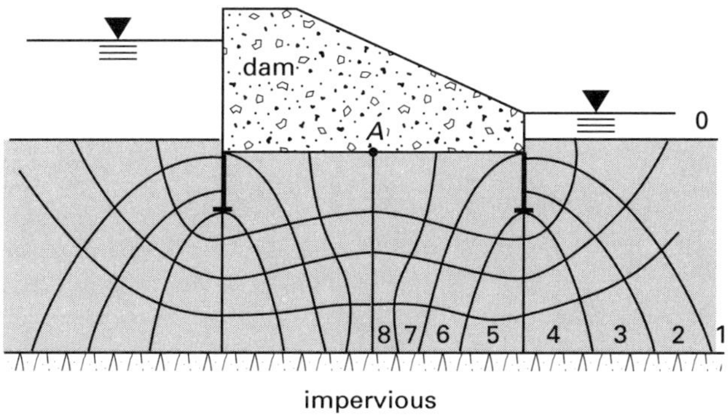

(b) Use Eq. 21.33 to solve for the flow rate.

$$Q = KH\left(\frac{N_f}{N_p}\right)$$

From the flow net as drawn, $N_f = 4$ and $N_p = 16$.

$$H = 10 \text{ m} - 1.5 \text{ m} = 8.5 \text{ m}$$

$$Q = \left(1.5 \ \frac{\text{cm}}{\text{min}}\right)\left(\frac{1 \text{ m}}{100 \text{ cm}}\right)(8.5 \text{ m})\left(\tfrac{4}{16}\right)\left(1 \ \frac{\text{m}}{\text{m}}\right)$$

$$= \boxed{0.031875 \text{ m}^3/\text{min} \quad [\text{per meter of width}]}$$

The answer is (A).

(c) From Eq. 21.36,

$$p_u = \left(\left(\frac{j}{N_p}\right)H + z\right)\rho_w g$$

$$= \left(\left(\tfrac{8}{16}\right)(8.5 \text{ m}) + 1.5 \text{ m}\right)\left(1000 \ \frac{\text{kg}}{\text{m}^3}\right)\left(9.81 \ \frac{\text{m}}{\text{s}^2}\right)$$

$$= \boxed{56\,407 \text{ Pa} \quad (56.4 \text{ kPa})}$$

The answer is (C).

4.

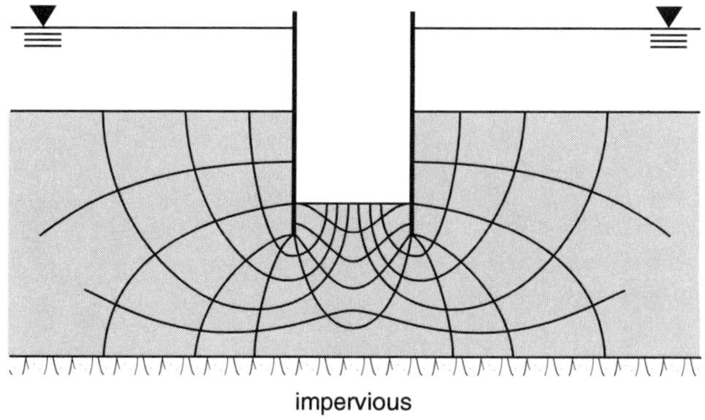

5. (a)

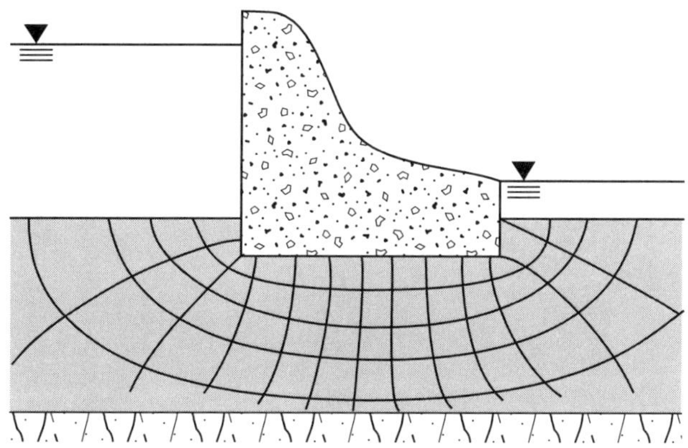

(b) From Eq. 21.33, with 4 flow paths and 13 equipotential drops,

$$Q = \left(0.15 \ \frac{\text{ft}}{\text{hr}}\right)(9.0 \text{ ft})\left(\tfrac{4}{13}\right)$$

$$= \boxed{0.42 \text{ ft}^3/\text{hr-ft width}}$$

The answer is (C).

22 Inorganic Chemistry

PRACTICE PROBLEMS

Empirical Formula Development

1. The gravimetric analysis of a compound is 40% carbon, 6.7% hydrogen, and 53.3% oxygen. What is the simplest formula for the compound?

(A) HCO
(B) HCO_2
(C) CH_2O
(D) CHO_2

2. A municipal water supply has the ionic concentrations listed.

Al^{+++}	0.5 mg/L
Ca^{++}	80.2 mg/L
Cl^{-}	85.9 mg/L
CO_2	19 mg/L
Fe^{++}	1.0 mg/L
HCO_3^-	185 mg/L
Mg^{++}	24.3 mg/L
Na^{+}	46.0 mg/L
SO_4^{--}	125 mg/L

(a) What is the total water hardness in mg/L? (b) How much lime ($Ca(OH)_2$) is required to remove the carbonate hardness?

3. The following concentrations of inorganic compounds are found during a routine analysis of a municipal water supply. Some of the water is treated with lime, and some is passed through a zeolite process. The zeolite has an exchange capacity of 10,000 grains of hardness per cubic foot (22.9 kg/m^3). 0.5 lbm of salt per 1000 grains of $CaCO_3$ hardness (3.5 kg of salt per kg of $CaCO_3$ hardness) are used to rejuvenate the zeolite.

$Ca(HCO_3)_2$	137 mg/L (as $CaCO_3$)
CO_2	0 mg/L
$MgSO_4$	72 mg/L (as $CaCO_3$)

(a) How much lime ($Ca(OH)_2$) is required to soften 1 million gallons (4 million liters) of this water to 100 mg/L as $CaCO_3$ if an excess of 30 mg/L is required for a complete reaction? (b) How much salt is required to rejuvenate the zeolite if 1 million gallons (4 million liters) are softened in that process?

SOLUTIONS

1. Calculate the mole ratios of the atoms by assuming there are 100 g of sample.

For 100 g of sample,

substance	mass	$\frac{m}{AW}$ = no. moles	mole ratio
C	40 g	$\frac{40}{12} = 3.33$	1
H	6.7 g	$\frac{6.7}{1} = 6.7$	2
O	53.3 g	$\frac{53.3}{16} = 3.33$	1

The empirical formula is $\boxed{CH_2O.}$

The answer is (C).

2. Use App. 22.C to determine hardness for both carbonates and noncarbonates.

$$\text{total hardness} = \sum \text{carbonates} + \sum \text{noncarbonates}$$

$$\text{(a) } Al^{+3}: \left(0.5\ \frac{mg}{L}\right)(5.56) = 2.78 \text{ mg/L as } CaCO_3$$

$$Ca^{+2}: \left(80.2\ \frac{mg}{L}\right)(2.5) = 200.5 \text{ mg/L as } CaCO_3$$

$$Fe^{+2}: \left(1.0\ \frac{mg}{L}\right)(1.79) = 1.79 \text{ mg/L as } CaCO_3$$

$$Mg^{+2}: \left(24.3\ \frac{mg}{L}\right)(4.10) = 99.6 \text{ mg/L as } CaCO_3$$

$$\text{total hardness} = \boxed{304.7 \text{ mg/L as } CaCO_3}$$

(Singly charged ions are excluded.)

(b) To remove the carbonate hardness, first remove the carbon dioxide.

$$CO_2: \left(19\ \frac{mg}{L}\ CO_2\right) \times \left(\frac{2.27 \text{ mg } CaCO_3}{L}\right)\left(\frac{1}{1.35}\right) = 31.9 \text{ mg/L } Ca(OH)_2$$

Water Treatment

$$HCO_3^{-1}:\ \left(185\ \frac{\text{mg}}{\text{L}}\ HCO_3^{-1}\right)\times\left(0.82\ \frac{\text{mg}}{\text{L}}\ CaCO_3\right)\left(\frac{1}{1.35}\right)$$
$$= 112.4\ \text{mg/L}\ Ca(OH)_2$$
$$\text{total} = 31.9\ \frac{\text{mg}}{\text{L}} + 112.4\ \frac{\text{mg}}{\text{L}}$$
$$= \boxed{144.3\ \text{mg/L of}\ Ca(OH)_2}$$

Notice that it does not matter if the HCO_3^- carbonate hardness comes from the Ca^{++}, Mg^{++}, Fe^{++}, or Al^{+++}. Adding lime will remove it. (There might still be some of these ions left over in the form of noncarbonate hardness.)

No soda ash is required because it is used to remove noncarbonate hardness.

3. Initial hardness:

$$\begin{aligned} Ca(HCO_3)_2 &= 137\ \text{mg/L}\ CaCO_3 \\ MgSO_4 &= 72\ \text{mg/L}\ CaCO_3 \\ \hline \text{hardness} &= 209\ \text{mg/L}\ CaCO_3 \end{aligned}$$

(a) Use lime to get the hardness down to 100 mg/L $CaCO_3$ knowing that an excess of 30 mg/L of $Ca(OH)_2$ is required.

The total hardness removed by $Ca(OH)_2$ is (209 mg/L $CaCO_3$ − 100 mg/L $CaCO_2$) 109 mg/L $CaCO_3$.

Use App. 22.C to convert from mg/L as $CaCO_3$ to mg/L as $Ca(OH)_2$.

$$\frac{\text{mg}}{\text{L}}\ Ca(OH)_2 = \left(109\ \frac{\text{mg}}{\text{L}}\ CaCO_3\right)\left(\frac{1}{1.35}\right)$$
$$= 80.7\ \text{mg/L as}\ Ca(OH)_2$$

Add an additional 30 mg/L of $Ca(OH)_2$. Use the conversion factor 1 mg/L = 8.345 lbm/10^6 gal.

$$\text{lbm}\ Ca(OH)_2 = \left(80.7\ \frac{\text{mg}}{\text{L}} + 30\ \frac{\text{mg}}{\text{L}}\right)\left(\frac{\frac{8.345\ \text{lbm}}{10^6\ \text{gal}}}{1\ \frac{\text{mg}}{\text{L}}}\right)$$
$$= \boxed{924\ \text{lbm}/10^6\ \text{gal}}$$

(b)

Customary U.S. Solution

Convert 109 mg/L as $CaCO_3$ to lbm/10^6 gal using the following conversion factor.

$$\left(109\ \frac{\text{mg}}{\text{L}}\right)\left(\frac{8.345\ \text{lbm}}{10^6\ \text{gal}}\right) = \frac{909.6\ \text{lbm hardness}}{10^6\ \text{gal}}$$

$$\left(\frac{909.6\ \text{lbm hardness}}{10^6\ \text{gal}}\right)\left(\frac{7000\ \text{grains}}{1\ \text{lbm}}\right)\left(\frac{0.5\ \text{lbm salt}}{1000\ \text{grains}}\right)$$
$$= \boxed{3184\ \text{lbm salt}/10^6\ \text{gal}}$$

SI Solution

$$\frac{\left(109\ \frac{\text{mg}}{\text{L}}\right)(4\times 10^6\text{L})\left(3.5\ \frac{\text{kg}}{\text{kg}}\right)}{10^6\ \frac{\text{mg}}{\text{kg}}} = \boxed{1526\ \text{kg salt}/4\ \text{ML}}$$

23 Water Supply Quality and Testing

PRACTICE PROBLEMS

Phosphorus

1. (*Time limit: one hour*) In a study of a small pond to determine phosphorus impact, the following information was collected.

pond size: 12 ac-ft (14.8×10^6 L)
watershed area: 4 ac (grassland)
average annual rainfall: 5 in
bioavailable P in runoff: 0.01 mg/L
runoff coefficient: 0.1

Biological processes in the pond biota convert phosphorus to a nonbioavailable form at the rate of 22% per year. Recycling of sediment phosphorus by rooted plants and by anaerobic conditions in the hypolimnion converts 12% per year of the nonbioavailable phosphorus back to bioavailable forms.

Runoff into the pond evaporates during the year, so no change in pond volume occurs.

(a) Starting from an initial condition of 0.1 mg/L bioavailable P, what is the P concentration after five annual cycles?

(A) 0.06 mg/L
(B) 0.11 mg/L
(C) 0.13 mg/L
(D) 0.18 mg/L

(b) What can be done to reduce the P accumulation?

(A) Add chemicals to precipitate phosphorus in the pond.
(B) Add chemicals to combine with phosphorus in the pond.
(C) Reduce use of phosphorus-based fertilizers in the surrounding fields.
(D) Use natural-based soaps and detergents in the home.

Alkalinity

2. (*Time limit: one hour*) Groundwater is used for a water supply. It is taken from the ground at 25°C. The initial properties of the water are as follows.

CO_2	60 mg/L (as $CaCO_3$)
alkalinity	200 mg/L (as $CaCO_3$)
pH	7.1

The water is treated for CO_2 removal by spraying into the atmosphere through a nozzle. The final CO_2 concentration is 5.6 mg/L (as $CaCO_3$) at 25°C. The first ionization constant of carbonic acid is 4.45×10^{-7}. What is the final pH of the water after spraying and recovery, assuming the alkalinity is unchanged in the process?

(A) 7.3
(B) 7.9
(C) 8.4
(D) 8.8

Solids

3. (*Time limit: one hour*) The solids content of a stream water sample is to be determined. The total solids content is determined by placing a portion of the sample into a porcelain evaporating dish, drying the sample at 105°C, and igniting the residue by placing the dried sample in a muffle furnace at 550°C. The following masses are recorded.

mass of empty dish: 50.326 g
mass of dish and sample: 118.400 g
mass of dish and dry solids: 50.437 g
mass of dish and ignited solids: 50.383 g

(a) The total solids content is

(A) 900 mg/L
(B) 1000 mg/L
(C) 1100 mg/L
(D) 1600 mg/L

(b) The total volatile solids content is

(A) 630 mg/L
(B) 710 mg/L
(C) 790 mg/L
(D) 830 mg/L

(c) The total fixed solids content is

(A) 270 mg/L
(B) 300 mg/L
(C) 420 mg/L
(D) 840 mg/L

The suspended solids content is determined by filtering a portion of the sample through a glass-fiber filter disk, drying the disk at 105°C, and igniting the residue by placing the dried sample in a muffle furnace at 550°C. The follow masses are recorded.

volume of sample: 30 mL
mass of filter disk: 0.1170 g
mass of disk and dry solids: 0.1278 g
mass of disk and ignited solids: 0.1248 g

(d) The total suspended solids content is

(A) 240 mg/L
(B) 360 mg/L
(C) 370 mg/L
(D) 820 mg/L

(e) The volatile suspended solids content is

(A) 100 mg/L
(B) 120 mg/L
(C) 230 mg/L
(D) 640 mg/L

(f) The fixed suspended solids content is

(A) 120 mg/L
(B) 180 mg/L
(C) 240 mg/L
(D) 260 mg/L

The dissolved solids content is determined by filtering a portion of the sample through a glass-fiber filter disk into a porcelain evaporating dish, drying the sample at 105°C, and igniting the residue by placing the dried sample in a muffle furnace at 550°C. The following masses are recorded.

volume of sample: 25 mL
mass of empty dish: 51.494 g
mass of dish and dry solids: 51.524 g
mass of dish and ignited solids: 51.506 g

(g) The total dissolved solids content is

(A) 1000 mg/L
(B) 1100 mg/L
(C) 1200 mg/L
(D) 1400 mg/L

(h) The volatile dissolved solids content is

(A) 680 mg/L
(B) 720 mg/L
(C) 810 mg/L
(D) 900 mg/L

(i) The fixed dissolved solids content is

(A) 230 mg/L
(B) 300 mg/L
(C) 410 mg/L
(D) 480 mg/L

(j) We would expect the total solids determined in part (a) to be equal to the sum of the total suspended solids determined in (d) plus the total dissolved solids determined in (g). Does the total solids content equal the sum of the suspended and dissolved solids?

(A) yes
(B) no, probably due to poor laboratory technique
(C) no, probably due to rounding of measured values
(D) no, probably because the samples were representative but not identical

Hardness

4. (*Time limit: one hour*) The laboratory analysis of a water sample is as follows. All concentrations are "as substance."

Ca^{++}	74.0 mg/L
Mg^{++}	18.3 mg/L
Na^{+}	27.6 mg/L
K^{+}	39.1 mg/L
pH	7.8
HCO_3^-	274.5 mg/L
SO_4^{--}	72.0 mg/L
Cl^-	49.7 mg/L

(a) The hardness of the water in terms of mg/L of calcium carbonate equivalent is

(A) 1.3 mg/L as $CaCO_3$
(B) 4.7 mg/L as $CaCO_3$
(C) 66 mg/L as $CaCO_3$
(D) 260 mg/L as $CaCO_3$

(b) Based on the laboratory analysis, which of the following can be said?

(A) It is surprising that no carbonates were found in the sample.
(B) The cations in solution, when converted to milliequivalents, will equal the anions, when converted to milliequivalents.
(C) The large amount of bicarbonate in the solution tends to make the water acidic.
(D) None of the above can be said.

(c) Assuming that hypothetical compounds are formed proportionally to the relative concentrations of ions, the hypothetical concentration of calcium bicarbonate in the sample is

(A) 3.7 meq/L
(B) 4.5 meq/L
(C) 4.7 meq/L
(D) 74 meq/L

(d) The hypothetical concentration of magnesium bicarbonate is

(A) 0.8 meq/L
(B) 1.5 meq/L
(C) 1.6 meq/L
(D) 5.2 meq/L

(e) The hypothetical concentration of magnesium sulfate is

(A) 0.7 meq/L
(B) 1.5 meq/L
(C) 4.5 meq/L
(D) 5.2 meq/L

(f) The hypothetical concentration of sodium sulfate is

(A) 0.8 meq/L
(B) 1.2 meq/L
(C) 1.5 meq/L
(D) 6.4 meq/L

(g) The hypothetical concentration of sodium chloride is

(A) 0.2 meq/L
(B) 0.4 meq/L
(C) 1.2 meq/L
(D) 1.4 meq/L

(h) The amount of lime (CaO) necessary in a lime softening process to remove the hardness caused by calcium bicarbonate is

(A) 17 mg/L
(B) 28 mg/L
(C) 95 mg/L
(D) 100 mg/L

(i) To remove hardness caused by magnesium bicarbonate, it is necessary to raise the pH by adding 35 mg/L of CaO in excess of the stoichiometric requirements. The amount of lime (CaO) necessary to remove the carbonate hardness caused by magnesium bicarbonate is

(A) 12 mg/L
(B) 45 mg/L
(C) 56 mg/L
(D) 80 mg/L

(j) The water is subsequently recarbonated to reduce the pH. Assume that by this softening process calcium hardness can be reduced to 30 mg/L and magnesium hardness can be reduced to 10 mg/L, both measured in terms of equivalent calcium carbonate. The amount of hardness that remains in the water is

(A) 10 mg/L as $CaCO_3$
(B) 30 mg/L as $CaCO_3$
(C) 45 mg/L as $CaCO_3$
(D) 75 mg/L as $CaCO_3$

SOLUTIONS

1. (a) annual rainfall = (4 ac)(5 in) = 20 ac-in

$$\begin{aligned}\text{runoff to pond} &= (0.1)(20\text{ ac-in})\\ &\quad\times\left(1.028\times 10^5\ \frac{\text{L}}{\text{ac-in}}\right)\\ &= 205{,}600\text{ L}\\ \begin{array}{r}\text{bioavailable P}\\ \text{reaching pond}\end{array} &= (205{,}600\text{ L})\left(0.01\ \frac{\text{mg}}{\text{L}}\right)\\ &= 2056\text{ mg}\\ \begin{array}{r}\text{initial bioavailable}\\ \text{P in pond}\end{array} &= (14.8\times 10^6\text{ L})\left(0.1\ \frac{\text{mg}}{\text{L}}\right)\\ &= 14.8\times 10^5\text{ mg}\end{aligned}$$

The total bioavailable P in the pond is

$$2056\text{ mg} + 14.8\times 10^5\text{ mg} = 14.82058\times 10^5\text{ mg}$$

Of this, a net percentage of 100% − 22% + 12% = 90% remains after biological processes and recycling.

$$(0.90)(14.82058\times 10^5\text{ mg}) = 13.33852\times 10^5\text{ mg}$$

The following table is prepared in a similar manner.

year	P at start of year	P from rainfall	total P available for activity	P at end of year
1	14.8×10^5	2056	14.82058×10^5	13.33852×10^5
2	13.33852×10^5	2056	13.35908×10^5	12.02317×10^5
3	12.02317×10^5	2056	12.04373×10^5	10.83936×10^5
4	10.83936×10^5	2056	10.85992×10^5	9.77393×10^5
5	9.77393×10^5	2056	9.79449×10^5	8.81504×10^5

The concentration after five years will be

$$\frac{8.81504\times 10^5\text{ mg}}{14.8\times 10^6\text{ L}} = 0.0596\text{ mg/L}$$

The answer is (A).

(b) To reduce the phosphorus accumulation in the pond, the arrival of additional bioavailable phosphorous must be reduced. This entails watershed management to reduce the amount of phosphorus applied as fertilizer and released through other sources. As eutrophication is a natural process accelerated by the availability of plant nutrients (nitrogen and phosphorus especially), reducing phosphorus can slow the process.

While there are chemical means to alter the forms of phosphorus to nonbioavailable states, this usually is not practical on a large scale and is potentially harmful in itself.

It is unlikely that the pond receives untreated discharge from local homes. Use of phosphate-rich detergents in the home will not affect the pond.

The answer is (C).

Water Treatment

2. With CO_2 present, all alkalinity is in the bicarbonate (HCO_3^-) form. Therefore, the equilibrium expression for the first ionization of carbonic acid may be used.

$$CO_2 + H_2O \rightarrow H_2CO_3 \leftrightarrow H^+ + HCO_3^-$$

$$K_1 = \frac{[H^+][HCO_3^-]}{[H_2CO_3]} = 4.45 \times 10^{-7}$$

$$[H_2CO_3] = [CO_2] = \frac{5.6\ \frac{mg}{L}}{\left(100\ \frac{g}{mol}\right)\left(1000\ \frac{mg}{g}\right)} = 5.6 \times 10^{-5}$$

$$[HCO_3^-] = \frac{200\ \frac{mg}{L}}{\left(100\ \frac{g}{mol}\right)\left(1000\ \frac{mg}{g}\right)} = 2 \times 10^{-3}$$

Solve for the hydrogen ion concentration.

$$[H^+] = \frac{K_1[CO_2]}{[H_2CO_3]} = \frac{(4.45 \times 10^{-7})(5.6 \times 10^{-5})}{2 \times 10^{-3}} = 1.246 \times 10^{-8}$$

$$pH = -\log[H^+] = -\log(1.246 \times 10^{-8}) = \boxed{7.9}$$

The answer is (B).

3. (a) The density of water is 1 g/mL. The volume of the tested sample is

$$\frac{118.4\ g - 50.326\ g}{1\ \frac{g}{mL}} = 68.1\ mL$$

$$TS = \frac{(50.437\ g - 50.326\ g)\left(1000\ \frac{mg}{g}\right)\left(1000\ \frac{mL}{L}\right)}{68.1\ mL} = \boxed{1630\ mg/L}$$

The answer is (D).

(b) $$TVS = \frac{(50.437\ g - 50.383\ g) \times \left(1000\ \frac{mg}{g}\right)\left(1000\ \frac{mL}{L}\right)}{68.1\ mL} = \boxed{793\ mg/L}$$

The answer is (C).

(c) $$TFS = 1630\ \frac{mg}{L} - 793\ \frac{mg}{L} = \boxed{837\ mg/L}$$

The answer is (D).

(d) $$TSS = \frac{(0.1278\ g - 0.1170\ g) \times \left(1000\ \frac{mg}{g}\right)\left(1000\ \frac{mL}{L}\right)}{30\ mL} = \boxed{360\ mg/L}$$

The answer is (B).

(e) $$VSS = \frac{(0.1278\ g - 0.1248\ g) \times \left(1000\ \frac{mg}{g}\right)\left(1000\ \frac{mL}{L}\right)}{30\ mL} = \boxed{100\ mg/L}$$

The answer is (A).

(f) $$FSS = 360\ \frac{mg}{L} - 100\ \frac{mg}{L} = \boxed{260\ mg/L}$$

The answer is (D).

(g) $$TDS = \frac{(51.524\ g - 51.494\ g) \times \left(1000\ \frac{mg}{g}\right)\left(1000\ \frac{mL}{L}\right)}{25\ mL} = \boxed{1200\ mg/L}$$

The answer is (C).

(h) $$VDS = \frac{(51.524\ g - 51.506\ g) \times \left(1000\ \frac{mg}{g}\right)\left(1000\ \frac{mL}{L}\right)}{25\ mL} = \boxed{720\ mg/L}$$

The answer is (B).

(i) $$FDS = 1200\ \frac{mg}{L} - 720\ \frac{mg}{L} = \boxed{480\ mg/L}$$

The answer is (D).

(j) If only the first two procedures were performed, the dissolved solids would be calculated as

$$\begin{aligned} \text{TDS} &= \text{TS} - \text{TSS} = 1630\ \frac{\text{mg}}{\text{L}} - 360\ \frac{\text{mg}}{\text{L}} \\ &= 1270\ \text{mg/L}\ \ (\text{vs. } 1200\ \text{mg/L}) \\ \text{VDS} &= \text{TVS} - \text{VSS} = 793\ \frac{\text{mg}}{\text{L}} - 100\ \frac{\text{mg}}{\text{L}} \\ &= 693\ \text{mg/L}\ \ (\text{vs. } 720\ \text{mg/L}) \\ \text{FDS} &= \text{TFS} - \text{FSS} = 837\ \frac{\text{mg}}{\text{L}} - 260\ \frac{\text{mg}}{\text{L}} \\ &= 577\ \text{mg/L}\ \ (\text{vs. } 480\ \text{mg/L}) \end{aligned}$$

The results of the three sets of tests are not the same. These differences are too great to be caused by improper rounding or faulty laboratory technique. This is probably the result of using samples that are not truly identical. For this reason, the suspended solids values are mathematically determined from the total solids and dissolved solids tests.

The answer is (D).

4. (a) Hardness is caused by multivalent cations: Ca^{++} and Mg^{++} in this example. Calculate the milliequivalents by dividing the measured concentration by the milliequivalent weight.

$$Ca^{++}\text{:}\ \ \frac{74.0\ \frac{\text{mg}}{\text{L}}}{20\ \frac{\text{mg}}{\text{meq}}} = 3.7\ \text{meq/L}$$

$$Mg^{++}\text{:}\ \ \frac{18.3\ \frac{\text{mg}}{\text{L}}}{12.2\ \frac{\text{mg}}{\text{meq}}} = 1.5\ \text{meq/L}$$

$$\text{total hardness} = 3.7\ \frac{\text{meq}}{\text{L}} + 1.5\ \frac{\text{meq}}{\text{L}} = 5.2\ \text{meq/L}$$

Determine the calcium carbonate equivalent by multiplying by the equivalent weight of calcium carbonate.

$$\begin{aligned} \text{hardness} &= \left(5.2\ \frac{\text{meq}}{\text{L}}\right)\left(50\ \frac{\text{mg}}{\text{meq}}\right) \\ &= \boxed{260\ \text{mg/L as } CaCO_3} \end{aligned}$$

The answer is (D).

(b) Alkalinity in the form of carbonate radical does not exist at a pH below 8.3. Bicarbonate is a form of alkalinity, and it neutralizes, not creates, acidity. To determine the cation/anion relationship, convert all of the concentrations to milliequivalents by dividing the measured concentrations by the milliequivalent weights.

$$Ca^{++}\text{:}\ \ \frac{74.0\ \frac{\text{mg}}{\text{L}}}{20\ \frac{\text{mg}}{\text{meq}}} = 3.7\ \text{meq/L}$$

$$Mg^{++}\text{:}\ \ \frac{18.3\ \frac{\text{mg}}{\text{L}}}{12.2\ \frac{\text{mg}}{\text{meq}}} = 1.5\ \text{meq/L}$$

$$Na^{+}\text{:}\ \ \frac{27.6\ \frac{\text{mg}}{\text{L}}}{23\ \frac{\text{mg}}{\text{meq}}} = 1.2\ \text{meq/L}$$

$$K^{+}\text{:}\ \ \frac{39.1\ \frac{\text{mg}}{\text{L}}}{39.1\ \frac{\text{mg}}{\text{meq}}} = 1.0\ \text{meq/L}$$

$$\begin{aligned} \text{total cations} &= 3.7\ \frac{\text{meq}}{\text{L}} + 1.5\ \frac{\text{meq}}{\text{L}} + 1.2\ \frac{\text{meq}}{\text{L}} \\ &\quad + 1.0\ \frac{\text{meq}}{\text{L}} \\ &= \boxed{7.4\ \text{meq/L}} \end{aligned}$$

$$HCO_3^{-}\text{:}\ \ \frac{274.5\ \frac{\text{mg}}{\text{L}}}{61\ \frac{\text{mg}}{\text{meq}}} = 4.5\ \text{meq/L}$$

$$SO_4^{--}\text{:}\ \ \frac{72.0\ \frac{\text{mg}}{\text{L}}}{48\ \frac{\text{mg}}{\text{meq}}} = 1.5\ \text{meq/L}$$

$$Cl^{-}\text{:}\ \ \frac{49.7\ \frac{\text{mg}}{\text{L}}}{35.5\ \frac{\text{mg}}{\text{meq}}} = 1.4\ \text{meq/L}$$

$$\begin{aligned} \text{total anions} &= 4.5\ \frac{\text{meq}}{\text{L}} + 1.5\ \frac{\text{meq}}{\text{L}} + 1.4\ \frac{\text{meq}}{\text{L}} \\ &= \boxed{7.4\ \text{meq/L}} \end{aligned}$$

The answer is (B).

(c) To determine the hypothetical compounds, construct a milliequivalent per liter bar graph for cations and anions, as shown.

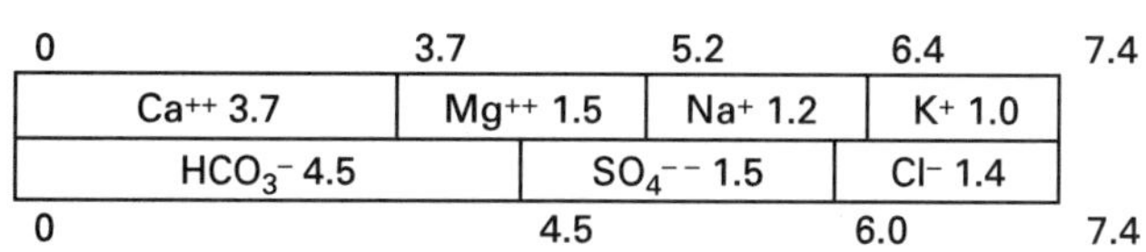

The hypothetical compounds are determined by moving from left to right from ions of greater concentration to ions of lesser concentration.

hypothetical	compound concentration (meq/L)	remarks
$Ca(HCO_3)_2$	3.7	Ca^{++} exhausted
$Mg(HCO_3)_2$	0.8	HCO_3^- exhausted
$MgSO_4$	0.7	Mg^{++} exhausted
Na_2SO_4	0.8	SO_4^{--} exhausted
NaCl	0.4	Na^+ exhausted
KCl	1.0	K^+ and Cl^- exhausted

The answer is (A).

(d) From part (c), the hypothetical concentration of magnesium bicarbonate is $\boxed{0.8 \text{ meq/L.}}$

The answer is (A).

(e) From part (c), the hypothetical concentration of magnesium sulfate is $\boxed{0.7 \text{ meq/L.}}$

The answer is (A).

(f) From part (c), the hypothetical concentration of sodium sulfate is $\boxed{0.8 \text{ meq/L.}}$

The answer is (A).

(g) From part (c), the hypothetical concentration of sodium chloride is $\boxed{0.4 \text{ meq/L.}}$

The answer is (B).

(h) The reactions are

$$CaO + H_2O \rightarrow Ca(OH)_2$$
$$Ca(HCO_3)_2 + Ca(OH)_2 \rightarrow CaCO_3 \downarrow + 2H_2O$$

One molecule of CaO forms one molecule of $Ca(OH)_2$, which in turn reacts with one molecule of $Ca(HCO_3)_2$. There are 3.7 meq/L of $Ca(HCO_3)_2$ in solution. The equivalent weight of CaO is $(40+16)/2 = 28$. Therefore, 3.7 meq/L of CaO are needed.

$$\left(3.7 \ \frac{\text{meq}}{\text{L}} \ \text{CaO}\right)\left(28 \ \frac{\text{mg}}{\text{meq}}\right) = \boxed{103.6 \text{ mg/L CaO}}$$

The answer is (D).

(i) The reactions are

$$CaO + H_2O \rightarrow Ca(OH)_2$$
$$Mg(HCO_3)_2 + 2Ca(OH)_2 \rightarrow 2CaCO_3 \downarrow + Mg(OH)_2 \downarrow + 2H_2O$$

Two molecules of $Ca(OH)_2$ are required for each molecule of $Mg(HCO_3)_2$.

$$\left(0.8 \ \frac{\text{meq}}{\text{L}} \ Mg(HCO_3)_2\right)(2)\left(28 \ \frac{\text{mg}}{\text{meq}}\right) + 35 \ \frac{\text{mg}}{\text{L}} \ \text{excess} = \boxed{79.8 \text{ mg/L}}$$

The answer is (D).

(j) The original hardness was in the hypothetical forms of calcium bicarbonate, magnesium bicarbonate, and magnesium sulfate. The bicarbonates have been removed in parts (h) and (i), leaving the residuals of 30 mg/L calcium hardness and 10 mg/L magnesium hardness. However, no attempt was made to remove the noncarbonate magnesium hardness represented by magnesium sulfate (this would have required the addition of soda ash and additional lime).

The residual hardness is as follows.

$$\text{calcium hardness:} \ \frac{30 \ \frac{\text{mg}}{\text{L}}}{50 \ \frac{\text{mg}}{\text{meq}}} = 0.6 \text{ meq/L}$$

$$\text{magnesium carbonate hardness:} \ \frac{10 \ \frac{\text{mg}}{\text{L}}}{50 \ \frac{\text{mg}}{\text{meq}}} = 0.2 \text{ meq/L}$$

From part (c), the magnesium noncarbonate hardness is 0.7 meq/L.

$$\begin{aligned}\text{residual hardness} &= 0.6 \ \frac{\text{meq}}{\text{L}} + 0.2 \ \frac{\text{meq}}{\text{L}} + 0.7 \ \frac{\text{meq}}{\text{L}} \\ &= 1.5 \text{ meq/L}\end{aligned}$$

$$\left(1.5 \ \frac{\text{meq}}{\text{L}}\right)\left(50 \ \frac{\text{mg}}{\text{meq}}\right) = \boxed{75 \text{ mg/L as } CaCO_3}$$

The answer is (D).

24 Water Supply Treatment and Distribution

PRACTICE PROBLEMS

Plain Sedimentation

1. A settling tank has an overflow rate of 100,000 gal/ft^2-day. Water carrying sediment of various sizes enters the tank. The sediment has the following distribution of settling velocities.

settling velocity (ft/min)	mass fraction remaining
10.0	0.54
5.0	0.45
2.0	0.35
1.0	0.20
0.75	0.10
0.50	0.03

(a) What is the gravimetric percentage of sediment particles completely removed?

(A) 32%
(B) 39%
(C) 47%
(D) 55%

(b) What is the total gravimetric percentage of all sediment particles removed?

(A) 35%
(B) 40%
(C) 50%
(D) 60%

2. A spherical sand particle has a specific gravity of 2.6 and a diameter of 1 mm. What is the settling velocity?

(A) 0.2 ft/sec
(B) 0.7 ft/sec
(C) 1.1 ft/sec
(D) 1.6 ft/sec

3. A mechanically cleaned circular clarifier is to be designed with the following characteristics.

flow rate	2.8 MGD
detention period	2 hr
surface loading	700 gal/ft^2-day

(a) What is the approximate diameter?

(A) 45 ft
(B) 60 ft
(C) 70 ft
(D) 90 ft

(b) What is the approximate depth?

(A) 6 ft
(B) 8 ft
(C) 12 ft
(D) 15 ft

(c) If the initial flow rate is reduced to 1.1 MGD, what is the surface loading?

(A) 190 gal/day-ft^2
(B) 230 gal/day-ft^2
(C) 250 gal/day-ft^2
(D) 280 gal/day-ft^2

(d) If the initial flow rate is reduced to 1.1 MGD, what is the average detention period?

(A) 4 hr
(B) 5 hr
(C) 6 hr
(D) 8 hr

4. (*Time limit: one hour*) A water treatment plant is designed to handle a total flow rate of 1.5 MGD. The current design includes two identical sedimentation basins with the following characteristics.

plan area	90 ft × 16 ft
depth	12 ft
total weir length	48 ft per basin
three-month sustained average low	70% of the design average daily flow
three-month sustained average high	200% of the design average daily flow

The basins must meet the following government standards.

minimum retention time	4.0 hrs
maximum weir load	20,000 gpd/ft
maximum velocity	0.5 ft/min

Do the basins meet the standards?

(A) Yes, specifications are met at both peak and low flows.
(B) No, specifications are not met at low flow.
(C) No, specifications are not met at high flow.
(D) No, specifications are not met at either high or low flow.

Mixing Physics

5. (*Time limit: one hour*) A flocculator tank with a volume of 200,000 ft^3 uses a paddle wheel to mix the coagulant in 60°F water. The operating characteristics are as follows.

mean velocity gradient	45 sec^{-1}
paddle drag coefficient	1.75
paddle tip velocity	2 ft/sec
relative water/paddle velocity	1.5 ft/sec

(a) What is the theoretical power required to drive the paddle?

(A) 12 hp
(B) 17 hp
(C) 60 hp
(D) 120 hp

(b) What is the drag force on the paddle?

(A) 450 lbf
(B) 910 lbf
(C) 5500 lbf
(D) 6400 lbf

(c) What is the required paddle area?

(A) 900 ft^2
(B) 1200 ft^2
(C) 1500 ft^2
(D) 1700 ft^2

Filtration

6. A water treatment plant has four square rapid sand filters. The flow rate is 4 gal/min-ft^2. Each filter has a treatment capacity of 600,000 gal/day. Each filter is backwashed once a day for 8 min. The rate of rise during washing is 24 in/min.

(a) What are the inside dimensions of each filter?

(A) 6 ft × 6 ft
(B) 8 ft × 8 ft
(C) 10 ft × 10 ft
(D) 12 ft × 12 ft

(b) What percentage of the filtered water is used for backwashing?

(A) 2%
(B) 4%
(C) 8%
(D) 10%

7. A water treatment plant has five identical rapid sand filters. Each is square in plan and has a capacity of 1.0 MGD. The application rate is 4 gal/min-ft^2, and the total wetted depth is 10 ft.

(a) What should be the cross-section dimensions of each filter?

(A) 7 ft × 7 ft
(B) 9 ft × 9 ft
(C) 10 ft × 10 ft
(D) 13 ft × 13 ft

(b) Each filter is backwashed every day for 5 min. The rate of rise of the backwash water is 2 ft/min. What percentage of the plant's filtered water will be used for backwashing?

(A) 1.3%
(B) 1.9%
(C) 2.4%
(D) 3.1%

Precipitation Softening

8. A town's water supply has the following ionic concentrations.

Al^{+++}	0.5 mg/L
Ca^{++}	80.2 mg/L
Cl^-	85.9 mg/L
CO_2	19 mg/L
CO_3^{--}	0
Fe^{++}	1.0 mg/L
Fl^-	0
HCO_3^-	185 mg/L
Mg^{++}	24.3 mg/L
Na^+	46.0 mg/L
NO_3^-	0
SO_4^{--}	125 mg/L

(a) What is the total hardness?

(A) 160 mg/L as $CaCO_3$
(B) 200 mg/L as $CaCO_3$
(C) 260 mg/L as $CaCO_3$
(D) 300 mg/L as $CaCO_3$

(b) How much slaked lime is required to combine with the carbonate hardness?

(A) 45 mg/L as substance
(B) 90 mg/L as substance
(C) 130 mg/L as substance
(D) 150 mg/L as substance

(c) How much soda ash is required to react with the carbonate hardness?

(A) none
(B) 15 mg/L as substance
(C) 35 mg/L as substance
(D) 60 mg/L as substance

9. A city's water supply contains the following ionic concentrations.

$Ca(HCO_3)_2$	137 mg/L as $CaCO_3$
$MgSO_4$	72 mg/L as $CaCO_3$
CO_2	0

(a) How much slaked lime is required to soften 1,000,000 gal of this water to a hardness of 100 mg/L if 30 mg/L of excess lime is used?

(A) 930 lbm
(B) 1200 lbm
(C) 1300 lbm
(D) 1700 lbm

(b) How much soda ash is required to soften 1,000,000 gal of this water to a hardness of 100 mg/L if 30 mg/L of excess lime is used?

(A) none
(B) 300 lbm
(C) 500 lbm
(D) 700 lbm

Zeolite Softening

10. The water described in Prob. 9 is to be softened using a zeolite process with the following characteristics: exchange capacity, 10,000 grains/ft^3; and salt requirement, 0.5 lbm per 1000 grains hardness removed. How much salt is required to soften the water to 100 mg/L hardness?

(A) 700 lbm/MG
(B) 1800 lbm/MG
(C) 2500 lbm/MG
(D) 3200 lbm/MG

11. The hardness of water from an underground aquifer is to be reduced from 245 mg/L to 80 mg/L using a zeolite process. The volumetric flow rate is 20,000 gal/day. The process has the following characteristics: resin exchange capacity, 20,000 grains/ft^3; and zeolite volume, 2 ft^3.

(a) What fraction of the water is bypassed around the process?

(A) 0.15
(B) 0.33
(C) 0.67
(D) 0.85

(b) What is the time between regenerations of the softener?

(A) 5 hr
(B) 16 hr
(C) 24 hr
(D) 30 hr

Demand

12. (*Time limit: one hour*) An area is expected to attain a population of 40,000 in 20 years. The cumulative per capita water demand for a peak day in the area is shown in the following diagram.

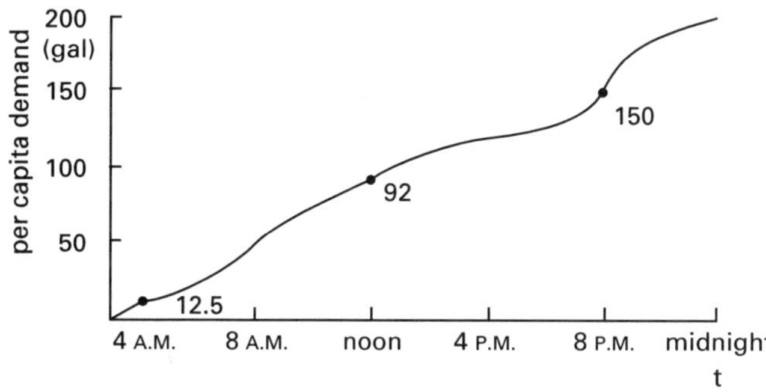

(a) What is the total per capita demand for a peak day?

(A) 160 gal
(B) 200 gal
(C) 240 gal
(D) 290 gal

(b) Assuming uniform operation over 24 hours, what storage volume is required in the treatment plant for all uses, including fire fighting demand?

(A) 1.1 MG
(B) 2.7 MG
(C) 3.2 MG
(D) 5.5 MG

(c) Assume that the pumping station runs uniformly from 4 A.M. until 8 A.M. to fill the storage tanks for the day. What storage is required to meet all uses, including fire fighting?

(A) 3.2 MG
(B) 3.8 MG
(C) 5.5 MG
(D) 8.7 MG

13. The water supply for a town of 15,000 people is taken from a river. The average consumption is 110 gpcd. The water has the following characteristics.

turbidity	20 to 100 NTU (varies)
total hardness	less than 60 mg/L as $CaCO_3$
coliform count	200 to 1000 per 100 mL (varies)

(a) What capacity should the distribution and treatment system have?

(A) 6000 gpm
(B) 7500 gpm
(C) 9500 gpm
(D) 12,000 gpm

Water Treatment

(b) If the application rate is 4 gal/min-ft^2, what total filter area is required?

(A) 290 ft^2
(B) 580 ft^2
(C) 910 ft^2
(D) 1100 ft^2

(c) Is softening required?

(A) No, 60 mg/L is soft water.
(B) No, softening would interfere with turbidity removal.
(C) Yes, the turbidity and coliform counts would also benefit.
(D) Yes, all municipal water should be softened.

(d) If a chlorine dose of 2 mg/L is required to obtain the desired chlorine residual, how much chlorine is required every 24 hours?

(A) 15 lbm
(B) 22 lbm
(C) 28 lbm
(D) 32 lbm

SOLUTIONS

1. (a)
$$v^* = \frac{\left(100{,}000\ \frac{\text{gal}}{\text{ft}^2\text{-day}}\right)\left(0.1337\ \frac{\text{ft}^3}{\text{gal}}\right)}{\left(24\ \frac{\text{hr}}{\text{day}}\right)\left(60\ \frac{\text{min}}{\text{hr}}\right)} = 9.28\ \text{ft/min}$$

Using interpolation, the mass fraction remaining is

$$0.45 + \left(\frac{9.28\ \frac{\text{ft}}{\text{min}} - 5.0\ \frac{\text{ft}}{\text{min}}}{10.0\ \frac{\text{ft}}{\text{min}} - 5.0\ \frac{\text{ft}}{\text{min}}}\right)(0.54 - 0.45) = 0.527 \text{ remains in flow}$$

$$1 - 0.527 = 0.473$$

47.3% completely removed

The answer is (C).

(b) Plot the mass fraction remaining versus the settling velocity.

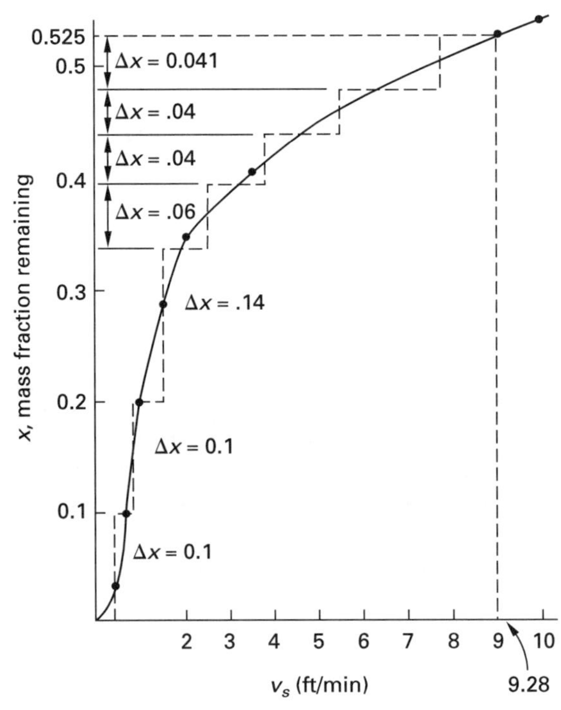

$$v_o = \left(\frac{100{,}000\ \text{gal}}{\text{ft}^2\text{-day}}\right)\left(\frac{1\ \text{ft}^3}{7.48\ \text{gal}}\right)\left(\frac{1\ \text{day}}{24\ \text{h}}\right)\left(\frac{1\ \text{hr}}{60\ \text{min}}\right) = 0.28\ \text{ft/min}$$

$$X_o \simeq 0.525 \quad (52.5\%)$$

Determine $\Delta x v_t$ by graphical integration.

Δx	v_t	$\Delta x v_t$
0.1	0.5	0.05
0.1	0.8	0.08
0.14	1.4	0.196
0.06	2.5	0.15
0.04	3.75	0.15
0.04	5.5	0.22
0.041	7.75	0.318
	total	1.164

The overall removal efficiency is

$$\begin{aligned} X &= 1 - X_o + \Sigma\left(\frac{\Delta x v_t}{v_o}\right) \\ &= 1 - 0.525 + \frac{1.164}{9.28} \\ &= 0.60 \end{aligned}$$

The answer is (D).

2. From Fig. 24.2, $v_s = \boxed{0.7 \text{ ft/sec}}$.

The answer is (B).

3. (a) The surface area is

$$A_{\text{surface}} = \frac{2.8 \times 10^6 \ \frac{\text{gal}}{\text{day}}}{700 \ \frac{\text{gal}}{\text{ft}^2\text{-day}}} = 4000 \text{ ft}^2$$

Since $A = \left(\frac{\pi}{4}\right) D^2$,

$$D = \sqrt{\frac{(4)(4000 \text{ ft}^2)}{\pi}} = \boxed{71.4 \text{ ft}}$$

The answer is (C).

(b) The volume is

$$V = \frac{\left(2.8 \times 10^6 \ \frac{\text{gal}}{\text{day}}\right)(2 \text{ hr})}{24 \ \frac{\text{hr}}{\text{day}}} = 2.333 \times 10^5 \text{ gal}$$

The depth is

$$\begin{aligned} d &= \frac{V}{A_{\text{surface}}} = \frac{(2.333 \times 10^5 \text{ gal})\left(0.1337 \ \frac{\text{ft}^3}{\text{gal}}\right)}{4000 \text{ ft}^2} \\ &= \boxed{7.8 \text{ ft}} \end{aligned}$$

The answer is (B).

(c)

$$v^* = \frac{1.1 \times 10^6 \ \frac{\text{gal}}{\text{day}}}{4000 \text{ ft}^2} = \boxed{275 \text{ gal/day-ft}^2}$$

The answer is (D).

(d) Use Eq. 24.6.

$$t = \frac{(2.333 \times 10^5 \text{ gal})\left(24 \ \frac{\text{hr}}{\text{day}}\right)}{1.1 \times 10^6 \ \frac{\text{gal}}{\text{day}}} = \boxed{5.09 \text{ hr}}$$

The answer is (B).

4. Assume there are two basins in parallel.

$$\text{volume per basin} = (90 \text{ ft})(16 \text{ ft})(12 \text{ ft}) = 17{,}280 \text{ ft}^3$$

(The freeboard is not given.)

The area per basin is

$$A = (90 \text{ ft})(16 \text{ ft}) = 1440 \text{ ft}^2$$

The three-month peak flow per basin is

$$(2)\left(\frac{1.5 \text{ MGD}}{2}\right) = 1.5 \text{ MGD}$$

$$(1.5 \text{ MGD})\left(1.547 \ \frac{\frac{\text{ft}^3}{\text{sec}}}{\text{MGD}}\right) = 2.32 \text{ ft}^3/\text{sec (cfs)}$$

The detention time at peak flow is given by Eq. 24.6.

$$t = \frac{V}{Q} = \frac{17{,}280 \text{ ft}^3}{\left(2.32 \ \frac{\text{ft}^3}{\text{sec}}\right)\left(60 \ \frac{\text{min}}{\text{hr}}\right)\left(60 \ \frac{\text{ft}^3}{\text{hr}}\right)} = 2.07 \text{ hr}$$

Since 2.07 hr < 4 hr, this is not acceptable.

The weir loading is

$$\frac{1.5 \times 10^6 \ \frac{\text{gal}}{\text{day}}}{48 \text{ ft}} = 31{,}250 \text{ gal/day-ft}$$

Since 31,250 > 20,000, this is not acceptable either.

The overflow rate is

$$v^* = \frac{Q}{A} = \frac{\left(2.32 \ \frac{\text{ft}^3}{\text{sec}}\right)\left(60 \ \frac{\text{sec}}{\text{min}}\right)}{1440 \text{ ft}^2} = 0.0967 \text{ ft/min}$$

Since 0.0967 < 0.5, this is acceptable.

At low flow,

$$\text{flow} = \frac{0.7}{2.0} = 0.35 \quad [35\% \text{ of high flow}]$$

$$t = \frac{2.07 \text{ hr}}{0.35} = 5.91 \text{ hr} \quad [\text{acceptable}]$$

$$\text{weir loading} = (0.35)\left(31{,}250 \ \frac{\text{gal}}{\text{day-ft}}\right) = 10{,}938 \ \frac{\text{gal}}{\text{day-ft}} \quad [\text{acceptable}]$$

$$v^* = (0.35)\left(0.1 \ \frac{\text{ft}}{\text{min}}\right) = 0.035 \text{ ft/min} \quad [\text{acceptable}]$$

The basins have been correctly designed for low flow but not for peak flow. One or more basins should be used.

The answer is (C).

5. For 60°F water, $\mu = 2.359 \times 10^{-5}$ lbf-sec/ft^2.

From Eq. 24.21,

$$P = \left(2.359 \times 10^{-5} \ \frac{\text{lbf-sec}}{\text{ft}^2}\right)\left(45 \ \frac{1}{\text{sec}}\right)^2 (200{,}000 \text{ ft}^3) = 9554 \text{ ft-lbf/sec}$$

(a)

$$\text{water horsepower} = \frac{9554 \ \frac{\text{ft-lbf}}{\text{sec}}}{550 \ \frac{\text{ft-lbf}}{\text{hp-sec}}} = \boxed{17.4 \text{ hp}}$$

The answer is (B).

(b) Since work = force × distance, then power = force × velocity.

$$D = \frac{P}{\text{v}} = \frac{9554 \ \frac{\text{ft-lbf}}{\text{sec}}}{1.5 \ \frac{\text{ft}}{\text{sec}}} = \boxed{6369 \text{ lbf}}$$

The answer is (D).

(c) Use Eq. 24.15.

$$A = \frac{(2)\left(32.2 \ \frac{\text{ft}}{\text{sec}^2}\right)(6369 \text{ lbf})}{(1.75)\left(62.4 \ \frac{\text{lbf}}{\text{ft}^3}\right)\left(1.5 \ \frac{\text{ft}}{\text{sec}}\right)^2} = \boxed{1669 \text{ ft}^2}$$

The answer is (D).

6. (a) The area is

$$A = \frac{600{,}000 \ \frac{\text{gal}}{\text{day}}}{\left(4 \ \frac{\text{gal}}{\text{min-ft}^2}\right)\left(24 \ \frac{\text{hr}}{\text{day}}\right)\left(60 \ \frac{\text{min}}{\text{hr}}\right)} = 104.2 \text{ ft}^2$$

$$\boxed{\text{use } 10 \text{ ft} \times 10 \text{ ft}}$$

The answer is (C).

(b) The volume is

$$V = \left(8 \ \frac{\text{min}}{\text{day}}\right)(100 \text{ ft}^2)\left(2 \ \frac{\text{ft}}{\text{min}}\right) = 1600 \text{ ft}^3/\text{day}$$

$$\left(1600 \ \frac{\text{ft}^3}{\text{day}}\right)\left(7.481 \ \frac{\text{gal}}{\text{ft}^3}\right) = 11{,}970 \text{ gal/day}$$

$$\frac{11{,}970 \ \frac{\text{gal}}{\text{day}}}{600{,}000 \ \frac{\text{gal}}{\text{day}}} = \boxed{0.01 \ (2\%)}$$

The answer is (A).

7. (a)

$$A = \frac{1 \times 10^6 \ \frac{\text{gal}}{\text{day}}}{\left(24 \ \frac{\text{hr}}{\text{day}}\right)\left(60 \ \frac{\text{min}}{\text{hr}}\right)\left(4 \ \frac{\text{gal}}{\text{min-ft}^2}\right)} = 173.6 \text{ ft}^2$$

$$\text{width} = \sqrt{173.6 \text{ ft}^2} = 13.2 \text{ ft}$$

$$\boxed{13.2 \text{ ft} \times 13.2 \text{ ft}}$$

The answer is (D).

(b) The water volume is

$$V = \left(5 \ \frac{\text{min}}{\text{day}}\right)(173.6 \text{ ft}^2)\left(2 \ \frac{\text{ft}}{\text{min}}\right)(5 \text{ filters}) = 8680 \text{ ft}^3/\text{day}$$

Convert to gallons.

$$\left(8680 \ \frac{\text{ft}^3}{\text{day}}\right)\left(7.481 \ \frac{\text{gal}}{\text{ft}^3}\right) = 64{,}935 \text{ gal/day}$$

$$\frac{64{,}935 \ \frac{\text{gal}}{\text{day}}}{5 \times 10^6 \ \frac{\text{gal}}{\text{day}}} = \boxed{0.013 \ (1.3\%)}$$

The answer is (A).

8. (a)

	mg/L as substance		factor from App. 22.C		
Ca^{++}:	80.2	×	2.5	=	200.5 mg/L
Mg^{++}:	24.3	×	4.1	=	99.63 mg/L
Fe^{++}:	1	×	1.79	=	1.79 mg/L
Al^{+++}:	0.5	×	5.56	=	2.78 mg/L

$$\text{hardness} = \boxed{304.7 \text{ mg/L}}$$

The answer is (D).

(b) To remove the carbonate hardness,

$$CO_2: \;\; 19 \; \frac{\text{mg}}{\text{L}} \times 2.27 = 43.13 \text{ mg/L as } CaCO_3$$

Add lime to remove the carbonate hardness. It does not matter whether the HCO_3^- comes from Mg^{++}, Ca^{++}, or Fe^{++}; adding lime will remove it.

There may be Mg^{++}, Ca^{++}, or Fe^{++} ions left over in the form of noncarbonate hardness, but the problem asked for carbonate hardness. Converting from mg/L of substance to mg/L as $CaCO_3$,

$$HCO_3^-: \;\; 185 \; \frac{\text{mg}}{\text{L}} \times 0.82 = 151.7 \text{ mg/L}$$

The total equivalents to be neutralized are

$$43.13 \; \frac{\text{mg}}{\text{L}} + 151.7 \; \frac{\text{mg}}{\text{L}} = 194.83 \text{ mg/L}$$

Convert $Ca(OH)_2$ using App. 22.C.

$$\frac{\text{mg}}{\text{L}} \text{ of } Ca(OH)_2 = \frac{194.83 \; \frac{\text{mg}}{\text{L}}}{1.35} = \boxed{144.3 \text{ mg/L as substance}}$$

The answer is (D).

(c) No soda ash is required since it is used to remove noncarbonate hardness.

The answer is (A).

9. (a) $Ca(HCO_3)_2$ and $MgSO_4$ both contribute to hardness. Since 100 mg/L of hardness is the goal, leave all $MgSO_4$ in the water. Take out 137 mg/L + 72 mg/L − 100 mg/L = 109 mg/L of $Ca(HCO_3)_2$. From App. 22.C (including the excess even though the reaction is not complete),

$$\text{pure } Ca(OH)_2 = 30 \; \frac{\text{mg}}{\text{L}} + \frac{109 \; \frac{\text{mg}}{\text{L}}}{1.35} = 110.74 \text{ mg/L}$$

$$\left(110.74 \; \frac{\text{mg}}{\text{L}}\right)\left(8.345 \; \frac{\text{lbm-L}}{\text{mg-MG}}\right) = \boxed{924 \text{ lbm/MG}}$$

The answer is (A).

(b) No soda ash is required.

The answer is (A).

10. The hardness removed is

$$137 \; \frac{\text{mg}}{\text{L}} + 72 \; \frac{\text{mg}}{\text{L}} - 100 \; \frac{\text{mg}}{\text{L}} = 109 \text{ mg/L}$$

$$\left(109 \; \frac{\text{mg}}{\text{L}}\right)\left(8.345 \; \frac{\text{lbm-L}}{\text{mg-MG}}\right) = 909.6 \text{ lbm hardness/MG}$$

$$\left(\frac{0.5 \text{ lbm}}{1000 \text{ gr}}\right)\left(909.6 \; \frac{\text{lbm}}{\text{MG}}\right)\left(7000 \; \frac{\text{gr}}{\text{lbm}}\right) = \boxed{3.18 \times 10^3 \text{ lbm/MG (3200 lbm/MG)}}$$

The answer is (D).

11. (a) A bypass process is required.

$$\text{fraction bypassed: } \frac{80 \; \frac{\text{mg}}{\text{L}}}{245 \; \frac{\text{mg}}{\text{L}}} = \boxed{0.326}$$

$$\text{fraction processed: } 1 - 0.326 = 0.674$$

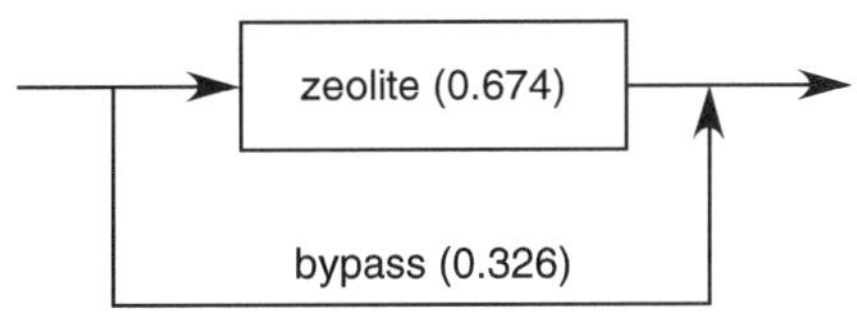

The answer is (B).

(b) The maximum hardness reduction is

$$\frac{(2 \text{ ft}^3)\left(20{,}000 \; \frac{\text{gr}}{\text{ft}^3}\right)}{7000 \; \frac{\text{gr}}{\text{lbm}}} = 5.71 \text{ lbm}$$

The hardness removal rate is

$$\frac{(0.674)\left(245 \; \frac{\text{mg}}{\text{L}}\right)\left(8.345 \times 10^{-6} \; \frac{\text{lbm}}{\text{gal}}\right)\left(20{,}000 \; \frac{\text{gal}}{\text{day}}\right)}{24 \; \frac{\text{hr}}{\text{day}}} = 1.15 \text{ lbm/hr}$$

$$t = \frac{5.71 \text{ lbm}}{1.15 \; \frac{\text{lbm}}{\text{hr}}} = \boxed{4.97 \text{ hr}}$$

The answer is (A).

12. (a) The daily demand is 200 gal.

The answer is (B).

(b)

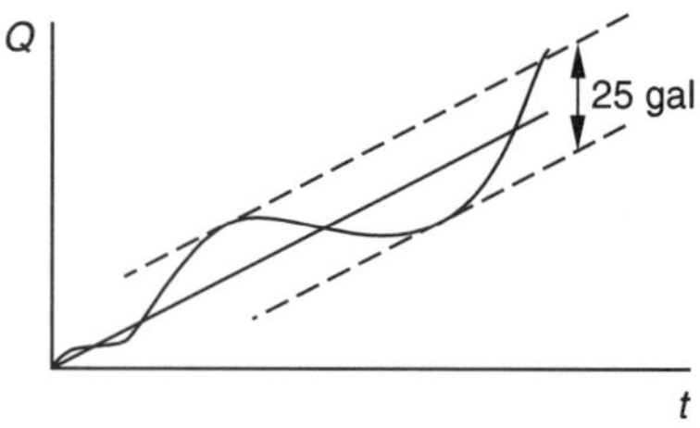

From the cumulative flow quantity, the storage requirement per capita-day is 25 gal.

The population use is

$$\left(25\ \frac{\text{gal}}{\text{day-person}}\right)(40{,}000\text{ people}) = 1{,}000{,}000\text{ gal/day}$$

From Eq. 24.53, the fire fighting requirement is

$$Q = 1020\sqrt{40}\left(1 - 0.01\sqrt{40}\right) = 6043\text{ gal/min}$$

The flow rate is 6043/1000 = 6 thousands of gallons, so maintain the flow for 6 hr. (Approximate ISO specifications.)

$$\begin{aligned}\text{capacity} &= 1{,}000{,}000\text{ gal} \\ &\quad + \left(6043\ \frac{\text{gal}}{\text{min}}\right)\left(60\ \frac{\text{min}}{\text{hr}}\right)(6\text{ hr}) \\ &= \boxed{3{,}175{,}000\text{ gal (3.2 MG)}}\end{aligned}$$

The answer is (C).

(c) The pump supplies demand from 4 A.M. until 8 A.M. The storage supplies demand from 8 A.M. until 4 A.M.

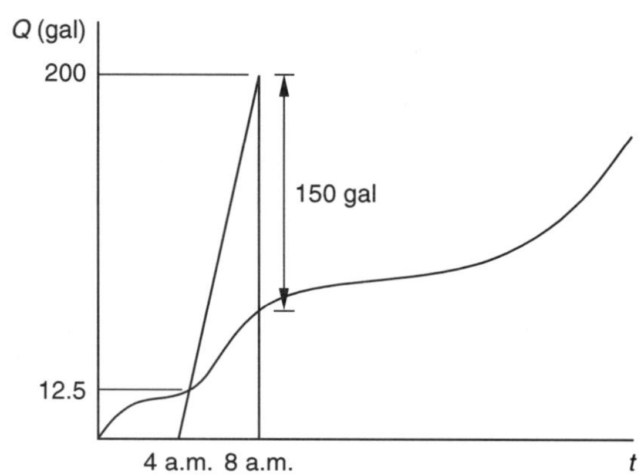

The population use is

$$(150\text{ gal} + 12.5\text{ gal})(40{,}000) = 6{,}500{,}000\text{ gal}$$

Now add fire fighting.

$$\begin{aligned}6{,}500{,}000\text{ gal} &+ \left(6043\ \frac{\text{gal}}{\text{min}}\right)\left(60\ \frac{\text{min}}{\text{hr}}\right)(6\text{ hr}) \\ &= \boxed{8{,}675{,}500\text{ gal (8.7 MG)}}\end{aligned}$$

The answer is (D).

13. (a) From Table 24.5, use 3 as the peak multiplier.

$$\frac{\left(110\ \frac{\text{gal}}{\text{day}}\right)(15{,}000)(3)}{\left(24\ \frac{\text{hr}}{\text{day}}\right)\left(60\ \frac{\text{min}}{\text{hr}}\right)} = 3437\text{ gal/min}$$

The fire fighting requirements are given by Eq. 24.53.

$$Q = 1020\sqrt{15}\left(1 - 0.01\sqrt{15}\right) = 3797\text{ gal/min}$$

The total maximum demand for which the distribution system should be designed is

$$3437\ \frac{\text{gal}}{\text{min}} + 3797\ \frac{\text{gal}}{\text{min}} = \boxed{7234\text{ gal/min}}$$

The answer is (B).

(b) The filter area should not be based on the maximum hourly rate since some of the demand during the peak hours can come from storage (clearwell or tanks). Also, fire requirements can bypass the filters if necessary.

$$\left(110\ \frac{\text{gal}}{\text{day-person}}\right)(15{,}000\text{ people}) = 1.65\times 10^6\text{ gal/day}$$

Using a flow rate of 4 gpm/ft^2, the required filter area is

$$A = \frac{1.65\times 10^6\ \frac{\text{gal}}{\text{day}}}{\left(4\ \frac{\text{gal}}{\text{min-ft}^2}\right)\left(24\ \frac{\text{hr}}{\text{day}}\right)\left(60\ \frac{\text{min}}{\text{hr}}\right)} = \boxed{286.5\text{ ft}^2}$$

The answer is (A).

(c) No, 60 mg/L is soft water.

The answer is (A).

(d) Disregarding the fire flow, the required average (not peak) daily chlorine mass is given by Eq. 24.11.

$$\frac{\left(110\ \frac{\text{gal}}{\text{day-person}}\right)(15{,}000\text{ people}) \times \left(2\ \frac{\text{mg}}{\text{L}}\right)\left(8.345\ \frac{\text{lbm-L}}{\text{MG-mg}}\right)}{10^6\ \frac{\text{gal}}{\text{MG}}} = \boxed{27.54\text{ lbm/day}}$$

The answer is (C).

25 Wastewater Quantity, Quality, and Testing

PRACTICE PROBLEMS

Sewer Velocities and Sizing

1. (*Time limit: one hour*) A town of 10,000 people (125 gpcd) has its own primary treatment plant.

(a) What mass of total solids should the treatment plant expect?

(A) 57 lbm/day
(B) 740 lbm/day
(C) 3800 lbm/day
(D) 8300 lbm/day

(b) If the town is 4 mi from the treatment plant and 400 ft above it in elevation, what minimum size pipe should be used between the town and the plant, assuming that the pipe flows full?

(A) 8 in
(B) 12 in
(C) 14 in
(D) 18 in

2. (*Time limit: one hour*) Your client has just completed a subdivision, and his sewage lines hook up to a collector that goes into a trunk. A problem has developed in the first manhole up from the trunk, and the collector pipe overflows periodically. An industrial plant is hooked directly into the trunk, and it is this plant's flow that is making your client's line back up. The subdivision is in a flood plain, and the sewer lines are very flat and cannot be steepened. (a) Describe two possible solutions to this problem. (b) Sketch plan views of your solutions.

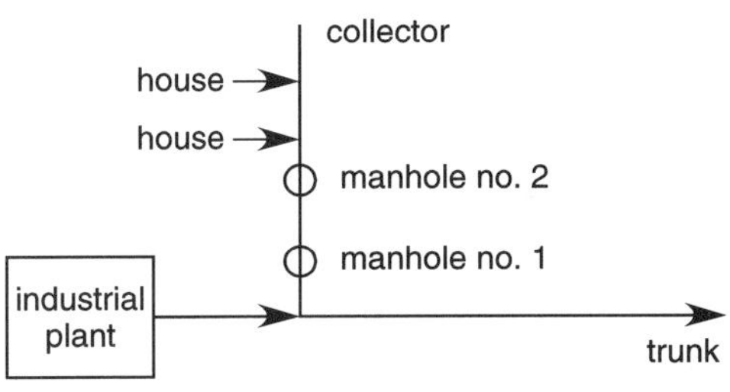

Dilution

3. Town A wants to discharge wastewater into a river upstream of town B. The state standard level for dissolved oxygen is 5 mg/L. Other data pertinent to this problem are given in the following table.

parameter	town A	river
flow (m^3/s)	0.280	0.877
ultimate BOD at 28°C (mg/L)	6.44	7.00
DO (mg/L)	1.00	6.00
K_d at 28°C, $days^{-1}$ (base-10)	N/A	0.199
K_r at 28°C, $days^{-1}$ (base-10)	N/A	0.370
velocity (m/s)	N/A	0.650
temperature (°C)	28	28

(a) Will the dissolved oxygen level of town B be reduced below the state standard level?

(A) yes; DO_{min} = 4.1 mg/L
(B) yes; DO_{min} = 4.8 mg/L
(C) no; DO_{min} = 5.3 mg/L
(D) no; DO_{min} = 5.9 mg/L

(b) At a distance of 5.79 km downstream, will the dissolved oxygen level be below the state standard level?

(A) yes; DO_{min} = 4.1 mg/L
(B) yes; DO_{min} = 4.8 mg/L
(C) no; DO_{min} = 5.3 mg/L
(D) no; DO_{min} = 5.9 mg/L

(c) At any point downstream, will the dissolved oxygen level be below the state standard level?

(A) yes; DO_{min} = 4.1 mg/L
(B) yes; DO_{min} = 4.7 mg/L
(C) no; DO_{min} = 5.1 mg/L
(D) no; DO_{min} = 5.5 mg/L

(d) Determine if the discharge will reduce the dissolved oxygen in the river below the state standard level if the winter river temperature drops to 12°C and all other characteristics remain the same.

(A) yes; DO_{min} = 4.5 mg/L
(B) yes; DO_{min} = 4.9 mg/L
(C) no; DO_{min} = 5.1 mg/L
(D) no; DO_{min} = 5.8 mg/L

4. To protect aquatic life, the limit on increase in temperature in a certain stream is 2°C at seasonal low flow. In addition, the stream must not contain more than 0.002 mg/L of un-ionized ammonia.

A manufacturing facility proposes to draw cooling water from the stream and return it to the stream at a higher temperature. They will also be discharging some un-ionized ammonia in the cooling water.

The critical low flow in the stream is 1200 m^3/s at a temperature of 20°C. The stream has no measurable un-ionized ammonia in its natural state. The facility intends to withdraw 60 m^3/s of cooling water and return it to the stream without loss of mass.

(a) What is the maximum temperature for the discharge water if the stream is not to be damaged?

(A) 40°C
(B) 60°C
(C) 80°C
(D) 90°C

(b) What is the maximum ammonia concentration for the discharge water if the stream is not to be damaged?

(A) 0.04 mg/L
(B) 0.06 mg/L
(C) 0.09 mg/L
(D) 0.13 mg/L

5. A rural town of 10,000 people discharges partially treated wastewater into a stream. The stream has the following characteristics.

minimum flow rate	120 ft^3/sec
velocity	3 mi/hr
minimum dissolved oxygen	7.5 mg/L at 15°C
temperature	15°C
BOD	0

The reoxygenation and deoxygenation coefficients are

reoxygenation coefficient of stream and effluent mixture:	0.2 day^{-1} at 20°C
deoxygenation coefficient:	0.15 day^{-1} at 20°C

The town's effluent has the following characteristics.

source	volume	BOD at 20°C	temp.
domestic	122 gpcd	0.191 lbm/cd	64°F
infiltration	116,000 gpd		51°F
industrial no. 1	180,000 gpd	800 mg/L	95°F
industrial no. 2	76,000 gpd	1700 mg/L	84°F

(a) What is the domestic waste BOD in mg/L at 20°C?

(A) 140 mg/L
(B) 160 mg/L
(C) 190 mg/L
(D) 220 mg/L

(b) What is the average effluent BOD in mg/L at 20°C just before discharge into the stream?

(A) 245 mg/L
(B) 270 mg/L
(C) 295 mg/L
(D) 315 mg/L

(c) What is the average temperature of the effluent just before discharge into the stream?

(A) 20°C
(B) 30°C
(C) 50°C
(D) 70°C

(d) If the wastewater does not contribute to dissolved oxygen at all, how far downstream is the theoretical point of minimum dissolved oxygen concentration?

(A) 30 mi
(B) 70 mi
(C) 80 mi
(D) 120 mi

(e) What is the theoretical minimum dissolved oxygen concentration in the stream?

(A) 4.7 mg/L
(B) 5.5 mg/L
(C) 7.2 mg/L
(D) 8.3 mg/L

6. A sewage treatment plant is being designed to handle both domestic and industrial wastewaters. The city population is 20,000, and an excess capacity factor of 15% is to be used for future domestic expansion. The wastewaters have the following characteristics.

source	volume	BOD
domestic	100 gpcd	0.18 lbm/cd
industrial no. 1	1.3 MGD	1100 mg/L
industrial no. 2	1.0 MGD	500 mg/L

(a) What is the design population equivalent for the plant?

(A) 60,000
(B) 75,000
(C) 90,000
(D) 110,000

(b) What is the plant's organic loading?

(A) 2300 lbm/day
(B) 8100 lbm/day
(C) 14,000 lbm/day
(D) 20,000 lbm/day

(c) What is the plant's hydraulic loading?

(A) 4.6 MGD
(B) 5.8 MGD
(C) 6.3 MGD
(D) 7.5 MGD

SOLUTIONS

1. Assume: 125 gpcd for average flow
800 mg/L total solids

(a)
$$\frac{(10{,}000)\left(125\ \frac{\text{gal}}{\text{day}}\right)\left(800\ \frac{\text{mg}}{\text{L}}\right)\left(8.345\ \frac{\text{lbm-L}}{\text{MG-mg}}\right)}{1\times 10^6\ \frac{\text{gal}}{\text{MG}}}$$
$$= \boxed{8345\ \text{lbm/day}}$$

The answer is (D).

(Minor variations in the assumptions should not affect the answer choice.)

(b)
$$S = \frac{400\ \text{ft}}{(4\ \text{mi})\left(5280\ \frac{\text{ft}}{\text{mi}}\right)} = 0.01894$$

Use Eq. 25.1.

$$\frac{Q_{\text{peak}}}{Q_{\text{ave}}} = \frac{18+\sqrt{P}}{4+\sqrt{P}} = \frac{18+\sqrt{10}}{4+\sqrt{10}} \approx 3.0$$

$$Q_{\text{peak}} = \frac{\left(125\ \frac{\text{gal}}{\text{day}}\right)(3)(10{,}000)\left(0.1337\ \frac{\text{ft}^3}{\text{gal}}\right)}{\left(24\ \frac{\text{hr}}{\text{day}}\right)\left(60\ \frac{\text{min}}{\text{hr}}\right)\left(60\ \frac{\text{sec}}{\text{min}}\right)} = 5.80\ \text{ft}^3/\text{sec}\quad(\text{cfs})$$

Assume $n = 0.013$. From Eq. 19.16,

$$D = 1.33\left(\frac{nQ}{\sqrt{S}}\right)^{\frac{3}{8}} = (1.33)\left[\frac{(0.013)\left(5.80\ \frac{\text{ft}^3}{\text{sec}}\right)}{\sqrt{0.01894}}\right]^{\frac{3}{8}} = 1.06\ \text{ft}\quad(12.7\ \text{in})$$

$$\boxed{14\ \text{in}}$$

(A 12 in pipe does not have the capacity of a 12.7 in pipe.)

The answer is (C).

2. (a) If the first manhole overflows, the piezometric head at the rim must be less than the head in the trunk. Further up at manhole no. 2, the increase in elevation is sufficient to raise rim no. 2, so the head in the trunk must be lowered.

Alternate solutions:

- relief storage (surge chambers) between trunk and manhole no. 1
- storage at plant and gradual release
- private trunk for plant
- private treatment and discharge for plant
- larger trunk capacity using larger or parallel pipes
- backflow preventors

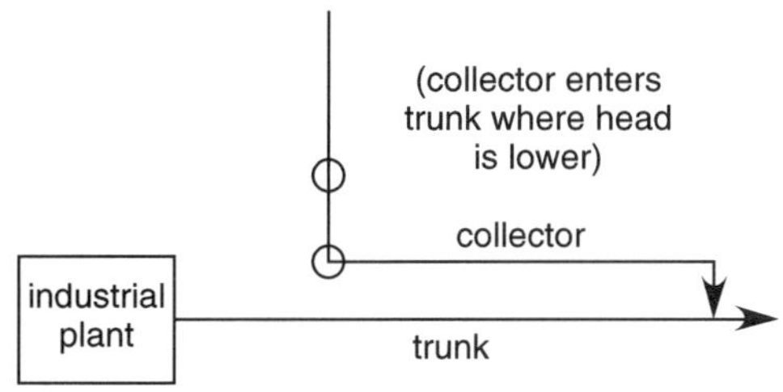

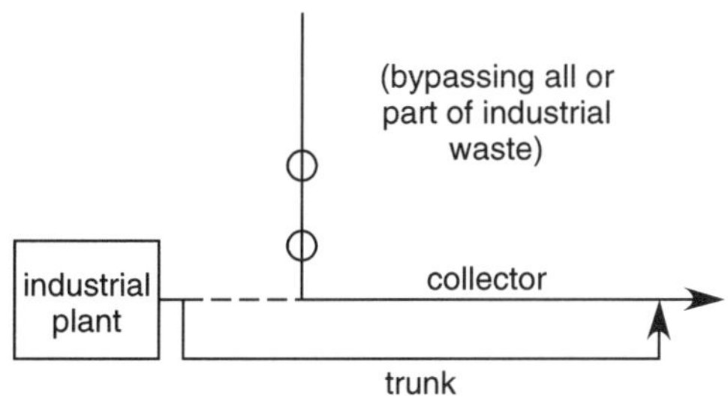

3. (a) Find the dissolved oxygen at town B from Eq. 25.33.

$$\text{DO} = \frac{Q_w\text{DO}_w + Q_r\text{DO}_r}{Q_w+Q_r} = \frac{\left(0.280\ \frac{\text{m}^3}{\text{s}}\right)\left(1\ \frac{\text{mg}}{\text{L}}\right)+\left(0.877\ \frac{\text{m}^3}{\text{s}}\right)\left(6\ \frac{\text{mg}}{\text{L}}\right)}{0.280\ \frac{\text{m}^3}{\text{s}}+0.877\ \frac{\text{m}^3}{\text{s}}} = \boxed{4.79\ \text{mg/L}}$$

So, immediately after mixing, the water will be reduced below the state standard of 5 mg/L.

The answer is (B).

(b) The composite ultimate BOD is

$$\text{BOD}_{u,28^\circ\text{C}} = \frac{Q_w\text{BOD}_w + Q_r\text{BOD}_r}{Q_w+Q_r} = \frac{\left(\begin{array}{c}\left(0.280\ \frac{\text{m}^3}{\text{s}}\right)\left(6.44\ \frac{\text{mg}}{\text{L}}\right)\\ +\left(0.877\ \frac{\text{m}^3}{\text{s}}\right)\left(7\ \frac{\text{mg}}{\text{L}}\right)\end{array}\right)}{0.280\ \frac{\text{m}^3}{\text{s}}+0.877\ \frac{\text{m}^3}{\text{s}}} = 6.8645\ \text{mg/L}$$

From App. 22.D, $DO_{sat} = 7.92$ mg/L at 28°C. The initial deficit is

$$\begin{aligned} D &= DO_{sat} - DO = 7.92\ \frac{mg}{L} - 4.79\ \frac{mg}{L} \\ &= 3.13\ mg/L \end{aligned}$$

Calculate travel time from town A to town B.

$$\begin{aligned} t &= \frac{(5.79\ km)\left(1000\ \frac{m}{km}\right)}{\left(0.650\ \frac{m}{s}\right)\left(86{,}400\ \frac{s}{day}\right)} \\ &= 0.1031\ day \end{aligned}$$

Calculate the deficit 5.79 km downstream. Use base-10 exponents because the K values are in base 10. Use Eq. 25.34.

$$\begin{aligned} D_t &= \left(\frac{K_d BOD_u}{K_r - K_d}\right)\left(10^{-K_d t} - 10^{-K_r t}\right) + D_0\left(10^{-K_r t}\right) \\ &= \frac{\left(0.199\ day^{-1}\right)\left(6.8645\ \frac{mg}{L}\right)}{0.370\ \frac{mg}{L} - 0.199\ \frac{mg}{L}} \\ &\quad \times \left(10^{-(0.199\ day^{-1})(0.1031\ day)} \right.\\ &\quad \left. -10^{-(0.370\ day^{-1})(0.1031\ day)}\right) \\ &\quad + \left(3.13\ \frac{mg}{L}\right)\left(10^{-(0.370\ day^{-1})(0.1031\ day)}\right) \\ &= 3.17\ mg/L \end{aligned}$$

Calculate the dissolved oxygen downstream.

$$\begin{aligned} DO_{town\ B} &= DO_{sat} - D \\ &= 7.92\ \frac{mg}{L} - 3.17\ \frac{mg}{L} \\ &= \boxed{4.75\ mg/L} \end{aligned}$$

This is below the state standard of 5 mg/L.

The answer is (B).

(c) Determine the critical time. Use Eq. 25.36.

$$\begin{aligned} t_c &= \left(\frac{1}{K_r - K_d}\right) \\ &\quad \times \log_{10}\left[\left(\frac{K_d BOD_u - K_r D_0 + K_d D_0}{K_d BOD_u}\right)\left(\frac{K_r}{K_d}\right)\right] \\ &= \left(\frac{1}{0.370\ day^{-1} - 0.199\ day^{-1}}\right) \\ &\quad \times \log_{10}\left[\left(\frac{\begin{array}{r}\left(0.199\ day^{-1}\right)\left(6.8645\ \frac{mg}{L}\right) \\ -\left(0.370\ day^{-1}\right)\left(3.13\ \frac{mg}{L}\right) \\ +\left(0.199\ day^{-1}\right)\left(3.13\ \frac{mg}{L}\right)\end{array}}{\left(0.199\ day^{-1}\right)\left(6.8645\ \frac{mg}{L}\right)}\right) \right.\\ &\quad \left. \times \left(\frac{0.370\ day^{-1}}{0.199\ day^{-1}}\right)\right] \\ &= 0.3122\ day \end{aligned}$$

Calculate critical deficit and dissolved oxygen. Use Eq. 25.37.

$$\begin{aligned} D_c &= \left(\frac{K_d BOD_u}{K_r}\right)10^{-K_d t} \\ &= \left(\frac{\left(0.199\ day^{-1}\right)\left(6.8645\ \frac{mg}{L}\right)}{0.370\ day^{-1}}\right) \\ &\quad \times \left(10^{-(0.199\ day^{-1})(0.3122\ day)}\right) \\ &= 3.20\ mg/L \\ DO &= DO_{sat} - D_c \\ &= 7.92\ \frac{mg}{L} - 3.20\ \frac{mg}{L} \\ &= \boxed{4.72\ mg/L} \end{aligned}$$

This is below the state standard of 5 mg/L.

The answer is (B).

(d) From Eq. 25.33, the composite dissolved oxygen is

$$\begin{aligned} \text{DO} &= \frac{Q_w\text{DO}_w + Q_r\text{DO}_r}{Q_w + Q_r} \\ &= \frac{\left(0.280\ \frac{\text{m}^3}{\text{s}}\right)\left(1\ \frac{\text{mg}}{\text{L}}\right) + \left(0.877\ \frac{\text{m}^3}{\text{s}}\right)\left(6\ \frac{\text{mg}}{\text{L}}\right)}{0.280\ \frac{\text{m}^3}{\text{s}} + 0.877\ \frac{\text{m}^3}{\text{s}}} \\ &= 4.79\ \text{mg/L} \quad \text{[no change]} \end{aligned}$$

At 12°C, the saturation dissolved oxygen is found from App. 22.D. $\text{DO}_{\text{sat},12^\circ\text{C}} = 10.83$ mg/L.

Calculate the initial deficit from Eq. 25.33.

$$\begin{aligned} D_o &= 10.83\ \frac{\text{mg}}{\text{L}} - 4.79\ \frac{\text{mg}}{\text{L}} \\ &= 6.04\ \text{mg/L} \end{aligned}$$

Calculate the temperature of the river water and wastewater mixture.

$$\begin{aligned} T &= \frac{Q_wT_w + Q_rT_r}{Q_w + Q_r} \\ &= \frac{\left(0.280\ \frac{\text{m}^3}{\text{s}}\right)(28^\circ\text{C}) + \left(0.877\ \frac{\text{m}^3}{\text{s}}\right)(12^\circ\text{C})}{0.280\ \frac{\text{m}^3}{\text{s}} + 0.877\ \frac{\text{m}^3}{\text{s}}} \\ &= 15.87^\circ\text{C} \end{aligned}$$

Calculate K_d at this temperature. Two different values of θ_d are used for the two temperature ranges. Use Eq. 25.27.

$$\begin{aligned} K_{d,T_1} &= K_{d,T_2}\theta^{T_1-T_2} \\ K_{d,20^\circ} &= (0.199\ \text{day}^{-1})(1.056)^{20^\circ\text{C}-28^\circ\text{C}} \\ &= 0.1287\ \text{day}^{-1} \\ K_{d,T_1} &= K_{d,T_2}\theta^{T_1-T_2} \\ K_{d,15.87^\circ\text{C}} &= (0.1287\ \text{day}^{-1})(1.135)^{15.87^\circ\text{C}-20^\circ\text{C}} \\ &= 0.07628\ \text{day}^{-1} \end{aligned}$$

Similarly, use Eq. 25.22 to correct K_r.

$$\begin{aligned} K_{r,T_1} &= K_{r,T_2}(1.024)^{T_1-T_2} \\ K_{r,15.87^\circ\text{C}} &= K_{r,28^\circ\text{C}}(1.024)^{15.87^\circ\text{C}-28^\circ\text{C}} \\ &= (0.370\ \text{day}^{-1})(1.024)^{15.87^\circ\text{C}-28^\circ\text{C}} \\ &= 0.2775\ \text{day}^{-1} \end{aligned}$$

Correct BOD_u to the new temperature. Use Eq. 25.31 twice.

$$\begin{aligned} \text{BOD}_{u,20^\circ\text{C}} &= \frac{\text{BOD}_{u,28^\circ\text{C}}}{(0.02)(28^\circ\text{C}) + 0.6} \\ \text{BOD}_{u,15.87^\circ\text{C}} &= (\text{BOD}_{u,20^\circ\text{C}})\big((0.02)(15.87^\circ\text{C}) + 0.6\big) \\ &= \frac{\left(6.8645\ \frac{\text{mg}}{\text{L}}\right)\big((0.02)(15.87^\circ\text{C}) + 0.06\big)}{(0.02)(28^\circ\text{C}) + 0.6} \\ &= 5.43\ \text{mg/L} \end{aligned}$$

Calculate the deficit at town B. Use Eq. 25.34.

$$\begin{aligned} D_\text{B} &= \left(\frac{K_d\text{BOD}_u}{K_r - K_d}\right)(10^{-K_dt} - 10^{-K_rt}) + D_0(10^{-K_rt}) \\ &= \left(\frac{(0.07628\ \text{day}^{-1})\left(5.43\ \frac{\text{mg}}{\text{L}}\right)}{0.2775\ \text{day}^{-1} - 0.07628\ \text{day}^{-1}}\right) \\ &\quad \times \left(10^{-(0.07628\ \text{day}^{-1})(0.1031\ \text{day})} - 10^{-(0.2775\ \text{day}^{-1})(0.1031\ \text{day})}\right) \\ &\quad + \left(6.04\ \frac{\text{mg}}{\text{L}}\right)\left(10^{-(0.2775\ \text{day}^{-1})(0.1031\ \text{day})}\right) \\ &= 5.75\ \text{mg/L} \end{aligned}$$

The dissolved oxygen at town B is

$$\begin{aligned} \text{DO}_\text{B} &= \text{DO}_\text{sat} - D_\text{B} \\ &= 10.83\ \frac{\text{mg}}{\text{L}} - 5.75\ \frac{\text{mg}}{\text{L}} \\ &= \boxed{5.08\ \text{mg/L}} \end{aligned}$$

This meets the state standard of 5 mg/L.

The answer is (C).

4. (a)

$$\begin{aligned} T &= 20^\circ\text{C} + 2^\circ\text{C} = 22^\circ\text{C} \\ T_1 &= 20^\circ\text{C} \\ Q_1 + Q_2 &= 1200\ \text{m}^3/\text{s} \\ Q_2 &= 60\ \text{m}^3/\text{s} \\ Q_1 &= 1200\ \frac{\text{m}^3}{\text{s}} - 60\ \frac{\text{m}^3}{\text{s}} = 1140\ \text{m}^3/\text{s} \\ T &= \frac{T_1Q_1 + T_2Q_2}{Q_1 + Q_2} \\ 22^\circ\text{C} &= \frac{(20^\circ\text{C})\left(1140\ \frac{\text{m}^3}{\text{s}}\right) + T_2\left(60\ \frac{\text{m}^3}{\text{s}}\right)}{1200\ \frac{\text{m}^3}{\text{s}}} \\ T_2 &= \boxed{60^\circ\text{C}} \end{aligned}$$

The answer is (B).

(b)

$$\begin{aligned} C &= 0 + 0.002\ \frac{\text{mg}}{\text{L}} \\ C_1 &= 0 \\ 0.002\ \frac{\text{mg}}{\text{L}} &= \frac{(0)\left(1140\ \frac{\text{m}^3}{\text{s}}\right) + C_2\left(60\ \frac{\text{m}^3}{\text{s}}\right)}{1200\ \frac{\text{m}^3}{\text{s}}} \\ &= \boxed{0.04\ \text{mg/L}} \end{aligned}$$

The answer is (A).

Water Treatment

5. (a) The domestic BOD concentration is

$$\frac{\left(0.191\ \frac{\text{lbm}}{\text{cd}}\right)\left(10^6\ \frac{\text{gal}}{\text{MG}}\right)\left(0.1198\ \frac{\text{mg-MG}}{\text{L-lbm}}\right)}{122\ \frac{\text{gal}}{\text{cd}}}$$

$$= \boxed{187.6\ \text{mg/L}}$$

The answer is (C).

(b) Use Eq. 25.33.

$$\frac{\left(122\ \frac{\text{gal}}{\text{day}}\right)(10{,}000)\left(187.6\ \frac{\text{mg}}{\text{L}}\right) + \left(16{,}000\ \frac{\text{gal}}{\text{day}}\right)(0) + \left(180{,}000\ \frac{\text{gal}}{\text{day}}\right)\left(800\ \frac{\text{mg}}{\text{L}}\right) + \left(76{,}000\ \frac{\text{gal}}{\text{day}}\right)\left(1700\ \frac{\text{mg}}{\text{L}}\right)}{\left(122\ \frac{\text{gal}}{\text{day}}\right)(10{,}000) + 116{,}000\ \frac{\text{gal}}{\text{day}} + 180{,}000\ \frac{\text{gal}}{\text{day}} + 76{,}000\ \frac{\text{gal}}{\text{day}}}$$

$$= \boxed{315.4\ \text{mg/L}}$$

The answer is (D).

(c) Use Eq. 25.33.

$$\frac{\left(122\ \frac{\text{gal}}{\text{day}}\right)(10{,}000)(64^\circ\text{F}) + \left(116{,}000\ \frac{\text{gal}}{\text{day}}\right)(51^\circ\text{F}) + \left(180{,}000\ \frac{\text{gal}}{\text{day}}\right)(95^\circ\text{F}) + \left(76{,}000\ \frac{\text{gal}}{\text{day}}\right)(84^\circ\text{F})}{\left(122\ \frac{\text{gal}}{\text{day}}\right)(10{,}000) + 116{,}000\ \frac{\text{gal}}{\text{day}} + 180{,}000\ \frac{\text{gal}}{\text{day}} + 76{,}000\ \frac{\text{gal}}{\text{day}}}$$

$$= \boxed{67.5^\circ\text{F}\ (19.7^\circ\text{C})}$$

The answer is (A).

(d) The total discharge into the river is

$$\left[\left(122\ \frac{\text{gal}}{\text{day}}\right)(10{,}000) + 116{,}000\ \frac{\text{gal}}{\text{day}} + 180{,}000\ \frac{\text{gal}}{\text{day}} + 76{,}000\ \frac{\text{gal}}{\text{day}}\right]\left(1.547 \times 10^{-6}\ \frac{\text{ft}^3\text{-day}}{\text{sec-gal}}\right)$$

$$= 2.46\ \text{ft}^3/\text{sec}\ \ (\text{cfs})$$

step 1: Find the stream conditions immediately after mixing.

$$\text{BOD}_{5,20^\circ\text{C}} = \frac{\left(2.46\ \frac{\text{ft}^3}{\text{sec}}\right)\left(315.4\ \frac{\text{mg}}{\text{L}}\right) + \left(120\ \frac{\text{ft}^3}{\text{sec}}\right)(0)}{2.46\ \frac{\text{ft}^3}{\text{sec}} + 120\ \frac{\text{ft}^3}{\text{sec}}} = 6.34\ \text{mg/L}$$

$$\text{DO} = \frac{\left(2.46\ \frac{\text{ft}^3}{\text{sec}}\right)(0) + \left(120\ \frac{\text{ft}^3}{\text{sec}}\right)\left(7.5\ \frac{\text{mg}}{\text{L}}\right)}{2.46\ \frac{\text{ft}^3}{\text{sec}} + 120\ \frac{\text{ft}^3}{\text{sec}}} = 7.35\ \text{mg/L}$$

$$T = \frac{\left(2.46\ \frac{\text{ft}^3}{\text{sec}}\right)(19.7^\circ\text{C}) + \left(120\ \frac{\text{ft}^3}{\text{sec}}\right)(15^\circ\text{C})}{2.46\ \frac{\text{ft}^3}{\text{sec}} + 120\ \frac{\text{ft}^3}{\text{sec}}} = 15.1^\circ\text{C}$$

step 2: Calculate the rate constants at 15.1°C. Use Eq. 25.27.

$$K_{d,15.1^\circ\text{C}} = (1.135)^{15.1^\circ\text{C}-20^\circ\text{C}}(0.15\ \text{day}^{-1}) = 0.0807\ \text{day}^{-1}$$

$$K_{r,15.1^\circ\text{C}} = (1.024)^{15.1^\circ\text{C}-20^\circ\text{C}}(0.2\ \text{day}^{-1}) = 0.178\ \text{day}^{-1}$$

step 3: Estimate BOD_u. Use Eq. 25.29.

$$\text{BOD}_{u,20^\circ\text{C}} = \frac{6.34\ \frac{\text{mg}}{\text{L}}}{1 - 10^{-(0.1\ \text{day}^{-1})(5\ \text{days})}} = 9.27\ \text{mg/L}$$

Use Eq. 25.31 to convert BOD_u to 15.1°C.

$$\text{BOD}_{u,15.1^\circ\text{C}} = \text{BOD}_{u,20^\circ\text{C}}(0.02T_{^\circ\text{C}} + 0.6) = \left(9.27\ \frac{\text{mg}}{\text{L}}\right)\left((0.02)(15.1^\circ\text{C}) + 0.6\right) = 8.36\ \text{mg/L}$$

step 4: From App. 22.D at 15°C, saturated DO = 10.15 mg/L. Since the actual is 7.35 mg/L, the deficit is

$$D_0 = 10.15\ \frac{\text{mg}}{\text{L}} - 7.35\ \frac{\text{mg}}{\text{L}} = 2.8\ \text{mg/L}$$

step 5: Calculate t_c. Use Eq. 25.36.

$$t_c = \left(\frac{1}{0.178\ \text{day}^{-1} - 0.0807\ \text{day}^{-1}}\right) \times \log_{10}\left[\frac{\left(0.0807\ \text{day}^{-1}\right)\left(8.36\ \frac{\text{mg}}{\text{L}}\right) - \left(0.178\ \text{day}^{-1}\right)\left(2.8\ \frac{\text{mg}}{\text{L}}\right) + \left(0.0807\ \text{day}^{-1}\right)\left(2.8\ \frac{\text{mg}}{\text{L}}\right)}{\left(0.0807\ \text{day}^{-1}\right)\left(8.36\ \frac{\text{mg}}{\text{L}}\right)} \times \left(\frac{0.178\ \text{day}^{-1}}{0.0807\ \text{day}^{-1}}\right)\right] = 1.22\ \text{days}$$

step 6: The distance downstream is

$$(1.22\ \text{days})\left(3\ \frac{\text{mi}}{\text{hr}}\right)\left(24\ \frac{\text{hr}}{\text{day}}\right) = \boxed{87.8\ \text{mi}}$$

The answer is (B).

(e) *step 7:* Use Eq. 25.37.

$$D_c = \left[\frac{\left(0.0807\ \text{day}^{-1}\right)\left(8.36\ \frac{\text{mg}}{\text{L}}\right)}{0.178\ \text{day}^{-1}}\right] \times 10^{-(0.0807\ \text{day}^{-1})(1.22\ \text{days})} = 3.02\ \text{mg/L}$$

step 8:

$$\text{DO}_{\text{min}} = 10.2\ \frac{\text{mg}}{\text{L}} - 3.02\ \frac{\text{mg}}{\text{L}} = \boxed{7.18\ \text{mg/L}}$$

The answer is (C).

6. (a) Do not apply the population expansion factor to the industrial effluents. Use Eq. 25.2.

$$P_e = (20)(1.15) + \frac{\left(1100\ \frac{\text{mg}}{\text{L}}\right)\left(1.3\times 10^6\ \frac{\text{gal}}{\text{day}}\right) \times \left(8.345\times 10^{-9}\ \frac{\text{lbm-L}}{\text{MG-mg}}\right)}{0.20\ \frac{\text{lbm}}{\text{day}}} + \frac{(500)\left(1.0\times 10^6\ \frac{\text{gal}}{\text{day}}\right) \times \left(8.345\times 10^{-9}\ \frac{\text{lbm-L}}{\text{MG-mg}}\right)}{0.20\ \frac{\text{lbm}}{\text{day}}} = \boxed{103.5\ [\text{thousands of people}]}$$

The answer is (D).

(b) Since the plant loading is requested, the organic loading can be given in lbm/day. Compare to Eq. 29.10 for the filter loading.

$$L_{\text{BOD}} = (112{,}500)\left(0.18\ \frac{\text{lbm}}{\text{day}}\right) = \boxed{20{,}250\ \text{lbm/day}}$$

The answer is (D).

(c)

$$L_H = (20{,}000)(1.15)\left(100\ \frac{\text{gal}}{\text{day}}\right) + 1.3\times 10^6\ \frac{\text{gal}}{\text{day}} + 1.0\times 10^6\ \frac{\text{gal}}{\text{day}} = \boxed{4.6\times 10^6\ \text{gal/day (4.6 MGD)}}$$

The answer is (A).

Water Treatment

26 Wastewater Treatment

PRACTICE PROBLEMS

Lagoons

1. A cheese factory located in a normally warm state has liquid waste with the following characteristics.

waste no. 1	
volume	10,000 gal/day
BOD	1000 mg/L
waste no. 2	
volume	25,000 gal/day
BOD	250 mg/L

The factory will use a 4 ft deep on-site lagoon to stabilize the waste.

(a) What is the total BOD loading?

(A) 60 lbm/day
(B) 85 lbm/day
(C) 110 lbm/day
(D) 140 lbm/day

(b) What lagoon size is required?

(A) 2.7 ac
(B) 6.8 ac
(C) 8.3 ac
(D) 10.9 ac

(c) What is the detention time?

(A) 5 wk
(B) 14 wk
(C) 36 wk
(D) 52 wk

Trickling Filters

2. It is estimated that the BOD of raw sewage received at a treatment plant serving a population of 20,000 will be 300 mg/L. It is estimated that the per capita BOD loading is 0.17 lbm/day. 30% of the influent BOD is removed by settling. One single-stage high-rate trickling filter is to be used to reduce the plant effluent to 50 mg/L. Recirculation is from the filter effluent to the primary settling influent. *Ten States' Standards* is in effect.

(a) What is the design flow rate?

(A) 0.9 MGD
(B) 1.1 MGD
(C) 1.4 MGD
(D) 2.0 MGD

(b) Assuming a design flow rate of 1.35 MGD, what is the total organic load on the filter?

(A) 1700 lbm/day
(B) 2100 lbm/day
(C) 2400 lbm/day
(D) 3400 lbm/day

(c) Assuming an incoming volume of 1.35 MGD and an organic loading of 60 lbm/day-1000 ft^3, what flow should be recirculated?

(A) 0.6 MGD
(B) 0.9 MGD
(C) 1.2 MGD
(D) 2.2 MGD

(d) What is the overall plant efficiency?

(A) 45%
(B) 76%
(C) 83%
(D) 91%

3. The average wastewater flow from a community of 20,000 is 125 gpcd. The 5 day, 20°C BOD is 250 mg/L. The suspended solids content is 300 mg/L. A final plant effluent of 50 mg/L of BOD is to be achieved through the use of settling tanks and two identical trickling filters operating in parallel. The settling tanks are to be designed to a standard of 1000 gpd/ft^2. The trickling filters are to be 6 ft deep and designed for an organic loading of 50.8 lbm/day-1000 ft^3. There is no recirculation.

(a) What settling tank surface area is required?

(A) 2500 ft^2
(B) 3000 ft^2
(C) 3500 ft^2
(D) 4500 ft^2

(b) What settling tank diameter is required?

(A) 40 ft
(B) 50 ft
(C) 65 ft
(D) 80 ft

(c) Estimate the BOD removal in the settling tanks.

(A) 15%
(B) 30%
(C) 45%
(D) 60%

(d) What is the trickling filter diameter?

(A) 60 ft
(B) 70 ft
(C) 80 ft
(D) 90 ft

4. Wastewater from a city with a population of 40,000 has an average daily flow of 4.4 MGD. The sewage has the following characteristics.

BOD_5 at 20°C	160 mg/L
COD	800 mg/L
total solids	900 mg/L
suspended solids	180 mg/L
volatile solids	320 mg/L
settleable solids	8 mg/L
pH	7.8

The wastewater is to be treated with primary settling and secondary trickling filtration. The settling basins are to be circular, 8 ft deep, and designed to a standard overflow rate of 1000 gal/day-ft^2.

(a) Assuming two identical basins operating in parallel, what should be the diameter of each sedimentation basin in order to remove about 30% of the BOD?

(A) 45 ft
(B) 53 ft
(C) 77 ft
(D) 86 ft

(b) What is the detention time?

(A) 1.4 hr
(B) 1.8 hr
(C) 2.3 hr
(D) 3.1 hr

(c) What is the weir loading?

(A) 8200 gal/day-ft
(B) 11,000 gal/day-ft
(C) 13,000 gal/day-ft
(D) 15,000 gal/day-ft

5. (*Time limit: one hour*) A small community has a projected average flow of 1 MGD. Incoming wastewater has the following properties: BOD, 250 mg/L; grit specific gravity, 2.65; and total suspended solids, 400 mg/L. The community wants to have a wastewater treatment plant consisting of a single aerated grit chamber, a single primary clarifier, two identical circular trickling filters, and a single secondary clarifier. Recirculation from the second clarifier to the entrance of the trickling filter will be 100% of the average flow. There is no equalization basin. The final effluent is to have a BOD of 30 mg/L.

(a) What is the peak design flow at the grit chamber?

(A) 1.0 MGD
(B) 1.5 MGD
(C) 2.0 MGD
(D) 2.5 MGD

(b) Determine the aerated grit chamber width assuming a 3 min detention time, 20 ft length, and a width:depth ratio of 1.25.

(A) 4 ft
(B) 6 ft
(C) 8 ft
(D) 10 ft

(c) Determine the approximate air requirements for the grit chamber in order to capture approximately 95% of the grit.

(A) 60 cfm
(B) 140 cfm
(C) 280 cfm
(D) 420 cfm

(d) If the clarifier is 12 ft deep, what should its diameter be?

(A) 50 ft
(B) 56 ft
(C) 63 ft
(D) 81 ft

(e) What is the peak flow experienced by the trickling filters?

(A) 1.0 MGD
(B) 2.0 MGD
(C) 3.0 MGD
(D) 4.0 MGD

(f) Assume the primary clarifier removes 30% of the incoming BOD, and the filters see the average flow only. Determine the diameter of the trickling filters assuming a 6 ft deep rock bed.

(A) 45 ft
(B) 55 ft
(C) 65 ft
(D) 85 ft

(g) Determine the diameter of the final clarifier.

(A) 40 ft
(B) 50 ft
(C) 60 ft
(D) 80 ft

Recirculating Biological Contactors

6. (*Time limit: one hour*) 1.5 MGD of wastewater with a BOD of 250 mg/L is processed by a high-rate rock trickling filter followed by recirculating biological contactor (RBC) processing. The trickling filter is 75 ft in diameter and 6 ft deep and was designed to NRC standards. The BOD removal efficiency of the RBC process is given by the following equation. (k has a value of 2.45 gal/day-ft^2, and Q (in units of gal/day) does not include recirculation.)

$$\eta_{\text{BOD}} = \frac{1}{\left(1+\frac{kA}{Q}\right)^3}$$

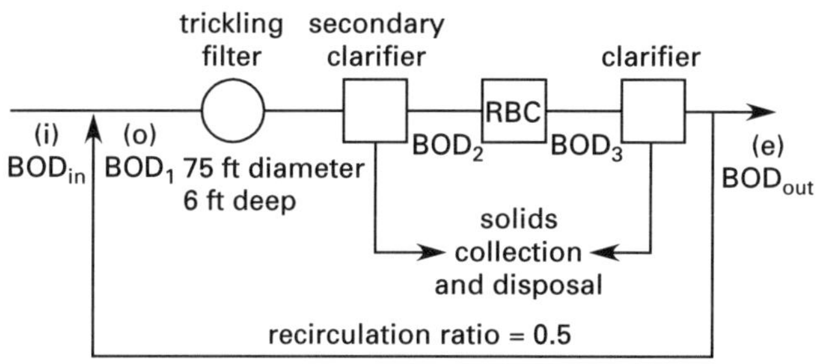

(a) What exposed RBC surface area is required to achieve an effluent BOD$_{out}$ of 30 mg/L?

(A) 4.2×10^4 ft^2
(B) 8.4×10^4 ft^2
(C) 1.3×10^5 ft^2
(D) 1.6×10^5 ft^2

(b) The recirculation pick-up point is relocated from after the final clarifier to after the trickling filter. The efficiency of the RBC process is 50%. Determine the recirculation ratio such that BOD$_{out}$ is 30 mg/L.

(A) 25%
(B) 50%
(C) 75%
(D) 100%

(c) If BOD$_2$ = 85 mg/L, BOD$_{out}$ = 30 mg/L, the yield is 0.4 lbm/lbm BOD removed, and the sludge specific gravity is essentially 1.0, what is the approximate sludge volume produced from the clarifiers?

(A) 4 ft^3/day
(B) 12 ft^3/day
(C) 21 ft^3/day
(D) 35 ft^3/day

SOLUTIONS

1. (a) The total BOD is

$$\left(1000\ \frac{\text{mg}}{\text{L}}\right)\left(8.345\ \frac{\text{lbm-L}}{\text{MG-mg}}\right)\left(\frac{10{,}000\ \frac{\text{gal}}{\text{day}}}{1{,}000{,}000\ \frac{\text{gal}}{\text{MG}}}\right) + \left(250\ \frac{\text{mg}}{\text{L}}\right)\left(8.345\ \frac{\text{lbm-L}}{\text{MG-mg}}\right)\left(\frac{25{,}000\ \frac{\text{gal}}{\text{day}}}{1{,}000{,}000\ \frac{\text{gal}}{\text{MG}}}\right)$$

$$= \boxed{135.6\ \text{lbm/day}\ (1.4\ \text{MGD})}$$

The answer is (D).

(b) Assume 20 lbm BOD/ac-day for a nonaerated stabilization pond. From Eq. 26.4, the required area is

$$A = \frac{Q}{\text{v}^*} = \frac{135.6\ \frac{\text{lbm}}{\text{day}}}{20\ \frac{\text{lbm}}{\text{ac-day}}} = \boxed{6.8\ \text{ac}}$$

The answer is (B).

(c) Use Eq. 26.5.

$$t_d = \frac{\text{V}}{Q} = \frac{(6.8\ \text{ac})\left(43{,}560\ \frac{\text{ft}^2}{\text{ac}}\right)(4\ \text{ft})}{\left(35{,}000\ \frac{\text{gal}}{\text{day}}\right)\left(1.547\times 10^{-6}\ \frac{\text{ft}^3\text{-day}}{\text{sec-gal}}\right)\left(3600\ \frac{\text{sec}}{\text{hr}}\right)} = \boxed{6078\ \text{hr}\ \ (36.2\ \text{weeks})}$$

The answer is (C).

2. (a)
$$Q = \frac{\left(0.17\ \frac{\text{lbm}}{\text{capita-day}}\right)(20{,}000\ \text{people})}{\left(8.345\times 10^{-6}\ \frac{\text{lbm-L}}{\text{gal-mg}}\right)\left(300\ \frac{\text{mg}}{\text{L}}\right)} = \boxed{1.358\times 10^6\ \text{gal/day}}$$

The answer is (C).

(*Ten States' Standards* specifies 100 gpcd in the absence of other information. In that case, Q = (100 gal/capita-day)(20,000 people) = 2×10^6 gal/day.)

(b) The BOD load leaving the primary clarifier and entering the filter is found from Eq. 26.12.

$$\text{BOD}_i = (1 - 0.30)\left(300\ \frac{\text{mg}}{\text{L}}\right) = 210\ \text{mg/L}$$

$$L_{\text{BOD}} = (1.35\ \text{MGD})\left(210\ \frac{\text{mg}}{\text{L}}\right)\left(8.345\ \frac{\text{lbm-L}}{\text{MG-mg}}\right) = \boxed{2366\ \text{lbm/day}}$$

The answer is (C).

(c) The efficiency of the filter and secondary clarifier is

$$\eta = \frac{210\ \frac{\text{mg}}{\text{L}} - 50\ \frac{\text{mg}}{\text{L}}}{210\ \frac{\text{mg}}{\text{L}}} = 0.762\ (76.2\%)$$

Use Eq. 26.13.

$$0.762 = \cfrac{1}{1 + 0.0561\sqrt{\cfrac{60\ \frac{\text{lbm}}{\text{day-1000 ft}^3}}{F}}}$$

$$F = 1.936$$

Use Eq. 26.15.

$$1.936 = \frac{1+R}{\left(1+(0.1)(R)\right)^2}$$

$$R = 1.6$$

Use Eq. 26.10.

$$Q_r = (1.6)(1.35\ \text{MGD}) = \boxed{2.16\ \text{MGD}}$$

The answer is (D).

(d) Use Eq. 26.9.

$$\eta = \frac{300\ \frac{\text{mg}}{\text{L}} - 50\ \frac{\text{mg}}{\text{L}}}{300\ \frac{\text{mg}}{\text{L}}} = \boxed{0.833\ (83.3\%)}$$

The answer is (C).

3. Use Eq. 26.4. Disregarding variations in peak flow, the average design volume is

$$\frac{\left(125\ \frac{\text{gal}}{\text{capita-day}}\right)(20{,}000\ \text{people})}{1{,}000{,}000\ \frac{\text{gal}}{\text{MG}}} = 2.5\ \text{MGD}$$

(a) The settling tank surface area is

$$\frac{2.5\times 10^6\ \frac{\text{gal}}{\text{day}}}{1000\ \frac{\text{gal}}{\text{day-ft}^2}} = \boxed{2500\ \text{ft}^2}$$

The answer is (A).

(b) The required diameter when using two tanks in parallel is

$$D = \sqrt{\frac{(4)(2500\ \text{ft}^2)}{2\pi}} = \boxed{39.9\ \text{ft (use 40 ft) each}}$$

The answer is (A).

(c) A 30% removal is typical.

The answer is (B).

(d) BOD entering the filter is

$$(1-0.30)\left(250\ \frac{\text{mg}}{\text{L}}\right) = 175\ \text{MG/L}$$

The filter efficiency is

$$\eta = \frac{175\ \frac{\text{mg}}{\text{L}} - 50\ \frac{\text{mg}}{\text{L}}}{175\ \frac{\text{mg}}{\text{L}}} = 0.71$$

From Fig. 26.4 with 71% efficiency and $R = 0$, $L_{\text{BOD}} = 60\ \text{lbm/day-1000 ft}^3$.

The total load is found from Eq. 26.12 rearranged in terms of V_1.

$$\frac{(2.5\ \text{MGD})\left(175\ \frac{\text{MG}}{\text{L}}\right) \times \left(8.345\ \frac{\text{lbm-L}}{\text{MG-MG}}\right)\left(1000\ \frac{\text{ft}^3}{1000\ \text{ft}^3}\right)}{\left(60\ \frac{\text{lbm}}{\text{day-1000 ft}^3}\right)(2\ \text{filters})}$$

$$= 30{,}424\ \text{ft}^3/\text{filter}$$

With a depth of 6 ft, the required surface area per filter is

$$A = \frac{V}{\text{depth}} = \frac{30{,}424\ \text{ft}^3}{6\ \text{ft}} = 5071\ \text{ft}^2$$

If two filters are used in parallel, the required diameter per filter is

$$D = \sqrt{\frac{(4)(5071\ \text{ft}^2)}{2\pi}} = \boxed{56.8\ \text{ft}}$$

The answer is (A).

4. (a) There is nothing particularly special about a basin that removes 30% BOD. Choose two basins in parallel, each working with half of the total flow. Choose an overflow rate of 1000 gpd/ft^2. The area per basin is given by Eq. 26.4.

$$A = \frac{4.4\times 10^6\ \frac{\text{gal}}{\text{day}}}{(2)\left(1000\ \frac{\text{gal}}{\text{day-ft}^2}\right)} = 2200\ \text{ft}^2$$

$$d = \sqrt{\left(\frac{4}{\pi}\right)(2200\ \text{ft}^2)} = \boxed{52.9\ \text{ft}}$$

The answer is (B).

(b) The detention time is given by Eq. 26.5.

$$t = \frac{(2200\ \text{ft}^2)(8\ \text{ft})}{\left(2.2\ \frac{\text{MGD}}{\text{tank}}\right)\left(1.547\ \frac{\frac{\text{ft}^3}{\text{sec}}}{\text{MGD}}\right)\left(3600\ \frac{\text{sec}}{\text{hr}}\right)}$$

$$= \boxed{1.436\ \text{hr}}$$

The answer is (A).

(c)

$$\begin{aligned}\text{circumference} &= \pi(52.9\ \text{ft}) \\ &= 166.2\ \text{ft}\end{aligned}$$

$$\text{weir loading} = \frac{2.2\times 10^6\ \frac{\text{gal}}{\text{day}}}{166.2\ \text{ft}} = \boxed{13{,}240\ \text{gal/day-ft (gpd/ft)}}$$

The answer is (C).

5.

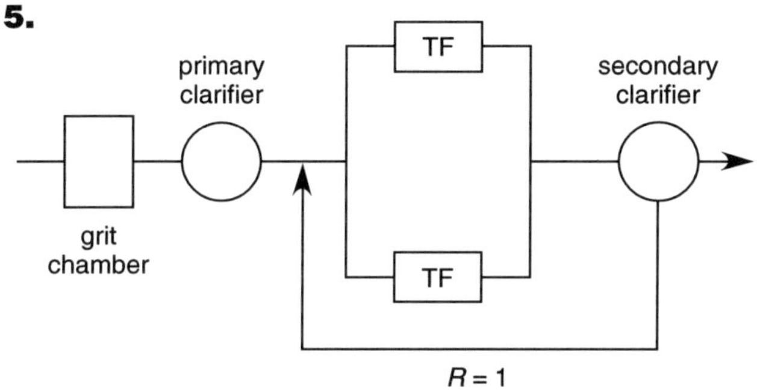

(a) Assume a peak flow multiplier of 2 since the population size is unknown. (Alternatively, a population "equivalent" can be used to estimate the population, and then Eq. 28.1 can be used.) The peak flow is

$$(2)(1\ \text{MGD}) = \boxed{2\ \text{MGD}}$$

The answer is (C).

(b) The peak flow rate per second is

$$Q_{\text{peak}} = \frac{(2)(1\ \text{MGD})\left(10^6\ \frac{\text{gal}}{\text{MG}}\right)\left(0.1337\ \frac{\text{ft}^3}{\text{gal}}\right)}{\left(24\ \frac{\text{hr}}{\text{day}}\right)\left(60\ \frac{\text{min}}{\text{hr}}\right)\left(60\ \frac{\text{sec}}{\text{min}}\right)} = 3.095\ \text{ft}^3/\text{sec}$$

With a detention time of 3 min, the volume of the grit chamber would be

$$V = Qt = \left(3.095\ \frac{\text{ft}^3}{\text{sec}}\right)\left(60\ \frac{\text{sec}}{\text{min}}\right)(3\ \text{min}) = 557.1\ \text{ft}^3$$

$$557.1\ \text{ft}^3 = (20\ \text{ft})(1.25)(\text{depth})^2$$

$$\text{water depth} = 4.72\ \text{ft}\quad \left[\text{round to } 4\tfrac{3}{4}\ \text{ft}\right]$$

$$\text{chamber width} = (1.25)(4.72\ \text{ft}) = \boxed{5.9\ \text{ft}\quad [\text{round to 6 ft}]}$$

The answer is (B).

(c) This is a shallow chamber. Use 3 cfm/ft.

$$\left(3\ \frac{\text{ft}^3}{\text{min-ft}}\right)(20\ \text{ft}) = \boxed{60\ \text{ft}^3/\text{min (cfm) of air}}$$

The answer is (A).

(d) (Note that the *Ten States' Standards* requires two basins.) Surface loading is the primary design parameter. Choose a surface loading of 1000 gal/day-ft^2. The volume is

$$A = \frac{Q}{v^*} = \frac{(2\ \text{MGD})\left(10^6\ \frac{\text{gal}}{\text{day-MGD}}\right)}{1000\ \frac{\text{gal}}{\text{day-ft}^2}} = 2000\ \text{ft}^2$$

$$D = \sqrt{\frac{4A}{\pi}} = \sqrt{\frac{(4)(2000\ \text{ft}^2)}{\pi}} = \boxed{50.5\ \text{ft}}$$

The answer is (A).

(e) The volume of the primary clarifier is fixed by the weir height. Therefore, the clarifier does not provide any storage (i.e., no damping of the flow rates). The fluctuations in flow will be experienced by the trickling filters.

$$Q_p + R = 2\ \text{MGD} + 1\ \text{MGD} = 3\ \text{MGD}$$

The answer is (C).

(f) Assume the primary sedimentation basin removes 30% of the BOD. BOD incoming to the trickle filter process is given by Eq. 26.12.

$$(1 - 0.30)\left(250\ \frac{\text{mg}}{\text{L}}\right) = 175\ \text{mg/L}$$

$$\left(175\ \frac{\text{mg}}{\text{L}}\right)\left(8.345\ \frac{\text{lbm-L}}{\text{MG-mg}}\right)(1\ \text{MGD}) = 1460\ \text{lbm/day}$$

Assume the primary basin damps out variations in flow, and peak multipliers do not apply. The required process efficiency is given by Eq. 26.9.

$$\eta = \frac{\text{BOD}_{\text{in}} - \text{BOD}_{\text{out}}}{\text{BOD}_{\text{in}}} = \frac{175\ \frac{\text{mg}}{\text{L}} - 30\ \frac{\text{mg}}{\text{L}}}{175\ \frac{\text{mg}}{\text{L}}} = 0.829\ (82.9\%)$$

From Eq. 26.15 with $w = 0.1$ and $R = 1$,

$$F = \frac{1+R}{(1+wR)^2} = \frac{1+1}{\left(1+(0.1)(1)\right)^2} = 1.65$$

From Eq. 26.13,

$$\eta = \frac{1}{1+0.0561\sqrt{\frac{L_{\text{BOD}}}{F}}}$$

$$0.83 = \frac{1}{1+0.0561\sqrt{\frac{L_{\text{BOD}}}{1.65}}}$$

$$L_{\text{BOD}} \approx 22 \text{ lbm/day-1000 ft}^3$$

$$\text{filter volume} = \frac{\left(1460 \ \frac{\text{lbm}}{\text{day}}\right)\left(1000 \ \frac{\text{ft}^3}{1000 \text{ ft}^3}\right)}{\left(22 \ \frac{\text{lbm}}{\text{day-1000 ft}^3}\right)(2 \text{ filters})} = 33{,}180 \text{ ft}^3$$

$$D = \sqrt{\frac{(4)(33{,}180 \text{ ft}^3)}{\pi(6 \text{ ft})}} = \boxed{83.9 \text{ ft } (85 \text{ ft})}$$

Use two 85 ft diameter, 6 ft deep filters.

The answer is (D).

(g) For the final clarifier, the maximum overflow rate is 800 gpd/ft^2, and the minimum depth is 7 ft.

$$A = \frac{1\times 10^6 \ \frac{\text{gal}}{\text{day}}}{800 \ \frac{\text{gal}}{\text{day-ft}^2}} = 1250 \text{ ft}^2$$

$$D = \sqrt{\frac{(4)(1250 \text{ ft}^2)}{\pi}} = \boxed{39.9 \text{ ft } (40 \text{ ft})}$$

Use a 40 ft diameter, 7 ft deep basin.

The answer is (A).

6. (a) Assume the last clarifier removes 30% of the remaining BOD.

$$\text{BOD}_3 = \frac{30 \ \frac{\text{mg}}{\text{L}}}{1-0.3} = 42.857 \text{ mg/L}$$

There is no recirculation that matches the NRC model. The NRC model "recirculation" is from the trickling filter discharge directly back to the entrance to the filter. This problem's recirculation is from several processes beyond the trickling filter. The recirculation increases the BOD loading and dilutes the influent.

The BOD loading to the trickling filter must include recirculation and is given by Eq. 26.12.

$$L_{\text{BOD}} = \frac{(1.5 \text{ MGD})\left(250 \ \frac{\text{mg}}{\text{L}}\right) \times \left(8.345 \ \frac{\text{lbm-L}}{\text{MG-mg}}\right)\left(1000 \ \frac{\text{ft}^3}{1000 \text{ ft}^3}\right)}{\pi\left(\frac{75 \text{ ft}}{2}\right)^2 (6 \text{ ft})} + \frac{(0.5)(1.5 \text{ MGD}\left(30 \ \frac{\text{mg}}{\text{L}}\right) \times \left(8.345 \ \frac{\text{lbm-L}}{\text{MG-mg}}\right)\left(1000 \ \frac{\text{ft}^3}{1000 \text{ ft}^3}\right)}{\pi\left(\frac{75 \text{ ft}}{2}\right)^2 (6 \text{ ft})}$$

$$= 118.058 \ \frac{\text{lbm}}{\text{day-1000 ft}^3} + 7.083 \ \frac{\text{lbm}}{\text{day-1000 ft}^3} = 125.141 \text{ lbm/day-1000 ft}^3$$

Since $R = 0$, $F = 1$ [Eq. 26.15].

The filter/clarifier efficiency is given by Eq. 26.13.

$$\eta = \frac{1}{1+0.0561\sqrt{\frac{125.141 \ \frac{\text{lbm}}{\text{day-1000 ft}^3}}{1.00}}} = 0.6144 \ (61.44\%)$$

$$\text{BOD}_2 = (1-0.6144)\left(250 \ \frac{\text{mg}}{\text{L}}\right) = 96.4 \text{ mg/L}$$

The removal fraction in the RBC must be

$$\frac{96.4 \ \frac{\text{mg}}{\text{L}} - 42.857 \ \frac{\text{mg}}{\text{L}}}{96.4 \ \frac{\text{mg}}{\text{L}}} = 0.5554$$

Solving the given performance equation,

$$0.5554 = \frac{1}{\left(1+\frac{2.45A}{1.5\times 10^6}\right)^3}$$

$$A = \boxed{1.32 \times 10^5 \text{ ft}^2}$$

The answer is (C).

(b) This is one of the recirculation modes to which the NRC model applies. Solve the problem backward to get $\text{BOD}_{\text{out}} = 30$ mg/L.

$$\text{BOD}_3 = \frac{30\ \frac{\text{mg}}{\text{L}}}{1 - 0.3} = 42.857\ \text{mg/L}$$

$$\text{BOD}_2 = \frac{42.857\ \frac{\text{mg}}{\text{L}}}{1 - 0.50} = 85.714\ \text{mg/L}$$

$$\eta_{\text{trickling filter and clarifier}} = \frac{250\ \frac{\text{mg}}{\text{L}} - 85.714\ \frac{\text{mg}}{\text{L}}}{250\ \frac{\text{mg}}{\text{L}}} = 0.6571\ (65.71\%)$$

The efficiency in the NRC model includes the effect of the clarifier, even though the recirculation occurs before the clarifier.

In this configuration, the BOD loading does not include the effects of recirculation, as the NRC model places a higher emphasis on organic loading than on hydraulic loading. From part (a), $L_{\text{BOD}} = 118.058$ lbm/day-1000 ft^3.

From Eq. 26.13,

$$F = \frac{L_{\text{BOD}}}{\left(\frac{1-\eta}{0.0561\eta}\right)^2} = \frac{118.058\ \frac{\text{lbm}}{\text{day-1000 ft}^2}}{\left[\frac{1 - 0.6571}{(0.0561)(0.6571)}\right]^2} = 1.3644$$

$$1.3644 = \frac{1+R}{(1+0.1R)^2}$$

$$R = \boxed{0.51\ (51\%)}$$

The answer is (B).

(c) Approximate the BOD removal of the secondary (first in-line) clarifier.

$$\text{BOD}_{\text{entering}} = \frac{\text{BOD}_2}{1-\eta} = \frac{85\ \frac{\text{mg}}{\text{L}}}{1 - 0.30} = 121.4\ \text{mg/L}$$

[assuming 30% removal]

The sludge production is

$$\left(0.4\ \frac{\text{lbm}}{\text{lbm}}\right)\left(\left(121.4\ \frac{\text{mg}}{\text{L}} - 85\ \frac{\text{mg}}{\text{L}}\right) + \left(43\ \frac{\text{mg}}{\text{L}} - 30\ \frac{\text{mg}}{\text{L}}\right)\right)\left(8.345\ \frac{\text{lbm-L}}{\text{MG-mg}}\right)(1.5\,\text{MGD}) = 247.3\ \text{lbm/day}$$

The sludge volume is

$$V = \frac{247.3\ \frac{\text{lbm}}{\text{day}}}{62.4\ \frac{\text{lbm}}{\text{ft}^3}} = \boxed{3.96\ \text{ft}^3/\text{day}}$$

The answer is (A).

Water Treatment

27 Activated Sludge and Sludge Processing

PRACTICE PROBLEMS

Sludge Quantities

1. 33 m^3/day of thickened sludge with a suspended solids content of 3.8% and 13 m^3/day of anaerobic digester sludge with a suspended solids content of 7.8 are produced in a wastewater treatment plant.

(a) What would be the effect of using a filter press to increase the solids content of the thickened sludge to 24%?

(A) 4000 m^3/yr
(B) 8000 m^3/yr
(C) 10,000 m^3/yr
(D) 15,000 m^3/yr

(b) What volume of digester sludge must be disposed of from sand drying beds that increase the solids concentration to 35%?

(A) 500 m^3/yr
(B) 1000 m^3/yr
(C) 2000 m^3/yr
(D) 4000 m^3/yr

2. An activated sludge plant processes 10 MGD of wastewater with 240 mg/L BOD and 225 mg/L suspended solids. 70% of the suspended solids are inorganic. The discharge from the final clarifier contains 15 mg/L BOD (all organic) and 20 mg/L suspended solids (all inorganic). Primary clarification removes 60% of the suspended solids and 35% of the BOD. The BOD reduction in the primary clarifier does not contribute to sludge production. The sludge produced has a specific gravity of 1.02 and a solids content of 6%. The cell yield (conversion of BOD reduction to biological solids) is 60%. The final clarifier does not reduce BOD.

(a) What is the daily mass of dry sludge solids produced?

(A) 2500 lbm/day
(B) 5000 lbm/day
(C) 10,000 lbm/day
(D) 20,000 lbm/day

(b) Assuming that the sludge is completely dried and compressed solid, what is the daily sludge volume?

(A) 200 ft^3/day
(B) 500 ft^3/day
(C) 1000 ft^3/day
(D) 1600 ft^3/day

(c) Assuming that the final moisture content in the sludge is 70% and air voids increase the volume by 10%, what is the daily sludge volume?

(A) 200 ft^3/day
(B) 500 ft^3/day
(C) 1000 ft^3/day
(D) 1600 ft^3/day

SOLUTIONS

1. (a) Sludge volume is inversely proportional to the solids content. The sludge volume from the filter press is

$$V_2 = V_1 \left(\frac{\text{SS}_1}{\text{SS}_2}\right) = \left(33\ \frac{\text{m}^3}{\text{day}}\right)\left(\frac{0.038}{0.24}\right) = 5.23\ \text{m}^3/\text{day}$$

The yearly decrease in volume is

$$\left(33\ \frac{\text{m}^3}{\text{day}} - 5.23\ \frac{\text{m}^3}{\text{day}}\right)\left(365\ \frac{\text{day}}{\text{yr}}\right) = \boxed{10{,}136\ \text{m}^3/\text{yr}}$$

The answer is (C).

(b) The sludge volume from the sand drying bed is

$$V_2 = V_1 \left(\frac{\text{SS}_1}{\text{SS}_2}\right) = \left(13\ \frac{\text{m}^3}{\text{day}}\right)\left(\frac{0.078}{0.35}\right) = 2.90\ \text{m}^3/\text{day}$$

The yearly disposal volume is

$$\left(2.9\ \frac{\text{m}^3}{\text{day}}\right)\left(365\ \frac{\text{day}}{\text{yr}}\right) = \boxed{1059\ \text{m}^3/\text{yr}}$$

The answer is (B).

2. The BOD and SS are not mutually exclusive. Some of the SS is organic in nature, and this shows up as BOD.

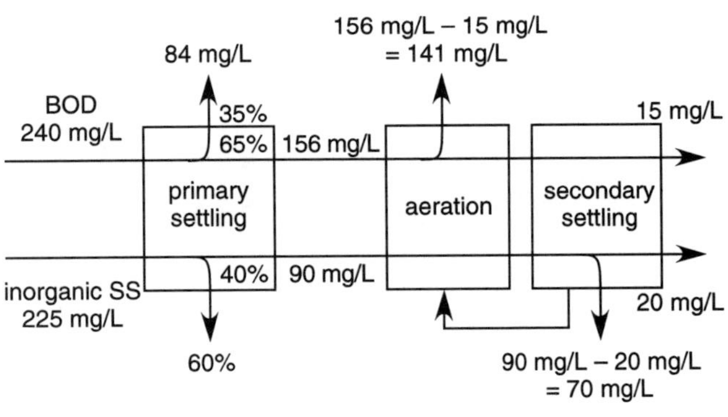

(a) The total dry weight of SS removed in all processes is

$$\left(225\ \frac{\text{mg}}{\text{L}} - 20\ \frac{\text{mg}}{\text{L}}\right)\left(8.345\ \frac{\text{lbm-L}}{\text{MG-mg}}\right)\left(10\ \frac{\text{MG}}{\text{day}}\right) = 17{,}107\ \text{lbm/day}$$

The BOD entering the secondary process is

$$(1 - 0.35)\left(240\ \frac{\text{mg}}{\text{L}}\right) = 156\ \text{mg/L}$$

Solids from BOD reduction in the secondary process are

$$Y(\Delta\text{BOD})Q = (0.60)\left(156\ \frac{\text{mg}}{\text{L}} - 15\ \frac{\text{mg}}{\text{L}}\right) \times \left(10\ \frac{\text{MG}}{\text{day}}\right)\left(8.345\ \frac{\text{lbm-L}}{\text{MG-mg}}\right) = 7060\ \text{lbm/day}$$

The total dry sludge mass is

$$17{,}107\ \frac{\text{lbm}}{\text{day}} + 7060\ \frac{\text{lbm}}{\text{day}} = \boxed{24{,}167\ \text{lbm/day}}$$

The answer is (D).

(b) The specific gravity of the sludge solids can be found from Eq. 27.47.

$$\frac{1}{\text{SG}} = \frac{1-s}{1} + \frac{s}{\text{SG}_{\text{solids}}}$$

$$\frac{1}{1.02} = \frac{1-0.06}{1} + \frac{0.06}{\text{SG}_{\text{solids}}}$$

$$\text{SG}_{\text{solids}} = 1.485$$

The density of the solids is

$$\rho_{\text{solids}} = (1.485)\left(62.4\ \frac{\text{lbm}}{\text{ft}^3}\right) = 92.66\ \text{lbm/ft}^3$$

The solid dry sludge volume is

$$V_{\text{solid}} = \frac{m}{\rho} = \frac{18{,}576\ \frac{\text{lbm}}{\text{day}}}{92.66\ \frac{\text{lbm}}{\text{ft}^3}} = \boxed{200.5\ \text{ft}^3/\text{day}}$$

The answer is (A).

(c) The disposal volume is

$$V_t = V_{\text{solid}} + V_{\text{water}}$$

$$V_{\text{solid}} = \frac{m_{\text{solid}}}{\rho}$$

$$V_{\text{water}} = \frac{m_{\text{water}}}{62.4\ \frac{\text{lbm}}{\text{ft}^3}} = \frac{0.70 m_t}{62.4\ \frac{\text{lbm}}{\text{ft}^3}} = \frac{0.70 m_{\text{solids}}}{\left(62.4\ \frac{\text{lbm}}{\text{ft}^3}\right)(1-0.70)}$$

$$V \approx (1.10)\left[\frac{24{,}167\ \frac{\text{lbm}}{\text{day}}}{92.66\ \frac{\text{lbm}}{\text{ft}^3}} + \frac{\left(24{,}167\ \frac{\text{lbm}}{\text{day}}\right)(0.70)}{\left(62.4\ \frac{\text{lbm}}{\text{ft}^3}\right)(1-0.70)}\right] = \boxed{1280.9\ \text{ft}^3/\text{day}}$$

The answer is (C).

28 Protection of Wetlands

PRACTICE PROBLEMS

1. Which of the following correctly describe the main components of wetlands?

I. Wetlands have water at the surface or in the root zone.

II. Wetlands support vegetation that is adapted to wet conditions.

III. Wetlands only have fresh water.

IV. Wetlands are flooded at least 6 months of the year.

V. Wetlands have unique soil conditions that usually differ from adjacent uplands.

(A) I, III, and IV
(B) II, III, and V
(C) I, II, and V
(D) III, IV, and V

2. The term "jurisdictional wetlands" means

(A) lands that meet the U.S. Fish and Wildlife scientific definition of wetlands
(B) lands that are regulated primarily by the EPA
(C) lands that meet the legal definition for regulation under Section 404 of the Clean Water Act
(D) lands that have met a court test

3. Palustrine wetlands do not include

(A) tidal wetlands
(B) nontidal wetlands
(C) marshes
(D) swamps

4. Riparian forested wetlands are

I. dry for portions of the growing season

II. periodically flooded by adjacent streams and rivers

III. usually of low productivity due to lack of nutrients

IV. referred to as bottomland hardwood forests in the southeast

V. have standing water for most of the growing season

(A) I, II, and III
(B) I, III, and V
(C) II, III, and IV
(D) I, II, and IV

5. The most important factor in determining the character, functions, and values of a wetland is

(A) habitat
(B) hydrology
(C) salinity
(D) wildlife

6. Hydroperiod is

(A) the seasonal pattern of the rise and fall of the surface and subsurface water level
(B) the depth of water at the start of the growing season
(C) not affected by tides
(D) always constant from year to year

7. The water budget for wetlands does not include which of the following terms?

(A) precipitation
(B) groundwater inflow and outflow
(C) evapotranspiration
(D) average water depth

8. Which of the following is not true concerning a pulsing hydroperiod?

(A) It is indicative of unproductive wetlands.
(B) It is associated with input of nutrients.
(C) It is typically associated with riparian wetlands.
(D) It is a key factor in defining the character of a wetland.

9. Which of the following is not true about the biochemistry of wetlands?

(A) Wetlands can be sinks, sources, or transformers of chemicals.

(B) Wetlands are always high nutrient systems.

(C) Wetlands are considered open systems when there is substantial movement of nutrients across the ecosystem boundary.

(D) Eutrophic wetlands are rich in mineral and organic nutrients.

10. Which of the following changes typically would have the greatest impact on the character, functions, and values of wetlands?

(A) adverse change in hydrology

(B) oil and gas exploration

(C) pipeline construction

(D) marina construction

11. Which of the following is not applicable to wetlands protection?

(A) the Rivers and Harbors Act of 1899

(B) Section 404 of the Clean Water Act

(C) the National Environmental Protection Act

(D) the Safe Drinking Water Act

12. The least environmentally damaging practicable alternative refers to

(A) the alternative for which a permit can be issued by the Corps of Engineers for work in wetlands

(B) the least cost alternative that is proposed by the developer for a project in wetlands

(C) the alternative that must be selected to meet the requirements of the Swampbuster Act

(D) the alternative that the U.S. Fish and Wildlife Service must approve in the absence of a finding of no significant impact

SOLUTIONS

1. *The answer is (C).*

2. *The answer is (C).*

3. *The answer is (A).*

4. *The answer is (D).*

5. *The answer is (B).*

6. *The answer is (A).*

7. From Eq. 23.1

$$\frac{\Delta V}{\Delta t} = P_n + S_i + G_i - \text{ET} - S_o - G_o \pm T$$

The answer is (D).

8. *The answer is (A).*

9. *The answer is (B).*

10. *The answer is (A).*

11. *The answer is (D).*

12. *The answer is (A).*

29 Thermodynamic Properties of Substances

PRACTICE PROBLEMS

1. What is the molar enthalpy of 250°F (120°C) steam with a quality of 92%?

2. What is the ratio of specific heats for air at 600°F (300°C)?

3. What is the density of helium at 600°F (300°C) and one standard atmosphere?

SOLUTIONS

1. *Customary U.S. Solution*

From App. 29.A, for 250°F steam, the enthalpy of saturated liquid, h_f, is 218.6 Btu/lbm. The heat of vaporization, h_{fg}, is 945.6 Btu/lbm. The enthalpy is given by Eq. 29.40.

$$\begin{aligned} h &= h_f + xh_{fg} \\ &= 218.6\ \frac{\text{Btu}}{\text{lbm}} + (0.92)\left(945.6\ \frac{\text{Btu}}{\text{lbm}}\right) \\ &= 1088.6\ \text{Btu/lbm} \end{aligned}$$

The molecular weight of water is 18 lbm/lbmol. The molar enthalpy is given by Eq. 29.14.

$$\begin{aligned} H &= (\text{MW})h \\ &= \left(18\ \frac{\text{lbm}}{\text{lbmol}}\right)\left(1088.6\ \frac{\text{Btu}}{\text{lbm}}\right) \\ &= \boxed{19{,}595\ \text{Btu/lbmol}} \end{aligned}$$

SI Solution

From App. 29.N, for 120°C steam, the enthalpy of saturated liquid, h_f, is 503.71 kJ/kg. The heat of vaporization, h_{fg}, is 2202.6 kJ/kg. The enthalpy is given by Eq. 29.40.

$$\begin{aligned} h &= h_f + xh_{fg} \\ &= 503.71\ \frac{\text{kJ}}{\text{kg}} + (0.92)\left(2202.6\ \frac{\text{kJ}}{\text{kg}}\right) \\ &= 2530.1\ \text{kJ/kg} \end{aligned}$$

The molecular weight of water is 18 kg/kmol. Molar enthalpy is given by Eq. 29.14.

$$\begin{aligned} H &= (\text{MW})h \\ &= \left(18\ \frac{\text{kg}}{\text{kmol}}\right)\left(2530.1\ \frac{\text{kJ}}{\text{kg}}\right) \\ &= \boxed{45\,542\ \text{kJ/kmol}} \end{aligned}$$

2. *Customary U.S. Solution*

The absolute temperature is

$$600°\text{F} + 460 = 1060°\text{R}$$

From Table 22.1, the specific heat at constant pressure for air at 1060°R is $c_p = 0.250$ Btu/lbm-°R.

From Eq. 29.48, the specific gas constant is

$$R = \frac{R^*}{\text{MW}} = \frac{1545.33 \ \frac{\text{ft-lbf}}{\text{lbmol-°R}}}{28.967 \ \frac{\text{lbm}}{\text{lbmol}}}$$
$$= 53.35 \text{ ft-lbf/lbm-°R}$$

From Eq. 29.95(b),

$$c_v = c_p - \frac{R}{J}$$
$$= 0.250 \ \frac{\text{Btu}}{\text{lbm-°R}} - \frac{53.35 \ \frac{\text{ft-lbf}}{\text{lbm-°R}}}{778 \ \frac{\text{ft-lbf}}{\text{Btu}}}$$
$$= 0.1814 \text{ Btu/lbm-°R}$$

The ratio of specific heats is given by Eq. 29.28.

$$k = \frac{c_p}{c_v} = \frac{0.250 \ \frac{\text{Btu}}{\text{lbm-°R}}}{0.1814 \ \frac{\text{Btu}}{\text{lbm-°R}}}$$
$$= \boxed{1.378}$$

SI Solution

From Table 22.1, specific heat at constant pressure for air is 0.250 Btu/lbm-°R. From the table footnote, the SI specific heat at constant pressure for air is

$$c_p = \left(0.250 \ \frac{\text{Btu}}{\text{lbm-°R}}\right)\left(4.187 \ \frac{\frac{\text{kJ}}{\text{kg·K}}}{\frac{\text{Btu}}{\text{lbm-°R}}}\right)$$
$$= 1.047 \text{ kJ/kg·K}$$

From Eq. 29.48, the specific gas constant is

$$R = \frac{R^*}{\text{MW}} = \frac{8314.3 \ \frac{\text{J}}{\text{kmol·K}}}{28.967 \ \frac{\text{kg}}{\text{kmol}}}$$
$$= 287.0 \text{ J/kg·K}$$

From Eq. 29.95(a),

$$c_v = c_p - R$$
$$= \left(1.047 \ \frac{\text{kJ}}{\text{kg·K}}\right)\left(1000 \ \frac{\text{J}}{\text{kJ}}\right) - 287.0 \ \frac{\text{J}}{\text{kg·K}}$$
$$= 760 \text{ J/kg·K}$$

The ratio of specific heats is given by Eq. 29.28.

$$k = \frac{c_p}{c_v} = \frac{\left(1.047 \ \frac{\text{kJ}}{\text{kg·K}}\right)\left(1000 \ \frac{\text{J}}{\text{kJ}}\right)}{760 \ \frac{\text{J}}{\text{kg·K}}}$$
$$= \boxed{1.377}$$

3. *Customary U.S. Solution*

From Eq. 29.48, the specific gas constant is

$$R = \frac{R^*}{\text{MW}} = \frac{1545.33 \ \frac{\text{ft-lbf}}{\text{lbmol-°R}}}{4 \ \frac{\text{lbm}}{\text{lbmol}}}$$
$$= 386.3 \text{ ft-lbf/lbm-°R}$$

The absolute temperature is

$$600°\text{F} + 460 = 1060°\text{R}$$

From Eq. 29.50, the density of helium is

$$\rho = \frac{p}{RT}$$
$$= \frac{\left(14.7 \ \frac{\text{lbf}}{\text{in}^2}\right)\left(144 \ \frac{\text{in}^2}{\text{ft}^2}\right)}{\left(386.3 \ \frac{\text{ft-lbf}}{\text{lbm-°R}}\right)(1060°\text{R})}$$
$$= \boxed{0.00517 \text{ lbm/ft}^3}$$

SI Solution

From Eq. 29.48, the specific gas constant is

$$R = \frac{R^*}{\text{MW}} = \frac{8314.3 \ \frac{\text{J}}{\text{kmol·K}}}{4 \ \frac{\text{kg}}{\text{kmol}}}$$
$$= 2079 \text{ J/kg·K}$$

The absolute temperature is

$$300°\text{C} + 273 = 573\text{K}$$

From Eq. 29.50, the density of helium is

$$\rho = \frac{p}{RT}$$
$$= \frac{1.013 \times 10^5 \text{ Pa}}{\left(2079 \ \frac{\text{J}}{\text{kg·K}}\right)(573\text{K})}$$
$$= \boxed{0.0850 \text{ kg/m}^3}$$

30 Changes in Thermodynamic Properties

PRACTICE PROBLEMS

1. Cast iron is heated from 80 to 780°F (27 to 416°C). What heat is required per unit mass?

2. The ventilation rate in a building is 3×10^5 ft^3/hr (2.4 m^3/s). The air is heated from 35 to 75°F (2 to 24°C) by water whose temperature decreases from 180 to 150°F (82 to 66°C). What is the water flow rate in gal/min (L/s)?

3. 8.0 ft^3 (0.25 m^3) of 180°F, 14.7 psia (82°C, 101.3 kPa) air are cooled to 100°F (38°C) in a constant-pressure process. What work is done?

4. What is the availability of an isentropic process using steam with a quality of 95% and operating between 300 psia and 50 psia (2 MPa and 0.35 MPa)?

5. (*Time limit: one hour*) A closed air heater receives 540°F, 100 psia (280°C, 700 kPa) air and heats it to 1540°F (840°C). The outside temperature is 100°F (40°C). The pressure of the air drops 20 psi (150 kPa) as it passes through the heater. What is the percentage loss in available energy due to the pressure drop?

6. (*Time limit: one hour*) Xenon gas at 20 psia and 70°F (150 kPa and 21°C) is compressed to 3800 psia and 70°F (25 MPa and 21°C) by a compressor/heat exchanger combination. The compressed gas is stored in a 100 ft^3 (3 m^3) rigid tank initially charged with xenon gas at 20 psia (150 kPa). (a) What is the mass of the xenon gas initially in the tank? (b) What is the average mass flow rate of xenon gas into the tank if the compressor fills the tank in exactly one hour? (c) If filling takes exactly one hour and electricity costs $0.045 per kW-hr, what is the cost of filling the tank?

7. (*Time limit: one hour*) The mass of an insulated 20 ft^3 (0.6 m^3) steel tank is 40 lbm (20 kg). The steel has a specific heat of 0.11 Btu/lbm-°R (0.46 kJ/kg·K). The tank is placed in a room where the surrounding air is 70°F and 14.7 psia (21°C and 101.3 kPa). After the tank is evacuated to 1 psia and 70°F (7 kPa and 21°C), a valve is suddenly opened, allowing the tank to fill with room air. The air enters the tank in a well-mixed, turbulent condition. Find the air temperature inside the tank immediately after filling and after the gas and tank have reached thermal equilibrium, but before any heat loss to the tank or room occurs.

SOLUTIONS

1. *Customary U.S. Solution*

From Table 29.2, the approximate value of specific heat for cast iron is $c_p = 0.10$ Btu/lbm-°F.

The heat required per unit mass is

$$\begin{aligned} q &= c_p(T_2 - T_1) \\ &= \left(0.10\ \frac{\text{Btu}}{\text{lbm-°F}}\right)(780\text{°F} - 80\text{°F}) \\ &= \boxed{70\ \text{Btu/lbm}} \end{aligned}$$

SI Solution

From Table 29.2, the approximate value of specific heat of cast iron is $c_p = 0.42$ kJ/kg·K.

The heat required per unit mass is

$$\begin{aligned} q &= c_p(T_2 - T_1) \\ &= \left(0.42\ \frac{\text{kJ}}{\text{kg·K}}\right)(416\text{°C} - 27\text{°C}) \\ &= \boxed{163.4\ \text{kJ/kg}} \end{aligned}$$

2. *Customary U.S. Solution*

First calculate the mass flow rate of air to be heated by using the ideal gas law. (Usually the air mass would be evaluated at the entering conditions. This problem is ambiguous. The building conditions are used because a building ventilation rate was specified.)

$$\dot{m}_{\text{air}} = \frac{p\dot{V}}{RT}$$

The absolute temperature is

$$T = 75\text{°F} + 460 = 535\text{°R}$$

$$\begin{aligned} \dot{m}_{\text{air}} &= \frac{\left(14.7\ \frac{\text{lbf}}{\text{in}^2}\right)\left(144\ \frac{\text{in}^2}{\text{ft}^2}\right)\left(3 \times 10^5\ \frac{\text{ft}^3}{\text{hr}}\right)}{\left(53.3\ \frac{\text{ft-lbf}}{\text{lbm-°R}}\right)(535\text{°R})} \\ &= 2.227 \times 10^4\ \text{lbm/hr} \end{aligned}$$

Ventilation

The heat lost by the water is equal to the heat gained by the air.

$$\dot{m}_w c_{p,w}(T_{1,w} - T_{2,w}) = \dot{m}_{\text{air}} c_{p,\text{air}}(T_{2,\text{air}} - T_{1,\text{air}})$$

$$\dot{m}_w \left(1\ \frac{\text{Btu}}{\text{lbm-°F}}\right)(180\text{°F} - 150\text{°F}) = \left(2.227 \times 10^4\ \frac{\text{lbm}}{\text{hr}}\right)\left(0.241\ \frac{\text{Btu}}{\text{lbm-°F}}\right)(75\text{°F} - 35\text{°F})$$

$$\dot{m}_w = 7156.1\ \text{lbm/hr}$$

From App. 14.A, the density of water at 165°F is approximately 61 lbm/ft^3. The water volume flow rate is

$$\dot{V}_w = \frac{\dot{m}}{\rho} = \frac{\left(7156.1\ \frac{\text{lbm}}{\text{hr}}\right)\left(\frac{1\ \text{hr}}{3600\ \text{sec}}\right)}{\left(61\ \frac{\text{lbm}}{\text{ft}^3}\right)\left(0.002228\ \frac{\text{ft}^3\text{-min}}{\text{sec-gal}}\right)} = \boxed{14.63\ \text{gal/min}}$$

SI Solution

First calculate the mass flow rate of air to be heated by using the ideal gas law. (Usually the air mass would be evaluated at the entering conditions. This problem is ambiguous. The building conditions are used because a building ventilation rate was specified.)

$$\dot{m}_{\text{air}} = \frac{p\dot{V}}{RT}$$

The absolute temperature is

$$T = 24\text{°C} + 273 = 297\text{K}$$

$$\dot{m}_{\text{air}} = \frac{(1.013 \times 10^5\ \text{Pa})\left(2.4\ \frac{\text{m}^3}{\text{s}}\right)}{\left(287\ \frac{\text{J}}{\text{kg·K}}\right)(297\text{K})} = 2.85\ \text{kg/s}$$

The heat lost by the water is equal to the heat gained by the air.

$$\dot{m}_w c_{p,w}(T_{1,w} - T_{2,w}) = \dot{m}_{\text{air}} c_{p,\text{air}}(T_{2,\text{air}} - T_{1,\text{air}})$$

$$\dot{m}_w \left(4.190\ \frac{\text{kJ}}{\text{kg·K}}\right)(82\text{°C} - 66\text{°C}) = \left(2.85\ \frac{\text{kg}}{\text{s}}\right)\left(1.005\ \frac{\text{kJ}}{\text{kg·K}}\right)(24\text{°C} - 2\text{°C})$$

$$\dot{m}_w = 0.940\ \text{kg/s}$$

From App. 35.B, the density of water at 74°C is approximately 976 kg/m^3. The water volume flow rate is

$$\dot{V}_w = \left(\frac{\dot{m}}{\rho}\right) = \left(0.940\ \frac{\text{kg}}{\text{s}}\right)\left(976\ \frac{\text{kg}}{\text{m}^3}\right)\left(1000\ \frac{\text{L}}{\text{m}^3}\right) = \boxed{0.963\ \text{L/s}}$$

3. *Customary U.S. Solution*

The mass of air is

$$m = \frac{p_1V_1}{RT_1} = \frac{\left(14.7\ \frac{\text{lbf}}{\text{in}^2}\right)\left(144\ \frac{\text{in}^2}{\text{ft}^2}\right)(8.0\ \text{ft}^3)}{\left(53.3\ \frac{\text{ft-lbf}}{\text{lbm-°R}}\right)(180\text{°F} + 460)} = 0.4964\ \text{lbm}$$

For a constant pressure process from Eq. 30.51, on a per unit mass basis,

$$W = R(T_2 - T_1)$$

The total work for m in lbm is

$$W = mR(T_2 - T_1) = (0.4964\ \text{lbm})\left(53.3\ \frac{\text{ft-lbf}}{\text{lbm-°R}}\right)(100\text{°F} - 180\text{°F}) = \boxed{-2116.6\ \text{ft-lbf}}$$

This is negative because work is done by the system.

SI Solution

The mass of air is

$$m = \frac{p_1V_1}{RT_1} = \frac{(101.3\ \text{kPa})\left(1000\ \frac{\text{Pa}}{\text{kPa}}\right)(0.25\ \text{m}^3)}{\left(287\ \frac{\text{J}}{\text{kg·K}}\right)(82\text{°C} + 273)} = 0.2486\ \text{kg}$$

For a constant pressure process from Eq. 30.51, on a per unit mass basis,

$$W = R(T_2 - T_1)$$

The total work for m in kg is

$$W = mR(T_2 - T_1) = (0.2486\ \text{kg})\left(287\ \frac{\text{J}}{\text{kg·K}}\right)(38^\circ\text{C} - 82^\circ\text{C}) = \boxed{-3139.3\ \text{J}}$$

4. *Customary U.S. Solution*

From App. 29.B, for 300 psia, the enthalpy of saturated liquid, h_f, is 394.1 Btu/lbm. The heat of vaporization, h_{fg}, is 809.8 Btu/lbm. The enthalpy is given by Eq. 29.40.

$$h_1 = h_f + xh_{fg} = 394.1\ \frac{\text{Btu}}{\text{lbm}} + (0.95)\left(809.8\ \frac{\text{Btu}}{\text{lbm}}\right) = 1163.4\ \text{Btu/lbm}$$

From the Mollier diagram, for an isentropic process from 300 psia to 50 psia, $h_2 = 1031$ Btu/lbm.

The availability is calculated from Eq. 30.164 using an isentropic process ($s_1 = s_2$) for unit mass.

$$\text{availability} = h_1 - h_2 = 1163.4\ \frac{\text{Btu}}{\text{lbm}} - 1031\ \frac{\text{Btu}}{\text{lbm}} = \boxed{132.4\ \text{Btu/lbm}}$$

SI Solution

From App. 29.O, for 2 MPa, the enthalpy of saturated liquid, h_f, is 908.79 kJ/kg. The heat of vaporization, h_{fg}, is 1890.7 kJ/kg. The enthalpy is given by Eq. 24.40.

$$h_1 = h_f + xh_{fg} = 908.79\ \frac{\text{kJ}}{\text{kg}} + (0.95)\left(1890.7\ \frac{\text{kJ}}{\text{kg}}\right) = 2705.0\ \text{kJ/kg}$$

From the Mollier diagram, for an isentropic process from 2 MPa to 0.35 MPa, $h_2 = 2405$ kJ/kg.

The availability is calculated from Eq. 30.164 using an isentropic process ($s_1 = s_2$) for unit mass.

$$\text{availability} = h_1 - h_2 = 2705\ \frac{\text{kJ}}{\text{kg}} - 2405\ \frac{\text{kJ}}{\text{kg}} = \boxed{300\ \text{kJ/kg}}$$

5. *Customary U.S. Solution*

The absolute temperature at the inlet of the air heater is

$$T_1 = 540^\circ\text{F} + 460 = 1000^\circ\text{R}$$

The absolute temperature at the outlet of the air heater is

$$T_2 = 1540^\circ\text{F} + 460 = 2000^\circ\text{R}$$

Since pressures are low and temperatures are high, use an air table.

From App. 29.F at 1000°R,

$$h_1 = 240.98\ \text{Btu/lbm}$$
$$\phi_1 = 0.75042\ \text{Btu/lbm-}^\circ\text{R}$$

From App. 29.F at 2000°R,

$$h_2 = 504.71\ \text{Btu/lbm}$$
$$\phi_2 = 0.93205\ \text{Btu/lbm-}^\circ\text{R}$$

The availability per unit mass is calculated from Eq. 30.164 using $T_L = 100^\circ\text{F} + 460 = 560^\circ\text{R}$.

$$W_{\text{max}} = h_1 - h_2 + T_L(s_2 - s_1)$$

For no pressure drop,

$$s_2 - s_1 = \phi_2 - \phi_1$$

$$W_{\text{max}} = h_1 - h_2 + T_L(\phi_2 - \phi_1) = 240.98\ \frac{\text{Btu}}{\text{lbm}} - 504.71\ \frac{\text{Btu}}{\text{lbm}} + (560^\circ\text{R})\left(0.93205\ \frac{\text{Btu}}{\text{lbm-}^\circ\text{R}} - 0.75042\ \frac{\text{Btu}}{\text{lbm-}^\circ\text{R}}\right) = -162.02\ \text{Btu/lbm}$$

With a pressure drop from 100 psia to 80 psia, from Eq. 29.39,

$$s_2 - s_1 = \phi_2 - \phi_1 - \left(\frac{R}{J}\right)\ln\left(\frac{p_2}{p_1}\right)$$

$$W_{\text{max},p\ \text{loss}} = h_1 - h_2 + T_L\left[\phi_2 - \phi_1 - \left(\frac{R}{J}\right)\ln\left(\frac{p_2}{p_1}\right)\right]$$

$$= 240.98\ \frac{\text{Btu}}{\text{lbm}} - 504.71\ \frac{\text{Btu}}{\text{lbm}} + (560^\circ\text{R}) \times \left[0.93205\ \frac{\text{Btu}}{\text{lbm-}^\circ\text{R}} - 0.75042\ \frac{\text{Btu}}{\text{lbm-}^\circ\text{R}} - \left(\frac{53.3\ \frac{\text{ft-lbf}}{\text{lbm-}^\circ\text{R}}}{778\ \frac{\text{ft-lbf}}{\text{Btu}}}\right)\ln\left(\frac{80\ \text{psia}}{100\ \text{psia}}\right)\right]$$

$$= -153.46\ \text{Btu/lbm}$$

Ventilation

The percentage loss in available energy is

$$\frac{W_{\max} - W_{\max,p\text{ loss}}}{W_{\max}} \times 100\%$$

$$= \frac{-162.02\ \frac{\text{Btu}}{\text{lbm}} - \left(-153.46\ \frac{\text{Btu}}{\text{lbm}}\right)}{-162.02\ \frac{\text{Btu}}{\text{lbm}}} \times 100\%$$

$$= \boxed{5.28\%}$$

SI Solution

The absolute temerature at the inlet of the air heater is

$$T_1 = 280°\text{C} + 273 = 553\text{K}$$

The absolute temperature at the outlet of the air heater is

$$T_2 = 840°\text{C} + 273 = 1113\text{K}$$

Since pressures are low and temperatures are high, use an air table.

From App. 29.S at 553K,

$$h_1 = 557.9\ \text{kJ/kg}$$
$$\phi_1 = 2.32372\ \text{kJ/kg·K}$$

From App. 29.S at 1113K,

$$h_2 = 1176.2\ \text{kJ/kg}$$
$$\phi_2 = 3.09092\ \text{kJ/kg·K}$$

The availability per unit mass is calculated from Eq. 30.164 using $T_L = 40°\text{C} + 273 = 313\text{K}$.

$$W_{\max} = h_1 - h_2 + T_L(s_2 - s_1)$$

For no pressure drop,

$$s_2 - s_1 = \phi_2 - \phi_1$$
$$W_{\max} = h_1 - h_2 + T_L(\phi_2 - \phi_1)$$
$$= 557.9\ \frac{\text{kJ}}{\text{kg}} - 1176.2\ \frac{\text{kJ}}{\text{kg}} + (313\text{K})\left(3.09092\ \frac{\text{kJ}}{\text{kg·K}} - 2.32372\ \frac{\text{kJ}}{\text{kg·K}}\right)$$
$$= -378.17\ \text{kJ/kg}$$

With a pressure drop from 700 kPa to 550 kPa, from Eq. 29.39,

$$s_2 - s_1 = \phi_2 - \phi_1 - R\ln\left(\frac{p_2}{p_1}\right)$$
$$W_{\max,p\text{ loss}} = h_1 - h_2 + T_L\left[\phi_2 - \phi_1 - R\ln\left(\frac{p_2}{p_1}\right)\right]$$
$$= 557.9\ \frac{\text{kJ}}{\text{kg}} - 1176.2\ \frac{\text{kJ}}{\text{kg}} + (313\text{K})\left[3.09092\ \frac{\text{kJ}}{\text{kg·K}} - 2.32372\ \frac{\text{kJ}}{\text{kg·K}} - \left(\frac{287\ \frac{\text{J}}{\text{kg·K}}}{1000\ \frac{\text{J}}{\text{kJ}}}\right)\ln\left(\frac{550\ \text{kPa}}{700\ \text{kPa}}\right)\right]$$
$$= -356.50\ \text{kJ/kg}$$

The percentage loss in available energy is

$$\frac{W_{\max} - W_{\max,p\text{ loss}}}{W_{\max}} \times 100\%$$

$$= \frac{-378.17\ \frac{\text{kJ}}{\text{kg}} - \left(-356.50\ \frac{\text{kJ}}{\text{kg}}\right)}{-378.17\ \frac{\text{kJ}}{\text{kg}}} \times 100\%$$

$$= \boxed{5.73\%}$$

6. *Customary U.S. Solution*

(a) Assume the tank is originally at 70°F. The absolute temperature is

$$T = 70°\text{F} + 460 = 530°\text{R}$$

From Table 29.7, R = 11.77 ft-lbf/lbm-°R. From Eq. 29.47,

$$m = \frac{pV}{RT}$$
$$= \frac{\left(20\ \frac{\text{lbf}}{\text{in}^2}\right)\left(144\ \frac{\text{in}^2}{\text{ft}^2}\right)(100\ \text{ft}^3)}{\left(11.77\ \frac{\text{ft-lbf}}{\text{lbm-°R}}\right)(530°\text{R})}$$
$$= \boxed{46.17\ \text{lbm}}$$

(b) From Table 29.4, the critical temperature and pressure of xenon are 521.9°R and 58.2 atm, respectively.

The reduced variables are

$$T_r = \frac{T}{T_c} = \frac{530°\text{R}}{521.9°\text{R}} = 1.02$$

$$p_r = \frac{p}{p_c} = \frac{3800 \text{ psia}}{(58.2 \text{ atm})\left(14.7 \frac{\text{psia}}{\text{atm}}\right)} = 4.44$$

From App. 29.Z, z is read as 0.61. Using Eq. 24.93,

$$m = \frac{pV}{zRT} = \frac{\left(3800 \frac{\text{lbf}}{\text{in}^2}\right)\left(144 \frac{\text{in}^2}{\text{ft}^2}\right)(100 \text{ ft}^3)}{(0.61)\left(11.77 \frac{\text{ft-lbf}}{\text{lbm-°R}}\right)(530°\text{R})}$$
$$= 14{,}380 \text{ lbm}$$

The average mass flow rate of xenon is

$$\dot{m} = \frac{14{,}380 \text{ lbm} - 46.17 \text{ lbm}}{1 \text{ hr}}$$
$$= \boxed{14{,}334 \text{ lbm/hr}}$$

(c) For isothermal compression, the work per unit mass is calculated from Eq. 30.79.

$$W = mRT \ln\left(\frac{p_1}{p_2}\right)$$
$$= \frac{(14{,}334 \text{ lbm})\left(11.77 \frac{\text{ft-lbf}}{\text{lbm-°R}}\right) \times (530°\text{R}) \ln\left(\frac{20 \text{ psia}}{3800 \text{ psia}}\right)}{\left(778 \frac{\text{ft-lbf}}{\text{Btu}}\right)\left(3413 \frac{\text{Btu}}{\text{kW·hr}}\right)}$$
$$= -176.7 \text{ kW-hr} \quad (\text{for 1 hr})$$

The cost of electricity is

$$\left(\frac{\$0.045}{\text{kW-hr}}\right)(176.7 \text{ kW-hr}) = \boxed{\$7.95}$$

SI Solution

(a) Assume the tank is originally at 21°C. The absolute temperature is

$$T = 21°\text{C} + 273 = 294\text{K}$$

From Table 29.7, $R = 63.32$ J/kg·K. From Eq. 29.47,

$$m = \frac{pV}{RT}$$
$$= \frac{(150 \text{ kPa})\left(1000 \frac{\text{Pa}}{\text{kPa}}\right)(3 \text{ m}^3)}{\left(63.32 \frac{\text{J}}{\text{kg·K}}\right)(294\text{K})}$$
$$= \boxed{24.17 \text{ kg}}$$

(b) From Table 29.4, the critical temperature and pressure of xenon are 289.9K and 58.2 atm, respectively. The reduced variables are

$$T_r = \frac{T}{T_c} = \frac{294\text{K}}{289.9\text{K}} = 1.01$$

$$p_r = \frac{p}{p_c} = \frac{25 \text{ MPa}}{(58.2 \text{ atm})\left(0.1013 \frac{\text{MPa}}{\text{atm}}\right)} = 4.24$$

From App. 29.Z, z is read as 0.59. Using Eq. 29.93,

$$m = \frac{pV}{zRT}$$
$$= \frac{(25 \text{ MPa})\left(10^6 \frac{\text{Pa}}{\text{MPa}}\right)(3 \text{ m}^3)}{(0.59)\left(63.32 \frac{\text{J}}{\text{kg·K}}\right)(294\text{K})}$$
$$= 6.828 \text{ kg}$$

The average mass flow rate of xenon is

$$\dot{m} = \frac{6828 \text{ kg} - 24.17 \text{ kg}}{(1 \text{ h})\left(3600 \frac{\text{s}}{\text{h}}\right)}$$
$$= \boxed{1.89 \text{ kg/s}}$$

(c) For isothermal compression, the work per unit mass is calculated from Eq. 30.79.

$$W = mRT \ln\left(\frac{p_1}{p_2}\right)$$
$$= (6828 \text{ kg} - 24.17 \text{ kg})\left(63.32 \frac{\text{J}}{\text{kg·K}}\right)(294\text{K})$$
$$\times \ln\left(\frac{150 \text{ kPa}}{(25 \text{ MPa})\left(1000 \frac{\text{kPa}}{\text{MPa}}\right)}\right)$$
$$\times \left(\frac{1 \text{ kJ}}{1000 \text{ J}}\right)\left(\frac{1 \text{ h}}{3600 \text{ s}}\right)$$
$$= -180 \text{ kW·h}$$

The cost of electricity is

$$\left(\frac{\$0.045}{\text{kW·h}}\right)(180 \text{ kW·h}) = \boxed{\$8.10}$$

7. Choose the control volume to include the air outside the tank that is pushed into the tank (subscript "e" for "entering"), as well as the tank volume.

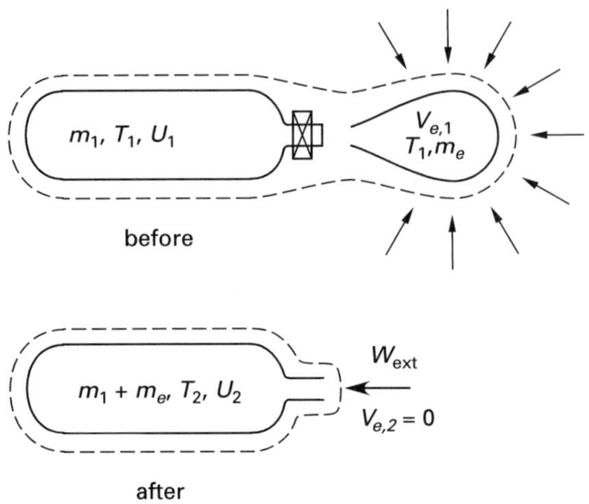

Customary U.S. Solution

The absolute temperature of the air in the tank when evacuated is

$$T_1 = 70°F + 460 = 530°R$$

From Table 29.7, $R = 53.3$ ft-lbf/lbm-°R. From Eq. 29.47,

$$\begin{aligned} m &= \frac{p_1 V_1}{RT_1} \\ &= \frac{\left(1\ \frac{\text{lbf}}{\text{in}^2}\right)\left(144\ \frac{\text{in}^2}{\text{ft}^2}\right)(20\ \text{ft}^3)}{\left(53.3\ \frac{\text{ft-lbf}}{\text{lbm-°R}}\right)(530°\text{R})} \\ &= 0.102\ \text{lbm} \end{aligned}$$

Assume $T_2 = 300°F$. The absolute temperature is

$$T_2 = 300°F + 460 = 760°R$$

From Eq. 29.47,

$$\begin{aligned} m_2 &= m_1 + m_e \\ &= \frac{p_2 V_2}{RT_2} \\ &= \frac{\left(14.7\ \frac{\text{lbf}}{\text{in}^2}\right)\left(144\ \frac{\text{in}^2}{\text{ft}^2}\right)(20\ \text{ft}^3)}{\left(53.3\ \frac{\text{ft-lbf}}{\text{lbm-°R}}\right)(760°\text{R})} \\ m_1 + m_e &\approx 1.045\ \text{lbm} \\ m_e &\approx 1.045\ \text{lbm} - m_1 \\ &= 1.045\ \text{lbm} - 0.102\ \text{lbm} \\ &= 0.943\ \text{lbm} \end{aligned}$$

From Eq. 29.47, the initial volume of the external air is

$$\begin{aligned} V_{e,1} &= \frac{mRT}{p} \\ &= \frac{(0.943\ \text{lbm})\left(53.3\ \frac{\text{ft-lbf}}{\text{lbm-°R}}\right)(530°\text{R})}{\left(14.7\ \frac{\text{lbf}}{\text{in}^2}\right)\left(144\ \frac{\text{in}^2}{\text{ft}^2}\right)} \\ &= 12.58\ \text{ft}^3 \end{aligned}$$

From Eq. 30.23, for a closed system,

$$Q = \Delta U + W$$

For an adiabatic system, $Q = 0$.

$$W_{ext} = \Delta U$$

For a constant pressure, closed system, from Eq. 30.50, the total work is

$$\begin{aligned} W_{ext} &= p(V_{e,2} - V_{e,1}) \\ &= \frac{\left(14.7\ \frac{\text{lbf}}{\text{in}^2}\right)\left(144\ \frac{\text{in}^2}{\text{ft}^2}\right)(0 - 12.58\ \text{ft}^3)}{778\ \frac{\text{ft-lbf}}{\text{Btu}}} \\ &= -34.23\ \text{Btu} \quad \left[\begin{array}{c}\text{surroundings do work}\\ \text{on the system}\end{array}\right] \end{aligned}$$

This energy is used to raise the temperature of the air and tank. Consider air as an ideal gas. The process inside the tank is not constant-pressure, as was the compression of the external air mass. However, the process is adiabatic, so $\Delta U = -W$. From Eq. 30.37, $\Delta U = c_v \Delta T$.

$$\begin{aligned} W_{ext} &= \big((m_1 + m_e)c_v\big)(T_2 - T_1) \\ 34.23\ \text{Btu} &= \left((1.045\ \text{lbm})\left(0.171\ \frac{\text{Btu}}{\text{lbm-°F}}\right) \times (T_2 - 70°\text{F})\right) \\ T_2 &= 261.6°\text{F} \end{aligned}$$

(A second iteration may be required.)

Now, allow the 261.6°F air and the 70°F tank to reach thermal equilibrium. The process will be constant-pressure.

$$\begin{aligned} q_{air} &= q_{tank} \\ c_{p,air} m_{air}(T_{air} - T_{equilibrium}) &= c_{p,tank} m_{tank}(T_{equilibrium} - T_1) \\ \left(0.24\ \frac{\text{Btu}}{\text{lbm-°F}}\right)(1.045\ \text{lbm})(261.6°\text{F} - T_{equilibrium}) &= \left(0.11\ \frac{\text{Btu}}{\text{lbm-°F}}\right)(40\ \text{lbm})(T_{equilibrium} - 70°\text{F}) \\ T_{equilibrium} &= \boxed{80.3°\text{F}} \end{aligned}$$

SI Solution

The absolute temperature of the air in the tank when evacuated is

$$T_1 = 21°\text{C} + 273 = 294\text{K}$$

From Table 29.7, $R = 287$ J/kg·K. From Eq. 29.47,

$$\begin{aligned} m &= \frac{p_1 V_1}{R T_1} \\ &= \frac{(7\ \text{kPa})\left(1000\ \frac{\text{Pa}}{\text{kPa}}\right)(0.6\ \text{m}^3)}{\left(287\ \frac{\text{J}}{\text{kg·K}}\right)(294\text{K})} \\ &= 0.0498\ \text{kg} \end{aligned}$$

Assume $T_2 = 127°\text{C}$. The absolute temperature is

$$T_2 = 127°\text{C} + 273 = 400\text{K}$$

From Eq. 29.47,

$$\begin{aligned} m_2 = m_1 + m_e &= \frac{p_2 V_2}{R T_2} \\ &= \frac{(101.3\ \text{kPa})\left(1000\ \frac{\text{Pa}}{\text{kPa}}\right)(0.6\ \text{m}^3)}{\left(287\ \frac{\text{J}}{\text{kg·K}}\right)(400\text{K})} \\ &= 0.5294\ \text{kg} \\ m_e &\approx 0.5294\ \text{kg} - m_1 \\ &= 0.5294\ \text{kg} - 0.0498\ \text{kg} \\ &= 0.4796\ \text{kg} \end{aligned}$$

From Eq. 29.47, the initial volume of the external air is

$$\begin{aligned} V_{e,1} &= \frac{mRT}{p} \\ &= \frac{(0.4796\ \text{kg})\left(287\ \frac{\text{J}}{\text{kg·K}}\right)(294\text{K})}{(101.3\ \text{kPa})\left(1000\ \frac{\text{Pa}}{\text{kPa}}\right)} \\ &= 0.3995\ \text{m}^3 \end{aligned}$$

From Eq. 30.23, for a closed system,

$$Q = \Delta U + W$$

For an adiabatic system, $Q = 0$.

$$W_{\text{ext}} = \Delta U$$

For a constant pressure, closed system, from Eq. 30.50, the total work is

$$\begin{aligned} W_{\text{ext}} &= p(V_{e,2} - V_{e,1}) \\ &= (101.3\ \text{kPa})\left(1000\ \frac{\text{Pa}}{\text{kPa}}\right)(0 - 0.3995\ \text{m}^3) \\ &= -40\,469\ \text{J} \quad \left[\begin{array}{c}\text{surroundings do work} \\ \text{on the system}\end{array}\right] \end{aligned}$$

This energy is used to raise the temperature of the air and tank. Consider air as an ideal gas. The process inside the tank is not constant-pressure, as was the compression of the external air mass. However, the process is adiabatic, so $\Delta U = -W$. From Eq. 30.37, $\Delta U = c_v \Delta T$.

$$\begin{aligned} W_{\text{ext}} &= \big((m_1 + m_e)c_v\big)(T_2 - T_1) \\ 40\,469\ \text{J} &= \left((0.5294\ \text{kg})\left(718\ \frac{\text{J}}{\text{kg·K}}\right)(T_2 - 21°\text{C})\right) \\ T_2 &= 127.5°\text{C} \end{aligned}$$

This is close enough to the assumed value of T_2 that a second iteration is not necessary.

Now, allow the 127.5°C air and the 21°C tank to reach thermal equilibrium. The process will be constant-pressure.

$$\begin{aligned} q_{\text{air}} &= q_{\text{tank}} \\ c_{p,\text{air}} m_{\text{air}}(T_{\text{air}} - T_{\text{equilibrium}}) &= c_{p,\text{tank}} m_{\text{tank}}(T_{\text{equilibrium}} - T_1) \\ \left(1005\ \frac{\text{J}}{\text{kg·K}}\right)(0.5294\ \text{kg})(127.5°\text{C} - T_{\text{equilibrium}}) &= \left(460\ \frac{\text{J}}{\text{kg·K}}\right)(20\ \text{kg})(T_{\text{equilibrium}} - 21°\text{C}) \\ T_{\text{equilibrium}} &= \boxed{26.8°\text{C}} \end{aligned}$$

Ventilation

31 HVAC: Psychrometrics

PRACTICE PROBLEMS

1. A room contains air at 80°F (27°C) dry-bulb and 67°F (19°C) wet-bulb. The total pressure is 1 atm. What are the (a) humidity, (b) enthalpy, and (c) specific heat?

2. If one row of cooling coils bypasses one-third of the air passing through it, what is the theoretical bypass factor for four rows of identical cooling coils in series?

3. 1000 ft^3/min (0.5 m^3/s) of air at 50°F (10°C) dry-bulb and 95% relative humidity are mixed with 1500 ft^3/min (0.75 m^3/s) of air at 76°F (24°C) and 45% relative humidity. What are the (a) dry-bulb temperature, (b) specific humidity, and (c) dew point of the mixture?

4. Air at 60°F (16°C) dry-bulb and 45°F (7°C) wet-bulb passes through an air washer with a humidifying efficiency of 70%. What are the (a) effective bypass factor of the system and (b) dry-bulb temperature of the air leaving the washer?

5. 95°F (35°C) dry-bulb, 75°F (24°C) wet-bulb air passes through a cooling tower and leaves at 85°F (29°C) dry-bulb and 90% relative humidity. What are the (a) enthalpy change and (b) change in moisture content per cubic foot (meter) of air?

6. An air washer receives 1800 ft^3/min (0.85 m^3/s) of air at 70°F (21°C) and 40% relative humidity and discharges the air at 75% relative humidity. A recirculating water spray with a constant temperature of 50°F (10°C) is used. (a) What mass of makeup water is required per minute? (b) What will be the condition of the discharged air?

7. Repeat Prob. 6 using saturated steam at atmospheric pressure in place of the 50°F (10°C) water spray.

8. During performances, a theater experiences a sensible heat load of 500,000 Btu/hr (150 kW) and a moisture load of 175 lbm/hr (80 kg/h). Air enters the theater at 65°F (18°C) and 55% relative humidity and is removed when it reaches 75°F (24°C) or 60% relative humidity, whichever comes first. (a) What is the ventilation rate in mass of air per hour? (b) What are the conditions of the air leaving the theater?

9. 500 ft^3/min (0.25 m^3/s) of air at 80°F (27°C) dry-bulb and 70% relative humidity are removed from a room. 150 ft^3/min (0.075 m^3/s) pass through an air conditioner and leave saturated at 50°F (10°C). The remaining 350 ft^3/min (0.175 m^3/s) bypass the air conditioner and mix with the conditioned air at 1 atm. What is the mixture's (a) temperature, (b) humidity ratio, and (c) relative humidity? (d) What is the heat load (in tons) of the air conditioner?

10. (*Time limit: one hour*) A dehumidifier takes 5000 ft^3/min (2.36 m^3/s) of air at 95°F (35°C) dry-bulb and 70% relative humidity and discharges it at 60°F (16°C) dry-bulb and 95% relative humidity. The dehumidifier uses an R-12 refrigeration cycle operating between 100°F (saturated) (38°C) and 50°F (10°C). (a) Locate the air entering and leaving points on the psychrometric chart. (b) Find the quantity of water removed from the air. (c) Find the quantity of heat removed from the air. (d) Draw the temperature-entropy and enthalpy-entropy diagrams for the refrigeration cycle. (e) Find the temperature, pressure, enthalpy, entropy, and specific volume for each endpoint of the refrigeration cycle.

11. (*Time limit: one hour*) 1500 ft^3/min (0.71 m^3/s) of saturated 25 psia (170 kPa) air are heated from 200 to 400°F (93 to 204°C) in a constant pressure, constant moisture drying process. (a) What is the final relative humidity? (b) What is the final specific humidity? (c) How much heat is required per unit mass of dry air? (d) What is the final dew point?

12. (*Time limit: one hour*) 410 lbm/hr (0.052 kg/s) of dry 800°F (427°C) air pass through a scrubber to reduce particulate emissions. To protect the elastomeric seals in the scrubber, the air temperature is reduced to 350°F (177°C) by passing the air through a spray of 80°F (27°C) water. The pressure in the spray chamber is 20 psia (140 kPa). (a) How much water is evaporated per hour? (b) What will be the relative humidity of the air leaving the spray chamber?

13. (*Time limit: one hour*) An evaporative counterflow air cooling tower removes 1×10^6 Btu/hr (290 kW) from a water flow. The temperature of the water is reduced from 120 to 110°F (49 to 43°C). Air enters the cooling tower at 91°F (33°C) and 60% relative humidity, and air leaves at 100°F (38°C) and 82% relative humidity. Calculate the (a) air flow rate and (b) quantity of makeup water.

SOLUTIONS

1. *Customary U.S. Solution*

(a) and (b) Locate the intersection of 80°F dry bulb and 67°F wet bulb on the psychrometric chart (App. 31.A). Read the value of humidity and enthalpy.

$$\omega = \boxed{0.0112 \text{ lbm moisture/lbm dry air}}$$

$$h = \boxed{31.5 \text{ Btu/lbm dry air}}$$

(c) c_p is gravimetrically weighted. c_p for air is 0.240 Btu/lbm-°F, and c_p for steam is approximately 0.40 Btu/lbm-°F.

$$G_{\text{air}} = \frac{1}{1 + 0.0112} = 0.989$$

$$G_{\text{steam}} = \frac{0.0112}{1 + 0.0112} = 0.011$$

$$\begin{aligned} c_{p,\text{mixture}} &= G_{\text{air}} c_{p,\text{air}} + G_{\text{steam}} c_{p,\text{steam}} \\ &= (0.989)\left(0.240 \ \frac{\text{Btu}}{\text{lbm-°F}}\right) \\ &\quad + (0.011)\left(0.40 \ \frac{\text{Btu}}{\text{lbm-°F}}\right) \\ &= \boxed{0.242 \text{ Btu/lbm-°F}} \end{aligned}$$

SI Solution

(a) and (b) Locate the intersection of 27°C dry bulb and 19°C wet bulb on the psychrometric chart (App. 31.B). Read the value of humidity and enthalpy.

$$\begin{aligned} \omega &= \left(10.5 \ \frac{\text{g}}{\text{kg dry air}}\right)\left(\frac{1 \text{ kg}}{1000 \text{ g}}\right) \\ &= \boxed{0.0105 \text{ kg/kg dry air}} \end{aligned}$$

$$h = \boxed{53.9 \text{ kJ/kg dry air}}$$

(c) c_p is gravimetrically weighted. c_p for air is 1.0048 kJ/kg·K, and c_p for steam is approximately 1.675 kJ/kg·K.

$$G_{\text{air}} = \frac{1}{1 + 0.0105} = 0.990$$

$$G_{\text{steam}} = \frac{0.0105}{1 + 0.0105} = 0.010$$

$$\begin{aligned} c_{p,\text{mixture}} &= G_{\text{air}} c_{p,\text{air}} + G_{\text{air}} c_{p,\text{steam}} \\ &= (0.990)\left(1.0048 \ \frac{\text{kJ}}{\text{kg·K}}\right) \\ &\quad + (0.010)\left(1.675 \ \frac{\text{kJ}}{\text{kg·K}}\right) \\ &= \boxed{1.0115 \text{ kJ/kg·K}} \end{aligned}$$

2.

$$\begin{aligned} \text{BF}_{n \text{ rows}} &= (\text{BF}_{1 \text{ row}})^n \\ &= \left(\tfrac{1}{3}\right)^4 = \boxed{0.0123} \end{aligned}$$

3.

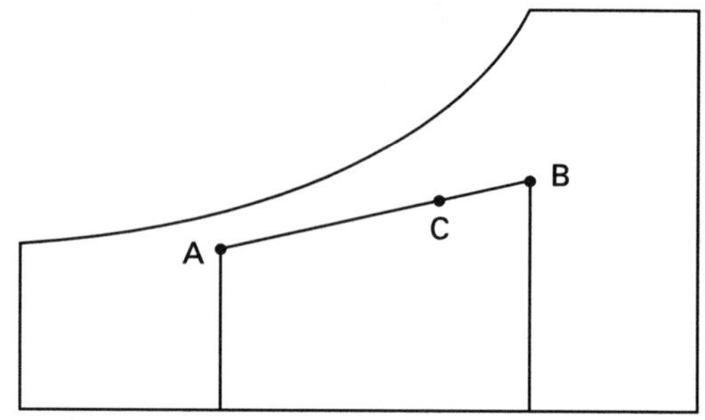

Customary U.S. Solution

(a) Locate the two points on the psychrometric chart and draw a line between them.

Reading from the chart (specific volumes),

$$\upsilon_{\text{A}} = 13.0 \text{ ft}^3/\text{lbm}$$
$$\upsilon_{\text{B}} = 13.7 \text{ ft}^3/\text{lbm}$$

The density at each point is

$$\rho_{\text{A}} = \frac{1}{\upsilon_{\text{A}}} = \frac{1}{13.0 \ \frac{\text{ft}^3}{\text{lbm}}} = 0.0769 \text{ lbm/ft}^3$$

$$\rho_{\text{B}} = \frac{1}{\upsilon_{\text{B}}} = \frac{1}{13.7 \ \frac{\text{ft}^3}{\text{lbm}}} = 0.0730 \text{ lbm/ft}^3$$

The mass flow at each point is

$$\begin{aligned} \dot{m}_{\text{A}} &= \rho_{\text{A}} \dot{V}_{\text{A}} = \left(0.0769 \ \frac{\text{lbm}}{\text{ft}^3}\right)\left(1000 \ \frac{\text{ft}^3}{\text{min}}\right) \\ &= 76.9 \text{ lbm/min} \end{aligned}$$

$$\begin{aligned} \dot{m}_{\text{B}} &= \rho_{\text{B}} \dot{V}_{\text{B}} = \left(0.0730 \ \frac{\text{lbm}}{\text{ft}^3}\right)\left(1500 \ \frac{\text{ft}^3}{\text{min}}\right) \\ &= 109.5 \text{ lbm/min} \end{aligned}$$

The gravimetric fraction of flow A is

$$\frac{76.9 \ \frac{\text{lbm}}{\text{min}}}{76.9 \ \frac{\text{lbm}}{\text{min}} + 109.5 \ \frac{\text{lbm}}{\text{min}}} = 0.413$$

Since the scales are all linear,

$$\begin{aligned} 0.413 &= \frac{T_B - T_C}{T_B - T_A} \\ T_C &= T_B - (0.413)(T_B - T_A) \\ &= 76°\text{F} - (0.413)(76°\text{F} - 50°\text{F}) \\ &= \boxed{65.3°\text{F}} \end{aligned}$$

(b) $\omega = \boxed{0.0082 \text{ lbm moisture/lbm dry air}}$

(c) $T_{dp} = \boxed{51°F}$

SI Solution

(a) Locate the two points on the psychrometric chart and draw a line between them.

Reading from the chart (specific volumes),

$$\upsilon_A = 0.813 \text{ m}^3\text{/kg dry air}$$
$$\upsilon_B = 0.856 \text{ m}^3\text{/kg dry air}$$

The density at each point is

$$\rho_A = \frac{1}{\upsilon_A} = \frac{1}{0.813 \ \frac{\text{m}^3}{\text{kg}}} = 1.23 \text{ kg/m}^3$$

$$\rho_B = \frac{1}{\upsilon_B} = \frac{1}{0.856 \ \frac{\text{m}^3}{\text{kg}}} = 1.17 \text{ kg/m}^3$$

The mass flow at each point is

$$\dot{m}_A = \rho_A \dot{V}_A = \left(1.23 \ \frac{\text{kg}}{\text{m}^3}\right)\left(0.5 \ \frac{\text{m}^3}{\text{s}}\right) = 0.615 \text{ kg/s}$$

$$\dot{m}_B = \rho_B \dot{V}_B = \left(1.17 \ \frac{\text{kg}}{\text{m}^3}\right)\left(0.75 \ \frac{\text{m}^3}{\text{s}}\right) = 0.878 \text{ kg/s}$$

The gravimetric fraction of flow A is

$$\frac{0.615 \ \frac{\text{kg}}{\text{s}}}{0.615 \ \frac{\text{kg}}{\text{s}} + 0.878 \ \frac{\text{kg}}{\text{s}}} = 0.412$$

Since the scales are linear,

$$0.412 = \frac{T_B - T_C}{T_B - T_A}$$

$$\begin{aligned} T_C &= T_B - (0.412)(T_B - T_A) \\ &= 24°\text{C} - (0.412)(24°\text{C} - 10°\text{C}) \\ &= \boxed{18.2°\text{C}} \end{aligned}$$

(b)

$$\begin{aligned} \omega &= \left(8.0 \ \frac{\text{g}}{\text{kg dry air}}\right)\left(\frac{1 \text{ kg}}{1000 \text{ g}}\right) \\ &= \boxed{0.008 \text{ kg/kg dry air}} \end{aligned}$$

(c) $T_{dp} = \boxed{10.6°\text{C}}$

4. *Customary U.S. Solution*

(a) From Eq. 31.24, the bypass factor is

$$\begin{aligned} \text{BF} &= 1 - \eta_{sat} \\ &= 1 - 0.70 = \boxed{0.30} \end{aligned}$$

(b) From Eq. 31.34, the dry-bulb temperature of air leaving the washer can be determined.

$$\eta_{sat} = \frac{T_{db,in} - T_{db,out}}{T_{db,in} - T_{wb,in}}$$

$$0.70 = \frac{60°\text{F} - T_{db,out}}{60°\text{F} - 45°\text{F}}$$

$$T_{db,out} = \boxed{49.5°\text{F}}$$

SI Solution

(a) From Eq. 31.24, the bypass factor is

$$\begin{aligned} \text{BF} &= 1 - \eta_{sat} \\ &= 1 - 0.70 = \boxed{0.30} \end{aligned}$$

(b) From Eq. 31.34, the dry-bulb temperature of air leaving the washer can be determined.

$$\eta_{sat} = \frac{T_{db,in} - T_{db,out}}{T_{db,in} - T_{wb,in}}$$

$$0.70 = \frac{16°\text{C} - T_{db,out}}{16°\text{C} - 7°\text{C}}$$

$$T_{db,out} = \boxed{9.7°\text{C}}$$

5. *Customary U.S. Solution*

(a) Refer to the psychrometric chart (App. 31.A).

At point 1, properties of air at $T_{db} = 95°\text{F}$ and $T_{wb} = 75°\text{F}$ are

$$\omega_1 = 0.0141 \text{ lbm moisture/lbm air}$$
$$h_1 = 38.4 \text{ lbm air}$$
$$\upsilon_1 = 14.3 \text{ ft}^3\text{/lbm air}$$

At point 2, properties of air at $T_{db} = 85°\text{F}$ and 90% relative humidity are

$$\omega_2 = 0.0237 \text{ lbm moisture/lbm air}$$
$$h_2 = 46.6 \text{ Btu/lbm air}$$

Ventilation

The enthalpy change is

$$\frac{h_2 - h_1}{v_1} = \frac{46.6\ \frac{\text{Btu}}{\text{lbm air}} - 38.4\ \frac{\text{Btu}}{\text{lbm air}}}{14.3\ \frac{\text{ft}^3}{\text{lbm air}}} = \boxed{0.573\ \text{Btu/ft}^3\ \text{air}}$$

(b) The moisture added is

$$\frac{\omega_2 - \omega_1}{v_1} = \frac{0.0237\ \frac{\text{lbm moisture}}{\text{lbm air}} - 0.0141\ \frac{\text{lbm moisture}}{\text{lbm air}}}{14.3\ \frac{\text{ft}^3}{\text{lbm air}}} = \boxed{6.71 \times 10^{-4}\ \text{lbm/ft}^3\ \text{air}}$$

SI Solution

(a) Refer to the psychrometric chart (App. 31.B).

At point 1, properties of air at $T_{\text{db}} = 35°\text{C}$ and $T_{\text{wb}} = 24°\text{C}$ are

$$\omega_1 = 14.3\ \text{g/kg air}$$
$$h_1 = 71.8\ \text{kJ/kg air}$$
$$v_1 = 0.8893\ \text{m}^3/\text{kg air}$$

At point 2, properties of air at $T_{\text{db}} = 29°\text{C}$ and 90% relative humidity are

$$\omega_2 = 23.1\ \text{g/kg air}$$
$$h_2 = 88\ \text{kJ/kg air}$$

The enthalpy change is

$$\frac{h_2 - h_1}{v_1} = \frac{88\ \frac{\text{kJ}}{\text{kg air}} - 71.8\ \frac{\text{kJ}}{\text{kg air}}}{0.8893\ \frac{\text{m}^3}{\text{kg air}}} = \boxed{18.2\ \text{kJ/m}^3\ \text{air}}$$

(b) The moisture added is

$$\frac{\omega_2 - \omega_1}{v_1} = \frac{\left(23.1\ \frac{\text{g}}{\text{kg air}} - 14.3\ \frac{\text{kg}}{\text{kg air}}\right) \times \left(\frac{1\ \text{kg}}{1000\ \text{g}}\right)}{0.8893\ \frac{\text{m}^3}{\text{kg air}}} = \boxed{9.90 \times 10^{-3}\ \text{kg/m}^3\ \text{air}}$$

6.

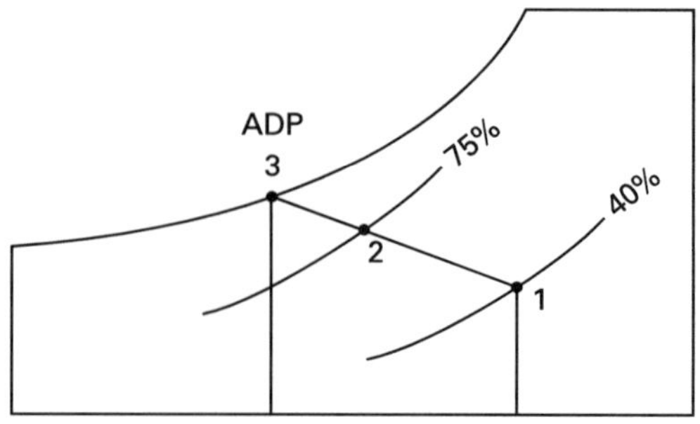

Customary U.S. Solution

(a) Refer to the psychrometric chart (App. 31.A).

At point 1, properties of air at $T_{\text{db}} = 70°\text{F}$ and $\phi = 40\%$ are

$$h_1 = 23.6\ \text{Btu/lbm air}$$
$$\omega_1 = 0.00623\ \text{lbm moisture/lbm air}$$
$$v_1 = 13.48\ \text{ft}^3/\text{lbm air}$$

The mass flow rate of incoming air is

$$\dot{m}_{a,1} = \frac{\dot{V}_1}{v_1} = \frac{1800\ \frac{\text{ft}^3}{\text{min}}}{13.48\ \frac{\text{ft}^3}{\text{lbm air}}} = 133.53\ \text{lbm air/min}$$

Locate point 1 on the psychrometric chart.

Notice that the temperature of the recirculating water is constant but not equal to the air's entering wet-bulb temperature. Therefore, this is not an adiabatic process.

Locate point 3 as 50°F saturated condition (water being sprayed) on the psychrometric chart.

Draw a line from point 1 to point 3. The intersection of this line with 75% relative humidity defines point 2 as

$$\boxed{\begin{aligned} h_2 &= 21.4\ \text{Btu/lbm air} \\ \omega_2 &= 0.0072\ \text{lbm moisture/lbm air} \\ T_{\text{db},2} &= 56°\text{F} \\ T_{\text{wb},2} &= 51.8°\text{F} \end{aligned}}$$

(b) The moisture (water) added is

$$\begin{aligned} \dot{m}_w &= \dot{m}_{a,1}(\omega_2 - \omega_1) \\ &= \left(133.53\ \frac{\text{lbm air}}{\text{min}}\right)\left(0.0072\ \frac{\text{lbm moisture}}{\text{lbm air}} - 0.00625\ \frac{\text{lbm moisture}}{\text{lbm air}}\right) \\ &= \boxed{0.127\ \text{lbm/min}} \end{aligned}$$

SI Solution

(a) Refer to the psychrometric chart (App. 31.B).

At point 1, properties of air at $T_{db} = 21°C$ and $\phi = 40\%$ are

$$h_1 = 36.75 \text{ kJ/kg air}$$
$$\omega_1 = 6.2 \text{ g moisture/kg air}$$
$$v_1 = 0.842 \text{ m}^3\text{/kg air}$$

The mass flow rate of incoming air is

$$\dot{m}_{a,1} = \frac{\dot{V}_1}{v_1} = \frac{0.85 \ \frac{\text{m}^3}{\text{s}}}{0.842 \ \frac{\text{m}^3}{\text{kg air}}} = 1.010 \text{ kg air/s}$$

Locate point 1 on the psychrometric chart.

Notice that the temperature of the recirculating water is constant but not equal to the air's entering wet-bulb temperature. Therefore, this is not an adiabatic process.

Locate point 3 as 10°C saturated condition (water being sprayed) on the psychrometric chart.

Draw a line from point 1 to point 3. The intersection of this line with 75% relative humidity defines point 2 as

$$\boxed{\begin{aligned} T_{2,db} &= 14.7°C \\ h_2 &= 34.1 \text{ kJ/kg air} \\ \omega_2 &= 7.6 \text{ g moisture/kg air} \end{aligned}}$$

(b) The water added is

$$\begin{aligned} \dot{m}_w &= \dot{m}_{a,1}(\omega_2 - \omega_1) \\ &= \left(1.010 \ \frac{\text{kg air}}{\text{s}}\right)\left(7.6 \ \frac{\text{g moisture}}{\text{kg air}} - 6.2 \ \frac{\text{g moisture}}{\text{kg air}}\right) \\ &\quad \times \left(\frac{1 \text{ kg}}{1000 \text{ g}}\right) \\ &= \boxed{0.00141 \text{ kg/s}} \end{aligned}$$

7. *Customary U.S. Solution*

From Prob. 6,

$$\omega_1 = 0.00623 \text{ lbm moisture/lbm air}$$
$$h_1 = 23.6 \text{ Btu/lbm air}$$
$$\dot{m}_{a,1} = 133.53 \text{ lbm air/min}$$

From the steam table (App. 29.B) for 1 atm steam, $h_{steam} = 1150.5$ Btu/lbm.

From the conservation of energy equation (Eq. 31.32),

$$\dot{m}_{a,1}h_1 + \dot{m}_{steam}h_{steam} = \dot{m}_{a,1}h_2$$

$$\left(133.53 \ \frac{\text{lbm air}}{\text{min}}\right)\left(23.6 \ \frac{\text{Btu}}{\text{lbm air}}\right) + \dot{m}_{steam}\left(1150.5 \ \frac{\text{Btu}}{\text{lbm}}\right) = \left(133.53 \ \frac{\text{lbm air}}{\text{min}}\right)h_2 \quad \textit{[Eq. 1]}$$

From conservation of mass for the water (Eq. 31.33),

$$\dot{m}_{a,1}\omega_1 + \dot{m}_{steam} = \dot{m}_{a,2}\omega_2$$

$$\left(133.53 \ \frac{\text{lbm air}}{\text{min}}\right)\left(0.00623 \ \frac{\text{lbm moisture}}{\text{lbm air}}\right) + \dot{m}_{steam} = \left(133.53 \ \frac{\text{lbm air}}{\text{min}}\right)\omega_2 \quad \textit{[Eq. 2]}$$

Since no single relationship exists between ω_2, $\dot{m}_{steam}$, and h_2, a trial and error solution is required. Once $\dot{m}_{steam}$ is selected, ω_2 and h_2 can be found from Eq. 1 and Eq. 2 as

$$h_2 = (8.616)\dot{m}_{steam} + 23.6$$
$$\omega_2 = 0.00623 + 0.00749\dot{m}_{steam}$$

Once h_2 and ω_2 are known, the relative humidity can be determined from the psychrometric chart. Continue the process until a relative humidity of 75% is achieved.

$\dot{m}_{steam}$ $\left(\frac{\text{lbm}}{\text{min}}\right)$	ω_2 $\left(\frac{\text{lbm moisture}}{\text{lbm air}}\right)$	h_2 $\left(\frac{\text{Btu}}{\text{lbm air}}\right)$	ϕ_2 (%)
0.3	0.00848	26.18	53
0.4	0.00923	27.05	58
0.5	0.00998	27.91	62
0.6	0.01072	28.77	66
0.7	0.01147	29.63	70
0.8	0.01222	30.49	74
0.82	0.01237	30.67	74.8
0.83	0.01245	30.75	75.3
0.825	0.01241	30.71	75.0

$$\boxed{\begin{aligned} \dot{m}_{steam} &= 0.825 \text{ lbm/min} \\ \omega_2 &= 0.01241 \text{ lbm moisture/lbm air} \\ h_2 &= 30.71 \text{ Btu/lbm air} \\ T_{db} &= 71.5°F \\ T_{wb} &= 65.9°F \end{aligned}}$$

SI Solution

From Prob. 6,

$$\omega_1 = 6.2 \text{ g moisture/kg air}$$
$$h_1 = 36.75 \text{ kJ/kg air}$$
$$\dot{m}_{a,1} = 1.010 \text{ kg air/s}$$

From the steam table (App. 29.O), for 1 atm steam, $h_{steam} = 2675.5$ kJ/kg.

From the conservation of energy equation (Eq. 31.32),

$$\dot{m}_{a,1}h_1 + \dot{m}_{steam}h_{steam} = \dot{m}_{a,1}h_2$$

$$\left(1.010 \ \frac{\text{kg air}}{\text{s}}\right)\left(36.75 \ \frac{\text{kJ}}{\text{kg air}}\right) + \dot{m}_{steam}\left(2675.5 \ \frac{\text{kJ}}{\text{kg}}\right) = \left(1.010 \ \frac{\text{kg air}}{\text{s}}\right)h_2 \quad \textit{[Eq. 1]}$$

Ventilation

From conservation of mass of water (Eq. 31.33),

$$\dot{m}_{a,1}\omega_1 + \dot{m}_{\text{steam}} = \dot{m}_{a,2}\omega_2$$

$$\left(1.010\ \frac{\text{kg air}}{\text{s}}\right)\left(6.2\ \frac{\text{g moisture}}{\text{kg air}}\right)\left(\frac{1\ \text{kg}}{1000\ \text{kg}}\right) + \dot{m}_{\text{steam}} = \left(1.010\ \frac{\text{kg air}}{\text{s}}\right)\omega_2 \quad \textit{[Eq. 2]}$$

Since no single relationship exists between ω_2, $\dot{m}_{\text{steam}}$, and h_2, a trial and error solution is required. Once $\dot{m}_{\text{steam}}$ is selected, ω_2 and h_2 can be found from Eq. 1 and Eq. 2 as

$$h_2 = 36.75 + 2649.0\dot{m}_{\text{steam}}$$
$$\omega_2 = 0.0062 + 0.99\dot{m}_{\text{steam}}$$

Once h_2 and ω_2 are known, the relative humidity can be determined from the psychrometric chart. Continue the process until a relative humidity of 75% is achieved.

$\dot{m}_{\text{steam}}$ $\left(\frac{\text{kg}}{\text{s}}\right)$	ω_2 $\left(\frac{\text{kg moisture}}{\text{kg air}}\right)$	h_2 $\left(\frac{\text{kJ}}{\text{kg air}}\right)$	ϕ_2 (%)
0.005	0.0112	50.00	69.5
0.0055	0.0116	51.32	74.5
0.0056	0.0117	51.58	75.0

$$\begin{aligned}\dot{m}_{\text{steam}} &= 0.0056\ \text{kg/s}\\ \omega_2 &= \left(0.0117\ \frac{\text{kg moisture}}{\text{kg air}}\right)\left(1000\ \frac{\text{g}}{\text{kg}}\right)\\ &= 11.7\ \text{g moisture/kg air}\\ T_{\text{db}} &= 21.3^\circ\text{C}\\ T_{\text{wb}} &= 18.2^\circ\text{C}\end{aligned}$$

8. *Customary U.S. Solution*

(a) From the psychrometric chart (App. 31.A), for incoming air at 65°F and 55% relative humidity, $\omega_1 = 0.0072$ lbm moisture/lbm air.

With sensible heating as a limiting factor, calculate the mass flow rate of air entering the theater from Eq. 31.26 (ventilation rate).

$$\dot{q} = \dot{m}_a(c_{p,\text{air}} + \omega c_{p,\text{moisture}})(T_2 - T_1)$$

$$500{,}000\ \frac{\text{Btu}}{\text{hr}} = \dot{m}_a\left[0.240\ \frac{\text{Btu}}{\text{lbm-}^\circ\text{F}} + \left(0.0072\ \frac{\text{lbm moisture}}{\text{lbm air}}\right) \times \left(0.444\ \frac{\text{Btu}}{\text{lbm-}^\circ\text{F}}\right)\right](75^\circ\text{F} - 65^\circ\text{F})$$

$$\dot{m}_a = \boxed{2.056 \times 10^5\ \text{lbm air/hr}}$$

(b) Assume that this air absorbs all the moisture. Then, the final humidity ratio is given by

$$\dot{m}_w = \dot{m}_a(\omega_2 - \omega_1)$$

$$\omega_2 = \left(\frac{\dot{m}_w}{\dot{m}_a}\right) + \omega_1 = \frac{175\ \frac{\text{lbm moisture}}{\text{hr}}}{2.056 \times 10^5\ \frac{\text{lbm air}}{\text{hr}}} + 0.0072\ \frac{\text{lbm moisture}}{\text{lbm air}}$$
$$= 0.00805\ \text{lbm moisture/lbm air}$$

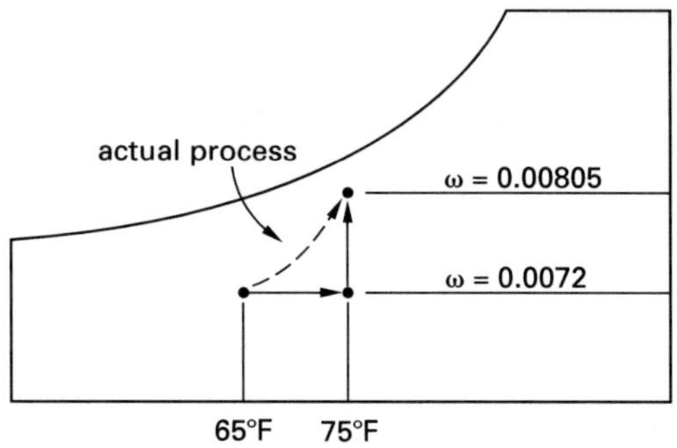

The final conditions are

$$T_{\text{db}} = \boxed{75^\circ\text{F}\ \ \text{[given]}}$$

$$\omega_2 = \boxed{0.00805\ \text{lbm moisture/lbm air}}$$

From the psychrometric chart (App. 31.A), the relative humidity is 44%. This is below 60%.

SI Solution

(a) From the psychrometric chart (App. 31.B), for incoming air at 18°C and 55% relative humidity, $\omega_1 = 7.1$ g moisture/kg air.

With sensible heating as a limiting factor, calculate the mass flow rate of air entering the theater from Eq. 31.26 (ventilation rate).

$$\dot{q} = \dot{m}_a(c_{p,\text{air}} + \omega c_{p,\text{moisture}}) \times (T_2 - T_1)$$

$$(150\,\text{kW})\left(1000\ \frac{\text{W}}{\text{kW}}\right) = \dot{m}_a\left[\left(1.005\ \frac{\text{kJ}}{\text{kg}\cdot{}^\circ\text{C}}\right)\left(1000\ \frac{\text{J}}{\text{kJ}}\right) + \left(1.805\ \frac{\text{kJ}}{\text{kg}\cdot{}^\circ\text{C}}\right)\left(1000\ \frac{\text{J}}{\text{kJ}}\right) \times \left(7.1\ \frac{\text{g moisture}}{\text{kg air}}\right)\left(\frac{1\ \text{kg}}{1000\ \text{g}}\right)\right] \times (24^\circ\text{C} - 18^\circ\text{C})$$

$$\dot{m}_a = \boxed{24.56\,\text{kg/s}}$$

(b) Assume that this air absorbs all the moisture. Then, the final humidity ratio is given by

$$\dot{m}_w = \dot{m}_a(\omega_2 - \omega_1)$$

$$\omega_2 = \frac{\dot{m}_w}{\dot{m}_a} + \omega_1$$

$$= \frac{\left(80\ \frac{\text{kg}}{\text{h}}\right)\left(\frac{1\ \text{h}}{3600\ \text{s}}\right)}{24.56\ \frac{\text{kg}}{\text{s}}}$$

$$+ \left(7.0\ \frac{\text{g moisture}}{\text{kg air}}\right)\left(\frac{1\ \text{kg}}{1000\ \text{g}}\right)$$

$$= 0.00790\ \text{kg moisture/kg air}$$

The final conditions are

$$T_{\text{db}} = \boxed{24°\text{C}}$$

$$\omega_2 = \left(0.00790\ \frac{\text{kg moisture}}{\text{kg air}}\right)\left(1000\ \frac{\text{g}}{\text{kg}}\right)$$

$$= \boxed{7.9\ \text{g moisture/kg air}}$$

From the psychrometric chart (App. 31.B), the relative humidity is 44%. This is below 60%.

9.

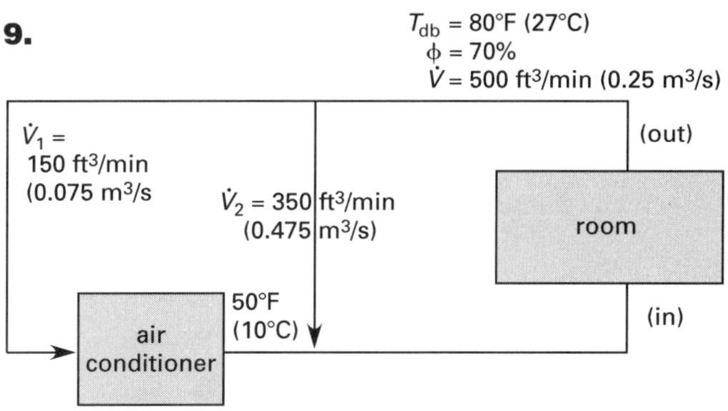

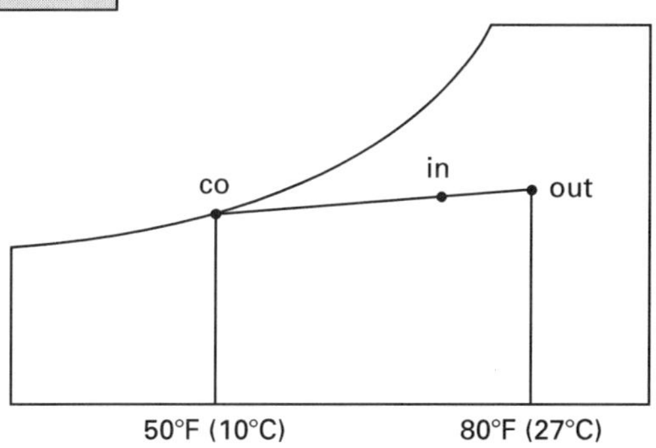

Customary U.S. Solution

Locate point (out) ($T_{\text{db}} = 80°\text{F}$ and $\phi = 70\%$) and point (co) (saturated at 50°F) on the psychrometric chart. At point (out) from App. 31.A,

$$v_{\text{out}} = 13.95\ \text{ft}^3/\text{lbm air}$$

$$h_{\text{out}} = 36.2\ \text{Btu/lbm air}$$

At point (co), $h_{\text{co}} = 20.3$ Btu/lbm air.

The air mass flow rate through the air conditioner is

$$\dot{m}_1 = \frac{\dot{V}_1}{v_1} = \frac{\dot{V}_1}{v_{\text{out}}} = \frac{150\ \frac{\text{ft}^3}{\text{min}}}{13.95\ \frac{\text{ft}^3}{\text{lbm air}}}$$

$$= 10.75\ \text{lbm air/min}$$

The mass flow rate of the bypass air is

$$\dot{m}_2 = \frac{\dot{V}_2}{v} = \frac{350\ \frac{\text{ft}^3}{\text{min}}}{13.95\ \frac{\text{ft}^3}{\text{lbm air}}} = 25.09\ \text{lbm air/min}$$

The percentage of bypass air is

$$x = \frac{25.09\ \frac{\text{lbm air}}{\text{min}}}{10.75\ \frac{\text{lbm air}}{\text{min}} + 25.09\ \frac{\text{lbm air}}{\text{min}}} = 0.70\quad(70\%)$$

Using the lever rule and the fact that all of the temperature scales are linear,

$$T_{\text{db,in}} = T_{\text{co}} + (0.70)(T_{\text{out}} - T_{\text{co}})$$

$$= 50°\text{F} + (0.70)(80°\text{F} - 50°\text{F})$$

$$= 71°\text{F}$$

At that point,

(a) $T_{\text{db,in}} = \boxed{71°\text{F}}$

(b) $\omega_{\text{in}} = \boxed{0.0132\ \text{lbm moisture/lbm air}}$

(c) $\phi_{\text{in}} = \boxed{81\%}$

(d) The air conditioner capacity is given by

$$\dot{Q} = \dot{m}_{\text{air}}(h_{t,2} - h_{t,1}) = \dot{m}_1(h_{\text{out}} - h_{\text{co}})$$

$$= \left(10.75\ \frac{\text{lbm air}}{\text{min}}\right)\left(36.2\ \frac{\text{Btu}}{\text{lbm air}} - 20.3\ \frac{\text{Btu}}{\text{lbm air}}\right)$$

$$\times \left(\frac{1\ \text{ton}}{200\ \frac{\text{Btu}}{\text{min}}}\right)$$

$$= \boxed{0.85\ \text{ton}}$$

SI Solution

Locate point (out) ($T_{\text{db}} = 27°\text{C}$, $\phi = 70\%$) and point (co) (saturated at 10°C) on the psychrometric chart. At point (out) from App. 31.B,

$$v_{\text{out}} = 0.872\ \text{m}^3/\text{kg air}$$

$$h_{\text{out}} = 67.3\ \text{kJ/kg air}$$

At point (co) from App. 31.B, $h_{co} = 29.26$ kJ/kg air.

At mass flow rate through the air conditioner,

$$\dot{m}_1 = \frac{\dot{V}_1}{v} = \frac{0.075\ \frac{\text{m}^3}{\text{s}}}{0.872\ \frac{\text{m}^3}{\text{kg air}}} = 0.0860 \text{ kg air/s}$$

The flow rate of bypass air is

$$\dot{m}_2 = \frac{\dot{V}_2}{v} = \frac{0.175\ \frac{\text{m}^3}{\text{s}}}{0.872\ \frac{\text{m}^3}{\text{kg air}}} = 0.2007 \text{ kg air/s}$$

The percentage of bypass air is

$$x = \frac{0.2007\ \frac{\text{kg air}}{\text{s}}}{0.0860\ \frac{\text{kg air}}{\text{s}} + 0.2007\ \frac{\text{kg air}}{\text{s}}} = 0.70 \quad (70\%)$$

Using the lever rule and the fact that all of the temperature scales are linear,

$$\begin{aligned} T_{\text{db,in}} &= T_{\text{co}} + (0.70)(T_{\text{out}} - T_{\text{co}}) \\ &= 10^\circ\text{C} + (0.70)(27^\circ\text{C} - 10^\circ\text{C}) \\ &= 21.9^\circ\text{C} \end{aligned}$$

At that point,

(a) $T_{\text{db,in}} = \boxed{21.9^\circ\text{C}}$

(b) $\omega_{\text{in}} = \boxed{13.4 \text{ g moisture/kg air}}$

(c) $\phi_{\text{in}} = \boxed{81\%}$

(d) The air conditioner capacity is given by

$$\begin{aligned} \dot{Q} &= \dot{m}_{\text{air}}(h_{t,2} - h_{t,1}) = \dot{m}_1(h_{\text{out}} - h_{\text{co}}) \\ &= \left(0.0860\ \frac{\text{kg air}}{\text{s}}\right)\left(67.3\ \frac{\text{kJ}}{\text{kg air}} - 29.26\ \frac{\text{kJ}}{\text{kg air}}\right) \\ &\quad \times \left(0.2843\ \frac{\text{ton}}{\text{kW}}\right) \\ &= \boxed{0.93 \text{ ton}} \end{aligned}$$

10.

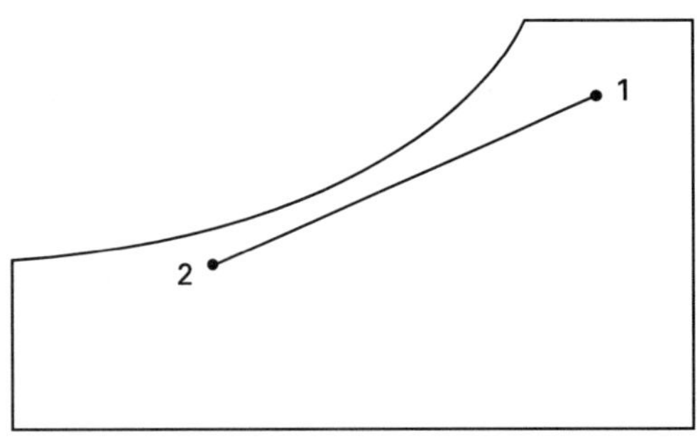

Customary U.S. Solution

(a) At point 1, from the psychrometric chart (App. 31.A) at $T_{\text{db}} = 95^\circ$F and $\phi = 70\%$,

$$\boxed{\begin{aligned} h_1 &= 50.7 \text{ Btu/lbm air} \\ v_1 &= 14.56 \text{ ft}^3\text{/lbm air} \\ \omega_1 &= 0.0253 \text{ lbm water/lbm air} \end{aligned}}$$

At point 2, from the psychrometric chart (App. 31.A) at $T_{\text{db}} = 60^\circ$F and $\phi = 95\%$,

$$\boxed{\begin{aligned} h_2 &= 25.8 \text{ Btu/lbm air} \\ \omega_2 &= 0.0105 \text{ lbm water/lbm air} \end{aligned}}$$

(b) The air mass flow rate is

$$\dot{m}_a = \frac{\dot{V}}{v_1} = \frac{5000\ \frac{\text{ft}^3}{\text{min}}}{14.56\ \frac{\text{ft}^3}{\text{lbm air}}} = 343.4 \text{ lbm air/min}$$

From Eq. 31.27, the water removed is

$$\begin{aligned} \dot{m}_w &= \dot{m}_a(\omega_1 - \omega_2) \\ &= \left(343.4\ \frac{\text{lbm air}}{\text{min}}\right) \\ &\quad \times \left(0.0253\ \frac{\text{lbm water}}{\text{lbm air}} - 0.0105\ \frac{\text{lbm water}}{\text{lbm air}}\right) \\ &= \boxed{5.08 \text{ lbm water/min}} \end{aligned}$$

(c) From Eq. 31.28, the quantity of heat removed is

$$\begin{aligned} \dot{q} &= \dot{m}_a(h_1 - h_2) \\ &= \left(343.4\ \frac{\text{lbm air}}{\text{min}}\right)\left(50.7\ \frac{\text{Btu}}{\text{lbm air}} - 25.8\ \frac{\text{Btu}}{\text{lbm air}}\right) \\ &= \boxed{8551 \text{ Btu/min}} \end{aligned}$$

(d) Considering an R-12 refrigeration cycle operating at saturated condition at 100°F, the T-s and h-s diagrams are as follows.

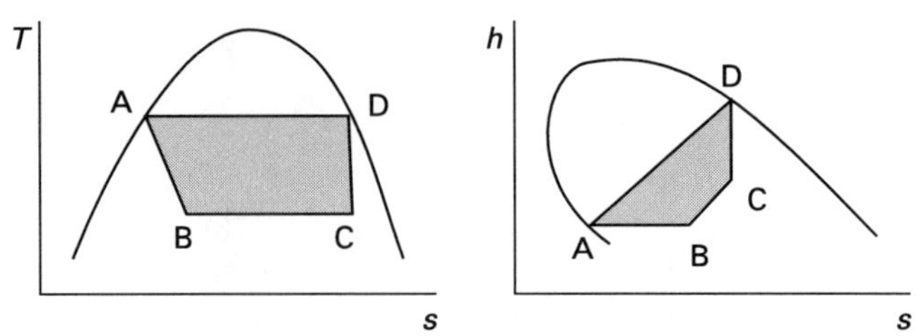

Use App. 29.G for saturated conditions.

(e) At A,

$$T = 100°\text{F} \quad \text{[given]}$$
$$p = p_{\text{sat}} \text{ at } 100°\text{F} = 131.6 \text{ psia}$$
$$h_{\text{A}} = h_f \text{ at } 100°\text{F} = 31.16 \text{ Btu/lbm}$$
$$s_{\text{A}} = s_f \text{ at } 100°\text{F} = 0.06316 \text{ Btu/lbm-°R}$$
$$v_{\text{A}} = v_f \text{ at } 100°\text{F} = 0.0127 \text{ ft}^3\text{/lbm}$$

At point B,

$$T = 50°\text{F} \quad \text{[given]}$$
$$p = p_{\text{sat}} \text{ at } 50°\text{F} = 61.39 \text{ psia}$$
$$h_{\text{B}} = h_{\text{a}} = 31.16 \text{ Btu/lbm}$$
$$h_{f,\text{B}} = 19.27 \text{ Btu/lbm}$$
$$h_{fg,\text{B}} = 64.51 \text{ Btu/lbm}$$
$$x = \frac{h_B - h_{f,\text{B}}}{h_{fg,\text{B}}} = \frac{31.16 \ \frac{\text{Btu}}{\text{lbm}} - 19.27 \ \frac{\text{Btu}}{\text{lbm}}}{64.51 \ \frac{\text{Btu}}{\text{lbm}}} = 0.184$$

$$s_{\text{B}} = s_{f,\text{B}} + x s_{fg,\text{B}} = 0.04126 \ \frac{\text{Btu}}{\text{lbm-°R}} + (0.184)\left(0.12659 \ \frac{\text{Btu}}{\text{lbm-°R}}\right) = 0.06455 \text{ Btu/lbm-°R}$$

$$v_{\text{B}} = v_{f,\text{B}} + x(v_g - v_f) = 0.0118 \ \frac{\text{ft}^3}{\text{lbm}} + (0.184)\left(0.673 \ \frac{\text{ft}^3}{\text{lbm}} - 0.0118 \ \frac{\text{ft}^3}{\text{lbm}}\right) = 0.1335 \text{ ft}^3\text{/lbm}$$

At point D,

$$T = 100°\text{F} \quad \text{[given]}$$
$$p = p_{\text{sat}} \text{ at } 100°\text{F} = 131.6 \text{ psia}$$
$$h_{\text{D}} = h_g \text{ at } 100°\text{F} = 88.62 \text{ Btu/lbm}$$
$$s_{\text{D}} = s_g \text{ at } 100°\text{F} = 0.16584 \text{ Btu/lbm-°R}$$
$$v_{\text{D}} = v_g \text{ at } 100°\text{F} = 0.319 \text{ ft}^3\text{/lbm}$$

At point C,

$$T = 50°\text{F} \quad \text{[given]}$$
$$p = p_{\text{sat}} \text{ at } 50°\text{F} = 61.39 \text{ psia}$$
$$s_{\text{C}} = s_{\text{D}} = 0.16584 \text{ Btu/lbm-°R}$$
$$s_{f,\text{C}} = 0.04126 \text{ Btu/lbm-°R}$$
$$s_{fg,\text{C}} = 0.12659 \text{ Btu/lbm-°R}$$

$$x = \frac{s_{\text{C}} - s_{f,\text{C}}}{s_{fg,\text{C}}} = \frac{0.16584 \ \frac{\text{Btu}}{\text{lbm-°R}} - 0.04126 \ \frac{\text{Btu}}{\text{lbm-°R}}}{0.12659 \ \frac{\text{Btu}}{\text{lbm-°R}}} = 0.984$$

$$h_{\text{C}} = h_{f,\text{C}} + x h_{fg,\text{C}} = 19.27 \ \frac{\text{Btu}}{\text{lbm}} + (0.984)\left(64.51 \ \frac{\text{Btu}}{\text{lbm}}\right) = 82.75 \text{ Btu/lbm}$$

$$v_{\text{C}} = v_{f,\text{C}} + x(v_{g,\text{C}} - v_{f,\text{C}}) = 0.0118 \ \frac{\text{ft}^3}{\text{lbm}} + (0.984)\left(0.673 \ \frac{\text{ft}^3}{\text{lbm}} - 0.0118 \ \frac{\text{ft}^3}{\text{lbm}}\right) = 0.662 \text{ ft}^3\text{/lbm}$$

SI Solution

(a) The psychrometric chart is shown in the customary U.S. solution. At point 1, from the psychrometric chart (App. 31.B) at $T_{\text{db}} = 35°\text{C}$ and $\phi = 70\%$,

$$\boxed{\begin{aligned} h_1 &= 99.9 \text{ kJ/kg air} \\ v_1 &= 0.91 \text{ m}^3\text{/kg air} \\ \omega_1 &= \left(25.3 \ \frac{\text{g moisture}}{\text{kg air}}\right)\left(\frac{1 \text{ kg}}{1000 \text{ g}}\right) \\ &= 0.0253 \text{ kg moisture/kg air} \end{aligned}}$$

At point 2, from the psychrometric chart (App. 31.B) at $T_{\text{db}} = 16°\text{C}$ and $\phi = 95\%$,

$$\boxed{\begin{aligned} h_2 &= 43.4 \text{ kJ/kg air} \\ \omega_2 &= \left(10.8 \ \frac{\text{g moisture}}{\text{kg air}}\right)\left(\frac{1 \text{ kg}}{1000 \text{ g}}\right) \\ &= 0.0108 \text{ kg moisture/kg air} \end{aligned}}$$

(b) The air mass flow rate is

$$\dot{m}_a = \frac{\dot{V}_1}{v_1} = \frac{\dot{V}_1}{v_{\text{out}}} = \frac{2.36 \ \frac{\text{m}^3}{\text{s}}}{0.91 \ \frac{\text{m}^3}{\text{kg air}}} = 2.593 \text{ kg air/s}$$

The water removed from Eq. 31.27 is

$$\dot{m}_w = \dot{m}_a(\omega_1 - \omega_2) = \left(2.593 \ \frac{\text{kg air}}{\text{s}}\right)\left(0.0253 \ \frac{\text{kg moisture}}{\text{kg air}} - 0.0108 \ \frac{\text{kg moisture}}{\text{kg air}}\right) = \boxed{0.0376 \text{ kg moisture/s}}$$

Ventilation

(c) From Eq. 31.28, the quantity of heat removed is

$$\begin{aligned}\dot{q} &= \dot{m}_a(h_1 - h_2)\\ &= \left(2.593\ \frac{\text{kg air}}{\text{s}}\right)\left(99.9\ \frac{\text{kJ}}{\text{kg air}} - 43.4\ \frac{\text{kJ}}{\text{kg air}}\right)\\ &= \boxed{146.5\ \text{kW}}\end{aligned}$$

(d) The T-s and h-s diagrams are shown in the customary U.S. solution.

(e) Use App. 29.T for saturated conditions.

At point A,

$$\begin{aligned}T &= 38^\circ\text{C}\quad\text{[given]}\\ p &= p_{\text{sat}}\ \text{at}\ 38^\circ\text{C} = 0.91324\ \text{MPa}\\ h_\text{A} &= h_f\ \text{at}\ 38^\circ\text{C} = 237.23\ \text{kJ/kg}\\ s_\text{A} &= s_f\ \text{at}\ 38^\circ\text{C} = 1.1259\ \text{kJ/kg}\cdot{}^\circ\text{C}\\ v_\text{A} &= v_f\ \text{at}\ 38^\circ\text{C} = \frac{1}{\rho_f} = \frac{1}{1261.9\ \frac{\text{kg}}{\text{m}^3}}\\ &= 0.0007925\ \text{m}^3/\text{kg}\end{aligned}$$

At point B,

$$\begin{aligned}T &= 10^\circ\text{C}\quad\text{[given]}\\ p &= p_{\text{sat}}\ \text{at}\ 10^\circ\text{C} = 0.42356\ \text{MPa}\\ h_\text{B} &= h_\text{A} = 237.23\ \text{kJ/kg}\\ h_{f,\text{B}} &= 209.48\ \text{kJ/kg}\\ h_{g,\text{B}} &= 356.79\ \text{kJ/kg}\end{aligned}$$

$$\begin{aligned}x &= \frac{h_\text{B} - h_{f,\text{B}}}{h_{g,\text{B}} - h_{f,\text{B}}} = \frac{237.23\ \frac{\text{kJ}}{\text{kg}} - 209.48\ \frac{\text{kJ}}{\text{kg}}}{356.79\ \frac{\text{kJ}}{\text{kg}} - 209.48\ \frac{\text{kJ}}{\text{kg}}}\\ &= 0.188\end{aligned}$$

$$\begin{aligned}s_\text{B} &= s_{f,\text{B}} + x(s_{g,\text{B}} - s_{f,\text{B}})\\ &= 1.0338\ \frac{\text{kJ}}{\text{kg}\cdot\text{K}}\\ &\quad + (0.188)\left(1.5541\ \frac{\text{kJ}}{\text{kg}\cdot\text{K}} - 1.0338\ \frac{\text{kJ}}{\text{kg}\cdot\text{K}}\right)\\ &= 1.1316\ \text{kJ/kg}\cdot\text{K}\end{aligned}$$

$$\begin{aligned}v_\text{B} &= v_{f,\text{B}} + x(v_{g,\text{B}} - v_{f,\text{B}})\\ v_{f,\text{B}} &= \frac{1}{\rho_{f,\text{B}}} = \frac{1}{1363.0\ \frac{\text{kg}}{\text{m}^3}} = 0.0007337\ \text{m}^3/\text{kg}\\ v_{g,\text{B}} &= 0.04119\ \text{m}^3/\text{kg}\\ v_\text{B} &= 0.0007337\ \frac{\text{m}^3}{\text{kg}}\\ &\quad + (0.188)\left(0.04119\ \frac{\text{m}^3}{\text{kg}} - 0.0007337\ \frac{\text{m}^3}{\text{kg}}\right)\\ &= 0.008339\ \text{m}^3/\text{kg}\end{aligned}$$

At point D,

$$\begin{aligned}T &= 38^\circ\text{C}\quad\text{[given]}\\ p &= p_{\text{sat}}\ \text{at}\ 38^\circ\text{C} = 0.42356\ \text{MPa}\\ h_\text{D} &= h_g\ \text{at}\ 38^\circ\text{C} = 367.95\ \text{kJ/kg}\\ s_\text{D} &= s_g\ \text{at}\ 38^\circ\text{C} = 1.5461\ \text{kJ/kg}\cdot\text{K}\\ v_\text{D} &= v_g\ \text{at}\ 38^\circ\text{C} = 0.01931\ \text{m}^3/\text{kg}\end{aligned}$$

At point C,

$$\begin{aligned}T &= 10^\circ\text{C}\\ p &= p_{\text{sat}}\ \text{at}\ 10^\circ\text{C} = 0.42356\ \text{MPa}\\ s_\text{C} &= s_\text{D} = 1.5461\ \text{kJ/kg}\cdot\text{K}\\ s_{f,\text{C}} &= 1.0338\ \text{kJ/kg}\cdot\text{K}\\ s_{g,\text{C}} &= 1.5541\ \text{kJ/kg}\cdot\text{K}\end{aligned}$$

$$\begin{aligned}x &= \frac{s_\text{C} - s_{f,\text{C}}}{s_{g,\text{C}} - s_{f,\text{C}}} = \frac{1.5461\ \frac{\text{kJ}}{\text{kg}\cdot\text{K}} - 1.0338\ \frac{\text{kJ}}{\text{kg}\cdot\text{K}}}{1.5541\ \frac{\text{kJ}}{\text{kg}\cdot\text{K}} - 1.0338\ \frac{\text{kJ}}{\text{kg}\cdot\text{K}}}\\ &= 0.985\end{aligned}$$

$$\begin{aligned}h_\text{C} &= h_{f,\text{C}} + x(h_{g,\text{C}} - h_{f,\text{C}})\\ &= 209.48\ \frac{\text{kJ}}{\text{kg}} + (0.985)\left(356.79\ \frac{\text{kJ}}{\text{kg}} - 209.48\ \frac{\text{kJ}}{\text{kg}}\right)\\ &= 354.58\ \text{kJ/kg}\end{aligned}$$

$$\begin{aligned}v_\text{C} &= v_{f,\text{C}} + x(v_{g,\text{C}} - v_{f,\text{C}})\\ v_{g,\text{C}} &= 0.04119\ \text{m}^3/\text{kg}\\ v_\text{C} &= 0.0007337\ \frac{\text{m}^3}{\text{kg}}\\ &\quad + (0.985)\left(0.04119\ \frac{\text{m}^3}{\text{kg}} - 0.0007337\ \frac{\text{m}^3}{\text{kg}}\right)\\ &= 0.04058\ \text{m}^3/\text{kg}\end{aligned}$$

11. *Customary U.S. Solution*

(a) The saturation pressure at 200°F from App. 29.A is $p_{\text{sat},1} = 11.529$ psia.

Since air is saturated (100% relative humidity), the water vapor pressure is equal to the saturation pressure.

$$p_{w,1} = p_{\text{sat},1} = 11.529\ \text{psia}$$

The partial pressure of the air is

$$\begin{aligned}p_{a,1} &= p_1 - p_{w,1} = 25\ \text{psia} - 11.529\ \text{psia}\\ &= 13.471\ \text{psia}\end{aligned}$$

From Table 29.7, the specific gas constants are

$$\begin{aligned}R_w &= 85.78\ \text{ft-lbf/lbm-}^\circ\text{R}\\ R_\text{air} &= 53.35\ \text{ft-lbf/lbm-}^\circ\text{R}\end{aligned}$$

The mass of water vapor from the ideal gas law is

$$\dot{m}_{w,1} = \frac{p_{w,1}\dot{V}}{R_w T} = \frac{\left(11.529\ \frac{\text{lbf}}{\text{in}^2}\right)\left(144\ \frac{\text{in}^2}{\text{ft}^2}\right)\left(1500\ \frac{\text{ft}^3}{\text{min}}\right)}{\left(85.78\ \frac{\text{ft-lbf}}{\text{lbm-}^\circ\text{R}}\right)(200^\circ\text{F} + 460)} = 43.99\ \text{lbm/min water}$$

The mass of air from the ideal gas law is

$$\dot{m}_{a,1} = \frac{p_{a,1}\dot{V}}{R_{\text{air}} T} = \frac{\left(13.471\ \frac{\text{lbf}}{\text{in}^2}\right)\left(144\ \frac{\text{in}^2}{\text{ft}^2}\right)\left(1500\ \frac{\text{ft}^3}{\text{min}}\right)}{\left(53.35\ \frac{\text{ft-lbf}}{\text{lbm-}^\circ\text{R}}\right)(200^\circ\text{F} + 460)} = 82.64\ \text{lbm/min air}$$

The humidity ratio is

$$\omega_1 = \frac{\dot{m}_{w,1}}{\dot{m}_{a,1}} = \frac{43.99\ \frac{\text{lbm water}}{\text{min}}}{82.64\ \frac{\text{lbm air}}{\text{min}}} = 0.532\ \text{lbm water/lbm air}$$

Since it is a constant pressure, constant moisture drying process, mole fractions and partial pressures do not change.

$$p_{w,2} = p_{w,1} = 11.529\ \text{psia}$$

The saturation pressure at 400°F from App. 29.A is $p_{\text{sat},2} = 247.1$ psia.

The relative humidity at state 2 is

$$\phi_2 = \frac{p_{w,2}}{p_{\text{sat},2}} = \frac{11.529\ \text{psia}}{247.1\ \text{psia}} = \boxed{0.0467\ \ (4.67\%)}$$

(b) The specific humidity remains constant.

$$\omega_2 = \omega_1 = \boxed{0.532\ \text{lbm water/lbm air}}$$

(c) The heat required consists of two parts.

Obtain enthalpy for air from App. 29.F.

The absolute temperatures are

$$T_1 = 200^\circ\text{F} + 460 = 660^\circ\text{R}$$
$$T_2 = 400^\circ\text{F} + 460 = 860^\circ\text{R}$$
$$h_1 = 157.92\ \text{Btu/lbm}$$
$$h_2 = 206.46\ \text{Btu/lbm}$$

The heat absorbed by the air is

$$q_1 = h_2 - h_1 = 206.46\ \frac{\text{Btu}}{\text{lbm}} - 157.92\ \frac{\text{Btu}}{\text{lbm}} = 48.54\ \text{Btu/lbm dry air}$$

(There will be a small error if constant specific heat is used instead.)

For water, use the Mollier diagram. From App. 29.E, h_1 at 200°F and 11.529 psia is 1146 Btu/lbm (almost saturated).

Follow a constant 11.529 psia pressure curve up to 400°F.

$$h_2 = 1240\ \text{Btu/lbm}$$

(There will be a small error if Eq. 31.19(b) is used instead.)

The heat absorbed by the steam is

$$q_2 = \omega(h_2 - h_1) = \left(0.532\ \frac{\text{lbm water}}{\text{lbm air}}\right)\left(1240\ \frac{\text{Btu}}{\text{lbm}} - 1146\ \frac{\text{Btu}}{\text{lbm}}\right) = 50.01\ \text{Btu/lbm air}$$

The total heat absorbed is

$$q_{\text{total}} = q_1 + q_2 = 48.54\ \frac{\text{Btu}}{\text{lbm air}} + 50.01\ \frac{\text{Btu}}{\text{lbm air}} = \boxed{98.55\ \text{Btu/lbm air}}$$

(d) The dew point is the temperature at which water starts to condense out in a constant pressure process. Following the constant 11.529 psia pressure line back to the saturation line, $\boxed{T_{\text{dp}} = 200^\circ\text{F}.}$

SI Solution

(a) From App. 29.N, the saturation pressure at 93°C is

$$p_{\text{sat},1} = (0.7879\ \text{bar})\left(100\ \frac{\text{kPa}}{\text{bar}}\right) = 78.79\ \text{kPa}$$

Since air is saturated (100% relative humidity), water vapor pressure is equal to saturation pressure.

$$p_{w,1} = p_{\text{sat},1} = 78.79\ \text{kPa}$$

The partial pressure of air is

$$p_{a,1} = p_1 - p_{w,1} = 170\ \text{kPa} - 78.79\ \text{kPa} = 91.21\ \text{kPa}$$

From Table 29.7, the specific gas constants are

$$R_w = 461.50\ \text{J/kg·K}$$
$$R_{\text{air}} = 287.03\ \text{J/kg·K}$$

Ventilation

The mass of water vapor from the ideal gas law is

$$\dot{m}_{w,1} = \frac{p_{w,1}\dot{V}}{R_w T} = \frac{(78.79 \text{ kPa})\left(1000 \frac{\text{Pa}}{\text{kPa}}\right)\left(0.71 \frac{\text{m}^3}{\text{s}}\right)}{\left(461.50 \frac{\text{J}}{\text{kg·K}}\right)(93°\text{C} + 273)} = 0.3312 \text{ kg/s water}$$

The mass of air from the ideal gas law is

$$\dot{m}_{a,1} = \frac{p_{a,1}\dot{V}}{R_{\text{air}} T} = \frac{(91.21 \text{ kPa})\left(1000 \frac{\text{Pa}}{\text{kPa}}\right)\left(0.71 \frac{\text{m}^3}{\text{s}}\right)}{\left(287.03 \frac{\text{J}}{\text{kg·K}}\right)(93°\text{C} + 273)} = 0.6164 \text{ kg/s air}$$

The humidity ratio is

$$\omega_1 = \frac{\dot{m}_{w,1}}{\dot{m}_{a,1}} = \frac{0.3312 \frac{\text{kg water}}{\text{s}}}{0.6164 \frac{\text{kg air}}{\text{s}}} = 0.537 \text{ kg water/kg air}$$

Since it is a constant pressure, constant moisture drying process, mole fractions and partial pressure do not change.

$$p_{w,2} = p_{w,1} = 78.79 \text{ kPa}$$

From App. 29.N, the saturation pressure at 204°C is

$$p_{\text{sat},2} = (16.95 \text{ bar})\left(100 \frac{\text{kPa}}{\text{bar}}\right) = 1695 \text{ kPa}$$

The relative humidity at state 2 is

$$\phi_2 = \frac{p_{w,2}}{p_{\text{sat},2}} = \frac{78.79 \text{ kPa}}{1695 \text{ kPa}} = \boxed{0.0465 \ (4.65\%)}$$

(b) The specific humidity remains constant.

$$\omega_2 = \omega_1 = \boxed{0.537 \text{ kg water/kg air}}$$

(c) The heat required consists of two parts.

Obtain enthalpy for air from App. 29.S.

The absolute temperatures are

$$T_1 = 93°\text{C} + 273 = 366\text{K}$$
$$T_2 = 204°\text{C} + 273 = 477\text{K}$$
$$h_1 = 366.63 \text{ kJ/kg}$$
$$h_2 = 479.42 \text{ kJ/kg}$$

The heat absorbed by air is

$$q_1 = h_2 - h_1 = 479.42 \frac{\text{kJ}}{\text{kg}} - 366.63 \frac{\text{kJ}}{\text{kg}} = 112.79 \text{ kJ/kg air}$$

(There will be a small error if constant specific heat is used instead.)

For water use the Mollier diagram. From App. 29.R, h_1 at 93°C and 78.79 kPa is 2670 kJ/kg (almost saturated).

Follow a constant 78.79 kPa pressure curve up to 204°C.

$$h_2 = 2890 \text{ kJ/kg}$$

(There will be a small error if Eq. 31.19(a) is used instead.)

The heat absorbed by steam is

$$q_2 = \omega(h_2 - h_1) = \left(0.537 \frac{\text{kg water}}{\text{kg air}}\right)\left(2890 \frac{\text{kJ}}{\text{kg}} - 2670 \frac{\text{kJ}}{\text{kg}}\right) = 118.14 \text{ kJ/kg air}$$

The total heat absorbed is

$$q_{\text{total}} = q_1 + q_2 = 112.79 \frac{\text{kJ}}{\text{kg air}} + 118.14 \frac{\text{kJ}}{\text{kg air}} = \boxed{230.93 \text{ kJ/kg air}}$$

(d) The dew point is the temperature at which water starts to condense out in a constant pressure process. Following the constant 78.79 kPa pressure line back to the saturation line, $\boxed{T_{\text{dp}} \approx 93°\text{C}.}$

12. *Customary U.S. Solution*

(a) The absolute air temperatures are

$$T_1 = 800°\text{F} + 460 = 1260°\text{R}$$
$$T_2 = 350°\text{F} + 460 = 810°\text{R}$$

At low pressures, use air tables. From App. 29.F,

$$h_1 = 306.65 \text{ Btu/lbm}$$
$$h_2 = 194.25 \text{ Btu/lbm}$$

From App. 29.A, the enthalpy of water at 80°F is $h_{w,1} = 48.09$ Btu/lbm.

From Eq. 31.19(b), the enthalpy of steam at 350°F is

$$h_{w,2} \approx \left(0.444 \frac{\text{Btu}}{\text{lbm-°F}}\right)(350°\text{F}) + 1061 \frac{\text{Btu}}{\text{lbm}} = 1216.4 \text{ Btu/lbm}$$

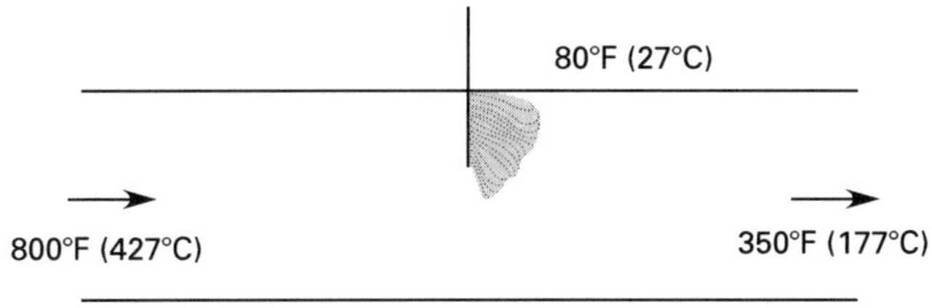

Air temperature is reduced from 800 to 350°F, and this energy is used to change water at 80°F to steam at 350°F. From the energy balance equation,

$$\dot{m}_w(h_{w,2} - h_{w,1}) = \dot{m}_a(h_1 - h_2)$$

$$\dot{m}_w = \frac{\dot{m}_a(h_1 - h_2)}{h_{w,2} - h_{w,1}}$$

$$= \frac{\left(410\ \frac{\text{lbm}}{\text{hr}}\right)\left(306.65\ \frac{\text{Btu}}{\text{lbm}} - 194.25\ \frac{\text{Btu}}{\text{lbm}}\right)}{1216.4\ \frac{\text{Btu}}{\text{lbm}} - 48.09\ \frac{\text{Btu}}{\text{lbm}}}$$

$$= \boxed{39.4\ \text{lbm/hr water}}$$

(b) The number of moles of water evaporated (in the air mixture) is

$$\dot{n}_w = \frac{39.4\ \frac{\text{lbm}}{\text{hr}}}{18.016\ \frac{\text{lbm}}{\text{lbmol}}} = 2.19\ \text{lbmol/hr}$$

The number of moles of air at the exit in the mixture is

$$\dot{n}_a = \frac{410\ \frac{\text{lbm}}{\text{hr}}}{28.967\ \frac{\text{lbm}}{\text{lbmol}}} = 14.15\ \text{lbmol/hr}$$

The mole fraction of water in the mixture is

$$x_w = \frac{\dot{n}_w}{\dot{n}_a + \dot{n}_w}$$

$$= \frac{2.19\ \frac{\text{lbmol}}{\text{hr}}}{14.15\ \frac{\text{lbmol}}{\text{hr}} + 2.19\ \frac{\text{lbmol}}{\text{hr}}} = 0.134$$

Partial pressure of water vapor is

$$p_w = x(p_{\text{chamber}}) = (0.134)(20\ \text{psia}) = 2.68\ \text{psia}$$

From App. 29.A, the saturation pressure at 350°F is $p_{\text{sat}} = 134.53$ psia.

The relative humidity is

$$\phi = \frac{p_w}{p_{\text{sat}}} = \frac{2.68\ \text{psia}}{134.53\ \text{psia}} = \boxed{0.020\ \ (2.0\%)}$$

SI Solution

(a) The absolute temperatures are

$$T_1 = 427°\text{C} + 273 = 700\text{K}$$
$$T_2 = 177°\text{C} + 273 = 450\text{K}$$

Air tables can be used at low pressures. From App. 29.S,

$$h_1 = 713.27\ \text{kJ/kg}$$
$$h_2 = 451.80\ \text{kJ/kg}$$

From App. 29.N, the enthalpy of water at 27°C is $h_{w,1} = 113.25$ kJ/kg.

From Eq. 31.19(a), the enthalpy of steam at 177°C is

$$h_{w,2} = \left(1.805\ \frac{\text{kJ}}{\text{kg}\cdot°\text{C}}\right)(177°\text{C}) + 2501\ \frac{\text{kJ}}{\text{kg}}$$
$$= 2820.5\ \text{kJ/kg}$$

Air temperature is reduced from 427 to 177°C, and this energy is used to change water at 27°C to steam at 177°C. From the energy balance equation,

$$\dot{m}_w = \frac{\dot{m}_a(h_1 - h_2)}{h_{w,2} - h_{w,1}}$$

$$= \frac{\left(0.052\ \frac{\text{kg}}{\text{s}}\right)\left(713.27\ \frac{\text{kJ}}{\text{kg}} - 451.80\ \frac{\text{kJ}}{\text{kg}}\right)}{2820.5\ \frac{\text{kJ}}{\text{kg}} - 113.25\ \frac{\text{kJ}}{\text{kg}}}$$

$$= \boxed{0.00502\ \text{kg/s}}$$

(b) The number of moles of water evaporated (in the air mixture) is

$$\dot{n}_w = \frac{0.00502\ \frac{\text{kg}}{\text{s}}}{18.016\ \frac{\text{kg}}{\text{kmol}}} = 2.79 \times 10^{-4}\ \text{kmol/s}$$

The number of moles of air in the mixture at the exit is

$$\dot{n}_a = \frac{0.052\ \frac{\text{kg}}{\text{s}}}{28.967\ \frac{\text{kg}}{\text{kmol}}} = 1.80 \times 10^{-3}\ \text{kmol/s}$$

The mole fraction of water in the mixture is

$$x_w = \frac{\dot{n}_w}{\dot{n}_a + \dot{n}_w}$$

$$= \frac{2.79 \times 10^{-4}\ \frac{\text{kmol}}{\text{s}}}{1.80 \times 10^{-3}\ \frac{\text{kmol}}{\text{s}} + 2.79 \times 10^{-4}\ \frac{\text{kmol}}{\text{s}}}$$

$$= 0.134$$

The partial pressure of water vapor is

$$p_w = xp_{\text{chamber}} = (0.134)(140 \text{ kPa}) = 18.76 \text{ kPa}$$

From App. 29.N, the saturation pressure at 177°C is

$$p_{\text{sat}} = (9.389 \text{ bar})\left(100 \ \frac{\text{kPa}}{\text{bar}}\right) = 938.9 \text{ kPa}$$

The relative humidity is

$$\phi = \frac{p_w}{p_{\text{sat}}} = \frac{18.76 \text{ kPa}}{938.9 \text{ kPa}} = \boxed{0.020 \ (2.0\%)}$$

13.

Customary U.S. Solution

(a) The cooled water flow rate is given by

$$Q = \dot{m}_w c_p \Delta T$$

$$1 \times 10^6 \text{ Btu} = \dot{m}_w \left(1.0 \ \frac{\text{Btu}}{\text{lbm-°F}}\right)(120°\text{F} - 110°\text{F})$$

$$\dot{m}_w = 1 \times 10^5 \text{ lbm/hr}$$

From the psychrometric chart (App. 31.A), for air in at $T_{\text{db}} = 91°\text{F}$ and $\phi = 60\%$,

$$h_{\text{in}} \approx 42.7 \text{ Btu/lbm}$$

$$\omega_{\text{in}} = 0.0190 \text{ lbm moisture/lbm air}$$

For air out, the normal psychrometric chart (offscale) cannot be used, so use Eq. 31.11 to find the humidity ratio and Eqs. 31.17, 31.18(b), and 31.19(b) to calculate the enthalpy of air. (App. 31.D could also be used as a simpler solution.)

From App. 29.A, the saturated steam pressure at 100°F is 0.9503 psia.

$$p_w = \phi p_{\text{sat}} = (0.82)(0.9503 \text{ psia}) = 0.7792 \text{ psia}$$

From Eq. 31.1,

$$p_a = p - p_w = 14.696 \text{ psia} - 0.7792 \text{ psia} = 13.9168 \text{ psia}$$

From Eq. 31.11, the humidity ratio for air out is

$$\phi = 1.608\omega_{\text{out}}\left(\frac{p_a}{p_{\text{sat}}}\right)$$

$$0.82 = 1.608\omega_{\text{out}}\left(\frac{13.9168 \text{ psia}}{0.9503 \text{ psia}}\right)$$

$$\omega_{\text{out}} = 0.0348 \text{ lbm moisture/lbm air}$$

From Eqs. 31.17, 31.18(b), and 31.19(b), the enthalpy of air out is

$$\begin{aligned} h_2 &= h_a + \omega_2 h_w \\ &= \left(0.240 \ \frac{\text{Btu}}{\text{lbm-°F}}\right) T_{2,°\text{F}} + \omega_2 \\ &\quad \times \left(\left(0.444 \ \frac{\text{Btu}}{\text{lbm-°F}}\right) T_{2,°\text{F}} + 1061 \ \frac{\text{Btu}}{\text{lbm}}\right) \\ &= \left(0.240 \ \frac{\text{Btu}}{\text{lbm-°F}}\right)(100°\text{F}) \\ &\quad + \left(0.0348 \ \frac{\text{lbm moisture}}{\text{lbm air}}\right)\left(\left(0.444 \ \frac{\text{Btu}}{\text{lbm-°F}}\right) \right. \\ &\quad \left. \times (100°\text{F}) + 1061 \ \frac{\text{Btu}}{\text{lbm}}\right) \\ &= 62.47 \text{ Btu/lbm air} \end{aligned}$$

The mass flow rate of air can be determined from Eq. 31.26.

$$q = \dot{m}_a(h_2 - h_1)$$

$$\dot{m}_{\text{air}} = \frac{q}{h_2 - h_1} = \frac{1 \times 10^6 \ \frac{\text{Btu}}{\text{hr}}}{62.47 \ \frac{\text{Btu}}{\text{lbm air}} - 42.7 \ \frac{\text{Btu}}{\text{lbm air}}} = \boxed{5.058 \times 10^4 \text{ lbm air/hr}}$$

(b) From conservation of water vapor,

$$\omega_1 \dot{m}_{\text{air}} + \dot{m}_{\text{make-up}} = \omega_2 \dot{m}_{\text{air}}$$

$$\dot{m}_{\text{make-up}} = \dot{m}_{\text{air}}(\omega_{\text{out}} - \omega_{\text{in}})$$

$$= \left(5.058 \times 10^4 \ \frac{\text{lbm air}}{\text{hr}}\right) \times \left(0.0348 \ \frac{\text{lbm moisture}}{\text{lbm air}} - 0.0191 \ \frac{\text{lbm moisture}}{\text{lbm air}}\right) = \boxed{794 \text{ lbm water/hr}}$$

SI Solution

(a) The cooled water flow rate is given by

$$Q = \dot{m}_w c_p \Delta T$$

$$290 \text{ kW} = \dot{m}_w \left(4.187 \ \frac{\text{kJ}}{\text{kg·°C}}\right)(49°\text{C} - 43°\text{C})$$

$$\dot{m}_w = 11.54 \text{ kg/s}$$

From the psychrometric chart (App. 31.B), for air in at $T_{db} = 33°C$ and $\phi = 60\%$,

$$h_{in} = 82.3 \text{ kJ/kg air}$$

$$\omega_{in} = \left(19.2 \,\frac{\text{g moisture}}{\text{kg air}}\right)\left(\frac{1 \text{ kg}}{1000 \text{ g}}\right) = 0.0192 \text{ kg moisture/kg air}$$

For air out, the psychrometric chart (off scale) cannot be used, so use Eq. 31.11 to find the humidity ratio and Eqs. 31.17, 31.18(a), and 31.19(a) to calculate enthalpy of air.

From App. 29.N, the saturated steam pressure at 38°C is 0.06632 bars.

$$\begin{aligned} p_w &= \phi p_{sat} \\ &= (0.82)(0.06632 \text{ bar}) \\ &= 0.05438 \text{ bar} \end{aligned}$$

From Eq. 31.1,

$$\begin{aligned} p_a &= p - p_w \\ &= 1 \text{ bar} - 0.05438 \text{ bar} \\ &= 0.94562 \text{ bar} \end{aligned}$$

From Eq. 31.11, the humidity ratio for air out is

$$\begin{aligned} \phi &= 1.608\omega_{out}\left(\frac{p_a}{p_{sat}}\right) \\ 0.82 &= 1.608\omega_{out}\left(\frac{0.94562 \text{ bar}}{0.06632 \text{ bar}}\right) \\ \omega_{out} &= 0.0358 \text{ kg moisture/kg air} \end{aligned}$$

From Eqs. 31.17, 31.18(a), and 31.19(a), the enthalpy of air out is

$$\begin{aligned} h_2 &= h_a + \omega_2 h_w \\ &= \left(1.005 \,\frac{\text{kJ}}{\text{kg}\cdot°\text{C}}\right) T_{°C} + \omega_{out}\left(\left(1.805 \,\frac{\text{kJ}}{\text{kg}\cdot°\text{C}}\right) T_{°C} + 2501 \,\frac{\text{kJ}}{\text{kg}}\right) \\ &= \left(1.005 \,\frac{\text{kJ}}{\text{kg}\cdot°\text{C}}\right)(38°\text{C}) + \left(0.0358 \,\frac{\text{kg moisture}}{\text{kg air}}\right) \\ &\quad \times \left(\left(1.805 \,\frac{\text{kJ}}{\text{kg}\cdot°\text{C}}\right)(38°\text{C}) + 2501 \,\frac{\text{kJ}}{\text{kg}}\right) \\ &= 130.2 \text{ kJ/kg air} \end{aligned}$$

The mass flow rate of air can be determined from Eq. 31.26.

$$q = \dot{m}_a(h_2 - h_1)$$

$$\dot{m}_{air} = \frac{q}{h_2 - h_1} = \frac{290 \text{ kW}}{130.2 \,\frac{\text{kJ}}{\text{kg air}} - 82.3 \,\frac{\text{kJ}}{\text{kg air}}} = \boxed{6.054 \text{ kg air/s}}$$

(b) From conservation of water vapor,

$$\omega_1 \dot{m}_{air} + \dot{m}_{make\text{-}up} = \omega_2 \dot{m}_{air}$$

$$\dot{m}_{make\text{-}up} = \dot{m}_{air}(\omega_{out} - \omega_{in})$$

$$\begin{aligned} &= \left(6.054 \,\frac{\text{kg air}}{\text{s}}\right) \\ &\quad \times \left(0.0358 \,\frac{\text{kg moisture}}{\text{kg air}} - 0.0192 \,\frac{\text{kg moisture}}{\text{kg air}}\right) \\ &= \boxed{0.100 \text{ kg water/s}} \end{aligned}$$

Ventilation

32 HVAC: Ventilation and Humidification Requirements

PRACTICE PROBLEMS

1. An office room has floor dimensions of 60 ft by 95 ft (18 m by 29 m) and a ceiling height of 10 ft (3 m). 45 people occupy the office, and half of them smoke. (a) Calculate the ventilation rate based on six air changes per hour. (b) State your assumptions and determine the ventilation rate based on occupancy.

2. 150 ppm of methanol (TLV = 200 ppm; MW = 32.04; SG = 0.792) and 285 ppm of methylene chloride (TLV = 500 ppm; MW = 84.94; SG = 1.336) are found in the air in a plating booth. Two pints of each are evaporated per hour. Use a mixing safety factor (i.e., a K value) of 6. What ventilation rate is required?

3. An auditorium is designed to seat 4500 people. The ventilation rate is 60 ft^3/min (1.68 m^3/min) per person of outside air. The outside temperature is 0°F (−18°C) dry-bulb, and the outside pressure is 14.6 psia (100.6 kPa). Air leaves the auditorium at 70°F (21°C) dry-bulb. There is no recirculation. The furnace has been sized assuming an internal heat gain of 1,250,000 Btu/hr (370 kW) and the outside conditions described. (a) At what temperature should the air enter the auditorium? (b) How much heat should be supplied to the ventilation air? (c) Has the furnace been sized properly?

4. A room is maintained at design conditions of 75°F (23.9°C) dry-bulb and 50% relative humidity. The air outside is at 95°F (35°C) dry-bulb and 75°F (23.9°C) wet-bulb. Conditioned air enters the room and increases 20°F (−11.1°C) in temperature before being removed from the room. The sensible and latent loads are 200,000 Btu/hr (60 kW) and 50,000 Btu/hr (15 kW), respectively. 2000 ft^3/min (57 m^3/min) of outside air are used. What are the (a) apparatus dew point and (b) quantity of air flowing through the air conditioner?

SOLUTIONS

1. *Customary U.S. Solution*

(a) The office volume is

$$V = (60 \text{ ft})(95 \text{ ft})(10 \text{ ft}) = 57{,}000 \text{ ft}^3$$

Based on six air changes per hour, the flow rate is

$$\dot{V} = \left(57{,}000 \ \frac{\text{ft}^3}{\text{air change}}\right)\left(6 \ \frac{\text{air changes}}{\text{hr}}\right)\left(\frac{1 \text{ hr}}{60 \text{ min}}\right) = \boxed{5700 \text{ ft}^3/\text{min}}$$

(b) For preferred ventilation based on Sec. 32-2, "Ventilation Standards," assume the following.

- The ventilation rate for a nonsmoking area is 20 ft^3/min.
- The ventilation rate for a smoking area ranges from 30 to 60 ft^3/min, with an average value of 45 ft^3/min.

The preferred ventilation rate is

$$\dot{V} = \left(\tfrac{1}{2}\right)(45 \text{ persons})\left(\frac{20 \ \frac{\text{ft}^3}{\text{min}}}{1 \text{ person}}\right) + \left(\tfrac{1}{2}\right)(45 \text{ persons})\left(\frac{45 \ \frac{\text{ft}^3}{\text{min}}}{1 \text{ person}}\right) = \boxed{1463 \text{ ft}^3/\text{min}}$$

SI Solution

(a) The office volume is

$$(18 \text{ m})(29 \text{ m})(3 \text{ m}) = 1566 \text{ m}^3$$

Based on six air changes per hour, the flow rate is

$$\dot{V} = \left(1566 \ \frac{\text{m}^3}{\text{air change}}\right)\left(6 \ \frac{\text{air changes}}{\text{h}}\right)\left(\frac{1 \text{ h}}{60 \text{ min}}\right) = \boxed{156.6 \text{ m}^3/\text{min}}$$

(b) For preferred ventilation based on Sec. 32-2, "Ventilation Standards," assume the following.

- The ventilation rate for a nonsmoking area is 0.57 m^3/min.
- The ventilation rate for a smoking area ranges from 0.84 to 1.68 m^3/min, with an average value of 1.26 m^3/min.

The preferred ventilation rate is

$$\dot{V} = \left(\tfrac{1}{2}\right)(45\text{ persons})\left(\frac{0.57\ \frac{\text{m}^3}{\text{min}}}{1\text{ person}}\right) + \left(\tfrac{1}{2}\right)(45\text{ persons})\left(\frac{1.26\ \frac{\text{m}^3}{\text{min}}}{1\text{ person}}\right) = \boxed{41.2\ \text{m}^3/\text{min}}$$

2. From Eq. 32.12(b), the ventilation rate required is

$$\dot{V}_{\text{cfm}} = \frac{(4.03\times 10^8)K(\text{SG})R_{\text{pints/min}}}{(\text{MW})\text{TLV}_{\text{ppm}}}$$

For the methanol,

$$\dot{V}_{\text{cfm}} = \frac{(4.03\times 10^8)(6)(0.792)\left(2\ \frac{\text{pints}}{\text{h}}\right)\left(\frac{1\text{ h}}{60\text{ min}}\right)}{(32.04)(200\text{ ppm})} = 9962\ \text{ft}^3/\text{min}$$

For the methylene chloride,

$$\dot{V}_{\text{cfm}} = \frac{(4.03\times 10^8)(6)(1.336)\left(2\ \frac{\text{pints}}{\text{hr}}\right)\left(\frac{1\text{ hr}}{60\text{ min}}\right)}{(84.94)(500\text{ ppm})} = 2535\ \text{ft}^3/\text{min}$$

The total ventilation rate is

$$9962\ \frac{\text{ft}^3}{\text{min}} + 2535\ \frac{\text{ft}^3}{\text{min}} = \boxed{12{,}497\ \text{ft}^3/\text{min}}$$

3. *Customary U.S. Solution*

(a) The total ventilation rate is

$$\dot{V} = \left(\frac{60\ \frac{\text{ft}^3}{\text{min}}}{1\text{ person}}\right)(4500\text{ persons})\left(60\ \frac{\text{min}}{\text{hr}}\right) = 1.62\times 10^7\ \text{ft}^3/\text{hr}$$

The absolute temperature of outside air is

$$T = 0°\text{F} + 460 = 460°\text{R}$$

From the ideal gas law, the density of outside air is

$$\rho = \frac{p}{RT} = \frac{\left(14.6\ \frac{\text{lbf}}{\text{in}^2}\right)\left(144\ \frac{\text{in}^2}{\text{ft}^2}\right)}{\left(53.35\ \frac{\text{ft-lbf}}{\text{lbm-°R}}\right)(460°\text{R})} = 0.08567\ \text{lbm/ft}^3$$

The mass flow rate is

$$\dot{m} = \dot{V}\rho = \left(1.62\times 10^7\ \frac{\text{ft}^3}{\text{hr}}\right)\left(0.08567\ \frac{\text{lbm}}{\text{ft}^3}\right) = 1.388\times 10^6\ \text{lbm/hr}$$

Assume no latent heat (no moisture) at 0°F.

From Table 32.1, the sensible heat generated by each person seated in the theater is 225 Btu/hr. Therefore,

$$Q_{\text{in from people}} = \left(\frac{225\ \frac{\text{Btu}}{\text{hr}}}{\text{person}}\right)(4500\text{ persons}) = 1.01\times 10^6\ \text{Btu/hr}$$

The air leaves the auditorium at 70°F. The specific heat is 0.240 Btu/lbm-°F (remains fairly constant). Since Q is known, the air temperature entering the auditorium can be calculated by

$$Q = \dot{m}c_p(T_{\text{out,air}} - T_{\text{in,air}})$$

$$T_{\text{in,air}} = T_{\text{out,air}} - \frac{Q}{\dot{m}c_p} = 70°\text{F} - \frac{1.01\times 10^6\ \frac{\text{Btu}}{\text{hr}}}{\left(1.388\times 10^6\ \frac{\text{lbm}}{\text{hr}}\right)\left(0.240\ \frac{\text{Btu}}{\text{lbm-°F}}\right)} = \boxed{67.0°\text{F}}$$

(b) The heat needed to heat dry ventilation air from 0°F to 67°F is

$$Q = \dot{m}c_p\Delta T = \left(1.388\times 10^6\ \frac{\text{lbm}}{\text{hr}}\right)\left(0.240\ \frac{\text{Btu}}{\text{lbm-°F}}\right)(67°\text{F} - 0°\text{F}) = \boxed{2.23\times 10^7\ \text{Btu/hr}}$$

(c) Since 1.01×10^6 Btu/hr $< 1.25\times 10^6$ Btu/hr, the furnace is too small.

SI Solution

(a) The total ventilation rate is

$$\dot{V} = \left(\frac{1.68\ \frac{\text{m}^3}{\text{min}}}{\text{person}}\right)(4500\ \text{persons})\left(60\ \frac{\text{min}}{\text{h}}\right)$$
$$= 4.54 \times 10^5\ \text{m}^3/\text{h}$$

The absolute temperature of outside air is

$$T = -18°\text{C} + 273 = 255\text{K}$$

From the ideal gas law, the density of outside air is

$$\rho = \frac{p}{RT} = \frac{(100.6\ \text{kPa})\left(1000\ \frac{\text{Pa}}{\text{kPa}}\right)}{\left(287.03\ \frac{\text{J}}{\text{kg·K}}\right)(255\text{K})}$$
$$= 1.374\ \text{kg/m}^3$$

The mass flow rate is

$$\dot{m} = \rho\dot{V} = \left(1.374\ \frac{\text{kg}}{\text{m}^3}\right)\left(4.54 \times 10^5\ \frac{\text{m}^3}{\text{h}}\right)$$
$$= 6.238 \times 10^5\ \text{kg/h}$$

Assume no latent heat (no moisture) at −18°C.

From Table 32.1 and the table footnote, the sensible heat generated by people seated in the theater is

$$\left(225\ \frac{\text{Btu}}{\text{h}}\right)\left(0.293\ \frac{\text{W}}{\frac{\text{Btu}}{\text{h}}}\right) = 65.93\ \text{W}$$

Therefore,

$$Q_{\text{in from people}} = \left(65.93\ \frac{\text{W}}{\text{persons}}\right)(4500\ \text{persons}) \times \left(\frac{1\ \text{kW}}{1000\ \text{W}}\right)$$
$$= 296.7\ \text{kW}$$

The air leaves the auditorium at 21°C. The specific heat ≈ 1.0048 kJ/kg·K (remains fairly constant). The air temperature entering the auditorium can be found from known Q as

$$Q = \dot{m}c_p(T_{\text{out,air}} - T_{\text{in,air}})$$
$$T_{\text{in,air}} = T_{\text{out,air}} - \frac{Q}{\dot{m}c_p}$$
$$= 21°\text{C} - \frac{296.7\ \text{kW}}{\left(6.238 \times 10^5\ \frac{\text{kg}}{\text{h}}\right) \times \left(\frac{1\ \text{h}}{3600\ \text{s}}\right)\left(1.0048\ \frac{\text{kJ}}{\text{kg·K}}\right)}$$
$$= \boxed{19.3°\text{C}}$$

(b) The heat needed to heat dry ventilation air from −18°C to 19.3°C is

$$Q = \dot{m}c_p\Delta T$$
$$= \left(6.238 \times 10^5\ \frac{\text{kg}}{\text{h}}\right)\left(\frac{1\ \text{h}}{3600\ \text{s}}\right)\left(1.0048\ \frac{\text{kJ}}{\text{kg·K}}\right) \times (19.3°\text{C} - (-18°\text{C}))$$
$$= \boxed{6494\ \text{kW}}$$

(c) Since 296.7 kW < 370 kW, the furnace is too small.

4. *Customary U.S. Solution*

(a) From Eq. 38.22, the room sensible heat ratio is

$$\text{RSHR} = \frac{q_s}{q_t} = \frac{q_s}{q_s + q_l}$$
$$= \frac{200{,}000\ \frac{\text{Btu}}{\text{hr}}}{200{,}000\ \frac{\text{Btu}}{\text{hr}} + 50{,}000\ \frac{\text{Btu}}{\text{hr}}}$$
$$= 0.8$$

Locate the point i for design conditions of 75°F dry bulb and 50% relative humidity. Draw a condition line with a slope of 0.8 through point i.

$$\text{ADP} = \boxed{50.8°\text{F}} \quad \left[\begin{array}{c}\text{intersection of room condi-}\\ \text{tion and saturation line}\end{array}\right]$$
$$T_w = \text{ADP}$$
$$T_{\text{db,in}} = 75°\text{F} - 20°\text{F} = 55°\text{F}$$

The answer is (B).

(b) From Eq. 32.6(b), the volumetric flow rate of air entering the room is

$$\dot{V}_{\text{in,cfm}} = \frac{\dot{q}_{\text{s,Btu/hr}}}{\left(1.08\ \frac{\text{Btu-min}}{\text{ft}^3\text{-hr-°F}}\right)(T_{\text{i,°F}} - T_{\text{in,°F}})}$$
$$= \frac{200{,}000\ \frac{\text{Btu}}{\text{hr}}}{\left(1.08\ \frac{\text{Btu-min}}{\text{ft}^3\text{-hr-°F}}\right)(75°\text{F} - 55°\text{F})}$$
$$= 9259\ \text{ft}^3/\text{min}$$

This is a mixing problem.

Using the lever rule, and since the temperature scales are all linear, the fraction of air passing through the conditioner is

$$\frac{T_i - T_{\text{in}}}{T_i - T_{\text{co}}} = \frac{75°\text{F} - 55°\text{F}}{75°\text{F} - 50.8°\text{F}} = 0.826$$
$$\dot{V}_1 = (0.826)\left(9259\ \frac{\text{ft}^3}{\text{min}}\right) = \boxed{7648\ \text{ft}^3/\text{min}}$$

The answer is (C).

Ventilation

SI Solution

(a) From Eq. 38.22, the room sensible heat ratio is

$$\text{RSHR} = \frac{q_s}{q_s + q_l} = \frac{60\ \text{kW}}{60\ \text{kW} + 15\ \text{kW}} = 0.8$$

Locate the point i for design conditions of 23.9°C dry bulb and 50% relative humidity. Draw a condition line with a slope of 0.8 through point i.

$$\text{ADP} = \boxed{10^\circ\text{C}} \quad \left[\begin{array}{c}\text{intersection of room condi-}\\ \text{tion and saturation line}\end{array}\right]$$

$$T_w = \text{ADP}$$

$$T_{\text{db,in}} = 23.9^\circ\text{C} - 11.1^\circ\text{C} = 12.8^\circ\text{C}$$

The answer is (B).

(b) From Eq. 32.6(a),

$$\dot{V}_{\text{in,m}^3/\text{min}} = \frac{\dot{q}_{\text{s,kW}}}{\left(0.02\ \frac{\text{kJ·min}}{\text{m}^3\text{·s·°C}}\right)(T_{\text{i,°C}} - T_{\text{in,°C}})}$$

$$= \frac{60\ \text{kW}}{\left(0.02\ \frac{\text{kJ-min}}{\text{m}^3\text{·s·°C}}\right)(23.9^\circ\text{C} - 12.8^\circ\text{C})}$$

$$= 270.3\ \text{m}^3/\text{min}$$

This is a mixing problem.

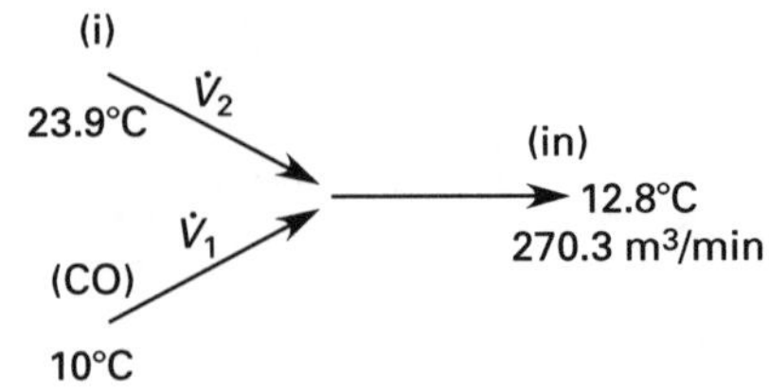

Using the lever rule, and since the temperature scales are all linear, the fraction of air passing through the conditioner is

$$\frac{T_i - T_{\text{in}}}{T_i - T_{\text{co}}} = \frac{23.9^\circ\text{C} - 12.8^\circ\text{C}}{23.9^\circ\text{C} - 10^\circ\text{C}}$$

$$= 0.799$$

$$\dot{V}_1 = (0.799)\left(270.3\ \frac{\text{m}^3}{\text{min}}\right)$$

$$= \boxed{216.0\ \text{m}^3/\text{min}}$$

The answer is (C).

33 HVAC: Heating Load

PRACTICE PROBLEMS

1. A conditioned room contains 12,000 W of fluorescent lights (with older coil ballasts) and twelve 90% efficient, 10 hp motors operating at 80% of their rated capacities. The lights are pendant-mounted on chains from the ceiling. The motors drive various pieces of machinery located in the conditioned space. What is the internal heat gain?

2. A building is located in the city of New York. There are 4772 (2651 in SI) degree days during the October 15 to May 15 period for this area. The building is heated by fuel oil whose heating value is 153,600 Btu/gal (42 800 MJ/m^3) and which costs \$0.15/gal (\$0.034/L). The calculated design heat loss is 3.5×10^6 Btu/hr (1 MW) based on 70°F (21.1°C) indoor and 0°F (−17.8°C) outdoor design temperatures. The furnace has an efficiency of 70%. What is the approximate cost for heating this building during the winter (October 15 through May 15)?

3. A flat roof consists of 1½ in (38 mm) of insulation installed over 3 in (76 mm) of soft pine sheathing, and a ¾ in (19 mm) acoustical ceiling suspended 4 in (100 mm) below the pine. The outdoor wind velocity is 15 mi/hr (24 km/h). The indoor design temperature is 80°F (26.7°C). The outdoor design temperature is 95°F (35°C). What is the overall coefficient of heat transfer?

4. A 12 ft by 12 ft (3.6 m by 3.6 m) floor is constructed as a concrete slab with two exposed edges. The slab edge heat loss coefficient for these two edges is 0.55 Btu/ft-hr-°F (0.95 W/m·°C). The other two edges form part of a heated basement wall. The indoor design temperature is 70°F (21.1°C); the outdoor design temperature is −10°F (−23.3°C). Use the slab edge method to determine the heat loss from the slab.

5. An office in a remodeled historic building has floor dimensions of 100 ft by 40 ft (30 m by 12 m) and a ceiling height of 10 ft (3 m). One of the 40 ft (12 m) walls is shared with an adjacent heated space. The three remaining walls have two 4 ft (1.2 m) wide by 6 ft (1.8 m) high, double glass with ¼ in (6.4 mm) air space, weatherstripped, double-hung windows per 20 ft (6 m). The crack coefficient for the windows is 32 ft^3/hr-ft (3.0 m^3/h·m). The basement and second floor are heated to 70°F (21.1°C). The wall coefficient is 0.2 Btu/ft^2-hr-°F (1.1 W/m^2·°C), and the outdoor wind velocity is 15 mi/hr (24 km/h). The indoor design temperature is 70°F (21.1°C); the outdoor design temperature is −10°F (−23.3°C). Including infiltration but disregarding ventilation air, what is the heating load?

6. A building is located in Newark, New Jersey. The heating season lasts for 245 days; the degree days are 5252 (2918 in SI); the outdoor design temperature is 0°F (−17.8°C). The temperature in the building is maintained at 70°F (21.1°C) between the hours of 8:30 A.M. and 5:30 P.M. During the rest of the day and the night, the temperature is allowed to drop to 50°F (10°C). The building is heated with coal that has a heating value of 13,000 Btu/lbm (30.2 MJ/kg). A heat loss of 650,000 Btu/hr (0.19 MW) has been calculated based on 70°F (21.1°C) indoor and 0°F (−17.8°C) outdoor temperatures. The furnace has an efficiency of 70%. What mass of coal is required each year?

7. The design heat loss of a building is 200,000 Btu/hr (60 kW) originally based on 70°F (21.1°C) indoor and 0°F(−17.8°C) outdoor design temperatures. At that location, there are 4200 (2333 in SI) degree days over the 210-day heating season. The building is occupied 24 hr/day. What is the percentage reduction in heating fuel if the thermostat is lowered from 70 to 68°F (21.1 to 20°C)?

8. (*Time limit: one hour*) A building is located where the annual heating season lasts 21 weeks. The indoor design temperature is 70°F (21.1°C). The gas furnace has an efficiency of 75%. Fuel costs \$0.25 per therm. The building is occupied only from 8 A.M. until 6 P.M., Monday through Friday. In the past, the building was maintained at 70°F (21.1°C) at all times and ventilated with one air change per hour. The thermostat is now being set back 12°F (6.7°C) and the ventilation reduced 50% during the unoccupied times. Infiltration through cracks and humidity changes are negligible. What are the annual savings?

internal volume:	801,000 ft^3 (22 700 m^3)
wall area:	11,040 ft^2 (993 m^2)

wall overall heat transfer coefficient:	0.15 Btu/ft^2-hr-°F (0.85 W/m^2·°C)
window area:	2760 ft^2 (260 m^2)
window overall heat transfer coefficient:	1.13 Btu/ft^2-hr-°F (6.42 W/m^2·°C)
roof area:	26,700 ft^2 (2480 m^2)
roof overall heat transfer coefficient:	0.05 Btu/ft^2-hr-°F (0.3 W/m^2·°C)
slab on grade:	690 lineal ft (210 m) of exposed slab edge
slab edge coefficient:	1.5 Btu/ft-hr-°F (2.6 W/m·°C)

SOLUTIONS

1. *Customary U.S. Solution*

Based on Sec. 33-10, for fluorescent lights, the rated wattage should be increased by 20–25% to account for ballast heating. Since the lights are pendant-mounted on chain from the ceiling, most of this heat enters the conditioned space. Assume rated wattage increases by 20%. From Eq. 33.7(b), the internal heat gain due to lights (SF = 1.2) is

$$\dot{q}_{\text{lights}} = \left(3413\ \frac{\text{Btu}}{\text{kW-hr}}\right)(\text{SF})\left(\frac{P}{\eta}\right)$$

$$\eta = 1 \quad [\text{given}]$$

$$\begin{aligned}\dot{q}_{\text{lights}} &= \left(3413\ \frac{\text{Btu}}{\text{kW-hr}}\right)(1.2) \\ &\quad\times\left(\frac{(12{,}000\ \text{W})\left(\frac{1\ \text{kW}}{1000\ \text{W}}\right)}{1}\right) \\ &= 4.9147\times10^4\ \text{Btu/hr}\end{aligned}$$

From Eq. 33.7(b), the internal heat gain due to motors (SF = 0.8) is

$$\dot{q}_{\text{motors}} = \left(2545\ \frac{\text{Btu}}{\text{hr-hp}}\right)(\text{SF})\left(\frac{P}{\eta}\right)$$

$$\eta = 0.90 \quad [\text{given}]$$

$$\begin{aligned}\dot{q}_{\text{motors}} &= (12\ \text{motors})\left(2545\ \frac{\text{Btu}}{\text{hr-hp}}\right)(0.8)\left(\frac{10\ \text{hp}}{0.90}\right) \\ &= 2.7147\times10^5\ \text{Btu/hr} \\ \dot{q}_{\text{total}} &= \dot{q}_{\text{lights}} + \dot{q}_{\text{motors}} \\ &= 4.9147\times10^4\ \frac{\text{Btu}}{\text{hr}} + 2.7147\times10^5\ \frac{\text{Btu}}{\text{hr}} \\ &= \boxed{3.206\times10^5\ \text{Btu/hr}}\end{aligned}$$

SI Solution

Based on the customary U.S. solution, SF = 1.2 for lights. From Eq. 33.7(a), the internal heat gain due to lights is

$$\eta = 1 \quad [\text{given}]$$

$$\begin{aligned}\dot{q}_{\text{lights}} &= (\text{SF})\left(\frac{P}{\eta}\right) \\ &= (1.2)\left(\frac{(12\,000\ \text{W})\left(\frac{1\ \text{kW}}{1000\ \text{W}}\right)}{1}\right) \\ &= 14.4\ \text{kW}\end{aligned}$$

From Eq. 33.7(a), the internal heat gain due to motors (12 motors) is

$$\dot{q}_{\text{motors}} = \left(0.7457\ \frac{\text{kW}}{\text{hp}}\right)(\text{SF})\left(\frac{P}{\eta}\right)$$

$$\text{SF} = 0.8 \quad \text{[given]}$$

$$\eta = 0.90 \quad \text{[given]}$$

$$\begin{aligned}\dot{q}_{\text{motors}} &= (12\ \text{motors})\left(0.7457\ \frac{\text{kW}}{\text{hp}}\right)(0.8)\left(\frac{10\ \text{hp}}{0.90}\right)\\ &= 79.5\ \text{kW}\end{aligned}$$

$$\begin{aligned}\dot{q}_{\text{total}} &= \dot{q}_{\text{lights}} + \dot{q}_{\text{motors}}\\ &= 14.4\ \text{kW} + 79.5\ \text{kW}\\ &= \boxed{93.9\ \text{kW}}\end{aligned}$$

2. *Customary U.S. Solution*

From Eq. 33.10(b), the fuel consumption (in gal/heating season) is

$$\frac{\left(24\ \frac{\text{hr}}{\text{day}}\right)\dot{q}_{\text{Btu/hr}}(\text{DD})}{(T_i - T_o)(\text{HV}_{\text{Btu/gal}})\eta_{\text{furnace}}}$$

$$\begin{aligned}&= \frac{\left(24\ \frac{\text{hr}}{\text{day}}\right)\left(3.5\times 10^6\ \frac{\text{Btu}}{\text{hr}}\right)(4772^\circ\text{F-days})}{(70^\circ\text{F} - 0^\circ\text{F})\left(153{,}600\ \frac{\text{Btu}}{\text{gal}}\right)(0.70)}\\ &= 53{,}260\ \text{gal}\end{aligned}$$

The total cost of 53,260 gal fuel at \$0.15/gal is

$$(53{,}260\ \text{gal})\left(\frac{\$0.15}{\text{gal}}\right) = \boxed{\$7989}$$

SI Solution

$$\begin{aligned}\text{HV}_{\text{kJ/L}} &= \left(42\,800\ \frac{\text{MJ}}{\text{m}^3}\right)\left(\frac{1000\ \text{kJ}}{1\ \text{MJ}}\right)\left(\frac{1\ \text{m}^3}{1000\ \text{L}}\right)\\ &= 42\,800\ \text{kJ/L}\end{aligned}$$

From Eq. 33.10(a), fuel consumption (in L/heating season) is

$$\frac{\left(86\,400\ \frac{\text{s}}{\text{day}}\right)\dot{q}(\text{DD})}{(T_c - T_o)(\text{HV})\eta_{\text{furnace}}}$$

$$\begin{aligned}&= \frac{\left(86\,400\ \frac{\text{s}}{\text{day}}\right)(1\ \text{MW})\left(\frac{1000\ \text{kW}}{1\ \text{MW}}\right) \times (2651^\circ\text{C-days})}{(21.1^\circ\text{C} - (-17.8^\circ\text{C}))\left(42\,800\ \frac{\text{kJ}}{\text{L}}\right)(0.70)}\\ &= 196\,531\ \text{L}\end{aligned}$$

The total cost at \$0.034/L is

$$(196\,531\ \text{L})\left(\frac{\$0.034}{\text{L}}\right) = \boxed{\$6682}$$

3. *Customary U.S. Solution*

From Table 33.3, the surface film coefficient for outdoor air (horizontal, heat flow down, 15 mph) is $h_o = 6\ \text{Btu/ft}^2\text{-hr-}^\circ\text{F}$. The film thermal resistance is

$$R_1 = \frac{1}{h_o} = \frac{1}{6\ \frac{\text{Btu}}{\text{ft}^2\text{-hr-}^\circ\text{F}}} = 0.167\ \text{ft}^2\text{-hr-}^\circ\text{F/Btu}$$

From App. 33.A, for roof insulation, the thermal resistance per inch is $2.78\ \text{ft}^2\text{-}^\circ\text{F-hr/Btu-in}$.

$$R_2 = \left(2.78\ \frac{\text{ft}^2\text{-}^\circ\text{F-hr}}{\text{Btu-in}}\right)(1.5\ \text{in}) = 4.17\ \text{ft}^2\text{-hr-}^\circ\text{F/Btu}$$

From App. 33.A, the thermal resistance per inch for soft wood is $1.25\ \text{ft}^2\text{-}^\circ\text{F-hr/Btu-in}$.

$$R_3 = \left(1.25\ \frac{\text{ft}^2\text{-}^\circ\text{F-hr}}{\text{Btu-in}}\right)(3\ \text{in}) = 3.75\ \text{ft}^2\text{-}^\circ\text{F-hr/Btu}$$

For the space between soft wood and an acoustic ceiling, the effective surface emissivity from Eq. 33.4 is

$$E = \frac{1}{\frac{1}{\epsilon_1} + \frac{1}{\epsilon_2} - 1}$$

$$\epsilon_1 = \epsilon_{\text{wood}} = 0.90 \quad \text{[Sec. 6]}$$

$$\epsilon_2 = \epsilon_{\text{paper}} = 0.95 \quad \text{[used for acoustic tile]}$$

$$E = \frac{1}{\frac{1}{0.90} + \frac{1}{0.95} - 1} = 0.86$$

From Table 33.2, the thermal conductance of 4 in horizontal planar air space at $E = 0.86$ with heat flow down is approximately $0.81\ \text{Btu/hr-ft}^2\text{-}^\circ\text{F}$. The thermal resistance is

$$R_4 = \frac{1}{0.81\ \frac{\text{Btu}}{\text{ft}^2\text{-hr-}^\circ\text{F}}} = 1.235\ \text{ft}^2\text{-hr-}^\circ\text{F/Btu}$$

From App. 33.A, the thermal resistance for 3/4 in acoustical tile is $R_5 = 1.78\ \text{ft}^2\text{-}^\circ\text{F-hr/Btu}$.

From Table 33.3, the surface film coefficient for indoor air (v = 0) (horizontal, heat flow down) is $h_i = 1.08\ \text{Btu/ft}^2\text{-hr-}^\circ\text{F}$. The film thermal resistance is

$$R_6 = \frac{1}{h_i} = \frac{1}{1.08\ \frac{\text{Btu}}{\text{ft}^2\text{-hr-}^\circ\text{F}}} = 0.926\ \text{ft}^2\text{-hr-}^\circ\text{F/Btu}$$

The total resistance is

$$\begin{aligned} R_{\text{total}} &= R_1 + R_2 + R_3 + R_4 + R_5 + R_6 \\ &= 0.167\ \frac{\text{ft}^2\text{-hr-}^\circ\text{F}}{\text{Btu}} + 4.17\ \frac{\text{ft}^2\text{-hr-}^\circ\text{F}}{\text{Btu}} \\ &\quad + 3.75\ \frac{\text{ft}^2\text{-hr-}^\circ\text{F}}{\text{Btu}} + 1.235\ \frac{\text{ft}^2\text{-hr-}^\circ\text{F}}{\text{Btu}} \\ &\quad + 1.78\ \frac{\text{ft}^2\text{-hr-}^\circ\text{F}}{\text{Btu}} + 0.926\ \frac{\text{ft}^2\text{-hr-}^\circ\text{F}}{\text{Btu}} \\ &= 12.03\ \text{ft}^2\text{-hr-}^\circ\text{F/Btu} \end{aligned}$$

The overall coefficient of heat transfer is

$$\begin{aligned} U &= \frac{1}{R_{\text{total}}} = \frac{1}{12.03\ \frac{\text{ft}^2\text{-hr-}^\circ\text{F}}{\text{Btu}}} \\ &= \boxed{0.0831\ \text{Btu/ft}^2\text{-hr-}^\circ\text{F}} \end{aligned}$$

SI Solution

From Table 33.3, the surface film coefficient for outdoor air (horizontal, heat flow down) at 15 mph is $h_o = 34.1\ \text{W/m}^2\cdot^\circ\text{C}$. The film thermal resistance is

$$R_1 = \frac{1}{h_o} = \frac{1}{34.1\ \frac{\text{W}}{\text{m}^2\cdot^\circ\text{C}}} = 0.0293\ \text{m}^2\cdot^\circ\text{C/W}$$

From App. 33.A for roof insulation, the thermal resistance per inch is 2.78 ft^2-°F-hr/Btu-in.

$$\begin{aligned} R_2 &= \left(2.78\ \frac{\text{ft}^2\text{-}^\circ\text{F-hr}}{\text{Btu-in}}\right)\left(6.93\ \frac{\frac{\text{m}\cdot^\circ\text{C}}{\text{W}}}{\frac{\text{ft}^2\text{-}^\circ\text{F-hr}}{\text{Btu-in}}}\right) \\ &\quad \times (38\ \text{mm})\left(\frac{1\ \text{m}}{1000\ \text{mm}}\right) \\ &= 0.732\ \text{m}^2\cdot^\circ\text{C/W} \end{aligned}$$

From App. 33.A, the thermal resistance per inch for soft wood is 1.25 ft^2-°F-hr/Btu-in.

$$\begin{aligned} R_3 &= \left(1.25\ \frac{\text{ft}^2\text{-}^\circ\text{F-hr}}{\text{Btu-in}}\right)\left(6.93\ \frac{\frac{\text{m}\cdot^\circ\text{C}}{\text{W}}}{\frac{\text{ft}^2\text{-}^\circ\text{F-hr}}{\text{Btu-in}}}\right) \\ &\quad \times (76\ \text{mm})\left(\frac{1\ \text{m}}{1000\ \text{mm}}\right) \\ &= 0.658\ \text{m}^2\cdot^\circ\text{C/W} \end{aligned}$$

From the customary U.S. solution, the effective surface emissivity between soft wood and an acoustic ceiling is $E = 0.86$.

From Table 33.2, the thermal conductance of 100 mm horizontal planar air space at $E = 0.86$ with heat flow down is approximately 4.6 W/m^2·°C. The thermal resistance is

$$R_4 = \frac{1}{4.6\ \frac{\text{W}}{\text{m}^2\cdot^\circ\text{C}}} = 0.217\ \text{m}^2\cdot^\circ\text{C/W}$$

From App. 33.A, the thermal resistance for 19 mm (3/4 in) acoustical tile is

$$\begin{aligned} R_5 &= \left(1.78\ \frac{\text{ft}^2\text{-}^\circ\text{F-hr}}{\text{Btu-in}}\right)\left(0.176\ \frac{\frac{\text{m}^2\cdot^\circ\text{C}}{\text{W}}}{\frac{\text{ft}^2\text{-}^\circ\text{F-hr}}{\text{Btu}}}\right) \\ &= 0.313\ \text{m}^2\cdot^\circ\text{C/W} \end{aligned}$$

From Table 33.3, the surface film coefficient for indoor air (v = 0) (horizontal, heat flow down) is $h_i = 6.13\ \text{W/m}^2\cdot^\circ\text{C}$. The film thermal resistance is

$$R_6 = \frac{1}{h_i} = \frac{1}{6.13\ \frac{\text{W}}{\text{m}^2\cdot^\circ\text{C}}} = 0.163\ \text{m}^2\cdot^\circ\text{C/W}$$

The total resistance is

$$\begin{aligned} R_{\text{total}} &= R_1 + R_2 + R_3 + R_4 + R_5 + R_6 \\ &= 0.0293\ \frac{\text{m}^2\cdot^\circ\text{C}}{\text{W}} + 0.732\ \frac{\text{m}^2\cdot^\circ\text{C}}{\text{W}} \\ &\quad + 0.658\ \frac{\text{m}^2\cdot^\circ\text{C}}{\text{W}} + 0.217\ \frac{\text{m}^2\cdot^\circ\text{C}}{\text{W}} \\ &\quad + 0.313\ \frac{\text{m}^2\cdot^\circ\text{C}}{\text{W}} + 0.163\ \frac{\text{m}^2\cdot^\circ\text{C}}{\text{W}} \\ &= 2.112\ \text{m}^2\cdot^\circ\text{C/W} \end{aligned}$$

The overall coefficient of heat transfer is

$$U = \frac{1}{R_{\text{total}}} = \frac{1}{2.112\ \frac{\text{m}^2\cdot^\circ\text{C}}{\text{W}}} = \boxed{0.473\ \text{W/m}^2\cdot^\circ\text{C}}$$

4. *Customary U.S. Solution*

From Eq. 33.5, the heat loss from the slab is

$$\dot{q} = pF(T_i - T_o)$$

The slab edge coefficient is given as $F = 0.55$ Btu/ft-hr-°F.

The perimeter is

$$\begin{aligned} p &= 12\ \text{ft} + 12\ \text{ft} \quad [\text{2 edges only}] \\ &= 24\ \text{ft} \end{aligned}$$

$$\begin{aligned} \dot{q} &= (24\ \text{ft})\left(0.55\ \frac{\text{Btu}}{\text{ft-hr-}^\circ\text{F}}\right)(70^\circ\text{F} - (-10^\circ\text{F})) \\ &= \boxed{1056\ \text{Btu/hr}} \end{aligned}$$

SI Solution

From Eq. 33.5, the heat loss from the slab is

$$\dot{q} = pF(T_i - T_o)$$

The slab edge coefficient is given as $F = 0.95$ W/m·°C.

The perimeter is

$$\begin{aligned} p &= 3.6 \text{ m} + 3.6 \text{ m} \quad [\text{2 edges only}] \\ &= 7.2 \text{ m} \\ \dot{q} &= (7.2 \text{ m})\left(0.95 \ \frac{\text{W}}{\text{m}\cdot{}^\circ\text{C}}\right)\left(21.1^\circ\text{C} - (-23.3^\circ\text{C})\right) \\ &= \boxed{303.7 \text{ W}} \end{aligned}$$

5.

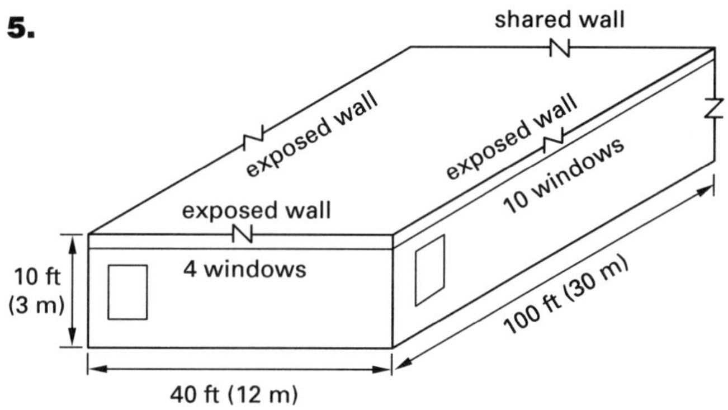

Customary U.S. Solution

For three unshared walls, the total exposed (walls + windows) area is

$$(10 \text{ ft})(40 \text{ ft} + 100 \text{ ft} + 100 \text{ ft}) = 2400 \text{ ft}^2$$

- There are two windows per 20 ft.
- The number of windows for a 40 ft wall is 4.
- The number of windows for each 100 ft wall is 10.
- The total number of windows is

$$4 + 10 + 10 = 24$$

- The total window area is

$$(24)(4 \text{ ft})(6 \text{ ft}) = 576 \text{ ft}^2$$

- The total wall area is

$$2400 \text{ ft}^2 - 576 \text{ ft}^2 = 1824 \text{ ft}^2$$

From Table 33.3, the outdoor film coefficient for a vertical wall in a 15 mph wind is $h_o = 6.00$ Btu/hr-ft^2-°F.

From Table 33.3, the indoor film coefficient for still air (vertical, heat flow horizontal) is $h_i = 1.46$ Btu/ft^2-hr-°F.

Notice that L/k was given in the problem statement, not k. The overall coefficient of heat transfer for the wall is

$$\begin{aligned} U_{\text{wall}} &= \frac{1}{\dfrac{1}{h_o} + \dfrac{\text{L}_{\text{wall}}}{k_{\text{wall}}} + \dfrac{1}{h_i}} \\ &= \frac{1}{\left(\begin{array}{c} \dfrac{1}{6.00 \ \dfrac{\text{Btu}}{\text{hr-ft}^2\text{-}^\circ\text{F}}} + \dfrac{1}{0.2 \ \dfrac{\text{Btu}}{\text{ft}^2\text{-hr-}^\circ\text{F}}} \\ + \dfrac{1}{1.46 \ \dfrac{\text{Btu}}{\text{ft}^2\text{-hr-}^\circ\text{F}}} \end{array}\right)} \\ &= 0.1709 \text{ Btu/hr-ft}^2\text{-}^\circ\text{F} \end{aligned}$$

The heat loss from the wall is

$$\begin{aligned} \dot{q}_{\text{wall}} &= U_{\text{wall}} A_{\text{wall}} \Delta T \\ &= \left(0.1709 \ \frac{\text{Btu}}{\text{hr-ft}^2\text{-}^\circ\text{F}}\right)(1824 \text{ ft}^2)\left(70^\circ\text{F} - (-10^\circ\text{F})\right) \\ &= 24{,}938 \text{ Btu/hr} \end{aligned}$$

From App. 33.A, for double, vertical (1/4 in air space) glass windows, the thermal resistance is 1.63 ft^2-°F-hr/Btu.

$$U_{\text{windows}} = \frac{1}{1.63 \ \dfrac{\text{ft}^2\text{-}^\circ\text{F-hr}}{\text{Btu}}} = 0.613 \text{ Btu/ft}^2\text{-}^\circ\text{F-hr}$$

The heat loss from the windows is

$$\begin{aligned} \dot{q}_{\text{windows}} &= U_{\text{windows}} A_{\text{windows}} \Delta T \\ &= \left(0.613 \ \frac{\text{Btu}}{\text{ft}^2\text{-}^\circ\text{F-hr}}\right)(576 \text{ ft}^2) \\ &\quad \times \left(70^\circ\text{F} - (-10^\circ\text{F})\right) \\ &= 28{,}247 \text{ Btu/hr} \end{aligned}$$

From Eq. 32.1, the infiltration flow rate is

$$\dot{V} = BL$$

The crack coefficient is given as $B = 32$ ft^3/hr-ft.

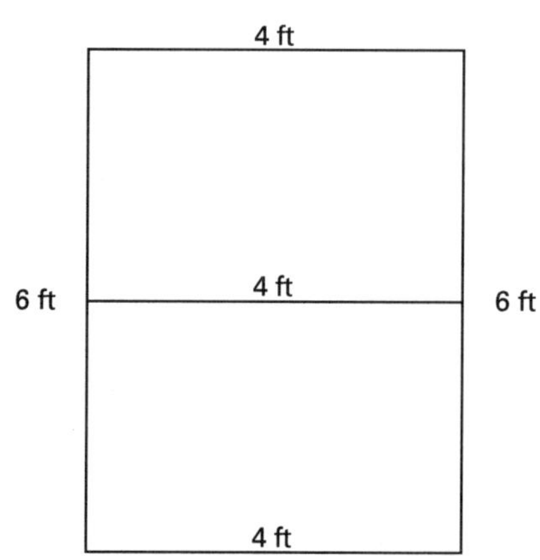

For double-hung windows, the crack length is

$$\begin{aligned} L &= 4 \text{ ft} + 4 \text{ ft} + 4 \text{ ft} + 6 \text{ ft} + 6 \text{ ft} \\ &= 24 \text{ ft} \\ \dot{V} &= \left(32 \ \frac{\text{ft}^3}{\text{hr-ft}}\right)(24 \text{ ft}) \\ &= 768 \text{ ft}^3/\text{hr per window} \end{aligned}$$

Although there are 24 windows, one wall of 10 is exfiltrating.

$$\dot{V} = (24 - 10)\left(768 \ \frac{\text{ft}^3}{\text{hr}}\right) = 10{,}752 \text{ ft}^3/\text{hr}$$

For atmospheric air, assume $\rho = 0.075$ lbm/ft^3 and $c_p = 0.24$ Btu/lbm-°F.

The mass flow rate of infiltrated air is

$$\begin{aligned} \dot{m} = \rho\dot{V} &= \left(0.075 \ \frac{\text{lbm}}{\text{ft}^3}\right)\left(10{,}752 \ \frac{\text{ft}^3}{\text{hr}}\right) \\ &= 806.4 \text{ lbm/hr} \end{aligned}$$

The heat loss due to infiltration is

$$\begin{aligned} \dot{q}_{\text{infiltration}} &= \dot{m}c_p\Delta T \\ &= \left(806.4 \ \frac{\text{lbm}}{\text{hr}}\right)\left(0.24 \ \frac{\text{Btu}}{\text{lbm-°F}}\right) \\ &\quad \times (70\text{°F} - (-10\text{°F})) \\ &= 15{,}483 \text{ Btu/hr} \end{aligned}$$

The total heat loss is

$$\begin{aligned} \dot{q}_{\text{total}} &= \dot{q}_{\text{wall}} + \dot{q}_{\text{windows}} + \dot{q}_{\text{infiltration}} \\ &= 24{,}938 \ \frac{\text{Btu}}{\text{hr}} + 28{,}247 \ \frac{\text{Btu}}{\text{hr}} + 15{,}483 \ \frac{\text{Btu}}{\text{hr}} \\ &= \boxed{68{,}668 \text{ Btu/hr}} \end{aligned}$$

SI Solution

For three unshared walls, the total exposed (walls + windows) area is

$$(3 \text{ m})(12 \text{ m} + 30 \text{ m} + 30 \text{ m}) = 216 \text{ m}^2$$

For 24 windows (from the customary U.S. solution), the total window area is

$$(24)(1.2 \text{ m})(1.8 \text{ m}) = 51.8 \text{ m}^2$$

The total wall area is

$$216 \text{ m}^2 - 51.8 \text{ m}^2 = 164.2 \text{ m}^2$$

From Table 33.3, the outdoor film coefficient for a vertical wall in a 15 mph wind is $h_o = 34.1$ W/m^2·°C.

From Table 33.3, the indoor film coefficient for still air (vertical, heat flow horizontal) is $h_i = 8.29$ W/m^2·°C.

Notice that L/k was given in the problem statement, not k. The overall coefficient of heat transfer for the wall is

$$\begin{aligned} U_{\text{wall}} &= \frac{1}{\frac{1}{h_o} + \frac{L_{\text{wall}}}{k_{\text{wall}}} + \frac{1}{h_i}} \\ &= \frac{1}{\frac{1}{34.1 \ \frac{\text{W}}{\text{m}^2\cdot\text{°C}}} + \frac{1}{1.1 \ \frac{\text{W}}{\text{m}^2\cdot\text{°C}}} + \frac{1}{8.29 \ \frac{\text{W}}{\text{m}^2\cdot\text{°C}}}} \\ &= 0.944 \text{ W/m}^2\cdot\text{°C} \end{aligned}$$

The heat loss from the wall is

$$\begin{aligned} \dot{q}_{\text{wall}} &= U_{\text{wall}}A_{\text{wall}}\Delta T \\ &= \left(0.944 \ \frac{\text{W}}{\text{m}^2\cdot\text{°C}}\right)(164.2 \text{ m}^2) \\ &\quad \times (21.1\text{°C} - (-23.3\text{°C})) \\ &= 6882 \text{ W} \end{aligned}$$

From the customary U.S. solution,

$$\begin{aligned} U_{\text{windows}} &= \left(0.613 \ \frac{\text{Btu}}{\text{ft}^2\text{-°F-hr}}\right)\left(5.68 \ \frac{\frac{\text{W}}{\text{m}^2\cdot\text{°C}}}{\frac{\text{Btu}}{\text{ft}^2\text{-°F-hr}}}\right) \\ &= 3.48 \text{ W/m}^2\cdot\text{°C} \end{aligned}$$

The heat loss from the windows is

$$\begin{aligned} \dot{q}_{\text{windows}} &= U_{\text{windows}}A_{\text{windows}}\Delta T \\ &= \left(3.48 \ \frac{\text{W}}{\text{m}^2\cdot\text{°C}}\right)(51.8 \text{ m}^2) \\ &\quad \times (21.1\text{°C} - (-23.3\text{°C})) \\ &= 8004 \text{ W} \end{aligned}$$

From Eq. 32.1, the infiltration flow rate is

$$\dot{V} = BL$$

The crack coefficient is given as $B = 3.0\ \text{m}^3/\text{h·m}$.

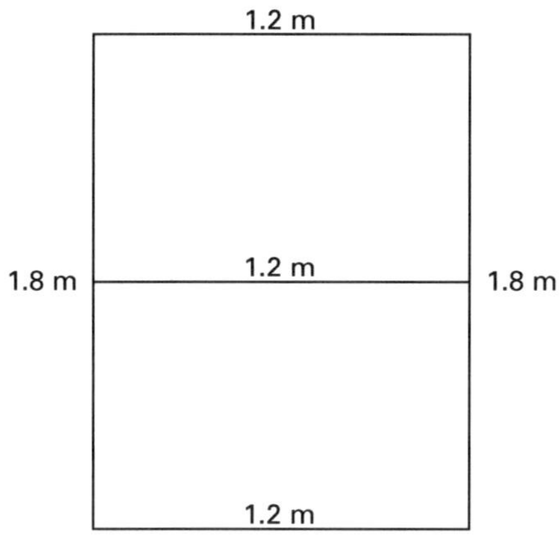

For double-hung windows, the crack length is

$$\begin{aligned} L &= 1.2\ \text{m} + 1.2\ \text{m} + 1.2\ \text{m} + 1.8\ \text{m} + 1.8\ \text{m} \\ &= 7.2\ \text{m} \end{aligned}$$

$$\dot{V} = \left(3.0\ \frac{\text{m}^3}{\text{h·m}}\right)(7.2\ \text{m}) = 21.6\ \text{m}^2/\text{h per window}$$

Although there are 24 windows, one wall of 10 is exfiltrating.

$$\dot{V} = (24 - 10)\left(21.6\ \frac{\text{m}^3}{\text{h}}\right) = 302.4\ \text{m}^3/\text{h}$$

For atmospheric air, assume $\rho = 1.20\ \text{kg/m}^3$ and $c_p = 1.005\ \text{kJ/kg·°C}$.

The mass flow rate of infiltrated air is

$$\begin{aligned} \dot{m} = \rho\dot{V} &= \left(1.2\ \frac{\text{kg}}{\text{m}^3}\right)\left(302.4\ \frac{\text{m}^3}{\text{h}}\right) \\ &= 362.9\ \text{kg/h} \end{aligned}$$

The heat loss due to infiltration is

$$\begin{aligned} \dot{q}_{\text{infiltration}} &= \dot{m}c_p\Delta T \\ &= \left(362.9\ \frac{\text{kg}}{\text{h}}\right)\left(1.005\ \frac{\text{kJ}}{\text{kg·°C}}\right)\left(1000\ \frac{\text{J}}{\text{kg}}\right) \\ &\quad \times \left(\frac{1\ \text{h}}{3600\ \text{s}}\right)(21.1°\text{C} - (-23.3°\text{C})) \\ &= 4498\ \text{W} \end{aligned}$$

The total heat loss is

$$\begin{aligned} \dot{q}_{\text{total}} &= \dot{q}_{\text{wall}} + \dot{q}_{\text{windows}} + \dot{q}_{\text{infiltration}} \\ &= 6882\ \text{W} + 8004\ \text{W} + 4498\ \text{W} \\ &= \boxed{19\,384\ \text{W}\ \ (19.4\ \text{kW})} \end{aligned}$$

6. *Customary U.S. Solution*

The heat loss per degree of temperature difference is

$$\begin{aligned} \frac{\dot{q}}{\Delta T} &= \frac{650{,}000\ \frac{\text{Btu}}{\text{hr}}}{70°\text{F} - 0°\text{F}} \\ &= 9286\ \text{Btu/hr-°F} \end{aligned}$$

From Eq. 33.9(b), the average temperature, $\overline{\overline{T}}$, over the entire heating season can be calculated.

$$\text{DD} = N(65°\text{F} - \overline{\overline{T}})$$

The number of heating degree days, DD, is 5252.

The number of days in the heating season, N, is 245.

$$\begin{aligned} 5252\ \text{days} &= (245\ \text{days})(65°\text{F} - \overline{\overline{T}}) \\ \overline{\overline{T}} &= 43.56°\text{F} \end{aligned}$$

The period between 8:30 A.M. and 5:30 P.M. is 9 hr. For this period, the temperature in the building is 70°F. For the remaining 15 hr, the temperature in the building is 50°F. The total winter heat loss is

$$\begin{aligned} &(245\ \text{days})\left(9286\ \frac{\text{Btu}}{\text{hr-°F}}\right) \\ &\quad \times \left[\left(9\ \frac{\text{hr}}{\text{day}}\right)(70°\text{F} - 43.56°\text{F}) + \left(15\ \frac{\text{hr}}{\text{day}}\right)(50°\text{F} - 43.56°\text{F})\right] \\ &\quad = 7.61 \times 10^8\ \text{Btu} \end{aligned}$$

The fuel consumption is

$$\frac{7.61 \times 10^8\ \text{Btu}}{\left(13{,}000\ \frac{\text{Btu}}{\text{lbm}}\right)(0.70)} = \boxed{83{,}626\ \text{lbm/yr}}$$

SI Solution

The heat loss per degree of temperature difference is

$$\begin{aligned} \frac{\dot{q}}{\Delta T} &= \frac{(0.19\ \text{MW})\left(10^6\ \frac{\text{W}}{\text{MW}}\right)}{21.1°\text{C} - (-17.8°\text{C})} \\ &= 4884\ \text{W/°C} \end{aligned}$$

From Eq. 33.9(a), the average temperature, $\overline{\overline{T}}$, over the entire heating season can be calculated.

$$\text{DD} = N(18°\text{C} - \overline{\overline{T}})$$

The number of heating Kelvin degree days, DD, is 2918.

Ventilation

The number of days in the entire heating season, N, is 245.

$$2918 \text{ days} = (245 \text{ days})(18^\circ\text{C} - \overline{\overline{T}})$$
$$\overline{\overline{T}} = 6.09^\circ\text{C}$$

The period between 8:30 A.M. and 5:30 P.M. is 9 h. For this period, the temperature in the building is 21.1°C. For the remaining 15 h, the temperature in the building is 10°C. The total winter heat loss is

$$\begin{aligned}&(245 \text{ days})\left(4884 \ \frac{\text{W}}{^\circ\text{C}}\right)\\ &\quad\times\left[\left(9 \ \frac{\text{h}}{\text{day}}\right)\left(3600 \ \frac{\text{s}}{\text{h}}\right)(21.1^\circ\text{C} - 6.09^\circ\text{C})\right.\\ &\quad\left.+\left(15 \ \frac{\text{h}}{\text{day}}\right)\left(3600 \ \frac{\text{s}}{\text{h}}\right)(10^\circ\text{C} - 6.09^\circ\text{C})\right]\left(\frac{1 \text{ MJ}}{10^6 \text{ J}}\right)\\ &\quad= 8.345 \times 10^5 \text{ MJ}\end{aligned}$$

The fuel consumption is

$$\frac{8.345 \times 10^5 \text{ MJ}}{\left(30.2 \ \frac{\text{MJ}}{\text{kg}}\right)(0.70)} = \boxed{39\,475 \text{ kg/yr}}$$

7. *Customary U.S. Solution*

The heat loss per degree of temperature difference is

$$\frac{\dot{q}}{\Delta T} = \frac{200{,}000 \ \frac{\text{Btu}}{\text{hr}}}{70^\circ\text{F} - 0^\circ\text{F}} = 2857 \text{ Btu/hr-}^\circ\text{F}$$

From Eq. 33.9(b), the average temperature, $\overline{\overline{T}}$, over the entire heating season can be calculated.

$$\text{DD} = N(65^\circ\text{F} - \overline{\overline{T}})$$

The number of heating degree days, DD, is 4200.

The number of days in the heating season, N, is 210.

$$4200 \text{ days} = (210 \text{ days})(65^\circ\text{F} - \overline{\overline{T}})$$
$$\overline{\overline{T}} = 45^\circ\text{F}$$

The total original winter heat loss based on 70°F indoor is

$$\begin{aligned}&(210 \text{ days})\left(24 \ \frac{\text{hr}}{\text{day}}\right)\left(2857 \ \frac{\text{Btu}}{\text{hr-}^\circ\text{F}}\right)\\ &\quad\times(70^\circ\text{F} - 45^\circ\text{F}) = 3.60 \times 10^8 \text{ Btu}\end{aligned}$$

The reduced heat loss based on 68°F indoor is

$$\begin{aligned}&(210 \text{ days})\left(24 \ \frac{\text{hr}}{\text{day}}\right)\left(2857 \ \frac{\text{Btu}}{\text{hr-}^\circ\text{F}}\right)\\ &\quad\times(68^\circ\text{F} - 45^\circ\text{F}) = 3.31 \times 10^8 \text{ Btu}\end{aligned}$$

The reduction is

$$\frac{3.60 \times 10^8 \text{ Btu} - 3.31 \times 10^8 \text{ Btu}}{3.60 \times 10^8 \text{ Btu}} = \boxed{0.081 \ \ (8.1\%)}$$

SI Solution

Since the problem only concerns the percentage reduction in heat loss, determine only the average temperature, $\overline{\overline{T}}$, as all other parameters except indoor temperature remain constant.

From Eq. 33.9(a), the average temperature, $\overline{\overline{T}}$, over the entire heating season can be calculated.

$$\text{DD} = N(18^\circ\text{C} - \overline{\overline{T}})$$

The number of heating Kelvin degree days, DD, is 2333.

The number of days in the entire heating season, N, is 210.

$$2333 \text{ days} = (210 \text{ days})(18^\circ\text{C} - \overline{\overline{T}})$$
$$\overline{\overline{T}} = 6.89^\circ\text{C}$$

The reduction is

$$\frac{\Delta T_{\text{original}} - \Delta T_{\text{reduced}}}{\Delta T_{\text{original}}}$$

$$\Delta T_{\text{original}} = 21.1^\circ\text{C} - 6.89^\circ\text{C} = 14.21^\circ\text{C}$$
$$\Delta T_{\text{reduced}} = 20^\circ\text{C} - 6.89^\circ\text{C} = 13.11^\circ\text{C}$$

The reduction is

$$\frac{14.21^\circ\text{C} - 13.11^\circ\text{C}}{14.21^\circ\text{C}} = \boxed{0.077 \ \ (7.7\%)}$$

8. *Customary U.S. Solution*

Prior to the thermostat change, the energy requirement for ventilation air is

$$\begin{aligned}q_{\text{air},1} &= m_1 c_p (T_{i,1} - T_o)\\ &= N_1 V t \rho c_p (T_{i,1} - T_o)\end{aligned}$$

After the change, the energy requirement is

$$\begin{aligned}q_{\text{air},2} &= m_2 c_p (T_{i,2} - T_o)\\ &= N_2 V t \rho c_p (T_{i,2} - T_o)\end{aligned}$$

The energy savings are

$$\begin{aligned}\Delta q_{\text{air}} &= q_{\text{air},1} - q_{\text{air},2} = N_1 V t \rho c_p (T_{i,1} - T_o)\\ &\quad - N_2 V t \rho c_p (T_{i,2} - T_o)\\ &= V t \rho c_p \big(N_1 (T_{i,1} - T_o) - N_2 (T_{i,2} - T_o)\big)\end{aligned}$$

During an entire week, the building is occupied 10 hr (from 8:00 A.M. to 6:00 P.M.) and unoccupied 14 hr each day for 5 days, and unoccupied 24 hr each day for 2 days (over the weekend).

The unoccupied time for a 21 week period over a year is

$$\begin{aligned} t &= \left(14 \ \frac{\text{hr}}{\text{day}}\right)\left(5 \ \frac{\text{days}}{\text{wk}}\right)\left(21 \ \frac{\text{wk}}{\text{yr}}\right) \\ &\quad + \left(24 \ \frac{\text{hr}}{\text{day}}\right)\left(2 \ \frac{\text{days}}{\text{wk}}\right)\left(21 \ \frac{\text{wk}}{\text{yr}}\right) \\ &= 2478 \ \text{hr/yr} \end{aligned}$$

With air at 70°F, $c_p \approx 0.24$ Btu/lbm-°F and $\rho \approx 0.075$ lbm/ft^3.

$$\begin{aligned} \Delta q_{\text{air}} &= \left(801{,}000 \ \frac{\text{ft}^3}{\text{air change}}\right)\left(2478 \ \frac{\text{hr}}{\text{yr}}\right)\left(0.075 \ \frac{\text{lbm}}{\text{ft}^3}\right) \\ &\quad \times \left(0.24 \ \frac{\text{Btu}}{\text{lbm-°F}}\right)\left(\left(1 \ \frac{\text{air change}}{\text{hr}}\right)\right. \\ &\quad \times (70°\text{F} - 30°\text{F}) - \left(0.5 \ \frac{\text{air change}}{\text{hr}}\right) \\ &\quad \left. \times \left((70°\text{F} - 12°\text{F}) - 30°\text{F}\right)\right) \\ &= 9.289 \times 10^8 \ \text{Btu/hr} \end{aligned}$$

The savings due to reduced heat losses from the roof, floor, and walls (for 12°F set back) are

$$\begin{aligned} \dot{q} &= UA\Delta T \\ &= \left[\left(0.15 \ \frac{\text{Btu}}{\text{ft}^2\text{-hr-°F}}\right)(11{,}040 \ \text{ft}^2) + \left(1.13 \ \frac{\text{Btu}}{\text{ft}^2\text{-hr-°F}}\right)\right. \\ &\quad \times (2760 \ \text{ft}^2) + \left(0.05 \ \frac{\text{Btu}}{\text{ft}^2\text{-hr-°F}}\right)(26{,}700 \ \text{ft}^2) \\ &\quad \left. + \left(1.5 \ \frac{\text{Btu}}{\text{ft-hr-°F}}\right)(690 \ \text{ft})\right](12°\text{F}) \\ &= 85{,}738 \ \text{Btu/hr} \\ q &= \dot{q}t = \left(85{,}738 \ \frac{\text{Btu}}{\text{hr}}\right)\left(2478 \ \frac{\text{hr}}{\text{yr}}\right) \\ &= 2.125 \times 10^8 \ \text{Btu/yr} \end{aligned}$$

The energy saved per year is

$$\begin{aligned} q_{\text{actual}} &= \frac{q_{\text{ideal}}}{\eta_{\text{furnace}}} = \frac{\Delta q_{\text{air}} + q}{\eta_{\text{furnace}}} \\ &= \frac{9.289 \times 10^8 \ \frac{\text{Btu}}{\text{yr}} + 2.125 \times 10^8 \ \frac{\text{Btu}}{\text{yr}}}{(0.75)\left(100{,}000 \ \frac{\text{Btu}}{\text{therm}}\right)} \\ &= 15{,}218 \ \text{therm/yr} \end{aligned}$$

The cost savings are

$$\left(15{,}218 \ \frac{\text{therm}}{\text{yr}}\right)\left(\frac{\$0.25}{\text{therm}}\right) = \boxed{\$3805 \text{ per year}}$$

SI Solution

Prior to the thermostat change, the energy requirement for ventilation air is

$$\begin{aligned} q_{\text{air},1} &= m_1 c_p (T_{i,1} - T_o) \\ &= N_1 V t \rho c_p (T_{i,1} - T_o) \end{aligned}$$

After the change, the energy requirement is

$$\begin{aligned} q_{\text{air},2} &= m_2 c_p (T_{i,2} - T_o) \\ &= N_2 V t \rho c_p (T_{i,2} - T_o) \end{aligned}$$

The energy savings are

$$\begin{aligned} \Delta q_{\text{air}} &= q_{\text{air},1} - q_{\text{air},2} = N_1 V t \rho c_p (T_{i,1} - T_o) \\ &\quad - N_2 V t \rho c_p (T_{i,2} - T_o) \\ &= V t \rho c_p \big(N_1 (T_{i,1} - T_o) - N_2 (T_{i,2} - T_o)\big) \end{aligned}$$

During an entire week, the building is occupied 10 hr (from 8:00 A.M. to 6:00 P.M.) and unoccupied 14 h each day for 5 days, and unoccupied 24 h each day for 2 days (over the weekend).

The unoccupied time for a 21 week period over a year is

$$\begin{aligned} t &= \left(14 \ \frac{\text{h}}{\text{day}}\right)\left(5 \ \frac{\text{days}}{\text{wk}}\right)\left(21 \ \frac{\text{wk}}{\text{yr}}\right) \\ &\quad + \left(24 \ \frac{\text{h}}{\text{day}}\right)\left(2 \ \frac{\text{days}}{\text{wk}}\right)\left(21 \ \frac{\text{wk}}{\text{yr}}\right) \\ &= 2478 \ \text{h/yr} \end{aligned}$$

With air at 21°C, $c_p \approx 1.005$ kJ/kg·°C and $\rho \approx 1.2$ kg/m^3.

$$\begin{aligned} \Delta q_{\text{air}} &= \left(22\,700 \ \frac{\text{m}^3}{\text{air change}}\right)\left(2478 \ \frac{\text{h}}{\text{yr}}\right)\left(1.2 \ \frac{\text{kg}}{\text{m}^3}\right) \\ &\quad \times \left(1.005 \ \frac{\text{kJ}}{\text{kg·°C}}\right)\left(\left(1 \ \frac{\text{air change}}{\text{h}}\right)\right. \\ &\quad \times (21.1°\text{C} - (-1°\text{C})) - \left(0.5 \ \frac{\text{air change}}{\text{h}}\right) \\ &\quad \left. \times \left((21.1°\text{C} - 6.7°\text{C}) - (-1°\text{C})\right)\right) \\ &= 9.769 \times 10^8 \ \text{kJ/yr} \end{aligned}$$

The savings due to reduced heat losses from the roof, floor, and walls for 6.7°C set back is

$$\begin{aligned}\dot{q} &= UA\Delta T \\ &= \left[\left(0.85\ \frac{\text{W}}{\text{m}^2\cdot{}^\circ\text{C}}\right)(993\ \text{m}^2) \right.\\ &\quad + \left(6.42\ \frac{\text{W}}{\text{m}^2\cdot{}^\circ\text{C}}\right)(260\ \text{m}^2) \\ &\quad + \left(0.3\ \frac{\text{W}}{\text{m}^2\cdot{}^\circ\text{C}}\right)(2480\ \text{m}^2) \\ &\quad \left. + \left(2.6\ \frac{\text{W}}{\text{m}\cdot{}^\circ\text{C}}\right)(210\ \text{m})\right](6.7^\circ\text{C}) \\ &= 25\,482\ \text{W}\end{aligned}$$

$$\begin{aligned}q = \dot{q}t &= (25\,482\ \text{W})\left(1\ \frac{\text{J}}{\text{s}\cdot\text{W}}\right)\left(2478\ \frac{\text{h}}{\text{yr}}\right) \\ &\quad \times\left(3600\ \frac{\text{s}}{\text{h}}\right)\left(\frac{1\ \text{kJ}}{1000\ \text{J}}\right) \\ &= 2.273\times 10^8\ \text{kJ/yr}\end{aligned}$$

The energy saved per year is

$$\begin{aligned}q_{\text{actual}} &= \frac{q_{\text{ideal}}}{\eta_{\text{furnace}}} = \frac{\Delta q_{\text{air}} + q}{\eta_{\text{furnace}}} \\ &= \frac{9.769\times 10^8\ \frac{\text{kJ}}{\text{yr}} + 2.273\times 10^8\ \frac{\text{kJ}}{\text{yr}}}{(0.75)\left(1000\ \frac{\text{kJ}}{\text{MJ}}\right)\left(105.506\ \frac{\text{MJ}}{\text{therm}}\right)} \\ &= 15\,218\ \text{therm/yr}\end{aligned}$$

The cost savings are

$$\left(15\,218\ \frac{\text{therm}}{\text{yr}}\right)\left(\frac{\$0.25}{\text{therm}}\right) = \boxed{\$3805\ \text{per year}}$$

34 HVAC: Cooling Load

PRACTICE PROBLEMS

1. (*Time limit: one hour*) The bypass air conditioning system shown has the following operating characteristics.

total air pressure: 14.7 psia (101.3 kPa)
supply temperature: 58°F (14.4°C) dry-bulb
sensible load: 200,000 Btu/hr (58.6 kW)
latent load: 450,000 grains/hr (29 kg/h)
outdoor air: 90°F (32.2°C) dry-bulb,
  76°F (24.4°C) wet-bulb
make-up air: 2000 ft^3/min (940 L/s)
condition of air leaving washer: saturated

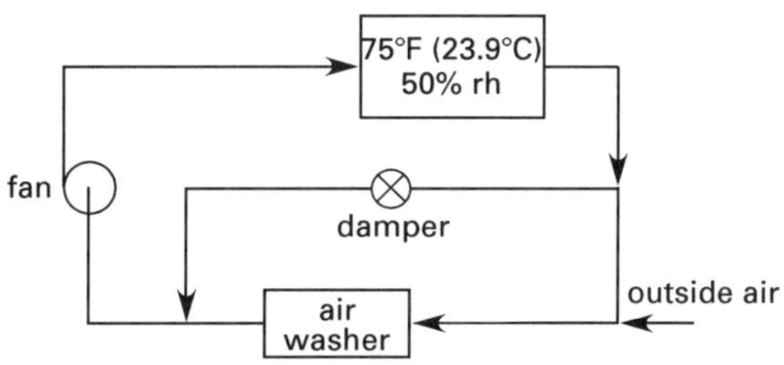

Find the (a) temperature of the air leaving the washer, (b) rate of supply air, and (c) moisture content of the supply air.

2. (*Time limit: one hour*) A building is located at 32°N latitude. The indoor design temperature is 78°F (25.6°C). The outdoor design temperature is 95°F (35°C). The daily temperature range is 22°F (12.2°C). Energy gains through the floor, from lights, and from occupants are insignificant. The building's construction is as follows.

walls:
- 1600 ft^2 (144 m^2) facing north
- 1400 ft^2 (126 m^2) facing south
- 1500 ft^2 (135 m^2) facing east
- 1400 ft^2 (126 m^2) facing west
- 4 in (100 mm) brick facing
- 3 in (75 mm) concrete block
- 1 in (25 mm) mineral wool (emissivity = 0.96)
- 2 in (50 mm) furring
- 3/8 in (9.5 mm) drywall gypsum (emissivity = 0.95)
- 1/2 in (12 mm) plaster

roof:
- 6000 ft^2 (540 m^2)
- 4 in concrete (100 mm)
- 2 in insulation (50 mm)
- insulation conductivity: 2.28 ft^2-°F-hr/Btu-in
- single felt layer
- 1 in (25 mm) air gap
- 1/2 in (12 mm) acoustical ceiling tile

windows:
- 100 ft^2 (9 m^2) facing east
- 1/4 in (6.4 mm) thick, single glazing
- cream-colored Venetian shades
- no exterior shading

(a) What is the cooling load for mid-July at 4:00 P.M. sun time? (b) Explain why this is or is not the peak cooling load.

SOLUTIONS

1. *Customary U.S. Solution*

(a) This is a standard bypass problem.

The indoor conditions are

$$T_i = 75°\text{F}$$
$$\phi_i = 50\%$$

The outdoor conditions are

$$T_{o,\text{db}} = 90°\text{F}$$
$$T_{o,\text{wb}} = 76°\text{F}$$

The ventilation rate is 2000 ft^3/min.

The loads are

$$q_s = 200{,}000 \text{ Btu/hr}$$
$$q_l = 450{,}000 \text{ gr/hr}$$

To find the sensible heat ratio, q_l must be expressed in Btu/hr.

From the psychrometric chart (App. 31.A), for the room conditions, $\omega = 0.0095$ lbm moisture/lbm dry air.

From Eq. 31.7,

$$\omega = (0.622)\left(\frac{p_w}{p_a}\right)$$
$$= (0.622)\left(\frac{p_w}{p - p_w}\right)$$
$$0.0095\ \frac{\text{lbm moisture}}{\text{lbm dry air}} = (0.622)\left(\frac{p_{w,\text{psia}}}{14.7 \text{ psia} - p_{w,\text{psia}}}\right)$$
$$p_w = 0.22 \text{ psia}$$

From steam tables (App. 29.A), for 0.22 psia,

$$h_{fg} = 1062 \text{ Btu/lbm}$$
$$q_l = \left(450{,}000\ \frac{\text{gr}}{\text{hr}}\right)\left(\frac{1 \text{ lbm}}{7000 \text{ gr}}\right)\left(1062\ \frac{\text{Btu}}{\text{lbm}}\right)$$
$$= 68{,}271 \text{ Btu/hr}$$

From Eq. 31.22, the sensible heat ratio is

$$\text{SHR} = \frac{q_s}{q_s + q_l} = \frac{200{,}000\ \frac{\text{Btu}}{\text{hr}}}{200{,}000\ \frac{\text{Btu}}{\text{hr}} + 68{,}271\ \frac{\text{Btu}}{\text{hr}}}$$
$$= 0.75$$

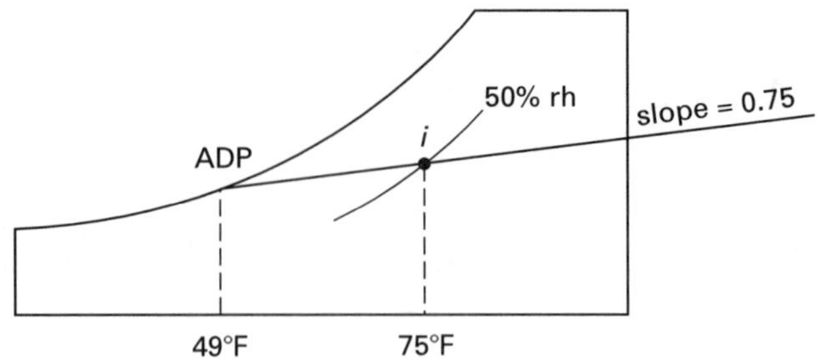

Locate 75°F and 50% relative humidity on the psychrometric chart (App. 31.A). Draw a line with a slope of 0.75 through this point.

The left-hand intersection shows ADP = 49°F. Since the air leaves the conditioner saturated,

$$\text{BF}_{\text{coil}} = 0$$
$$T_{\text{co}} = \boxed{49°\text{F}}$$

(b) Calculate the air flow through the room from Eq. 34.9(b).

$$\dot{V}_{\text{in}} = \frac{\dot{q}_{s,\text{Btu/hr}}}{\left(1.08\ \frac{\text{Btu-min}}{\text{ft}^3\text{-hr}}\right)(T_i - T_{\text{in}})}$$

$$\dot{V}_{\text{in,cfm}} = \frac{200{,}000\ \frac{\text{Btu}}{\text{hr}}}{\left(1.08\ \frac{\text{Btu-min}}{\text{ft}^3\text{-hr}}\right)(75°\text{F} - 58°\text{F})}$$
$$= \boxed{10{,}893 \text{ ft}^3/\text{min}}$$

(c) Since $T_{\text{in}} = 58°\text{F}$, locate this dry-bulb temperature on the condition line. Reading from the chart (App. 31.A),

$$\omega_{\text{in}} = \boxed{0.0081 \text{ lbm/lbm}}$$

SI Solution

(a) The indoor conditions are

$$T_i = 23.9°\text{C}$$
$$\phi_i = 50\%$$

The outdoor conditions are

$$T_{o,\text{db}} = 32.2°\text{C}$$
$$T_{o,\text{wb}} = 24.4°\text{C}$$

The ventilation rate is 940 L/s.

The loads are

$$q_s = 58.6 \text{ kW}$$
$$q_l = 29 \text{ kg/h}$$

To find the sensible heat ratio, q_l must be expressed in kW.

From the psychrometric chart (App. 31.B), for the room conditions, $\omega \approx 9.3$ g moisture/kg dry air (9.3×10^{-3} kg moisture/kg dry air).

From Eq. 31.7,

$$\omega = (0.622)\left(\frac{p_w}{p_a}\right)$$
$$= (0.622)\left(\frac{p_w}{p - p_w}\right)$$
$$9.3 \times 10^{-3}\ \frac{\text{kg moisture}}{\text{kg dry air}} = (0.622)\left(\frac{p_w}{101.3\ \text{kPa} - p_w}\right)$$
$$p_w = 1.49\ \text{kPa}$$

From steam tables (App. 29.N), for 1.49 kPa,

$$h_{fg} = 2470.7\ \text{kJ/kg}$$
$$q_l = \left(29\ \frac{\text{kg}}{\text{h}}\right)\left(\frac{1\ \text{h}}{3600\ \text{s}}\right)\left(2470.7\ \frac{\text{kJ}}{\text{kg}}\right) = 19.9\ \text{kW}$$

From Eq. 31.22, the sensible heat ratio is

$$\text{SHR} = \frac{q_s}{q_s + q_l} = \frac{58.6\ \text{kW}}{58.6\ \text{kW} + 19.9\ \text{kW}}$$
$$= 0.75$$

Locate 23.9°C and 50% relative humidity on the psychrometric chart (Fig. 31.B). Draw a line with a slope of 0.75 through this point.

The left-hand intersection shows ADP = 9°C. Since the air leaves the conditioner saturated,

$$\text{BF}_{\text{coil}} = 0$$
$$T_{\text{co}} = \boxed{9°\text{C}}$$

(b) Calculate the air flow through the room from Eq. 34.9(a).

$$\dot{V}_{\text{in,L/s}} = \frac{\dot{q}_{s,\text{W}}}{\left(1.20\ \frac{\text{W·s}}{\text{L·°C}}\right)(T_i - T_{\text{in}})}$$
$$= \frac{(58.6\ \text{kW})\left(1000\ \frac{\text{W}}{\text{kW}}\right)}{\left(1.20\ \frac{\text{W·s}}{\text{L·°C}}\right)(23.9°\text{C} - 14.4°\text{C})}$$
$$= \boxed{5140\ \text{L/s}}$$

(c) Since T_{in} = 14.4°C, locate this dry-bulb temperature on the condition line. Reading from the chart (31.B),

$$\omega_{\text{in}} = \boxed{8.1\ \text{g/kg dry air}}$$

2. (a) *step 1:* Determine thermal resistances for walls, roof, and windows.

For walls:

Assume 2 in furring to be 2 in air space.

From Eq. 33.4, the effective space emissivity is

$$\frac{1}{E} = \frac{1}{\epsilon_1} + \frac{1}{\epsilon_2} - 1$$
$$= \frac{1}{0.96} + \frac{1}{0.95} - 1$$
$$E = 0.91$$

From Table 33.2, for vertical air space orientation, the thermal conductance (by extrapolating) for $E = 0.91$ is

$$\frac{1}{C} = \frac{1}{1.07\ \frac{\text{Btu}}{\text{hr-ft}^2\text{-°F}}}$$
$$= 0.93\ \text{hr-ft}^2\text{-°F/Btu}$$

Use App. 33.A.

walls

type of construction	R, $\frac{\text{ft}^2\text{-°F-hr}}{\text{Btu}}$	notes
4 in brick spacing	0.44	
3 in concrete block	0.40	
1 in mineral wool	3.85	value given per in
2 in furring	0.93	see calculation above
3/8 in drywall gypsum	0.34	0.45 for 1/2 is given
1/2 in plaster (sand aggregate)	0.09	
surface outdoor (Table 33.3)	0.25	assume 7 1/2 mph air speed
surface indoor (Table 33.3)	0.68	still air, $U = 1.46\ \frac{\text{Btu}}{\text{ft}^2\text{-°F-hr}}$

roof

type of construction	R, $\frac{\text{ft}^2\text{-°F-hr}}{\text{Btu}}$	notes
4 in concrete	0.32	sand aggregate
2 in insulation	5.56	2.78/in
felt	0.06	
1 in air gap	0.87	$\epsilon = 0.90$ for buildup material
1/2 in accoustical ceiling tile	1.19	
surface outdoor (Table 33.3)	0.25	assume 7 1/2 mph air speed and downward heat flow ~ upward heat flow
surface indoor (Table 33.3)	0.93	still air, $U = 1.08\ \frac{\text{Btu}}{\text{ft}^2\text{-°F-hr}}$

Ventilation

windows		
type of construction	$R, \frac{\text{ft}^2\text{-°F-hr}}{\text{Btu}}$	notes
1/4 in thick, single glazing	0.66	without shades, $R = 0.88 \frac{\text{ft}^2\text{-°F-hr}}{\text{Btu}}$
cream colored venetian shades, no exterior shades		assume shading factor of 0.75

step 2: Determine the overall heat transfer coefficient for walls, roof, and windows using Eq. 33.3.

$$U = \frac{1}{\sum R_i}$$

For the walls,

$$U = \frac{1}{\begin{array}{l} 0.25 \frac{\text{ft}^2\text{-°F-hr}}{\text{Btu}} + 0.44 \frac{\text{ft}^2\text{-°F-hr}}{\text{Btu}} \\ + 0.40 \frac{\text{ft}^2\text{-°F-hr}}{\text{Btu}} + 3.85 \frac{\text{ft}^2\text{-°F-hr}}{\text{Btu}} \\ + 0.93 \frac{\text{ft}^2\text{-°F-hr}}{\text{Btu}} + 0.34 \frac{\text{ft}^2\text{-°F-hr}}{\text{Btu}} \\ + 0.09 \frac{\text{ft}^2\text{-°F-hr}}{\text{Btu}} + 0.68 \frac{\text{ft}^2\text{-°F-hr}}{\text{Btu}} \end{array}}$$

$$= 0.143 \text{ Btu/ft}^2\text{-°F-hr}$$

The cooling load is calculated from the equivalent temperature differences for the walls and roof. Tables of these values are not common; they are now often incorporated directly into HVAC computer application databases. Results from the remainder of this problem will vary with the temperature differences used.

Assume the following equivalent temperature differences for 4 in brick face walls at 4 P.M.

facing east: $\Delta T = 20°\text{F}$
facing west: $\Delta T = 20°\text{F}$
facing south: $\Delta T = 28°\text{F}$
facing north: $\Delta T = 17°\text{F}$

For 4 in brick facing at 4 P.M.,

$$\begin{aligned} Q_{\text{walls}} &= \left(0.143 \frac{\text{Btu}}{\text{ft}^2\text{-°F-hr}}\right) [(1600 \text{ ft}^2)(17°\text{F}) \\ &\quad + (1400 \text{ ft}^2)(28°\text{F}) + (1500 \text{ ft}^2)(20°\text{F}) \\ &\quad + (1400 \text{ ft}^2)(20°\text{F})] \\ &= 17{,}789 \text{ Btu/hr} \end{aligned}$$

For the roof,

$$U = \frac{1}{\begin{array}{l} 0.25 \frac{\text{ft}^2\text{-°F-hr}}{\text{Btu}} + 0.32 \frac{\text{ft}^2\text{-°F-hr}}{\text{Btu}} \\ + 5.56 \frac{\text{ft}^2\text{-°F-hr}}{\text{Btu}} + 0.06 \frac{\text{ft}^2\text{-°F-hr}}{\text{Btu}} \\ + 0.87 \frac{\text{ft}^2\text{-°F-hr}}{\text{Btu}} + 1.19 \frac{\text{ft}^2\text{-°F-hr}}{\text{Btu}} \\ + 0.93 \frac{\text{ft}^2\text{-°F-hr}}{\text{Btu}} \end{array}}$$

$$= 0.109 \text{ Btu/ft}^2\text{-°F-hr}$$

Assume an equivalent temperature difference for 4 in concrete roofing at 4 P.M. = 74°F. Therefore,

$$\begin{aligned} Q_{\text{roof}} &= \left(0.109 \frac{\text{Btu}}{\text{ft}^2\text{-°F-hr}}\right) (6000 \text{ ft}^2)(74°\text{F}) \\ &= 48{,}396 \text{ Btu/hr} \end{aligned}$$

For the windows:

Since all windows facing east will receive no direct sunlight at 4 P.M., consider $\Delta T = 95°\text{F} - 78°\text{F}$.

$$\begin{aligned} Q_{\text{windows}} &= \left(\frac{1}{0.66 \frac{\text{ft}^2\text{-°F-hr}}{\text{Btu}}}\right) (100 \text{ ft}^2)(95°\text{F} - 78°\text{F}) \\ &= 2576 \text{ Btu/hr} \end{aligned}$$

The total sensible transmission load (at 4 P.M.) is

$$\begin{aligned} Q_{\text{total}} &= 17{,}789 \frac{\text{Btu}}{\text{hr}} + 48{,}396 \frac{\text{Btu}}{\text{hr}} + 2576 \frac{\text{Btu}}{\text{hr}} \\ &= \boxed{68{,}761 \text{ Btu/hr}} \end{aligned}$$

(b) This may not be the peak load. Since Q_{walls} and Q_{roof} are the major contributors and ΔT equivalent temperature differences are maximum between 4 P.M. and 6 P.M., the peak cooling load will be maximum somewhere between 4 P.M. and 6 P.M. Consider Q_{walls} and Q_{roof} for 6 P.M.

At 6 P.M., the equivalent temperature differences for 4 in brick facing are assumed as

facing east: $\Delta T = 22°\text{F}$
facing west: $\Delta T = 35°\text{F}$
facing south: $\Delta T = 28°\text{F}$
facing north: $\Delta T = 20°\text{F}$

$$\begin{aligned}Q_{\text{walls}} &= \left(0.143\ \frac{\text{Btu}}{\text{ft}^2\text{-}^\circ\text{F-hr}}\right)\big[(1600\ \text{ft}^2)(20^\circ\text{F})\\ &\quad+\ (1400\ \text{ft}^2)(28^\circ\text{F}) + (1500\ \text{ft}^2)(22^\circ\text{F})\\ &\quad+\ (1400\ \text{ft}^2)(35^\circ\text{F})\big]\\ &= 21{,}908\ \text{Btu/hr}\end{aligned}$$

Assume the equivalent temperature difference for a 4 in concrete roof at 6 P.M. is 68°F.

$$\begin{aligned}Q_{\text{roof}} &= \left(0.109\ \frac{\text{Btu}}{\text{ft}^2\text{-}^\circ\text{F-hr}}\right)(6000\ \text{ft}^2)(68^\circ\text{F})\\ &= 44{,}472\ \text{Btu/hr}\end{aligned}$$

The total sensible transmission load at 6 P.M. is

$$\begin{aligned}Q_{\text{total}} &= 21{,}908\ \frac{\text{Btu}}{\text{hr}} + 44{,}472\ \frac{\text{Btu}}{\text{hr}} + 2576\ \frac{\text{Btu}}{\text{hr}}\\ &= \boxed{68{,}956\ \text{Btu/hr}}\end{aligned}$$

This is very close to the 4 P.M. load.

(An SI solution of this problem is not supported by this book.)

Ventilation

35 Air Conditioning Equipment

PRACTICE PROBLEMS

Chapter 35 of the *Environmental Engineering Reference Manual* contains no practice problems.

36 Fans and Ductwork

PRACTICE PROBLEMS

(Note: Round all duct dimensions to the next larger whole inch or multiples of 25 mm.)

1. A round 18 in (457 mm) duct is to be replaced by a rectangular duct with an aspect ratio of 4:1. What should be the dimensions of the rectangular duct?

2. Draw the system curve for a fan that moves 10,000 SFCM (4700 L/s) against a pressure of 4 in wg (1 kPa).

3. A 1/8 scale model fan is tested at 300 rpm. At standard air conditions and at that speed, the model fan moves 40 cfm (19 L/s) against 0.5 in wg (125 Pa). What is the air power if a full-sized fan operates at 5000 ft (1500 m) altitude at half the model's speed?

4. A duct system consists of 750 ft (230 m) of galvanized 20 in (508 mm) ducts, four round elbows (radius-to-diameter ratio of 1.5). Two 2 in (51 mm) diameter pipes pass perpendicularly through the duct at two locations. The flow rate is 6000 SCFM (4200 L/s). What is the friction loss?

5. 1500 SCFM (700 L/s) of air flows in an 18 in (457 mm) round duct. After a branch reduction of 300 SCFM (140 L/s), the fitting reduces to 14 in (356 mm) in the through direction. What is the static regain in the through direction?

6. (*Time limit: one hour*) A fan in a small theater delivers 1500 cfm (700 L/s) through the system shown. The duct is rectangular with an aspect ratio of 1.5:1. All elbows have a radius-to-diameter ratio of 1.5. The terminal pressure at each outlet must be 0.25 in wg (63 Pa) or higher. Takeoff fitting losses are to be disregarded. (a) Use the equal-friction method to size the system. (b) What is the static pressure at the fan? (c) What is the total pressure at the fan?

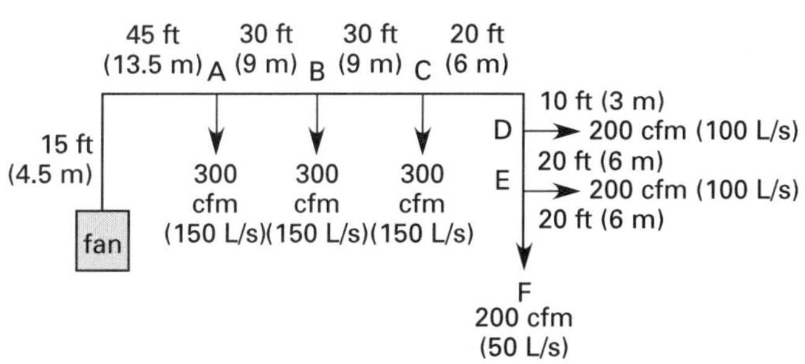

7. (*Time limit: one hour*) A fan in a theater delivers 300 cfm (140 L/s) of air through round ducts to each of the twelve outlets shown. The minimum terminal pressure at the outlets is 0.15 in wg (38 Pa). All elbows have a radius-to-diameter ratio of 1.25. Branch takeoff fitting losses are to be disregarded. (a) Use the equal-friction method to size the system. (b) What is the static pressure at the fan? (c) What is the total pressure at the fan?

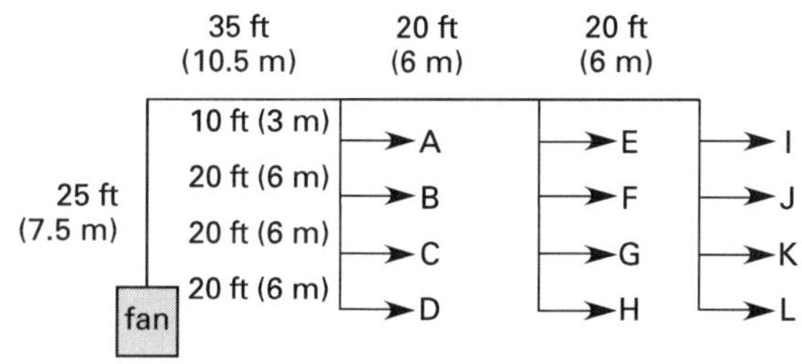

8. (*Time limit: one hour*) Size the longest run in Prob. 7 using the static regain method with a regain coefficient of 0.75. (Customary U.S. solution only required.)

9. (*Time limit: one hour*) 3000 cfm (1400 L/s) of air with a density of 0.075 lbm/ft^3 (1.2 kg/m^3) enters a 12 in (305 mm) diameter duct at section A. The four-piece 90° elbows have a radius-to-diameter ratio of 1.5 and an equivalent length of 14 diameters. The Darcy friction factor is 0.02 everywhere in the system. The static regain coefficient for this system is 0.65. (a) Use the static regain method to calculate the diameters at sections B and C. (b) Specify if dampers should be placed in sections A, B, and/or C. (c) What is the regain in sections B and C?

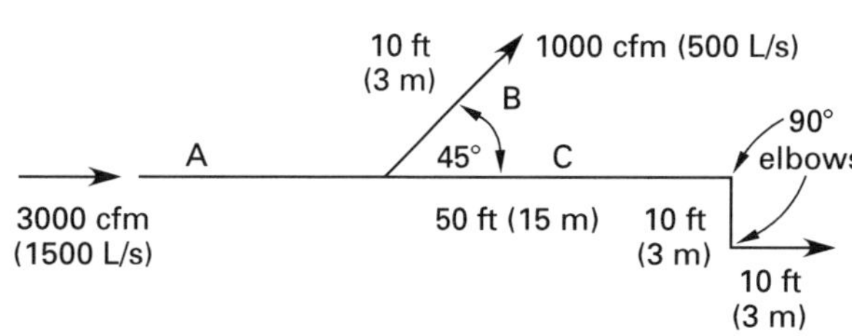

SOLUTIONS

1. *Customary U.S. Solution*

With aspect ratio $R = 4$, the short side is given by Eq. 36.31.

$$\text{short side} = \frac{D(1+R)^{\frac{1}{4}}}{1.3R^{\frac{5}{8}}} = \frac{(18 \text{ in})(1+4)^{\frac{1}{4}}}{(1.3)(4)^{\frac{5}{8}}} = 8.7 \text{ in} \approx \boxed{9.0 \text{ in}}$$

From Eq. 36.32, the long side is

$$\text{long side} = R(\text{short side}) = (4)(8.7 \text{ in}) = 34.8 \text{ in} \approx \boxed{35 \text{ in}}$$

SI Solution

With aspect ratio $R = 4$, the short side is given by Eq. 36.31.

$$\text{short side} = \frac{D(1+R)^{\frac{1}{4}}}{1.3R^{\frac{5}{8}}} = \frac{(457 \text{ mm})(1+4)^{\frac{1}{4}}}{(1.3)(4)^{\frac{5}{8}}} = \boxed{221 \text{ mm}}$$

From Eq. 36.32, the long side is

$$\text{long side} = R(\text{short side}) = (4)(221 \text{ mm}) = \boxed{884 \text{ mm}}$$

2. *Customary U.S. Solution*

Disregarding terminal pressure, the remainder of the points on the system curve can be found from Eq. 36.19.

$$\frac{p_2}{p_1} = \left(\frac{Q_2}{Q_1}\right)^2$$

$$Q_2 = Q_1\sqrt{\frac{p_2}{p_1}} = (10{,}000 \text{ SCFM})\sqrt{\frac{p_2}{4 \text{ in wg}}} = 5000\sqrt{p_2}$$

The following table can be generated from the preceding equation.

p_2 (in wg)	Q (SCFM)
1	5000
2	7071
3	8660
4	10,000
5	11,180
6	12,247

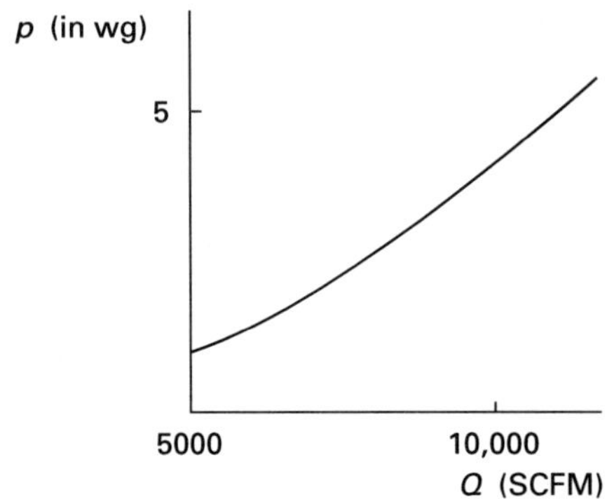

SI Solution

Disregarding terminal pressure, the remainder of the points on the system curve can be found from Eq. 36.19.

$$\frac{p_2}{p_1} = \left(\frac{Q_2}{Q_1}\right)^2$$

$$Q_2 = Q_1\sqrt{\frac{p_2}{p_1}} = \left(4700 \ \frac{\text{L}}{\text{s}}\right)\sqrt{\frac{p_2}{1 \text{ kPa}}} = 4700\sqrt{p_2}$$

The following table can be generated from the preceding equation.

p_2 (kPa)	Q (L/s)
1	4700
2	6647
3	8141
4	9400
5	10 510
6	11 513

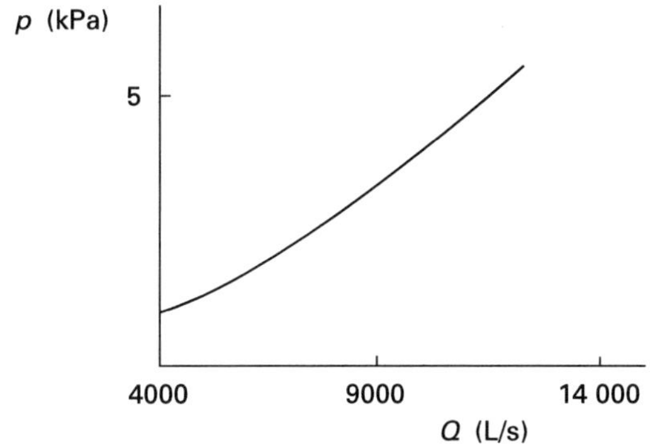

SI Solution

The air power is given by Eq. 36.11(a).

$$(\text{AkW})_1 = \frac{Q_{\text{L/s}}(\text{TP}_{\text{Pa}})}{10^6} = \frac{\left(19\ \frac{\text{L}}{\text{s}}\right)(125\ \text{Pa})}{10^6\ \frac{\text{L}\cdot\text{W}}{\text{m}^3\cdot\text{kW}}} = 2.375\times10^{-3}\ \text{kW}$$

In order to predict the performance of a dynamically similar fan, use Eq. 36.26.

$$\frac{(\text{AkW})_1}{(\text{AkW})_2} = \left(\frac{D_1}{D_2}\right)^5\left(\frac{n_1}{n_2}\right)^3\left(\frac{\gamma_1}{\gamma_2}\right)$$

$$(\text{AkW})_2 = (\text{AkW})_1\left(\frac{D_2}{D_1}\right)^5\left(\frac{n_2}{n_1}\right)^3\left(\frac{\gamma_2}{\gamma_1}\right)$$

$$\frac{D_2}{D_1} = \frac{1}{\frac{1}{8}} = 8$$

$$\frac{n_2}{n_1} = \tfrac{1}{2}$$

At standard air conditions, $\rho = 1.2\ \text{kg/m}^3$. Use standard atmospheric data from App. 18.A for 1500 m altitude.

$$\rho_{1525} = \rho_2 = \frac{p}{RT} = \frac{(0.8456\ \text{bar})\left(10^5\ \frac{\text{Pa}}{\text{bar}}\right)}{\left(287\ \frac{\text{J}}{\text{kg}\cdot\text{K}}\right)(278.4\text{K})} = 1.058\ \text{kg/m}^3$$

$$\frac{\gamma_2}{\gamma_1} = \frac{\rho_2 g}{\rho_1 g} = \frac{\rho_2}{\rho_1} = \frac{1.058\ \frac{\text{kg}}{\text{m}^3}}{1.2\ \frac{\text{kg}}{\text{m}^3}} = 0.88$$

$$(\text{AkW})_2 = (2.375\times10^{-3}\ \text{kW})(8)^5\left(\tfrac{1}{2}\right)^3(0.88) = \boxed{8.56\ \text{kW}}$$

3. *Customary U.S. Solution*

The air horsepower is given by Eq. 36.11(b).

$$(\text{AHP})_A = \frac{Q_{\text{cfm}}(\text{TP}_{\text{in wg}})}{6356} = \frac{(40\ \text{cfm})(0.5\ \text{in wg})}{6356\ \frac{\text{in-ft}^3}{\text{hp-min}}} = 3.147\times10^{-3}\ \text{hp}$$

In order to predict the performance of a dynamically similar fan, use Eq. 36.26.

$$\frac{(\text{AHP})_1}{(\text{AHP})_2} = \left(\frac{D_1}{D_2}\right)^5\left(\frac{n_1}{n_2}\right)^3\left(\frac{\gamma_1}{\gamma_2}\right)$$

$$(\text{AHP})_2 = (\text{AHP})_1\left(\frac{D_2}{D_1}\right)^5\left(\frac{n_2}{n_1}\right)^3\left(\frac{\gamma_2}{\gamma_1}\right)$$

$$\frac{D_2}{D_1} = \frac{1}{\frac{1}{8}} = 8$$

$$\frac{n_2}{n_1} = \tfrac{1}{2}$$

At standard air conditions, $\rho_1 = 0.075\ \text{lbm/ft}^3$. Use standard atmospheric data from App. 18.A for 5000 ft altitude.

$$\rho_{5000} = \rho_2 = \frac{p}{RT} = \frac{\left(12.225\ \frac{\text{lbf}}{\text{in}^2}\right)\left(144\ \frac{\text{in}^2}{\text{ft}^2}\right)}{\left(53.3\ \frac{\text{ft-lbf}}{\text{lbm-°R}}\right)(500.9\text{°R})} = 0.06594\ \text{lbm/ft}^3$$

$$\frac{\gamma_2}{\gamma_1} = \frac{\rho_2 g}{\rho_1 g} = \frac{\rho_2}{\rho_1} = \frac{0.06594\ \frac{\text{lbm}}{\text{ft}^3}}{0.075\ \frac{\text{lbm}}{\text{ft}^3}} = 0.8792$$

$$(\text{AHP})_2 = (3.147\times^{-3}\ \text{hp})(8)^5\left(\tfrac{1}{2}\right)^3(0.8792) = \boxed{11.33\ \text{hp}}$$

4. *Customary U.S. Solution*

From the standard friction loss in the standard duct chart (Fig. 36.4), the friction loss is 0.42 in of water per 100 ft of duct and v = 2700 ft/min. The friction loss due to the 750 ft of duct is

$$\text{FP}_{f,1} = (0.42\ \text{in wg})\left(\frac{750\ \text{ft}}{100\ \text{ft}}\right) = 3.15\ \text{in wg}$$

The equivalent length of each round elbow (radius-to-diameter ratio of 1.5) can be found from Table 36.4.

$$L_e = 12D$$

For four round elbows,

$$\begin{aligned} L_e &= (4)(12D) \\ &= \frac{(4)(12D)(20\ \text{in})}{12\ \frac{\text{in}}{\text{ft}}} = 80\ \text{ft} \end{aligned}$$

The friction loss due to the 80 ft of equivalent length of the duct is

$$\text{FP}_{f,2} = (0.42\ \text{in wg})\left(\frac{80\ \text{ft}}{100\ \text{ft}}\right) = 0.336\ \text{in wg}$$

For a cross duct,

$$\frac{D_1}{D_2} = \frac{2\ \text{in}}{20\ \text{in}} = 0.1$$

From Table 36.3, the loss coefficient is $K_{\text{up}} = 0.2$. For two cross pipes, the friction loss is given by Eq. 36.34.

$$\begin{aligned} \text{FP}_{f,3} &= (2)K_{\text{up}}\left(\frac{\text{v}}{4005}\right)^2 \\ &= (2)(0.2)\left(\frac{2700\ \frac{\text{ft}}{\text{min}}}{4005\ \frac{\text{ft}}{\text{min}}}\right)^2 = 0.182\ \text{in wg} \end{aligned}$$

The total loss is

$$\begin{aligned} \text{FP}_{f,1} + \text{FP}_{f,2} + \text{FP}_{f,3} &= 3.15\ \text{in wg} + 0.336\ \text{in wg} \\ &\quad + 0.182\ \text{in wg} \\ &= \boxed{3.668\ \text{in wg}} \end{aligned}$$

SI Solution

From the standard friction loss in the standard duct chart (Fig. 36.5), the friction loss is 9 Pa/m and v = 21.5 m/s. The friction loss due to the 230 m of the duct is

$$\text{FP}_{f,1} = \left(9\ \frac{\text{Pa}}{\text{m}}\right)(230\ \text{m}) = 2070\ \text{Pa}$$

The equivalent length of each round elbow (radius-to-diameter ratio of 1.5) can be found in Table 36.4.

$$L_e = 12D$$

For four round elbows,

$$\begin{aligned} L_e &= (4)(12D) \\ &= \frac{(4)(12D)(508\ \text{mm})}{\frac{1000\ \text{mm}}{\text{m}}} = 24.38\ \text{m} \end{aligned}$$

The friction loss due to the 24.38 m of equivalent length of duct is

$$\text{FP}_{f,2} = \left(9\ \frac{\text{Pa}}{\text{m}}\right)(24.38\ \text{m}) = 219.4\ \text{Pa}$$

For a cross duct,

$$\frac{D_1}{D_2} = \frac{51\ \text{mm}}{508\ \text{mm}} = 0.1$$

From Table 36.3, the loss coefficient K_{up} is 0.2. For two cross pipes, the friction loss is given by Eq. 36.34.

$$\begin{aligned} \text{FP}_{f,3} &= (2)K_{\text{up}}(\text{v}_{\text{m/s}})^2 \\ &= (2)(0.2)\left(21.5\ \frac{\text{m}}{\text{s}}\right)^2 = 184.9\ \text{Pa} \end{aligned}$$

The total loss is

$$\begin{aligned} \text{FP}_{f,1} + \text{FP}_{f,2} + \text{FP}_{f,3} &= 2070\ \text{Pa} + 219.4\ \text{Pa} \\ &\quad + 184.9\ \text{Pa} \\ &= \boxed{2474.3\ \text{Pa}} \end{aligned}$$

5. *Customary U.S. Solution*

For the 18 in duct,

$$\begin{aligned} A &= \frac{\pi}{4}D^2 = \left(\frac{\pi}{4}\right)\left(\frac{18\ \text{in}}{12\ \frac{\text{in}}{\text{ft}}}\right)^2 \\ &= 1.767\ \text{ft}^2 \end{aligned}$$

$$\begin{aligned} \text{v}_1 &= \frac{Q}{A} = \frac{1500\ \frac{\text{ft}^3}{\text{min}}}{1.767\ \text{ft}^2} \\ &= 848.9\ \text{ft/min} \end{aligned}$$

For the 14 in duct,

$$A = \frac{\pi}{4}D^2 = \left(\frac{\pi}{4}\right)\left(\frac{14\ \text{in}}{12\ \frac{\text{in}}{\text{ft}}}\right)^2 = 1.069\ \text{ft}^2$$

$$\begin{aligned} \text{v}_2 &= \frac{Q}{A} = \frac{1500\ \text{cfm} - 300\ \text{cfm}}{1.069\ \text{ft}^2} \\ &= 1122.5\ \text{ft/min} \end{aligned}$$

Since $v_2 > v_1$, there is no regain. The regain will be a static pressure loss. Use $R = 1.1$ in that case. From Eq. 36.39, the static pressure loss is

$$\begin{aligned}\text{SR}_{\text{actual}} &= R\left(\frac{v_{\text{up}}^2 - v_{\text{down}}^2}{(4005)^2}\right)\\ &= (1.1)\left(\frac{\left(848.9\ \frac{\text{ft}}{\text{min}}\right)^2 - \left(1122.5\ \frac{\text{ft}}{\text{min}}\right)^2}{(4005)^2}\right)\\ &= \boxed{-0.037 \text{ in wg loss}}\end{aligned}$$

SI Solution

For the 457 mm duct,

$$A = \frac{\pi}{4}D^2 = \left(\frac{\pi}{4}\right)\left(\frac{457 \text{ mm}}{1000\ \frac{\text{mm}}{\text{m}}}\right)^2 = 0.164 \text{ m}^2$$

$$\begin{aligned}v_1 = \frac{Q}{A} &= \frac{\left(700\ \frac{\text{L}}{\text{s}}\right)\left(\frac{1 \text{ m}^3}{1000 \text{ L}}\right)}{0.164 \text{ m}^2}\\ &= 4.27 \text{ m/s}\end{aligned}$$

For 356 mm duct,

$$A = \frac{\pi}{4}D^2 = \left(\frac{\pi}{4}\right)\left(\frac{356 \text{ mm}}{1000\ \frac{\text{mm}}{\text{m}}}\right)^2 = 0.1 \text{ m}^2$$

$$\begin{aligned}v_2 = \frac{Q}{A} &= \left(\frac{700\ \frac{\text{L}}{\text{s}} - 140\ \frac{\text{L}}{\text{s}}}{0.1 \text{ m}^2}\right)\left(\frac{1 \text{ m}^3}{1000 \text{ L}}\right)\\ &= 5.60 \text{ m/s}\end{aligned}$$

Since $v_2 > v_1$, there is no regain. The regain will be a static pressure loss. Use $R = 1.1$ in that case. From Eq. 36.5, the static pressure loss is

$$\begin{aligned}\text{SR} &= R\left(0.6\ \frac{\text{Pa·s}}{\text{m}}\right)(v_{\text{up}}^2 - v_{\text{down}}^2)\\ &= (1.1)\left(0.6\ \frac{\text{Pa·s}}{\text{m}}\right)\left(\left(4.27\ \frac{\text{m}}{\text{s}}\right)^2 - \left(5.60\ \frac{\text{m}}{\text{s}}\right)^2\right)\\ &= \boxed{-8.66 \text{ Pa loss}}\end{aligned}$$

6. *Customary U.S. Solution*

(a) *step 1:* From Table 36.7, choose the main duct velocity as 1600 fpm.

step 2: The total air flow from the fan is 1500 cfm. From Fig. 36.4, the main duct diameter is 13 in. The friction loss is 0.27 in wg per 100 ft.

step 3: After the first takeoff at A, the flow rate in section A–B is

$$1500 \text{ cfm} - 300 \text{ cfm} = 1200 \text{ cfm}$$

From Fig. 36.4, for 1200 cfm and 0.27 in wg per 100 ft, the diameter is 11.8 in (say 12 in) and the velocity is 1500 fpm. Similarly, the diameter and the velocity for other sections are obtained and listed in the following table.

section	Q (cfm)	D (in)	v (fpm)
fan–A	1500	13	1600
A–B	1200	12	1500
B–C	900	11	1400
C–D	600	9.5	1300
D–E	400	8	1150
E–F	200	6.3	975

These diameters are for round duct. The equivalent rectangular duct sides with an aspect ratio of 1.5 are found as follows.

$$a = \frac{D(1.5+1)^{0.25}}{(1.3)(1.5)^{0.625}} = 0.75D$$

$$b = Ra = (1.5)(0.75D) = 1.125D$$

Convert the diameters to sides a and b.

$$a_{\text{fan–A}} = (0.75)(13 \text{ in}) = 9.75 \text{ in}$$

$$b_{\text{fan–A}} = (1.125)(13 \text{ in}) = 14.63 \text{ in}$$

Use a 10×15 in duct.

The following table is prepared similarly, rounding up as appropriate.

section	D (in)	a (in)	b (in)
fan–A	13	10	15
A–B	12	9	14
B–C	11	9	13
C–D	9.5	8	11
D–E	8	6	9
E–F	6.3	5	8

(b) *step 4:* By inspection, the longest run is fan–F. From Table 36.4, the equivalent length of each bend is $12D$. For two elbows,

$$\begin{aligned}L_{e,\text{bend}} = 12D &= (12)\left(\frac{13 \text{ in} + 9.5 \text{ in}}{12\ \frac{\text{in}}{\text{ft}}}\right)\\ &= 22.5 \text{ ft}\end{aligned}$$

Ventilation

The equivalent length of the entire run is

15 ft	(fan to first bend)
45 ft	(first bend to point A)
30 ft	(section A–B)
30 ft	(section B–C)
20 ft	(point A to second bend)
22.5 ft	(equivalent length of two bends)
10 ft	(second bend to point D)
20 ft	(section D–E)
20 ft	(section E–F)
total: 212.5 ft	

The straight-through friction loss in the longest run is

$$\left(\frac{212.5\text{ ft}}{100\text{ ft}}\right)\left(0.27\ \frac{\text{in wg}}{100\text{ ft}}\right) = 0.57\text{ in wg}$$

The fan must be able to supply a static pressure of

$$\begin{aligned}\text{SP}_{\text{fan}} &= 0.57\text{ in wg} + 0.25\text{ in wg}\\ &= \boxed{0.82\text{ in wg}}\end{aligned}$$

(c) The total pressure supplied by the fan is

$$\begin{aligned}\text{TP}_{\text{fan}} &= 0.82\text{ in wg} + \left(\frac{1600\text{ fpm}}{4005}\right)^2\\ &= \boxed{0.98\text{ in wg}}\end{aligned}$$

SI Solution

(a) *step 1:* From Table 36.7, choose the main duct velocity as 1600 cfm. From the table footnote (*a*), the SI velocity is

$$\text{v}_{\text{main}} = (1600\text{ fpm})(0.00508) = 8.1\text{ m/s}$$

step 2: The total air flow from the fan is 700 L/s. From Fig. 36.5, the main duct diameter is approximately 340 mm. The friction loss is 2.3 Pa/m.

step 3: After the first takeoff at A, the flow rate in section A–B is

$$700\ \frac{\text{L}}{\text{s}} - 150\ \frac{\text{L}}{\text{s}} = 550\text{ L/s}$$

From Fig. 36.5 for 550 L/s and 2.3 Pa/m, the diameter is 300 mm and the velocity is 7.7 m/s. Similarly, the diameter and the velocity for other sections are obtained and listed in the following table.

section	*Q* (L/s)	*D* (mm)	v (m/s)
fan–A	700	340	8.1
A–B	550	300	7.7
B–C	400	275	7
C–D	250	225	6.3
D–E	150	180	5.5
E–F	50	125	4.4

These diameters are for round duct. The equivalent rectangular duct sides with an aspect ratio of 1.5 are found as follows.

$$a = \frac{D(1.5+1)^{0.25}}{(1.3)(1.5)^{0.625}} = 0.75D$$

$$b = Ra = (1.5)(0.75D) = 1.125D$$

Convert the diameters to sides a and b.

$$a_{\text{fan–A}} = (0.75)(340\text{ mm}) = 255\text{ mm}$$

$$b_{\text{fan–A}} = (1.125)(340\text{ mm}) = 383\text{ mm}$$

Use a 275 × 400 mm duct.

The following table is prepared similarly, rounding up as appropriate.

section	*D* (mm)	*a* (mm)	*b* (mm)
fan–A	340	275	400
A–B	300	225	350
B–C	275	225	325
C–D	225	175	275
D–E	180	150	225
E–F	125	100	150

(b) *step 4:* By inspection, the longest run is fan–F. From Table 36.4, the equivalent length of each bend is 12*D*. For two elbows,

$$\begin{aligned}L_{e,\text{bend}} &= 12D\\ &= (12)\left(\frac{340\text{ mm} + 225\text{ mm}}{1000\ \frac{\text{mm}}{\text{m}}}\right)\\ &= 6.8\text{ m}\end{aligned}$$

The equivalent length of the entire run is

4.5 m	(fan to first bend)
13.5 m	(first bend to point A)
9 m	(section A–B)
9 m	(section B–C)
6 m	(point C to second bend)
6.8 m	(equivalent length of two bends)
3 m	(second bend to point D)
6 m	(section D–E)
6 m	(section E–F)
total: 63.8 m	

The straight-through friction loss in the longest run is

$$(63.8\ \text{m})\left(2.3\ \frac{\text{Pa}}{\text{m}}\right) = 147\ \text{Pa}$$

The fan must be able to supply a static pressure of

$$\text{SP}_{\text{fan}} = 147\ \text{Pa} + 63\ \text{Pa} = \boxed{210\ \text{Pa}}$$

(c) The total pressure supplied by the fan is

$$\text{TP}_{\text{fan}} = 210\ \text{Pa} + \left(0.6\ \frac{\text{Pa}\cdot\text{s}^2}{\text{m}^2}\right)\left(8.1\ \frac{\text{m}}{\text{s}}\right)^2 = \boxed{249\ \text{Pa}}$$

7. *Customary U.S. Solution*

(a) *step 1:* From Table 36.7, choose the main duct velocity as 1600 fpm.

step 2: The total air flow from the fan is

$$(12)(300\ \text{cfm}) = 3600\ \text{cfm}$$

From Fig. 36.4, the main duct diameter is 20 in. The friction loss is 0.16 in wg per 100 ft.

step 3: After the first takeoff, the flow rate between first and second is

$$3600\ \text{cfm} - (4)(300\ \text{cfm}) = 2400\ \text{cfm}$$

From Fig. 36.4, for 2400 cfm and 0.16 in wg per 100 ft, the diameter is 17.2 in and the velocity is 1440 fpm. Similarly, the diameter and the velocity for other sections are obtained and listed in the following table.

section	Q (cfm)	D (in)	v (fpm)
fan–first	3600	20	1600
first–second	2400	17.2	1440
second–third	1200	13.2	1210
first–A	1200	13.2	1210
A–B	900	12	1150
B–C	600	10.2	1030
C–D	300	7.9	860
second–E	1200	13.2	1210
E–F	900	12	1150
F–G	600	10.2	1030
G–H	300	7.9	860
I–J	900	12	1150
J–K	600	10.2	1030
K–L	300	7.9	860

(b) *step 4:* By inspection, the longest run is fan–L. From Table 36.4, the equivalent length of each bend is $14.5D$ (interpolated). For two elbows,

$$L_{e,\text{bend}} = 14.5D = (14.5)\left(\frac{20\ \text{in} + 13.2\ \text{in}}{12\ \frac{\text{in}}{\text{ft}}}\right) = 40.1\ \text{ft}$$

The equivalent length of the entire run is

	25 ft	(fan to first bend)
	35 ft	(first bend to first section)
	20 ft	(first section to second section)
	20 ft	(second section to third section)
	10 ft	(third section to point I)
	40.1 ft	(equivalent length of two bends)
	20 ft	(point I to point J)
	20 ft	(point J to point K)
	20 ft	(point K to point L)
total:	210.1 ft	

The straight-through friction loss in the longest run is

$$\left(\frac{210.1\ \text{ft}}{100\ \text{ft}}\right)\left(0.16\ \frac{\text{in wg}}{100\ \text{ft}}\right) = 0.34\ \text{in wg}$$

The fan must be able to supply a static pressure of

$$\text{SP}_{\text{fan}} = 0.34\ \text{in wg} + 0.15\ \text{in wg} = \boxed{0.49\ \text{in wg}}$$

(c) The total pressure supplied by the fan is

$$\text{TP}_{\text{fan}} = 0.49\ \text{in wg} + \left(\frac{1600\ \text{fpm}}{4005}\right)^2 = \boxed{0.65\ \text{in wg}}$$

SI Solution

(a) *step 1:* From Table 36.7, choose the main duct velocity as 1600 fpm. From the table footnote (*a*), the SI velocity is

$$\text{v}_{\text{main}} = (1600\ \text{fpm})(0.00508) = 8.1\ \text{m/s}$$

step 2: The total air flow rate from the fan is

$$(12)\left(140\ \frac{\text{L}}{\text{s}}\right) = 1680\ \text{L/s}$$

From Fig. 36.5, the main duct diameter is 500 mm. The friction loss is 1.5 Pa/m.

Ventilation

step 3: After the first takeoff, the flow rate between first and second is

$$1680\ \frac{\text{L}}{\text{s}} - (4)\left(140\ \frac{\text{L}}{\text{s}}\right) = 1120\ \text{L/s}$$

From Fig. 36.5, for 1120 L/s and 1.5 Pa/m, the diameter is 440 mm and the velocity is 7.5 m/s. Similarly, the diameter and the velocity for other sections are obtained and listed in the following table.

section	Q (L/s)	D (mm)	v (m/s)
fan–first	1680	500	8.1
first–second	1120	440	7.5
second–third	560	335	6.4
first–A	560	335	6.4
A–B	420	305	5.9
B–C	280	260	5.4
C–D	140	195	4.5
second–E	560	335	6.4
E–F	420	305	5.9
F–G	280	260	5.4
G–H	140	195	4.5
I–J	420	305	5.9
J–K	280	260	5.4
K–L	140	195	4.5

(b) *step 4:* By inspection, the longest run is fan–L. From Table 36.4, the equivalent length of each bend is 14.5D (interpolated). For two elbows,

$$\begin{aligned} L_{e,\text{bend}} &= 14.5D \\ &= (14.5)\left(\frac{500\ \text{mm} + 335\ \text{mm}}{1000\ \frac{\text{mm}}{\text{m}}}\right) \\ &= 12.1\ \text{m} \end{aligned}$$

The equivalent length of the entire run is

7.5 m	(fan to first bend)
10.5 m	(first bend to first section)
6 m	(first section to second section)
6 m	(second section to third section)
3 m	(third section to point I)
12.1 m	(equivalent length of two bends)
6 m	(point I to point J)
6 m	(point J to point K)
6 m	(point K to point L)
total: 63.1 m	

The straight-through friction loss in the longest run is

$$(63.1\ \text{m})\left(1.5\ \frac{\text{Pa}}{\text{m}}\right) = 95\ \text{Pa}$$

The fan must be able to supply a static pressure of

$$\text{SP}_{\text{fan}} = 95\ \text{Pa} + 38\ \text{Pa} = \boxed{133\ \text{Pa}}$$

(c) The total pressure supplied by the fan is

$$\begin{aligned} \text{TP}_{\text{fan}} &= 133\ \text{Pa} + (0.6)\left(8.1\ \frac{\text{m}}{\text{s}}\right)^2 \\ &= \boxed{172\ \text{Pa}} \end{aligned}$$

8. *step 1:* From Table 36.7, choose the main duct velocity as 1600 fpm.

step 2: The area and diameter of the main duct are

$$A = \frac{Q}{\text{v}} = \frac{3600\ \text{cfm}}{1600\ \text{fpm}} = 2.25\ \text{ft}^2$$

$$\begin{aligned} D &= \sqrt{\frac{4A}{\pi}} = \sqrt{\frac{(4)(2.25\ \text{ft}^2)}{\pi}}\left(12\ \frac{\text{in}}{\text{ft}}\right) \\ &= 20.3\ \text{in} \end{aligned}$$

step 3: From Table 36.4, the equivalent length of each bend is 14.5D (interpolated). For the first elbow,

$$L_{e,\text{bend}} = 14.5D = (14.5)\left(\frac{20.3\ \text{in}}{12\ \frac{\text{in}}{\text{ft}}}\right) = 24.5\ \text{ft}$$

The equivalent length of the main duct from the fan to the first takeoff and bend is

$$L = 25\ \text{ft} + 24.5\ \text{ft} + 35\ \text{ft} = 84.5\ \text{ft}$$

step 4: From Fig. 36.4, the friction loss in the main run up to the branch takeoff is approximately 0.16 in wg per 100 ft. The actual friction loss is

$$\begin{aligned} \text{FP}_{\text{main}} &= (0.16\ \text{in wg})\left(\frac{84.5\ \text{ft}}{100\ \text{ft}}\right) \\ &= 0.135\ \text{in wg} \end{aligned}$$

step 5: After the first takeoff,

$$Q = 3600\ \text{cfm} - 1200\ \text{cfm} = 2400\ \text{cfm}$$

$$L = 20\ \text{ft}$$

$$\frac{L}{Q^{0.61}} = \frac{20\ \text{ft}}{(2400\ \text{cfm})^{0.61}} = 0.173$$

From Fig. 36.9, the velocity, v_2, is 1390 fpm.

step 6: Solve for the duct size from $A = Q/\text{v}$.

$$D_2 = \sqrt{\frac{4A}{\pi}} = \sqrt{\frac{4Q}{\pi \text{v}_2}} = \sqrt{\frac{(4)(2400 \text{ cfm})}{\pi(1390 \text{ fpm})}}\left(12 \frac{\text{in}}{\text{ft}}\right) = 18 \text{ in}$$

step 7: After the second takeoff,

$$Q = 2400 \text{ cfm} - 1200 \text{ cfm} = 1200 \text{ cfm}$$
$$L = 20 \text{ ft} + 14.5D + 10 \text{ ft} = 30 \text{ ft} + 14.5D$$

Since L contains diameter, D, as an unknown, this will require an iterative procedure. Using velocity $\text{v}_1 = 1390$ fpm before the takeoff and using an iterative procedure, $D = 15$ in and $\text{v} = 980$ fpm.

step 8: Proceeding similarly, the following table is developed for the remaining sizes.

section	L (ft)	Q (cfm)	$\frac{L}{Q^{0.61}}$	v (fpm)
I–J	20	900	0.32	830
J–K	20	600	0.40	660
K–L	20	300	0.62	500

step 9: Solve for the duct size from $A = Q/\text{v}$.

$$D = \sqrt{\frac{4A}{\pi}} = \sqrt{\frac{4Q}{\pi \text{v}}}$$

$$D_{\text{I–J}} = \sqrt{\frac{(4)(900 \text{ cfm})}{\pi(830 \text{ fpm})}}\left(12 \frac{\text{in}}{\text{ft}}\right) = 14 \text{ in}$$

$$D_{\text{J–K}} = \sqrt{\frac{(4)(600 \text{ cfm})}{\pi(660 \text{ fpm})}}\left(12 \frac{\text{in}}{\text{ft}}\right) = 13 \text{ in}$$

$$D_{\text{K–L}} = \sqrt{\frac{(4)(300 \text{ cfm})}{\pi(500 \text{ fpm})}}\left(12 \frac{\text{in}}{\text{ft}}\right) = 11 \text{ in}$$

step 10: The total pressure supplied by the fan is

$$0.135 \text{ in wg} + 0.15 \text{ in wg} = \boxed{0.285 \text{ in wg}}$$

9. *Customary U.S. Solution*

(a) and (b) No dampers are needed in duct A. The area of section A is

$$A = \frac{\pi}{4}D^2 = \left(\frac{\pi}{4}\right)\left(\frac{12 \text{ in}}{12 \frac{\text{in}}{\text{ft}}}\right)^2 = 0.7854 \text{ ft}^2$$

The velocity in section A is

$$\text{v}_\text{A} = \frac{Q}{A} = \frac{3000 \text{ cfm}}{(0.7854 \text{ ft}^2)\left(60 \frac{\text{sec}}{\text{min}}\right)} = \frac{3820 \text{ fpm}}{60 \frac{\text{sec}}{\text{min}}} = 63.66 \text{ ft/sec}$$

For a four-piece elbow with a radius-to-diameter ratio of 1.5, $L_e \approx 14D$.

Assume $D = 10$ in. The total equivalent length of run C is

$$L_e = 50 \text{ ft} + 10 \text{ ft} + 10 \text{ ft} + (2)\left[(14)\left(\frac{10 \text{ in}}{12 \frac{\text{in}}{\text{ft}}}\right)\right] = 93.3 \text{ ft}$$

For any diameter, D, in inches of section C, the velocity will be

$$\text{v}_\text{C} = \frac{Q}{A} = \frac{2000 \text{ cfm}}{\left(\frac{\pi}{4}\right)\left(\frac{D}{12 \frac{\text{in}}{\text{ft}}}\right)^2\left(60 \frac{\text{sec}}{\text{min}}\right)} = 6111.5/D^2$$

From Eq. 17.28, the friction loss in section C will be

$$h_{f,\text{C}} = \frac{fLv^2}{2Dg} = \frac{(0.02)(93.3 \text{ ft})\left(\frac{6111.5}{D^2}\right)^2}{(2)\left(\frac{D}{12 \frac{\text{in}}{\text{ft}}}\right)\left(32.2 \frac{\text{ft}}{\text{sec}^2}\right)} = \frac{1.3 \times 10^7}{D^5} \quad \text{[ft of air]}$$

Ventilation

The regain between A and C will be

$$h_{\text{regain}} = R\left(\frac{V_A^2 - V_G^2}{2g}\right)$$
$$= 0.65\left[\frac{\left(63.66\ \frac{\text{ft}}{\text{sec}}\right)^2 - \left(\frac{6111.5}{D^2}\right)^2}{(2)\left(32.2\ \frac{\text{ft}}{\text{sec}^2}\right)}\right]$$
$$= 40.9 - \frac{3.77\times 10^5}{D^4} \quad \text{[ft of air]}$$

The principle of static regain is that

$$h_{f,\text{C}} = h_{\text{regain}}$$
$$\frac{1.3\times 10^7}{D^5} = 40.9 - \frac{3.77\times 10^5}{D^4}$$

By trial and error, $D \approx 13.5$ in. Since the assumed value of D is different from the calculated value, this process should be repeated.

$$L_e = 50\text{ ft} + 10\text{ ft} + 10\text{ ft} + (2)\left[(14)\left(\frac{13.5\text{ in}}{12\ \frac{\text{in}}{\text{ft}}}\right)\right]$$
$$= 101.5\text{ ft}$$

From Eq. 17.28, the friction loss in section C will be

$$h_{f,\text{C}} = \frac{fL\text{v}^2}{2Dg}$$
$$= \frac{(0.02)(101.5\text{ ft})\left(\frac{6111.5}{D^2}\right)^2}{(2)\left(\frac{D}{12\ \frac{\text{in}}{\text{ft}}}\right)\left(32.2\ \frac{\text{ft}}{\text{sec}^2}\right)}$$
$$= \frac{1.41\times 10^7}{D^5} \quad \text{[ft of air]}$$

The principle of static regain is that

$$h_{f,\text{C}} = h_{\text{regain}}$$
$$\frac{1.41\times 10^7}{D^5} = 40.9 - \frac{3.77\times 10^5}{D^4}$$

By trial and error,

$$D_\text{C} = \boxed{13.63\text{ in}\quad \text{[say 14 in]}}$$

This results in a friction loss of

$$h_{f,\text{C}} = \frac{1.41\times 10^7}{(14\text{ in})^5} = 26.2\text{ ft of air}$$

Because the regain cancels friction loss, the pressure loss from A to C is zero. No dampers are needed in duct C. A damper is needed in section B, however.

For any diameter, D, in inches in section B, the velocity will be

$$\text{v}_\text{B} = \frac{Q}{A} = \frac{1000\text{ cfm}}{\left(\frac{\pi}{4}\right)\left(\frac{D}{12\ \frac{\text{in}}{\text{ft}}}\right)^2\left(60\ \frac{\text{sec}}{\text{min}}\right)}$$
$$= \frac{3055.8}{D^2} \quad \text{[ft/sec]}$$

The friction loss in section B for Eq. 17.28 will be

$$h_{f,\text{B}} = \frac{fL\text{v}^2}{2Dg}$$
$$= \frac{(0.02)(10\text{ ft})\left(\frac{3055.8}{D^2}\right)^2}{(2)\left(\frac{D}{12\ \frac{\text{in}}{\text{ft}}}\right)\left(32.2\ \frac{\text{ft}}{\text{sec}^2}\right)}$$
$$= \frac{3.48\times 10^5}{D^5} \quad \text{[ft of air]}$$

From Eq. 36.42, the friction loss in the branch takeoff between sections A and B will be

$$\Delta\text{TP}_{\text{A-B}} = K_{\text{br}}(\text{VP}_{\text{up}})$$

At this point, assume $\text{v}_\text{B}/\text{v}_\text{A} = 1.0$. Then from Table 36.6, for a 45° angle of takeoff, $K_{\text{br}} = 0.5$.

$$\Delta\text{TP}_{\text{A-B}} = (0.5)\left(\frac{3820\text{ fpm}}{4005}\right)^2 = 0.455\text{ in wg}$$

From Eq. 36.2,

$$p_{\text{psig}} = (0.455\text{ in wg})\left(0.0361\ \frac{\text{lbf}}{\text{in}^3}\right) = 0.01643\text{ psig}$$

From Eq. 36.1,

$$h_f = \frac{p_{\text{psig}}}{\gamma} = \frac{\left(0.01643\ \frac{\text{lbf}}{\text{in}^2}\right)\left(12\ \frac{\text{in}}{\text{ft}}\right)^2}{0.075\ \frac{\text{lbf}}{\text{ft}^3}}$$
$$= 31.5\text{ ft of air}$$

The regain between A and B will be

$$h_{\text{regain}} = R\left(\frac{\text{v}_\text{A}^2 - \text{v}_\text{B}^2}{2g}\right)$$
$$= (0.65)\left[\frac{\left(63.66\ \frac{\text{ft}}{\text{sec}}\right)^2 - \left(\frac{3055.8}{D^2}\right)^2}{(2)\left(32.2\ \frac{\text{ft}}{\text{sec}^2}\right)}\right]$$
$$= 40.9\text{ ft} - \frac{9.42\times 10^4}{D^4} \quad \text{[ft of air]}$$

Set the regain equal to the loss.

$$40.9\ \text{ft} - \frac{9.42\times 10^4}{D^4} = \frac{3.48\times 10^5}{D^5} + 31.5\ \text{ft}$$

By trial and error,

$$D_{\text{B}} = \boxed{10.77\ \text{in}\quad [\text{say } 11\ \text{in}]}$$

(c)

$$\begin{aligned} \text{v}_{\text{B}} &= \frac{3055.8}{D^2} \\ &= \frac{3055.8}{(11\ \text{in})^2} = 25.25\ \text{ft/sec} \\ \frac{\text{v}_{\text{B}}}{\text{v}_{\text{A}}} &= \frac{25.25\ \frac{\text{ft}}{\text{sec}}}{63.6\ \frac{\text{ft}}{\text{sec}}} \approx 0.4 \end{aligned}$$

From Table 36.6, for a 45° angle of takeoff, $K_{\text{br}} = 0.5$. Since the value of K_{br} remains the same, there is no need to repeat the preceding procedure. For section B, the friction canceled by the regain is

$$\begin{aligned} h_{f,\text{B}} &= 31.5\ \text{ft} + \frac{3.48\times 10^5}{D^5} \\ &= 31.5\ \text{ft} + \frac{3.48\times 10^5}{(11\ \text{in})^5} \\ &= \boxed{33.7\ \text{ft of air}} \end{aligned}$$

SI Solution

(a) and (b) No dampers are needed in duct A. The area of section A is

$$A = \frac{\pi}{4}D^2 = \left(\frac{\pi}{4}\right)\left(\frac{305\ \text{mm}}{1000\ \frac{\text{mm}}{\text{m}}}\right)^2 = 0.0731\ \text{m}^2$$

The velocity in section A is

$$\text{v}_{\text{A}} = \frac{Q}{A} = \frac{\left(1400\ \frac{\text{L}}{\text{s}}\right)\left(\frac{1\ \text{m}^3}{1000\ \text{L}}\right)}{0.0731\ \text{m}^2} = 19.15\ \text{m/s}$$

For a four-piece elbow with a radius-to-diameter ratio of 1.5, $L_e \approx 14D$.

Assume $D = 250$ mm. The total equivalent length of run C is

$$\begin{aligned} L_e &= 15\ \text{m} + 3\ \text{m} + 3\ \text{m} + (2)\left[(14)\left(\frac{250\ \text{mm}}{1000\ \frac{\text{mm}}{\text{m}}}\right)\right] \\ &= 28\ \text{m} \end{aligned}$$

For any diameter, D, in mm of section C, the velocity will be

$$\begin{aligned} \text{v}_{\text{C}} = \frac{Q}{A} &= \frac{\left(900\ \frac{\text{L}}{\text{s}}\right)\left(\frac{1\ \text{m}^3}{1000\ \text{L}}\right)}{\left(\frac{\pi}{4}\right)\left(\frac{D}{1000\ \frac{\text{mm}}{\text{m}}}\right)^2} \\ &= \frac{1.146\times 10^6}{\text{D}^2}\quad [\text{m/s}] \end{aligned}$$

From Eq. 17.28, the friction loss in section C will be

$$\begin{aligned} h_{f,\text{C}} &= \frac{fL\text{v}^2}{2Dg} \\ &= \frac{(0.02)(28\ \text{m})\left(\frac{1.146\times 10^6}{D^2}\right)^2}{(2)\left(\frac{D}{1000\ \frac{\text{mm}}{\text{m}}}\right)\left(9.81\ \frac{\text{m}}{\text{s}^2}\right)} \\ &= \frac{3.75\times 10^{13}}{D^5}\quad [\text{m of air}] \end{aligned}$$

The regain between A and C will be

$$\begin{aligned} h_{\text{regain}} &= R\left(\frac{\text{v}_{\text{A}}^2 - \text{v}_{\text{C}}^2}{2g}\right) \\ &= (0.65)\left[\frac{\left(19.15\ \frac{\text{m}}{\text{s}}\right)^2 - \left(\frac{1.146\times 10^6}{D^2}\right)^2}{(2)\left(9.81\ \frac{\text{m}}{\text{s}^2}\right)}\right] \\ &= 12.15\ \text{m} - \frac{4.35\times 10^{10}}{D^4}\quad [\text{m of air}] \end{aligned}$$

The principle of static regain is that

$$\begin{aligned} h_{f,\text{C}} &= h_{\text{regain}} \\ \frac{3.75\times 10^{13}}{D^5} &= 12.15\ \text{m} - \frac{4.35\times 10^{10}}{D^4} \end{aligned}$$

By trial and error, $D \approx 335$ mm. Since the assumed value of D is different from the calculated value, this process should be repeated.

$$\begin{aligned} L_e &= 15\ \text{m} + 3\ \text{m} + 3\ \text{m} + (2)\left[(14)\left(\frac{335\ \text{mm}}{1000\ \frac{\text{mm}}{\text{m}}}\right)\right] \\ &= 30.4\ \text{m} \end{aligned}$$

Ventilation

From Eq. 17.28, the friction loss in section C will be

$$h_{f,\mathrm{C}} = \frac{fL\mathrm{v}^2}{2Dg}$$
$$= \frac{(0.02)(30.4\ \mathrm{m})\left(\frac{1.146\times 10^6}{D^2}\right)^2}{(2)\left(\frac{D}{1000\ \frac{\mathrm{mm}}{\mathrm{m}}}\right)\left(9.81\ \frac{\mathrm{m}}{\mathrm{s}^2}\right)}$$
$$= \frac{4.07\times 10^{13}}{D^5}\quad [\text{m of air}]$$

The principle of static regain is that

$$h_{f,\mathrm{C}} = h_{\text{regain}}$$
$$\frac{4.07\times 10^{13}}{D^5} = 12.15 - \frac{4.35\times 10^{10}}{D^4}$$

By trial and error, $D = 340$ mm. This results in a friction loss of

$$h_{f,\mathrm{C}} = \frac{4.07\times 10^{13}}{(340\ \mathrm{mm})^5} = 8.96\ \text{m of air}$$

Because the regain cancels this friction loss, the pressure loss from A to C is zero. No dampers are needed in duct C. A damper is needed at B, however.

For any diameter, D, in mm in section B, the velocity will be

$$\mathrm{v}_B = \frac{Q}{A} = \frac{\left(500\ \frac{\mathrm{L}}{\mathrm{s}}\right)\left(\frac{1\ \mathrm{m}^3}{1000\ \mathrm{L}}\right)}{\left(\frac{\pi}{4}\right)\left(\frac{D}{1000\ \frac{\mathrm{mm}}{\mathrm{m}}}\right)^2}$$
$$= \frac{6.366\times 10^5}{\mathrm{D}^2}\quad [\mathrm{m/s}]$$

From Eq. 17.28, the friction loss in section B will be

$$h_{f,\mathrm{B}} = \frac{fL\mathrm{v}^2}{2Dg}$$
$$= \frac{(0.02)(3\ \mathrm{m})\left(\frac{6.366\times 10^5}{D^2}\right)^2}{(2)\left(\frac{D}{1000\ \frac{\mathrm{mm}}{\mathrm{m}}}\right)\left(9.81\ \frac{\mathrm{m}}{\mathrm{s}^2}\right)}$$
$$= \frac{1.239\times 10^{12}}{\mathrm{D}^5}\quad [\text{m of air}]$$

From Eq. 36.42, the friction loss in the branch takeoff between sections A and B will be

$$\Delta\mathrm{TP}_{\mathrm{A\text{-}B}} = K_{\mathrm{br}}(\mathrm{VP}_{\mathrm{up}})$$

At this point, assume $\mathrm{v_B}/\mathrm{v_A} = 1.0$. Then from Table 36.6, for a 45° angle of takeoff, $K_{\mathrm{br}} = 0.5$.

$$\Delta\mathrm{TP}_{\mathrm{A\text{-}B}} = (0.5)(0.6)\left(19.15\ \frac{\mathrm{m}}{\mathrm{s}}\right)^2$$
$$= 110\ \mathrm{Pa}$$

From Eq. 36.1,

$$h_f = \frac{p}{\rho g} = \frac{110\ \mathrm{Pa}}{\left(1.2\ \frac{\mathrm{kg}}{\mathrm{m}^3}\right)\left(9.81\ \frac{\mathrm{m}}{\mathrm{s}^2}\right)}$$
$$= 9.34\ \text{m of air}$$

The regain between A and B will be

$$h_{\text{regain}} = R\left(\frac{\mathrm{v_A^2 - v_B^2}}{2g}\right)$$
$$= (0.65)\left[\frac{\left(19.15\ \frac{\mathrm{m}}{\mathrm{s}}\right)^2 - \left(\frac{6.366\times 10^5}{D^2}\right)^2}{(2)\left(9.81\ \frac{\mathrm{m}}{\mathrm{s}^2}\right)}\right]$$
$$= \left(12.15\ \mathrm{m} - \frac{1.343\times 10^{10}}{D^4}\right)\quad [\text{m of air}]$$

Set the regain equal to the loss.

$$12.15\ \mathrm{m} - \frac{1.343\times 10^{10}}{D^4} = \frac{1.239\times 10^{12}}{D^5} + 9.34\ \mathrm{m}$$

By trial and error,

$$D_{\mathrm{B}} = 280\ \mathrm{mm}$$
$$\mathrm{v_B} = 8.12\ \mathrm{m/s}$$

(c)

$$\frac{\mathrm{v_B}}{\mathrm{v_A}} = \frac{8.12\ \frac{\mathrm{m}}{\mathrm{s}}}{19.15\ \frac{\mathrm{m}}{\mathrm{s}}} \approx 0.4$$

From Table 36.6, for a 45° angle of takeoff, $K_{\mathrm{br}} = 0.5$. Since the value of K_{br} remains the same, there is no need to repeat the preceding procedure. For section B, the friction canceled by the regain is

$$h_{f,\mathrm{B}} = 9.34\ \mathrm{m} + \frac{1.239\times 10^{12}}{D^5}$$
$$= 9.34\ \mathrm{m} + \frac{1.239\times 10^{12}}{(280\ \mathrm{mm})^5}$$
$$= \boxed{10.06\ \text{m of air}}$$

37 Fuels and Combustion

PRACTICE PROBLEMS

1. Methane with a heating value of 24,000 Btu/lbm (55.8 MJ/kg) is burned in a furnace with a 50% efficiency. How much water can be heated from 60 to 200°F (15 to 95°C) when 7 ft^3 (200 L) at gas industry standard conditions are burned?

2. 15 lbm/hr (6.8 kg/h) of propane (C_3H_8) are burned stoichiometrically in air. What volume of dry carbon dioxide (CO_2) is formed after cooling to 70°F (21°C) and 14.7 psia (101 kPa)?

3. In a particular installation, 30% by volume excess air at 15 psia (103 kPa) and 100°F (40°C) is needed for the combustion of methane. How much nitrogen passes through the furnace if methane is burned at the rate of 4000 ft^3/hr (31 L/s)?

4. How much air is required to completely burn one unit mass of a fuel that is 84% carbon, 15.3% hydrogen, 0.3% sulfur, and 0.4% nitrogen by weight?

5. Propane (C_3H_8) is burned with 20% excess air. What is the gravimetric percentage of carbon dioxide in the flue gas?

6. The ultimate analysis of a coal is 80% carbon, 4% hydrogen, 2% oxygen, and the rest ash. The flue gases are 60°F and 14.7 psia (15.6°C and 101.3 kPa) when sampled, and are 80% nitrogen, 12% carbon dioxide, 7% oxygen, and 1% carbon monoxide by volume. How much air is required to burn 1 lbm (1 kg) of coal under these conditions?

7. What is the approximate heating value of an oil with a specific gravity of 40° API?

8. The ultimate analysis of a coal is 75% carbon, 5% hydrogen, 3% oxygen, 2% nitrogen, and the rest ash. Atmospheric air is 60°F (16°C) and standard pressure. (a) What is the theoretical temperature of the combustion products? (b) Estimate the actual temperature of the combustion products. State your assumptions.

9. A fuel oil has the following ultimate analysis: 85.43% carbon, 11.31% hydrogen, 2.7% oxygen, 0.34% sulfur, and 0.22% nitrogen. The oil is burned with 60% excess air. Evaluate flue gas volumes at 600°F (320°C). (a) What volume of wet flue gases will be produced? (b) What volume of dry flue gases will be produced? (c) What will be the volumetric fraction of carbon dioxide?

10. The ultimate analysis of a coal is 51.45% carbon, 16.69% ash, 15.71% moisture, 7.28% oxygen, 4.02% hydrogen, 3.92% sulfur, and 0.93% nitrogen. 15,395 lbm (6923 kg) of the coal are burned, and 2816 lbm (1267 kg) of ash containing 20.9% carbon (by weight) are recovered. 13.3 lbm (kg) of dry gases are produced per pound (kilogram) of fuel burned. How much air was used per unit mass of fuel?

11. A coal is 65% carbon by weight. During combustion, 3% of the coal is lost in the ash pit. Combustion uses 9.87 lbm (kg) of air per pound (kg) of fuel. The flue gas analysis is 81.5% nitrogen, 9.5% carbon dioxide, and 9% oxygen. What is the percentage of excess air by mass?

12. A coal has an ultimate analysis of 67.34% carbon, 4.91% oxygen, 4.43% hydrogen, 4.28% sulfur, 1.08% nitrogen, and the rest ash. 3% of the carbon is lost during combustion. The flue gases are 81.9% nitrogen, 15.5% carbon dioxide, 1.6% carbon monoxide, and 1% oxygen by volume. What is the heat loss due to the formation of carbon monoxide?

13. A natural gas is 93% methane, 3.73% nitrogen, 1.82% hydrogen, 0.45% carbon monoxide, 0.35% oxygen, 0.25% ethylene, 0.22% carbon dioxide, and 0.18% hydrogen sulfide by volume. The gas is burned with 40% excess air. Atmospheric air is 60°F and standard atmospheric pressure. (a) What is the gas density? (b) What are the theoretical air requirements? (c) What is the wet flue gas analysis? (d) What is the dry flue gas analysis?

14. Coal enters a steam generator at 73°F (23°C). The coal has an ultimate analysis of 78.42% carbon, 8.25% oxygen, 5.68% ash, 5.56% hydrogen, 1.09% nitrogen, and 1.0% sulfur. The coal's heating value is 14,000 Btu/lbm (32.6 MJ/kg). 7.03% of the coal's weight is lost in the ash pit. The ash contains 31.5% carbon. Air at 67°F (19°C) wet bulb and 73°F (23°C) dry bulb is supplied. The flue gases consist of 80.08% nitrogen, 14.0% carbon dioxide, 5.5% oxygen, and 0.42% carbon monoxide. The flue gases are at a temperature of 575°F (300°C). Saturated water at 212°F (100°C) and 1.0 atm enters the steam generator. 11.12 lbm (kg) of water are evaporated in the boiler per pound (kilogram) of dry coal consumed. What is the complete heat balance, enumerating all combustion losses in the furnace?

15. Coal has a gravimetric analysis of 83% carbon, 5% hydrogen, 5% oxygen, and 7% noncombustible matter. 10% of the fired coal mass, including all of the noncombustible matter, is recovered in the ash pit. 26 lbm (26 kg) of air at 70°F (21°C) and 1 atmosphere are used per lbm (kg) of coal burned. When loaded into the furnace, the coal temperature is 60°F (15.6°C). The stack gas from coal combustion has a temperature of 550°F (290°C). What percentage of the combustion heat is carried away by the stack gases?

16. (*Time limit: one hour*) A utility boiler burns coal with an ultimate analysis of 76.56% carbon, 7.7% oxygen, 6.1% silicon, 5.5% hydrogen, 2.44% sulfur, and 1.7% nitrogen. 410 lbm/hr of refuse are removed with a composition of 30% carbon and 0% sulfur. All sulfur and the remaining carbon is burned. The power plant has the following characteristics.

- coal feed rate: 15,300 lbm/hr
- electric power rating: 17 MW
- generator efficiency: 95%
- steam generator efficiency: 86%
- cooling water rate: 225 ft^3/sec

(a) What is the emission rate of solid particulates in lbm/hr? (b) How much sulfur dioxide is produced per hour? (c) What is the temperature rise of the cooling water? (d) What efficiency must the flue gas particulate collectors have in order to meet a limit of 0.1 lbm of particulates per million Btu per hour (0.155 kg/MW)?

17. (*Time limit: one hour*) 250 SCFM (118 L/s) of propane are mixed with an oxidizer consisting of 60% oxygen and 40% nitrogen by volume in a proportion allowing 40% excess oxygen by weight. The maximum velocity for the two reactants when combined is 400 ft/min (2 m/s) at 14.7 psia (101 kPa) and 80°F (27°C). Maximum velocity for the products is 800 ft/min (4 m/s) at 8 psia (55 kPa) and 460°F (240°C). (a) Size the inlet pipe. (b) Size the stack. (c) What is the actual flow of oxygen? (d) What is the volume of flue gases? (e) What is the dew point of the flue gases?

18. (*Time limit: one hour*) An industrial process uses hot gas at 3600°R (1980°C) and 14.7 psia (101 kPa). It is proposed that propane be burned stoichiometrically in a mixture of nitrogen and oxygen. After passing through the process, gas will be exhausted through a duct, being cooled slowly to 100°F (38°C) and 14.7 psia (101 kPa) before discharge. The following data are available.

- The enthalpies of formation (at the standard reference temperature) are

$$C_3H_8(g) \quad \Delta H_f = +28{,}800 \text{ Btu/lbmol } (+67.0 \text{ GJ/kmol})$$

$$CO_2(g) \quad \Delta H_f = -169{,}300 \text{ Btu/lbmol } (-393.8 \text{ GJ/kmol})$$

$$H_2O(g) \quad \Delta H_f = -104{,}040 \text{ Btu/lbmol } (-242 \text{ GJ/kmol})$$

- The enthalpy increases from the standard reference temperature to 3600°R (1980°C) are

CO_2 :	39,791 Btu/lbmol	(92.6 GJ/kmol)
H_2O :	31,658 Btu/lbmol	(73.6 GJ/kmol)
N_2 :	24,471 Btu/lbmol	(56.9 GJ/kmol)

(a) What are the proportions by weight of the oxygen and nitrogen? (b) What is the amount of liquid water, if any, present in the stack gas? (c) How much water is removed from the stack gas?

19. (*Time limit: one hour*) An internal combustion engine is normally fuel-injected with gasoline (C_8H_{18}; lower heating value of 23,200 Btu/lbm; 54 MJ/kg). It is desired to switch to alcohol (C_2H_5OH; lower heating value of 11,930 Btu/lbm; 27.7 MJ/kg). Minimal changes to the engine are to be made. The indicated and mechanical efficiencies are unchanged. Stoichiometric air/fuel ratios are being used for initial calculations. (a) If the power output is to be unchanged, what must be the percentage change in specific fuel consumption? (b) If the fuel-injection velocity is to be unchanged, what is the percentage change in jet size? (c) If no changes are made to the engine, what will be the percentage change in power output?

SOLUTIONS

1. *Customary U.S. Solution*

For standard conditions, from Table 14.3 for the gas industry,

$$T = 60°\text{F} + 460 = 520°\text{R}$$
$$p = 14.73 \text{ psia}$$

From Eq. 29.48,

$$R = \frac{R^*}{\text{MW}}$$

From Table 29.7, for methane, MW = 16.043 lbm/lbmol. From Table 29.6, R^* = 1545.33 ft-lbf/lbmol-°R.

$$R = \frac{1545.33 \ \frac{\text{ft-lbf}}{\text{lbmol-°R}}}{16.043 \ \frac{\text{lbm}}{\text{lbmol}}} = 96.32 \text{ ft-lbf/lbm-°R}$$

From Eq. 29.47,

$$\begin{aligned} m &= \frac{pV}{RT} \\ &= \frac{\left(14.73 \ \frac{\text{lbf}}{\text{in}^2}\right)\left(144 \ \frac{\text{in}^2}{\text{ft}^2}\right)(7 \text{ ft}^3)}{\left(96.32 \ \frac{\text{ft-lbf}}{\text{lbm-°R}}\right)(520°\text{R})} \\ &= 0.296 \text{ lbm} \end{aligned}$$

The energy available from methane is

$$\begin{aligned} Q &= m(\text{HHV})\eta \\ &= (0.296 \text{ lbm})\left(24{,}000 \ \frac{\text{Btu}}{\text{lbm}}\right)(0.5) \\ &= 3552 \text{ Btu} \end{aligned}$$

This energy is used by water to heat from 60°F to 200°F.

$$Q = m_{\text{water}} c_p (T_2 - T_1)$$

$$m_{\text{water}} = \frac{3552 \text{ Btu}}{\left(1 \ \frac{\text{Btu}}{\text{lbm-°F}}\right)(200°\text{F} - 60°\text{F})} = \boxed{25.37 \text{ lbm}}$$

SI Solution

For standard conditions from Table 14.3 for the gas industry,

$$T = (60°\text{F} + 460)\left(\frac{1 \text{ K}}{1.8°\text{R}}\right) = 288.89\text{K}$$
$$p = (14.73 \text{ psia})\left(\frac{101.325 \text{ kPa}}{14.696 \text{ psia}}\right) = 101.56 \text{ kPa}$$

From Eq. 29.48,

$$R = \frac{R^*}{\text{MW}}$$

From Table 29.7, for methane, MW = 16.043 kg/kmol.

From Table 29.6, R^* = 8314.3 J/kmol·K.

$$R = \frac{8314.3 \ \frac{\text{J}}{\text{kmol·K}}}{16.043 \ \frac{\text{kg}}{\text{kmol}}} = 518.25 \text{ J/kg·K}$$

From Eq. 29.47,

$$\begin{aligned} m &= \frac{pV}{RT} \\ &= \frac{(101.56 \text{ kPa})\left(1000 \ \frac{\text{Pa}}{\text{kPa}}\right)(200 \text{ L})\left(\frac{1 \text{ m}^3}{1000 \text{ L}}\right)}{\left(518.25 \ \frac{\text{J}}{\text{kg·K}}\right)(288.89\text{K})} \\ &= 0.136 \text{ kg} \end{aligned}$$

The energy available from methane is

$$\begin{aligned} Q &= m(\text{HHV})\eta \\ &= (0.136 \text{ kg})\left(55.8 \ \frac{\text{MJ}}{\text{kg}}\right)\left(1000 \ \frac{\text{kJ}}{\text{MJ}}\right)(0.5) \\ &= 3794 \text{ kJ} \end{aligned}$$

This energy is used by water to heat it from 15°C to 95°C.

$$Q = m_{\text{water}} c_p (T_2 - T_1)$$

$$m_{\text{water}} = \frac{3794 \text{ kJ}}{\left(4.1868 \ \frac{\text{kJ}}{\text{kg·C}}\right)(95°\text{C} - 15°\text{C})} = \boxed{11.33 \text{ kg}}$$

2. *Customary U.S. Solution*

From Table 37.7,

	C_3H_8	+ $5O_2$	$\longrightarrow$ $3CO_2$	+ $4H_2O$
MW	44.097	(5)(32)	(3)(44.011)	
	44.097	160	132.033	

Combustion

The amount of carbon dioxide produced is 132.033 lbm/44.097 lbm propane. For 15 lbm/hr of propane, the amount of carbon dioxide produced is

$$\left(\frac{132.033\ \text{lbm}}{44.097\ \text{lbm}}\right)\left(15\ \frac{\text{lbm}}{\text{hr}}\right) = 44.91\ \text{lbm/hr}$$

From Eq. 29.48,

$$R = \frac{R^*}{\text{MW}} = \frac{1545.33\ \frac{\text{ft-lbf}}{\text{lbmol-°R}}}{44.097\ \frac{\text{lbm}}{\text{lbmol}}}$$
$$= 35.04\ \text{ft-lbf/lbm-°R}$$
$$T = 70°\text{F} + 460 = 530°\text{R}$$

From Eq. 29.47,

$$V = \frac{mRT}{p}$$
$$= \frac{\left(44.91\ \frac{\text{lbm}}{\text{hr}}\right)\left(35.04\ \frac{\text{ft-lbf}}{\text{lbm-°R}}\right)(530°\text{R})}{\left(14.7\ \frac{\text{lbf}}{\text{in}^2}\right)\left(144\ \frac{\text{in}^2}{\text{ft}^2}\right)}$$
$$= \boxed{394\ \text{ft}^3/\text{hr}}$$

SI Solution

From Table 37.7,

	C_3H_8	+	$5O_2$	$\longrightarrow$	$3CO_2$	+	$4H_2O$
MW	44.097		(5)(32)		(3)(44.011)		
	44.097		160		132.033		

The amount of carbon dioxide produced is 132.033 kg/44.097 kg propane. For 6.8 kg/h of propane, the amount of carbon dioxide produced is

$$\left(\frac{132.033\ \text{kg}}{44.097\ \text{kg}}\right)\left(6.8\ \frac{\text{kg}}{\text{h}}\right) = 20.36\ \text{kg/h}$$

From Eq. 29.48,

$$R = \frac{R^*}{\text{MW}} = \frac{8314.3\ \frac{\text{J}}{\text{kmol·K}}}{44.097\ \frac{\text{kg}}{\text{kmol}}}$$
$$= 188.55\ \text{J/kg·K}$$
$$T = 21°\text{C} + 273 = 294\text{K}$$

From Eq. 29.47,

$$V = \frac{mRT}{p}$$
$$= \frac{\left(20.36\ \frac{\text{kg}}{\text{h}}\right)\left(188.55\ \frac{\text{J}}{\text{kg·K}}\right)(294\text{K})}{(101\ \text{kPa})\left(1000\ \frac{\text{Pa}}{\text{kPa}}\right)}$$
$$= \boxed{11.17\ \text{m}^3/\text{h}}$$

3. The balanced chemical reaction equation is from Table 37.7.

$$CH_4 + 2O_2 \longrightarrow CO_2 + 2H_2O$$

With 30% excess air and considering there are 3.773 volumes of nitrogen for every volume of oxygen (Table 37.6), the reaction equation is

$$CH_4 + (1.3)(2)O_2 + (1.3)(2)(3.773)N_2 \longrightarrow CO_2 + 2H_2O + (1.3)(2)(3.773)N_2 + 0.6O_2$$

$$CH_4 + 2.6O_2 + 9.81N_2 \longrightarrow CO_2 + 2H_2O + 9.81N_2 + 0.6O_2$$

Customary U.S. Solution

The volume of nitrogen that accompanies 4000 ft^3/hr of entering methane is

$$V_{N_2} = \left(\frac{9.81\ \text{ft}^3\ N_2}{1\ \text{ft}^3\ CH_4}\right)\left(4000\ \frac{\text{ft}^3}{\text{hr}}\ CH_4\right)$$
$$= 39{,}240\ \text{ft}^3\ N_2/\text{hr}$$

This is the "partial volume" of nitrogen in the input stream.

From Eq. 29.48, for nitrogen,

$$R = \frac{R^*}{\text{MW}} = \frac{1545.33\ \frac{\text{ft-lbf}}{\text{lbmol-°R}}}{28.016\ \frac{\text{lbm}}{\text{lbmol}}}$$
$$= 55.16\ \text{ft-lbf/lbm-°R}$$

The absolute temperature is

$$T = 100°\text{F} + 460 = 560°\text{R}$$

From Eq. 29.47,

$$m_{N_2} = \frac{p_{N_2}V_{N_2}}{RT}$$
$$= \frac{\left(15\ \frac{\text{lbf}}{\text{in}^2}\right)\left(144\ \frac{\text{in}^2}{\text{ft}^2}\right)\left(39{,}240\ \frac{\text{ft}^3}{\text{hr}}\right)}{\left(55.16\ \frac{\text{ft-lbf}}{\text{lbm-°R}}\right)(560°\text{R})}$$
$$= \boxed{2744\ \text{lbm/hr}}$$

SI Solution

The volume of nitrogen that accompanies 31 L/s of entering methane is

$$\left(\frac{9.81\ \text{m}^3\ N_2}{1\ \text{m}^3\ CH_4}\right)\left(31\ \frac{\text{L}}{\text{s}}\right)\left(\frac{1\ \text{m}^3}{1000\ \text{L}}\right) = 0.3041\ \text{m}^3/\text{s}$$

This is the "partial volume" of nitrogen in the input stream.

From Eq. 29.48 for nitrogen,

$$R = \frac{R^*}{\text{MW}} = \frac{8314.3\ \frac{\text{J}}{\text{kmol·K}}}{28.016\ \frac{\text{kg}}{\text{kmol}}} = 296.8\ \text{J/kg·K}$$

The absolute temperature is

$$T = 40°\text{C} + 273 = 313\text{K}$$

From Eq. 29.47,

$$m_{N_2} = \frac{p_{N_2}V_{N_2}}{RT} = \frac{(103\ \text{kPa})\left(1000\ \frac{\text{Pa}}{\text{kPa}}\right)\left(0.3041\ \frac{\text{m}^3}{\text{s}}\right)}{\left(296.8\ \frac{\text{J}}{\text{kg·K}}\right)(313\text{K})} = \boxed{0.337\ \text{kg/s}}$$

4. From Table 37.7, combustion reactions are

$$\begin{array}{lccc} & \text{C} & + \ \text{O}_2 & \longrightarrow \text{CO}_2 \\ \text{MW} & 12 & 32 & \end{array}$$

The mass of oxygen required per unit mass of carbon is

$$\frac{32}{12} = 2.67$$

$$\begin{array}{lccc} & 2\text{H}_2 & + \ \text{O}_2 & \longrightarrow 2\text{H}_2\text{O} \\ \text{MW} & (2)(2) & 32 & \end{array}$$

The mass of oxygen required per unit mass of hydrogen is

$$\frac{32}{(2)(2)} = 8.0$$

$$\begin{array}{lccc} & \text{S} & + \ \text{O}_2 & \longrightarrow \text{SO}_2 \\ \text{MW} & 32.1 & 32 & \end{array}$$

The mass of oxygen required per unit mass of sulfur is

$$\frac{32}{32.1} = 1.0$$

Nitrogen does not burn.

The mass of oxygen required per unit mass of fuel is

$$(0.84)(2.67) + (0.153)(8) + (0.003)(1) = 3.47\ \text{unit of mass of O}_2\text{/unit mass fuel}$$

From Table 37.6, air is 0.2315 O_2/unit mass, so the air required is

$$\frac{3.47}{0.2315} = \boxed{15.0\ \text{lbm air/lbm fuel}}$$

[Customary U.S. Solution]

$$= \boxed{15.0\ \text{kg air/kg fuel}}$$

[SI Solution]

5. The balanced chemical reaction equation is

$$C_3H_8 + 5O_2 \longrightarrow 3CO_2 + 4H_2O$$

With 20% excess air, the oxygen volume is (1.2)(5) = 6.

$$C_3H_8 + 6O_2 \longrightarrow 3CO_2 + 4H_2O + O_2$$

From Table 37.6, there are 3.773 volumes of nitrogen for every volume of oxygen.

$$(6)(3.773) = 22.6$$

$$C_3H_8 + 6O_2 + 22.6\,N_2 \longrightarrow 3CO_2 + 4H_2O + O_2 + 22.6\,N_2$$

The percentage of carbon dioxide by weight in flue gas is

$$G_{CO_2} = \frac{(3)(44.011)}{(3)(44.011) + (4)(18.016) + 32 + (22.6)(28.016)} = \boxed{0.152\ \ (15.2\%)}$$

6. The actual air/fuel ratio can be estimated from the flue gas analysis and the fraction of carbon in fuel.

From Eq. 37.9,

$$\frac{m_{\text{air}}}{m_{\text{fuel}}} = \frac{3.04B_{N_2}G_C}{B_{CO_2} + B_{CO}} = \frac{(3.04)(80\%)(0.80)}{12\% + 1\%} = 14.97$$

The air required to burn 1 lbm of coal is $\boxed{14.97\ \text{lbm.}}$

(Customary U.S. Solution)

The air required to burn 1 kg of coal is $\boxed{14.97\ \text{kg.}}$

(SI Solution)

Alternative Solution

The use of Eq. 37.9 obscures the process of finding the air/fuel ratio. (The SI solution is similar but is not presented here.)

step 1: Find the mass of oxygen in the stack gases.

$$R_{CO_2} = 35.11 \text{ ft-lbf/lbm-°R}$$

$$R_{CO} = 55.17 \text{ ft-lbf/lbm-°R}$$

$$R_{O_2} = 48.29 \text{ ft-lbf/lbm-°R}$$

The partial densities are

$$\rho_{CO_2} = \frac{p}{RT} = \frac{(0.12)\left(14.7\ \frac{\text{lbf}}{\text{in}^2}\right)\left(144\ \frac{\text{in}^2}{\text{ft}^2}\right)}{\left(35.11\ \frac{\text{ft-lbf}}{\text{lbm-°R}}\right)(60°\text{F} + 460)}$$
$$= 1.391 \times 10^{-2} \text{ lbm/ft}^3$$

$$\rho_{CO} = \frac{(0.01)\left(14.7\ \frac{\text{lbf}}{\text{in}^2}\right)\left(144\ \frac{\text{in}^2}{\text{ft}^2}\right)}{\left(55.11\ \frac{\text{ft-lbf}}{\text{lbm-°R}}\right)(60°\text{F} + 460)}$$
$$= 7.387 \times 10^{-4} \text{ lbm/ft}^3$$

$$\rho_{O_2} = \frac{(0.07)\left(14.7\ \frac{\text{lbf}}{\text{in}^2}\right)\left(144\ \frac{\text{in}^2}{\text{ft}^2}\right)}{\left(48.29\ \frac{\text{ft-lbf}}{\text{lbm-°R}}\right)(60°\text{F} + 460)}$$
$$= 5.901 \times 10^{-3} \text{ lbm/ft}^3$$

The fraction of oxygen in the three components is

$$CO_2\text{:}\ \frac{32.0}{44} = 0.7273$$

$$CO\text{:}\ \frac{16}{28} = 0.5714$$

$$O_2\text{:}\ 1.00$$

In 100 ft^3 of stack gases, the total oxygen mass will be

$$(100 \text{ ft}^3)\Bigg[(0.7273)\left(1.391 \times 10^{-2}\ \frac{\text{lbm}}{\text{ft}^3}\right) + (0.5714)\left(7.387 \times 10^{-4}\ \frac{\text{lbm}}{\text{ft}^3}\right) + (1.00)\left(5.901 \times 10^{-3}\ \frac{\text{lbm}}{\text{ft}^3}\right)\Bigg] = 1.644 \text{ lbm}$$

step 2: Since air is 23.15% oxygen by weight, the mass of air per 100 ft^3 of stack gases is

$$\frac{1.644 \text{ lbm}}{0.2315} = 7.102 \text{ lbm}$$

step 3: Find the mass of carbon in the stack gases by a similar process.

$$CO_2\text{:}\ \frac{12}{44} = 0.2727$$

$$CO\text{:}\ \frac{12}{28} = 0.4286$$

$$(100 \text{ ft}^3)\Bigg[(0.2727)\left(1.391 \times 10^{-2}\ \frac{\text{lbm}}{\text{ft}^3}\right) + (0.4286)\left(7.387 \times 10^{-4}\ \frac{\text{lbm}}{\text{ft}^3}\right)\Bigg] = 0.4110 \text{ lbm}$$

step 4: The coal is 80% carbon, so the air per lbm of coal for combustion of the carbon is

$$\left(\frac{0.80\ \frac{\text{lbm carbon}}{\text{lbm coal}}}{0.4110\ \frac{\text{lbm carbon}}{100 \text{ ft}^3}}\right)(7.102 \text{ lbm}) = 13.824 \text{ lbm air/lbm coal}$$

step 5: This does not include air to burn hydrogen, since Orsat is a dry analysis.

The theoretical air for the hydrogen is given by Eq. 37.6.

$$R_{a/f,\text{H}} = \left(34.5\ \frac{\text{lbm}}{\text{lbm}}\right)\left(G_\text{H} - \frac{G_\text{O}}{8}\right)$$
$$= \left(34.05\ \frac{\text{lbm}}{\text{lbm}}\right)\left(0.04 - \frac{0.02}{8}\right)$$
$$= 1.277 \text{ lbm air/lbm fuel}$$

Ignoring any excess air for the hydrogen, the total air per pound of coal is

$$13.824\ \frac{\text{lbm air}}{\text{lbm coal}} + 1.277\ \frac{\text{lbm air}}{\text{lbm coal}} = \boxed{15.10 \text{ lbm air/lbm coal}}$$

7. From Eq. 14.11,

$$\text{SG} = \frac{141.5}{°\text{API} + 131.5}$$
$$= \frac{141.5}{40 + 131.5} = 0.825$$

Customary U.S. Solution

From Eq. 37.18(b),

$$\begin{aligned}\text{HHV} &= 22{,}320 - (3780)(\text{SG})^2\\ &= 22{,}320 - (3780)(0.825)^2\\ &= \boxed{19{,}747\ \text{Btu/lbm}}\end{aligned}$$

SI Solution

From Eq. 37.18(a),

$$\begin{aligned}\text{HHV} &= 51.92 - (8.792)(\text{SG})^2\\ &= 51.92 - (8.792)(0.825)^2\\ &= \boxed{45.94\ \text{MJ/kg}}\end{aligned}$$

8. *Customary U.S. Solution*

(a) *step 1:* From Eq. 37.16(b), substituting the lower heating value of hydrogen from App. 37.A, the lower heating value of coal is

$$\begin{aligned}\text{LHV} &= 14{,}093G_\text{C} + (51{,}623)\left(G_\text{H} - \frac{G_\text{O}}{8}\right)\\ &\quad + 3983G_\text{S}\\ &= (14{,}093)(0.75)\\ &\quad + (51{,}623)\left(0.05 - \frac{0.03}{8}\right) + (3983)(0)\\ &= 12{,}957\ \text{Btu/lbm}\end{aligned}$$

step 2: The gravimetric analysis of 1 lbm of coal is

carbon: 0.75 lbm

$$\begin{aligned}\text{free hydrogen: } &\left(G_\text{H,total} - \frac{G_O}{8}\right)\\ &= 0.05 - \frac{0.03}{8}\\ &= 0.0463\end{aligned}$$

The ratio of the molecular weight of water (18) to that of a hydrogen molecule (2) is 9.

water: $(9)(0.05 - 0.0463) = 0.0333$

nitrogen: 0.02

step 3: From Table 37.8, the theoretical stack gases per lbm coal for 0.75 lbm of coal are

$$\text{CO}_2 = (0.75)(3.667\ \text{lbm}) = 2.750\ \text{lbm}$$

$$\text{N}_2 = (0.75)(8.883\ \text{lbm}) = 6.662\ \text{lbm}$$

All products are calculated similarly (as the following table summarizes).

	CO_2	N_2	H_2O
from C:	2.750 lbm	6.662 lbm	
from H_2:		1.217 lbm	0.414 lbm
from H_2O:			0.0333 lbm
from O_2:	shows up in	CO_2 and H_2	
from N_2:		0.02 lbm	
total:	2.750 lbm	7.899 lbm	0.4473 lbm

step 4: Assume the combustion gases leave at 1000°F.

$$T_\text{ave} = \left(\tfrac{1}{2}\right)(60°\text{F} + 1000°\text{F}) = 530°\text{F}$$

$$T_\text{ave} = 530°\text{F} + 460 = 990°\text{R}$$

The specific heat values are given in Table 37.1.

$$\begin{aligned}c_{p\text{CO}_2} &= 0.251\ \text{Btu/lbm-°F}\\ c_{p\text{N}_2} &= 0.255\ \text{Btu/lbm-°F}\\ c_{p\text{H}_2\text{O}} &= 0.475\ \text{Btu/lbm-°F}\end{aligned}$$

The energy required to raise the combustion products for 1 lbm of coal 1°F is

$$\begin{aligned}&m_{\text{CO}_2}c_{p\text{CO}_2} + m_{\text{N}_2}c_{p\text{N}_2} + m_{\text{H}_2\text{O}}c_{p\text{H}_2\text{O}}\\ &= (2.750\ \text{lbm})\left(0.251\ \frac{\text{Btu}}{\text{lbm-°F}}\right)\\ &\quad + (7.899\ \text{lbm})\left(0.255\ \frac{\text{Btu}}{\text{lbm-°F}}\right)\\ &\quad + (0.4473\ \text{lbm})\left(0.475\ \frac{\text{Btu}}{\text{lbm-°F}}\right)\\ &= 2.92\ \text{Btu/°F}\end{aligned}$$

step 5: Assuming all combustion heat goes into the stack gases, the temperature is given by Eq. 37.19.

$$T_\text{max} = T_i + \frac{\text{lower heat of combustion}}{\text{energy required}}$$

$$= 60°\text{F} + \frac{12{,}957\ \frac{\text{Btu}}{\text{lbm}}}{2.92\ \frac{\text{Btu}}{\text{lbm-°F}}} = \boxed{4497°\text{F}}$$

[unreasonable]

(b) *step 6:* In reality, assuming 40% excess air and 75% of heat absorbed by the boiler, the excess air (based on 76.85% N_2 by weight) is

$$(0.40)\left(\frac{7.899\ \text{lbm}}{0.7685}\right) = 4.111\ \text{lbm}$$

From Table 37.1, $c_{p_\text{air}} = 0.249$ Btu/lbm-°R. Therefore,

$$T_\text{max} = 60°\text{F} + \frac{\left(12{,}957\ \frac{\text{Btu}}{\text{lbm}}\right)(1 - 0.75)}{2.92\ \frac{\text{Btu}}{°\text{F}} + (4.111\ \text{lbm})\left(0.249\ \frac{\text{Btu}}{\text{lbm-°R}}\right)}$$

$$= \boxed{881°\text{F}}$$

Combustion

SI Solution

(a) *step 1:* From Eq. 37.16(a), and substituting the lower heating value of hydrogen from App. 37.A, the lower heating value of coal is

$$\begin{aligned}\text{LHV} &= 32.78G_C + \left(51{,}623\ \frac{\text{Btu}}{\text{lbm}}\right)\left(2.326\ \frac{\text{kJ-lbm}}{\text{kg-Btu}}\right) \\ &\quad \times \left(\frac{1\ \text{MJ}}{1000\ \text{kJ}}\right)\left(G_H - \frac{G_O}{8}\right) \\ &\quad + 9.264\ \text{G}_S \\ &= (32.78)(0.75) + (120.1)\left(0.05 - \frac{0.03}{8}\right) \\ &\quad + (9.264)(0) \\ &= 30.14\ \text{MJ/kg}\end{aligned}$$

Steps 2 and 3 are the same as for the U.S. solution except that all masses are in kg.

step 4: Assume the combustion gases leave at 550°C.

$$\begin{aligned}T_{\text{ave}} &= \left(\tfrac{1}{2}\right)(16^\circ\text{C} + 550^\circ\text{C}) + 273 \\ &= 283^\circ\text{C} + 273 = 556\text{K}\end{aligned}$$

Specific heat values are given in Table 37.1. Using the footnote for SI units,

$$\begin{aligned}c_{p_{CO_2}} &= 1.051\ \text{kJ/kg·K} \\ c_{p_{N_2}} &= 1.068\ \text{kJ/kg·K} \\ c_{p_{H_2O}} &= 1.989\ \text{kJ/kg·K}\end{aligned}$$

The energy required to raise the combustion products for 1 kg of coal 1°C is

$$\begin{aligned}&m_{CO_2}c_{p_{CO_2}} + m_{N_2}c_{p_{N_2}} + m_{H_2O}c_{p_{H_2O}} \\ &= (2.750\ \text{kg})\left(1.051\ \frac{\text{kJ}}{\text{kg·K}}\right) \\ &\quad + (7.899\ \text{kg})\left(1.068\ \frac{\text{kJ}}{\text{kg·K}}\right) \\ &\quad + (0.4473\ \text{kg})\left(1.989\ \frac{\text{kJ}}{\text{kg·K}}\right) \\ &= 12.22\ \text{kJ/K}\end{aligned}$$

step 5: Assuming all combustion heat goes into the stack gases, the temperature is given by Eq. 37.19.

$$\begin{aligned}T_{\text{max}} &= T_i + \frac{\text{lower heat of combustion}}{\text{energy required}} \\ &= 16^\circ\text{C} + \frac{\left(30.14\ \frac{\text{MJ}}{\text{kg}}\right)\left(1000\ \frac{\text{kJ}}{\text{MJ}}\right)}{12.22\ \frac{\text{kJ}}{\text{K}}} \\ &= \boxed{2482^\circ\text{C}} \quad \text{[unreasonable]}\end{aligned}$$

(b) *step 6:* In reality, assuming 40% excess air and 75% of heat absorbed by the boiler, the excess air (based on 76.85% N_2 by weight) is

$$(0.40)\left(\frac{7.899\ \text{kg}}{0.7685}\right) = 4.111\ \text{kg}$$

From Table 37.1, using the table footnote, $c_{p_{\text{air}}} = 10\,043$ kJ/kg·K.

Therefore,

$$\begin{aligned}T_{\text{max}} &= 16^\circ\text{C} + \frac{\left(30.14\ \frac{\text{MJ}}{\text{kg}}\right) \times \left(1000\ \frac{\text{kJ}}{\text{MJ}}\right)(1 - 0.75)}{12.22\ \frac{\text{kJ}}{\text{K}} + (4.111\ \text{kg})\left(1.043\ \frac{\text{kJ}}{\text{kg·K}}\right)} \\ &= \boxed{472.5^\circ\text{C}}\end{aligned}$$

9. Assume the oxygen is in the form of moisture in the fuel.

The available oxygen is

$$\begin{aligned}G_{\text{H,free}} &= G_H - \frac{G_O}{8} \\ &= 0.1131 - \frac{0.027}{8} \\ &= 0.1097\end{aligned}$$

Customary U.S. Solution

step 1: From Table 37.8, find the stoichiometric oxygen required per lbm of fuel oil.

$$\begin{aligned}\text{C} \longrightarrow \text{CO}_2\text{: O}_2\ \text{required} &= (0.8543)(2.667\ \text{lbm}) \\ &= 2.2784\ \text{lbm} \\ \text{H}_2 \longrightarrow \text{H}_2\text{O: O}_2\ \text{required} &= (0.1097)(7.936\ \text{lbm}) \\ &= 0.8706\ \text{lbm} \\ \text{S} \longrightarrow \text{SO}_2\text{: O}_2\ \text{required} &= (0.0034)(0.998\ \text{lbm}) \\ &= 0.0034\ \text{lbm}\end{aligned}$$

The total amount of oxygen required per lbm of fuel oil is

$$\begin{aligned}&2.2784\ \text{lbm} + 0.8706\ \text{lbm} \\ &\quad + 0.0034\ \text{lbm} = 3.1524\ \text{lbm}\end{aligned}$$

step 2: With 60% excess air, the excess oxygen is

$$(0.6)(3.1524\ \text{lbm}) = 1.8914\ \text{lbm O}_2\text{/lbm fuel}$$

From Eq. 29.47, this oxygen occupies a volume of

$$V = \frac{mRT}{p}$$

At standard conditions,

$$p = 14.7 \text{ psia}$$
$$T = 60°\text{F} + 460 = 520°\text{R}$$

From Table 29.7, for oxygen, $R = 48.29$ ft-lbf/lbm-°R.

$$V = \frac{(1.8914 \text{ lbm})\left(48.29 \ \frac{\text{ft-lbf}}{\text{lbm-°R}}\right)(520°\text{R})}{\left(14.7 \ \frac{\text{lbf}}{\text{in}^2}\right)\left(144 \ \frac{\text{in}^2}{\text{ft}^2}\right)} = 22.44 \text{ ft}^3$$

step 3: The theoretical nitrogen based on Table 37.6 is

$$\left(\frac{3.1524 \text{ lbm}}{0.2315}\right)(0.7685) = 10.465 \text{ lbm N}_2/\text{lbm fuel}$$

The actual nitrogen with 60% excess air and nitrogen in the fuel is

$$\begin{aligned}(10.465 \text{ lbm})(1.6) + 0.0022 \text{ lbm} \\ = 16.746 \text{ lbm N}_2/\text{lbm fuel}\end{aligned}$$

From Eq. 29.47, this nitrogen occupies a volume of

$$V = \frac{mRT}{p}$$

From Table 29.7, R for nitrogen = 55.16 ft-lbf/lbm-°R.

$$V = \frac{(16.746 \text{ lbm})\left(55.16 \ \frac{\text{ft-lbf}}{\text{lbm-°R}}\right)(520°\text{R})}{\left(14.7 \ \frac{\text{lbf}}{\text{in}^2}\right)\left(144 \ \frac{\text{in}^2}{\text{ft}^2}\right)} = 226.91 \text{ ft}^3$$

step 4: From Table 37.8, the 60°F combustion product volumes per lbm of fuel will be

$$\begin{array}{lll} CO_2: & (0.8543)(31.63 \text{ ft}^3) & = 27.02 \text{ ft}^3 \\ H_2O: & (0.1131)(188.25 \text{ ft}^3) & = 21.29 \text{ ft}^3 \\ SO_2: & (0.0034)(11.84 \text{ ft}^3) & = 0.040 \text{ ft}^3 \\ N_2: & \text{from step 3} & = 226.91 \text{ ft}^3 \\ O_2: & \text{from step 2} & = 22.44 \text{ ft}^3 \\ \hline & \text{total} & = 297.7 \text{ ft}^3 \end{array}$$

(a) At 60°F, the wet volume will be $\boxed{297.7 \text{ ft}^3}$.

At 600°F, the wet volume is

$$V_{\text{wet},600°\text{F}} = (297.7 \text{ ft}^3)\left(\frac{600°\text{F} + 460}{60°\text{F} + 460}\right) = \boxed{606.9 \text{ ft}^3}$$

(b) At 60°F, the dry volume will be

$$297.7 \text{ ft}^3 - 21.29 \text{ ft}^3 = \boxed{276.4 \text{ ft}^3}$$

At 600°F, the dry volume is

$$V_{\text{dry},600°\text{F}} = (276.4 \text{ ft}^3)\left(\frac{600°\text{F} + 460}{60°\text{F} + 460}\right) = \boxed{563.4 \text{ ft}^3}$$

(c) The volumetric fraction of dry carbon dioxide is

$$\frac{27.02 \text{ ft}^3}{276.41 \text{ ft}^3} = \boxed{0.098 \ \ (9.8\%)}$$

SI Solution

step 1: From Table 37.8, find the stoichiometric oxygen required per lbm of fuel oil.

$$\begin{aligned} C \longrightarrow CO_2: \ O_2 \text{ required} &= (0.8543)(2.667 \text{ kg}) \\ &= 2.2784 \text{ kg} \\ H_2 \longrightarrow H_2O: \ O_2 \text{ required} &= (0.1097)(7.936 \text{ kg}) \\ &= 0.8706 \text{ kg} \\ S \longrightarrow SO_2: \ O_2 \text{ required} &= (0.0034)(0.998 \text{ kg}) \\ &= 0.0034 \text{ kg} \end{aligned}$$

The total amount of oxygen required per kg of fuel oil is

$$\begin{aligned} 2.2784 \text{ kg} + 0.8706 \text{ kg} \\ + 0.0034 \text{ kg} = 3.1524 \text{ kg} \end{aligned}$$

step 2: With 60% excess air, the excess oxygen is

$$(0.6)(3.1524 \text{ kg}) = 1.8914 \text{ kg O}_2/\text{kg fuel}$$

From Eq. 29.47, this oxygen occupies a volume of

$$V = \frac{mRT}{p}$$

At standard conditions,

$$p = 101.3 \text{ kPa}$$
$$T = 16°\text{C} + 273 = 289\text{K}$$

From Table 29.7, for oxygen, R = 259.82 J/kg·K.

$$V = \frac{(1.8914 \text{ kg})\left(259.82 \ \frac{\text{J}}{\text{kg·K}}\right)(289\text{K})}{(101.3 \text{ kPa})(1000 \text{ kPa})} = 1.402 \text{ m}^3$$

step 3: The theoretical nitrogen based on Table 37.6 is

$$\left(\frac{3.1524 \text{ kg}}{0.2315}\right)(0.7685) = 10.465 \text{ kg N}_2\text{/kg fuel}$$

The actual nitrogen with 60% excess air and nitrogen in the fuel is

$$(10.465 \text{ kg})(1.6) + 0.0022 \text{ kg} = 16.746 \text{ kg N}_2\text{/kg fuel}$$

From Eq. 29.47, this nitrogen occupies a volume of

$$V = \frac{mRT}{p}$$

From Table 29.7, R for nitrogen = 296.77 J/kg·K.

$$V = \frac{(16.746 \text{ kg})\left(296.77 \ \frac{\text{J}}{\text{kg·K}}\right)(289\text{K})}{(101.3 \text{ kPa})\left(1000 \ \frac{\text{Pa}}{\text{kPa}}\right)} = 14.178 \text{ m}^3$$

step 4: From Table 37.8, the 16°C combustion product volumes per kg of fuel will be

CO_2:	$(0.8543)(31.63 \text{ ft}^3)(0.06243)$	=	1.687 m^3
H_2O:	$(0.1131)(188.25 \text{ ft}^3)(0.06243)$	=	1.329 m^3
SO_2:	$(0.0034)(11.84 \text{ ft}^3)(0.06243)$	=	0.003 m^3
N_2:	from step 3	=	14.178 m^3
O_2:	from step 2	=	1.402 m^3
	total	=	18.599 m^3

(a) At 16°C, the wet volume will be $\boxed{18.599 \text{ m}^3}$.

At 320°C, the wet volume is

$$V_{\text{wet},320°\text{C}} = (18.599 \text{ m}^3)\left(\frac{320°\text{C} + 273}{16°\text{C} + 273}\right) = \boxed{38.16 \text{ m}^3}$$

(b) At 16°C, the dry volume will be

$$18.599 \text{ m}^3 - 1.329 \text{ m}^3 = \boxed{17.27 \text{ m}^3}$$

At 320°C, the dry volume is

$$V_{\text{dry},320°\text{C}} = (17.27 \text{ m}^3)\left(\frac{320°\text{C} + 273}{16°\text{C} + 273}\right) = \boxed{35.44 \text{ m}^3}$$

(c) The volumetric fraction of dry carbon dioxide is

$$\frac{1.687 \text{ m}^3}{17.27 \text{ m}^3} = \boxed{0.098 \ \ (9.8\%)}$$

10. *Customary U.S. Solution*

step 1: Based on the 15,395 lbm of coal burned producing 2816 lbm of ash containing 20.9% carbon by weight, the usable percentage of carbon per lbm of fuel is

$$0.5145 - \frac{(2816 \text{ lbm})(0.209)}{15{,}395 \text{ lbm}} = 0.4763$$

step 2: Since moisture is reported separately, assume all of the oxygen and hydrogen are free. (This is not ordinarily the case.) From Table 37.8, find the stoichiometric oxygen required per lbm of fuel.

$$\text{C} \longrightarrow \text{CO}_2\text{: O}_2 \text{ required} = (0.4763)(2.667 \text{ lbm}) = 1.2703 \text{ lbm}$$

$$\text{H}_2 \longrightarrow \text{H}_2\text{O: O}_2 \text{ required} = (0.0402)(7.936 \text{ lbm}) = 0.3190 \text{ lbm}$$

$$\text{S} \longrightarrow \text{SO}_2\text{: O}_2 \text{ required} = (0.0392)(0.998 \text{ lbm}) = 0.0391 \text{ lbm}$$

The total amount of O_2 required per lbm of fuel is

$$1.2703 \text{ lbm} + 0.3190 \text{ lbm} + 0.0391 \text{ lbm} - 0.0728 \text{ lbm} = 1.5556 \text{ lbm}$$

step 3: The theoretical air based on Table 37.6 is

$$\frac{1.5556 \text{ lbm}}{0.2315} = 6.720 \text{ lbm air/lbm fuel}$$

step 4: Ignoring fly ash, the theoretical dry products are given from Table 37.8.

$$\begin{aligned} CO_2&: (0.4763)(3.667 \text{ lbm}) &= 1.7466 \text{ lbm} \\ SO_2&: (0.0392)(1.998 \text{ lbm}) &= 0.0783 \text{ lbm} \\ N_2&: 0.0093 \text{ lbm} + (6.720 \text{ lbm}) \times (0.7685 \text{ lbm}) &= \underline{5.1736 \text{ lbm}} \\ & \text{total} &= 6.999 \text{ lbm} \end{aligned}$$

step 5: The excess air is

$$13.3 \text{ lbm} - 6.999 \text{ lbm} = 6.301 \text{ lbm}$$

step 6: The total air supplied per pound of fuel is

$$6.301 \text{ lbm} + 6.720 \text{ lbm} = \boxed{13.02 \text{ lbm}}$$

SI Solution

step 1: Based on the 6923 kg of coal burned producing 1267 kg of ash containing 20.9% carbon by weight, the usable percentage of carbon per kg of fuel is

$$0.5145 - \frac{(1267 \text{ kg})(0.209)}{6923 \text{ kg}} = 0.4763$$

step 2: From Table 37.8, find the stoichiometric oxygen required per kg of fuel.

$$\begin{aligned} C \longrightarrow CO_2\text{: } O_2 \text{ required} &= (0.4763)(2.667 \text{ kg}) \\ &= 1.2703 \text{ kg} \\ H_2 \longrightarrow H_2O\text{: } O_2 \text{ required} &= (0.0402)(7.936 \text{ kg}) \\ &= 0.3190 \text{ kg} \\ S \longrightarrow SO_2\text{: } O_2 \text{ required} &= (0.0392)(0.998 \text{ kg}) \\ &= 0.0391 \text{ kg} \end{aligned}$$

The total amount of O_2 required per kg of fuel is

$$\begin{aligned} &1.2703 \text{ kg} + 0.3190 \text{ kg} \\ &\quad + 0.0391 \text{ kg} - 0.0728 \text{ kg} = 1.5556 \text{ kg} \end{aligned}$$

step 3: The theoretical air based on Table 37.6 is

$$\frac{1.5556 \text{ kg}}{0.2315} = 6.720 \text{ kg air/kg fuel}$$

step 4: Ignoring fly ash, the theoretical dry products are given from Table 37.8.

$$\begin{aligned} CO_2&: (0.4763)(3.667 \text{ kg}) &= 1.7466 \text{ kg} \\ SO_2&: (0.0392)(1.998 \text{ kg}) &= 0.0783 \text{ kg} \\ N_2&: 0.0093 \text{ kg} + (6.720 \text{ kg}) \times (0.7685 \text{ kg}) &= \underline{5.1736 \text{ kg}} \\ & \text{total} &= 6.999 \text{ kg} \end{aligned}$$

step 5: The excess air is

$$13.3 \text{ kg} - 6.999 \text{ kg} = 6.301 \text{ kg}$$

step 6: The total air supplied is

$$6.301 \text{ kg} + 6.720 \text{ kg} = \boxed{13.02 \text{ kg}}$$

11. (The customary U.S. and SI solutions are essentially identical.)

From Eq. 37.9, the actual air/fuel ratio can be estimated as

$$R_{a/f,\text{actual}} = \frac{3.04 B_{N_2} G_C}{B_{CO_2} + B_{CO}}$$

A fraction of carbon is reduced due to the percentage of coal lost in the ash pit.

$$G_C = (1 - 0.03)(0.65) = 0.6305$$

$$\begin{aligned} R_{a/f,\text{actual}} &= \frac{(3.04)(0.815)(0.6305)}{0.095 + 0} \\ &= 16.44 \text{ lbm air/lbm fuel (kg air/kg fuel)} \end{aligned}$$

Combustion uses 9.87 lbm of air/lbm of fuel.

$$\begin{aligned} &= (9.87 \text{ lbm})(1 - 0.03) \\ &= 9.57 \text{ lbm of air/lbm of fuel} \end{aligned}$$

$$\begin{aligned} \% \text{ of excess air} &= \left(\frac{16.44 \text{ lbm} - 9.57 \text{ lbm}}{9.57 \text{ lbm}}\right)(100\%) \\ &= \boxed{71.8\%} \end{aligned}$$

12. From Eq. 37.23, the heat loss due to the formation of carbon monoxide is

$$q = \frac{(\text{HHV}_C - \text{HHV}_{CO}) G_C B_{CO}}{B_{CO_2} + B_{CO}}$$

From App. 37.A, the difference in the two heating values is

$$\begin{aligned} \text{HHV}_C - \text{HHV}_{CO} &= 14{,}093 \frac{\text{Btu}}{\text{lbm}} - 4347 \frac{\text{Btu}}{\text{lbm}} \\ &= 9746 \text{ Btu/lbm} \quad (22.67 \text{ MJ/kg}) \\ G_C &= (1 - 0.03)(0.6734) = 0.6532 \end{aligned}$$

Combustion

Customary U.S. Solution

$$q = \frac{\left(9746\ \frac{\text{Btu}}{\text{lbm}}\right)(0.6532)(1.6\%)}{15.5\% + 1.6\%} = \boxed{596\ \text{Btu/lbm}}$$

SI Solution

$$q = \frac{\left(22.67\ \frac{\text{MJ}}{\text{kg}}\right)(0.6532)(1.6\%)}{15.5\% + 1.6\%} = \boxed{1.39\ \text{MJ/kg}}$$

13. (a) For methane, $B = 0.93$.

From ideal gas laws (from Table 29.7, R for methane = 96.32 ft-lbf/lbm-°R),

$$\rho = \frac{p}{RT} = \frac{\left(14.7\ \frac{\text{lbf}}{\text{in}^2}\right)\left(144\ \frac{\text{in}^2}{\text{ft}^2}\right)}{\left(96.32\ \frac{\text{ft-lbf}}{\text{lbm-°R}}\right)(60°\text{F} + 460)} = 0.0422\ \text{lbm/ft}^3$$

From Table 37.9, $K = 9.55\ \text{ft}^3$ air/ft^3 fuel.

From Table 37.8,

products: 1 ft^3 CO_2, 2 ft^3 H_2O

From App. 37.A, HHV = 1013 Btu/ft^3.

Similar results for all the other fuel components are tabulated in the following table.

gas	B	ρ (lbm/ft^3)	ft^3 air	HHV (Btu/ft^3)	volumes of products: CO_2	H_2O	other
CH_4	0.93	0.0422	9.556	1013	1	2	–
N_2	0.0373	0.0738	–	–	–	–	1 N_2
CO	0.0045	0.0738	2.389	322	1	–	–
H_2	0.0182	0.0053	2.389	325	–	1	–
C_2H_4	0.0025	0.0739	14.33	1614	2	2	–
H_2S	0.0018	0.0900	7.167	647	–	1	1 SO_2
O_2	0.0035	0.0843	–	–	–	–	–
CO_2	0.0022	0.1160	–	–	1	–	–

From Eq. 29.78, the composite density is

$$\begin{aligned}\rho &= \sum B_i\rho_i \\ &= (0.93)\left(0.0422\ \frac{\text{lbm}}{\text{ft}^3}\right) + (0.0373)\left(0.0738\ \frac{\text{lbm}}{\text{ft}^3}\right) \\ &\quad + (0.0045)\left(0.0738\ \frac{\text{lbm}}{\text{ft}^3}\right) + (0.0182)\left(0.0053\ \frac{\text{lbm}}{\text{ft}^3}\right) \\ &\quad + (0.0025)\left(0.0739\ \frac{\text{lbm}}{\text{ft}^3}\right) + (0.0018)\left(0.0900\ \frac{\text{lbm}}{\text{ft}^3}\right) \\ &\quad + (0.0035)\left(0.0843\ \frac{\text{lbm}}{\text{ft}^3}\right) + (0.0022)\left(0.1160\ \frac{\text{lbm}}{\text{ft}^3}\right) \\ &= \boxed{0.0433\ \text{lbm/ft}^3}\end{aligned}$$

(b) The air is 20.9% oxygen by volume. The theoretical air requirements are

$$\begin{aligned}\sum B_i V_{\text{air,i}} - \frac{O_2\ \text{in fuel}}{0.209} &= (0.93)(9.556\ \text{ft}^3) + (0.0373)(0) \\ &\quad + (0.0045)(2.389\ \text{ft}^3) \\ &\quad + (0.0182)(2.389\ \text{ft}^3) \\ &\quad + (0.0025)(14.33\ \text{ft}^3) \\ &\quad + (0.0018)(7.167\ \text{ft}^3) \\ &\quad + (0.0035)(0) + (0.0022)(0) - \frac{0.0035}{0.209} \\ &= 8.990\ \text{ft}^3 - 0.01675\ \text{ft}^3 \\ &= \boxed{8.9733\ \text{ft}^3\ \text{air/ft}^3\ \text{fuel}}\end{aligned}$$

((c) and (d)) The theoretical oxygen will be

$$(8.9733\ \text{ft}^3)(0.209) = 1.875\ \text{ft}^3/\text{ft}^3$$

The excess oxygen will be

$$(0.4)(1.875\ \text{ft}^3) = 0.75\ \text{ft}^3/\text{ft}^3$$

Similarly, the total nitrogen in the stack gases is

$$(1.4)(0.791)(8.9733\ \text{ft}^3) + 0.034 = 9.971\ \text{ft}^3/\text{ft}^3\ \text{fuel}$$

The stack gases per ft^3 of fuel are

$$\begin{aligned}
\text{excess } O_2&: && = 0.75\ \text{ft}^3\\
\text{excess } N_2&: && = 9.971\ \text{ft}^3\\
\text{excess } SO_2&: && = 0.0018\ \text{ft}^3\\
\text{excess } CO_2&: (0.93)(1) + (0.0045)(1)\\
&\quad + (0.0025)(2) + (0.0022)(1) && = 0.942\ \text{ft}^3\\
\text{excess } H_2O&: (0.93)(2) + (0.0182)(1)\\
&\quad + (0.0025)(2) + (0.0018)(1) && = 1.885\ \text{ft}^3\\
&\text{total} && = 13.55\ \text{ft}^3
\end{aligned}$$

The total wet volume is 13.55 ft^3/ft^3 fuel.

The total dry volume is 11.66 ft^3/ft^3 fuel.

The volumetric analyses are

	O_2	N_2	SO_2	CO_2	H_2O
wet:	$\frac{0.75\ \text{ft}^3}{13.55\ \text{ft}^3}$	$\frac{9.971\ \text{ft}^3}{13.55\ \text{ft}^3}$	$\frac{0.0018\ \text{ft}^3}{13.55\ \text{ft}^3}$	$\frac{0.942\ \text{ft}^3}{13.55\ \text{ft}^3}$	$\frac{1.885}{13.55}$
	= 0.0554	0.736	–	0.070	0.139
dry:	$\frac{0.75\ \text{ft}^3}{11.66\ \text{ft}^3}$	$\frac{9.971\ \text{ft}^3}{11.66\ \text{ft}^3}$	$\frac{0.0018\ \text{ft}^3}{11.66\ \text{ft}^3}$	$\frac{0.942\ \text{ft}^3}{11.66\ \text{ft}^3}$	–
	= 0.0643	0.855	–	0.081	–

14. *Customary U.S. Solution*

step 1: From App. 29.B, the heat of vaporization at 212°F is $h_{fg} = 970.3$ Btu/lbm.

The heat absorbed in the boiler is

$$\begin{aligned}
m_{H_2O}h_{fg} &= (11.12\ \text{lbm } H_2O)\left(970.3\ \frac{\text{Btu}}{\text{lbm}}\right)\\
&= 10{,}789.7\ \text{Btu/lbm fuel}
\end{aligned}$$

step 2: The losses for heating stack gases can be found as follows.

The burned carbon per lbm of fuel is

$$0.7842 - (0.315)(0.0703) = 0.7621\ \text{lbm/lbm fuel}$$

The mass ratio of dry flue gases to solid fuel is given by Eq. 37.12.

$$\begin{aligned}
\frac{\text{mass of flue gas}}{\text{mass of solid fuel}} &= \frac{\begin{pmatrix}(11B_{CO_2} + 8B_{O_2}\\ + (7)(B_{CO} + B_{N_2}))\end{pmatrix}\left(G_C + \frac{G_S}{1.833}\right)}{(3)(B_{CO_2} + B_{CO})}\\
&= \frac{\begin{pmatrix}((11)(14.0) + (8)(5.5)\\ + (7)(0.42 + 80.08))\end{pmatrix}\left(0.7621 + \frac{0.01}{1.833}\right)}{(3)(14.0 + 0.42)}\\
&= 13.51\ \text{lbm stack gases/lbm fuel}
\end{aligned}$$

Properties of nitrogen frequently are assumed for dry flue gas. From Table 37.1, for nitrogen at an average temperature of (575°F + 73°F)/2 = 324°F (784°R), $c_p \approx 0.252$ Btu/lbm-°F.

The losses for heating stack gases are given by Eq. 37.20.

$$\begin{aligned}
q_1 &= m_{\text{flue gas}}c_p(T_{\text{flue gas}} - T_{\text{incoming air}})\\
&= (13.51\ \text{lbm})\left(0.252\ \frac{\text{Btu}}{\text{lbm-°F}}\right)(575°\text{F} - 67°\text{F})\\
&= 1729.5\ \text{Btu/lbm fuel}
\end{aligned}$$

The heat loss in the vapor formed during the combustion of hydrogen is given by Eq. 37.21.

$$q_2 = 8.94G_H(h_g - h_f)$$

Assume that the partial pressure of the water vapor is below 1 psia (the lowest App. 29.C goes).

h_g at 575°F and 1 psia can be found from superheat tables, App. 29.C.

$$h_g \approx 1324.2\ \text{Btu/lbm}$$

h_f at 73°F can be found from App. 29.A.

$$\begin{aligned}
h_f &= 41.09\ \text{Btu/lbm}\\
G_{H,\text{available}} &= G_{H,\text{total}} - \frac{G_O}{8}\\
&= 0.0556 - \frac{0.0825}{8}\\
&= 0.0453\\
q_2 &= (8.94)(0.0453)\left(1324.2\ \frac{\text{Btu}}{\text{lbm}} - 41.09\ \frac{\text{Btu}}{\text{lbm}}\right)\\
&= 519.6\ \text{Btu/lbm fuel}
\end{aligned}$$

Heat is also lost when it is absorbed by the moisture originally in the combustion air.

$$q_3 = \omega m_{\text{combustion air}}(h_g - h_g')$$

Assume that the partial pressure of the water vapor is below 1 psia (the lowest App. 29.C goes).

From App. 29.B, at 73°F and 0.4 psia, $h_g' \approx$ 1090 Btu/lbm. From the psychrometric chart,

$$\begin{aligned}
\omega &= 90\ \text{grains/lbm air}\\
&= 0.0129\ \text{lbm water/lbm air}
\end{aligned}$$

Considering the sulfur content, find the air/fuel ratio.

$$\frac{\text{lbm air}}{\text{lbm fuel}} = \frac{3.04 B_{N_2}\left(G_C + \frac{G_S}{1.833}\right)}{B_{CO_2} + B_{CO}}$$

$$= \frac{(3.04)(80.08\%)\left(0.7621 + \frac{0.01}{1.833}\right)}{14\% + 0.42\%}$$

$$= 12.96 \text{ lbm air/lbm fuel}$$

$$q_3 = \left(0.0129 \, \frac{\text{lbm water}}{\text{lbm air}}\right)\left(12.96 \, \frac{\text{lbm air}}{\text{lbm fuel}}\right) \times \left(1324.2 \, \frac{\text{Btu}}{\text{lbm}} - 1090 \, \frac{\text{Btu}}{\text{lbm}}\right)$$

$$= 39.2 \text{ Btu/lbm fuel}$$

The energy lost in incomplete combustion is given by Eq. 37.23.

$$q_4 = \frac{(\text{HHV}_C - \text{HHV}_{CO})\, G_C B_{CO}}{B_{CO_2} + B_{CO}}$$

$$= \frac{\left(9746 \, \frac{\text{Btu}}{\text{lbm}}\right)(0.7621)(0.42\%)}{14\% + 0.42\%}$$

$$= 216.3 \text{ Btu/lbm fuel}$$

The energy lost in unburned carbon is given by Eq. 37.24.

$$q_5 = \left(14{,}093 \, \frac{\text{Btu}}{\text{lbm}}\right) m_{\text{ash}} G_{\text{C,ash}}$$

$$= \left(14{,}093 \, \frac{\text{Btu}}{\text{lbm}}\right)(0.0703)(0.315)$$

$$= 312.1 \text{ Btu/lbm fuel}$$

The energy lost in radiation and unaccounted for is

$$14{,}000 \, \frac{\text{Btu}}{\text{lbm}} - 10{,}789.7 \, \frac{\text{Btu}}{\text{lbm}} - 1729.5 \, \frac{\text{Btu}}{\text{lbm}} - 519.6 \, \frac{\text{Btu}}{\text{lbm}} - 39.2 \, \frac{\text{Btu}}{\text{lbm}} - 216.3 \, \frac{\text{Btu}}{\text{lbm}} - 312.1 \, \frac{\text{Btu}}{\text{lbm}}$$

$$= \boxed{393.6 \text{ Btu/lbm fuel}}$$

SI Solution

step 1: From App. 29.N, the heat of vaporization at 100°C is $h_{fg} = 2257.0$ kJ/kg.

The heat absorbed in the boiler is

$$m_{H_2O} h_{fg} = (11.12 \text{ kg})\left(2257.0 \, \frac{\text{kJ}}{\text{kg}}\right) = 25\,098 \text{ kJ/kg fuel}$$

step 2: From step 2 of the U.S. solution,

$$\frac{\text{mass of flue gas}}{\text{mass of solid fuel}} = 13.51 \text{ kg stack gases/kg fuel}$$

Properties of nitrogen frequently are assumed for dry flue gas. For an average nitrogen temperature of (1/2)(300°C + 23°C) + 273 = 434.5 K (782°R), $c_p \approx 0.252$ Btu/lbm-°R.

c_p from Table 37.1 can be found (using the table footnote) as

$$c_p = \left(0.252 \, \frac{\text{Btu}}{\text{lbm-°R}}\right)\left(4.187 \, \frac{\frac{\text{kJ}}{\text{kg·K}}}{\frac{\text{Btu}}{\text{lbm-°R}}}\right)$$

$$= 1.055 \text{ kJ/kg·K}$$

The losses for heating stack gases are given by Eq. 37.20.

$$q_1 = m_{\text{flue gas}} c_p (T_{\text{flue gas}} - T_{\text{incoming air}})$$

$$= (13.51 \text{ kg})\left(1.055 \, \frac{\text{kJ}}{\text{kg·K}}\right)(300°\text{C} - 23°\text{C})$$

$$= 3948.1 \text{ kJ/kg fuel}$$

The heat loss in the vapor formed during the combustion of hydrogen is given by Eq. 37.21.

$$q_2 = 8.94 G_H (h_g - h_f)$$

Assume that the partial pressure of the water vapor is below 10 kPa (the lowest App. 29.P goes).

h_g at 300°C and 10 kPa can be found from superheat tables, App. 29.P.

$$h_g = 3076.5 \text{ kJ/kg}$$

h_f at 23°C can be found from App. 29.N.

$$h_f = 96.52 \text{ kJ/kg}$$

$$G_{\text{H,available}} = G_{\text{H,total}} - \frac{G_O}{8}$$

$$= 0.0556 - \frac{0.0825}{8}$$

$$= 0.0453$$

$$q_2 = (8.94)(0.0453)\left(3076.5 \, \frac{\text{kJ}}{\text{kg}} - 96.52 \, \frac{\text{kJ}}{\text{kg}}\right)$$

$$= 1206.8 \text{ kJ/kg fuel}$$

Heat is also lost when it is absorbed by the moisture originally in the combustion air.

$$q_3 = \omega m_{\text{combustion air}} (h_g - h'_g)$$

Assume that the partial pressure of the water vapor is low. At 23°C, from App. 29.O, $h'_g \approx$ 2500 kJ/kg.

From the psychrometric chart for 19°C wet bulb and 23°C dry bulb, ω = 12.2 g/kg dry air.

The air/fuel ratio from the customary U.S. solution is

$$\frac{\text{kg air}}{\text{kg fuel}} = 12.96 \text{ kg air/kg fuel}$$

$$\begin{aligned} q_3 &= \left(12.2 \ \frac{\text{g}}{\text{kg}}\right)\left(\frac{1 \text{ kg}}{1000 \text{ g}}\right)\left(12.96 \ \frac{\text{kg}}{\text{kg}}\right) \\ &\quad \times \left(3076.5 \ \frac{\text{kJ}}{\text{kg}} - 2500 \ \frac{\text{kJ}}{\text{kg}}\right) \\ &= 91.15 \text{ kJ/kg} \end{aligned}$$

Energy lost in incomplete combustion is given by Eq. 37.23.

$$\begin{aligned} q_4 &= \frac{(\text{HHV}_\text{C} - \text{HHV}_\text{CO})\,G_\text{C}B_\text{CO}}{B_{\text{CO}_2} + B_\text{CO}} \\ &= \left(\frac{\left(22.67 \ \frac{\text{MJ}}{\text{kg}}\right)(0.7621)(0.42)}{14 + 0.42}\right)\left(1000 \ \frac{\text{kJ}}{\text{MJ}}\right) \\ &= \left(0.5032 \ \frac{\text{MJ}}{\text{kg}}\right)\left(1000 \ \frac{\text{kJ}}{\text{MJ}}\right) \\ &= 503.2 \text{ kJ/kg fuel} \end{aligned}$$

The energy lost in unburned carbon is given by Eq. 37.24.

$$\begin{aligned} q_5 &= \left(32.8 \ \frac{\text{MJ}}{\text{kg}}\right) m_\text{ash} G_{\text{C,ash}} \\ &= \left(32.8 \ \frac{\text{MJ}}{\text{kg}}\right)\left(1000 \ \frac{\text{kJ}}{\text{MJ}}\right)(0.0703)(0.315) \\ &= 726.3 \text{ kJ/kg fuel} \end{aligned}$$

Energy lost in radiation and unaccounted for is

$$\begin{aligned} &\left(32.6 \ \frac{\text{MJ}}{\text{kg}}\right)\left(1000 \ \frac{\text{kJ}}{\text{MJ}}\right) - 25\,098 \ \frac{\text{kJ}}{\text{kg}} \\ &\quad - 3948.1 \ \frac{\text{kJ}}{\text{kg}} - 1206.8 \ \frac{\text{kJ}}{\text{kg}} - 91.15 \ \frac{\text{kJ}}{\text{kg}} \\ &\quad - 503.2 \ \frac{\text{kJ}}{\text{kg}} - 726.3 \ \frac{\text{kJ}}{\text{kg}} = \boxed{1026.5 \text{ kJ/kg}} \end{aligned}$$

15. *Customary U.S. Solution*

step 1: The incoming reactants on a per-pound basis are

0.07 lbm ash
0.05 lbm hydrogen
0.05 lbm oxygen
0.83 lbm carbon

This is an ultimate analysis. Assume that only the hydrogen that is not locked up with oxygen in the form of water is combustible. From Eq. 37.15, the available hydrogen is

$$\begin{aligned} G_{\text{H,available}} &= G_{\text{H,total}} - \frac{G_\text{O}}{8} \\ &= 0.05 \text{ lbm} - \frac{0.05 \text{ lbm}}{8} \\ &= 0.04375 \text{ lbm} \end{aligned}$$

The mass of water produced is the hydrogen mass plus eight times as much oxygen. The locked hydrogen is

$$0.05 \text{ lbm} - 0.04375 \text{ lbm} = 0.00625 \text{ lbm}$$

$$\begin{aligned} \text{lbm of moisture} &= G_\text{H} + G_\text{O} \\ &= G_\text{H} + 8G_\text{H} \\ &= 0.00625 \text{ lbm} + (8)(0.00625 \text{ lbm}) \\ &= 0.05625 \text{ lbm} \end{aligned}$$

The air is 23.15% oxygen by weight (Table 37.6), so other reactants for 26 lbm of air are

$$(0.2315)(26 \text{ lbm}) = 6.019 \text{ lbm } O_2$$

$$(0.7685)(26 \text{ lbm}) = 19.981 \text{ lbm } N_2$$

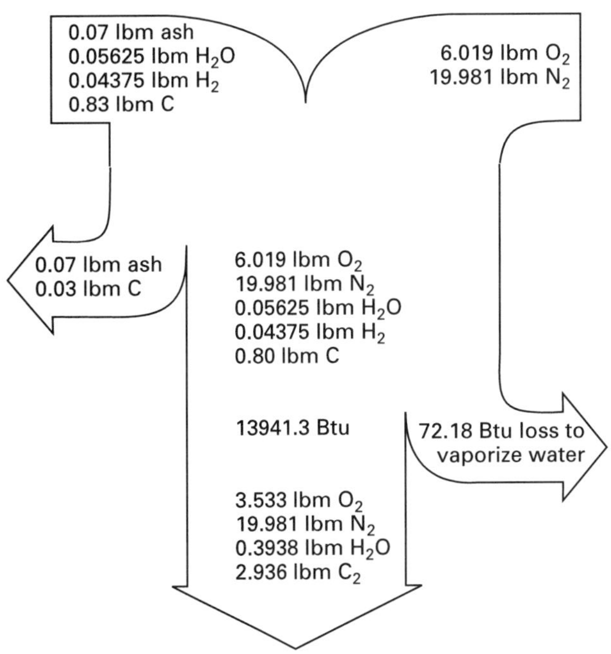

step 2: Ash pit material losses are 10% or 0.1 lbm, which includes all of the ash.

0.07 lbm ash (noncombustible matter)
0.03 lbm unburned carbon

step 3: Determine what remains.

6.019 lbm oxygen
19.981 lbm nitrogen
0.05625 lbm water
0.04375 lbm hydrogen
0.080 lbm carbon

step 4: Determine the energy loss in vaporizing the moisture.

$$q = (\text{moisture})(h_g - h_f)$$

From App. 29.C, h_g at 550°F and 14.7 psia is 1311.3 Btu/lbm.

From App. 29.A, h_f at 60°F is 28.08 Btu/lbm.

$$\begin{aligned} q &= (0.05625\ \text{lbm})\left(1311.3\ \frac{\text{Btu}}{\text{lbm}} - 28.08\ \frac{\text{Btu}}{\text{lbm}}\right) \\ &= 72.18\ \text{Btu} \end{aligned}$$

step 5: Calculate the heating value of the remaining fuel components using App. 37.A.

$$\begin{aligned} \text{HV}_\text{C} &= (0.80\ \text{lbm})\left(14{,}093\ \frac{\text{Btu}}{\text{lbm}}\right) \\ &= 11274.4\ \text{Btu} \\ \text{HV}_\text{H} &= (0.04375\ \text{lbm})\left(60{,}958\ \frac{\text{Btu}}{\text{lbm}}\right) \\ &= 2666.9\ \text{Btu} \end{aligned}$$

The heating value after the coal moisture is evaporated is

$$\begin{aligned} &11{,}274.4\ \text{Btu} + 2666.9\ \text{Btu} - 72.18\ \text{Btu} \\ &\qquad = 13{,}869\ \text{Btu} \end{aligned}$$

step 6: Using Table 37.8, determine the combustion products.

$$\begin{aligned} \text{oxygen required by carbon} &= (0.80)(2.667\ \text{lbm}) \\ &= 2.134\ \text{lbm} \\ \text{oxygen required by hydrogen} &= (0.04375)(7.936\ \text{lbm}) \\ &= 0.3472\ \text{lbm} \\ \text{carbon dioxide produced by carbon} &= (0.8)(3.667\ \text{lbm}) \\ &= 2.934\ \text{lbm} \\ \text{water produced by hydrogen} &= (0.04375)(8.936\ \text{lbm}) \\ &= 0.3910\ \text{lbm} \end{aligned}$$

The remaining oxygen is

$$6.019\ \text{lbm} - 2.134\ \text{lbm} - 0.3472\ \text{lbm} = 3.538\ \text{lbm}$$

step 7: The gaseous products must be heated from 70°F to 550°F. The average temperature is

$$\left(\tfrac{1}{2}\right)(70°\text{F} + 550°\text{F}) = 310°\text{F}\ (770°\text{R})$$

From Table 37.1, the specific heat of gaseous products is

gas	c_p $\left(\frac{\text{Btu}}{\text{lbm-°R}}\right)$
oxygen	0.228
nitrogen	0.252
water	0.460
carbon dioxide	0.225

$$\begin{aligned} Q_\text{heating} &= \left((3.538\ \text{lbm})\left(0.228\ \frac{\text{Btu}}{\text{lbm-°R}}\right)\right. \\ &\quad + (19.981\ \text{lbm})\left(0.252\ \frac{\text{Btu}}{\text{lbm-°R}}\right) \\ &\quad + (0.3910\ \text{lbm})\left(0.460\ \frac{\text{Btu}}{\text{lbm-°R}}\right) \\ &\quad \left. + (2.934\ \text{lbm})\left(0.225\ \frac{\text{Btu}}{\text{lbm-°R}}\right)\right) \\ &\quad \times (550°\text{F} - 70°\text{F}) \\ &= 3207.3\ \text{Btu} \end{aligned}$$

step 8: The percent loss is

$$\frac{3207.3\ \text{Btu} + 72.18\ \text{Btu}}{13{,}869\ \text{Btu}} = \boxed{0.236\ \ (23.6\%)}$$

SI Solution

Steps 1 through 3 are the same as for the customary U.S. solution except that everything is based on kg.

step 4: Determine the energy loss in the vaporizing moisture.

$$q = (\text{moisture})(h_g - h_f)$$

From App. 29.P, h_g at 290°C and 101.3 kPa is 3054.3 kJ/kg.

h_f at 15.6°C from App. 29.N is 65.51 kJ/kg.

$$\begin{aligned} q &= (0.05625\ \text{kg})\left(3054.3\ \frac{\text{kJ}}{\text{kg}} - 65.51\ \frac{\text{kJ}}{\text{kg}}\right) \\ &= 168.1\ \text{kJ} \end{aligned}$$

step 5: Calculate the heating value of the remaining fuel components using App. 37.A and the table footnote.

$$\mathrm{HV_C} = (0.80\ \mathrm{kg})\left(14\,093\ \frac{\mathrm{Btu}}{\mathrm{lbm}}\right)\left(2.326\ \frac{\frac{\mathrm{kJ}}{\mathrm{kg}}}{\frac{\mathrm{Btu}}{\mathrm{lbm}}}\right)$$

$$= 26\,224\ \mathrm{kJ}$$

$$\mathrm{HV_H} = (0.04375\ \mathrm{kg})\left(60\,958\ \frac{\mathrm{Btu}}{\mathrm{lbm}}\right)\left(2.326\ \frac{\frac{\mathrm{kJ}}{\mathrm{kg}}}{\frac{\mathrm{Btu}}{\mathrm{lbm}}}\right)$$

$$= 6203\ \mathrm{kJ}$$

The heating value after the coal moisture is evaporated is

$$26\,224\ \mathrm{kJ} + 6203\ \mathrm{kJ} - 168.1\ \mathrm{kJ} = 32\,259\ \mathrm{kJ}$$

step 6: This step is the same as for the customary U.S. solution except that all quantities are in kg.

step 7: The gaseous products must be heated from 21°C to 290°C. The average temperature is

$$\left(\tfrac{1}{2}\right)(21^\circ\mathrm{C} + 290^\circ\mathrm{C}) = 156^\circ\mathrm{C}\ (429^\circ\mathrm{K})$$

$$(429\mathrm{K})\left(1.8\ \frac{^\circ\mathrm{R}}{\mathrm{K}}\right) = 771^\circ\mathrm{R}$$

From Table 37.1, the specific heat of gaseous products is calculated using the table footnote.

gas	c_p $\left(\frac{\mathrm{kJ}}{\mathrm{kg\cdot K}}\right)$
oxygen	0.955
nitrogen	1.055
water	1.926
carbon dioxide	0.942

$$Q_{\mathrm{heating}} = \left((3.538\ \mathrm{kg})\left(0.955\ \frac{\mathrm{kJ}}{\mathrm{kg\cdot K}}\right) + (19.981\ \mathrm{lbm})\left(1.055\ \frac{\mathrm{kJ}}{\mathrm{kg\cdot K}}\right) + (0.3910\ \mathrm{kg})\left(1.926\ \frac{\mathrm{kJ}}{\mathrm{kg\cdot K}}\right) + (2.934\ \mathrm{kg})\left(0.942\ \frac{\mathrm{kJ}}{\mathrm{kg\cdot K}}\right)\right) \times (290^\circ\mathrm{C} - 21^\circ\mathrm{C})$$

$$= \boxed{7525.4\ \mathrm{kJ}}$$

step 8: The percentage loss is

$$\frac{7525.4\ \mathrm{kJ} + 168.1\ \mathrm{kJ}}{32\,259\ \mathrm{kJ}} = \boxed{0.238\ \ (23.8\%)}$$

16. (a) Silicon in ash is SiO_2 with a molecular weight of

$$28.09\ \frac{\mathrm{lbm}}{\mathrm{lbmol}} + (2)\left(16\ \frac{\mathrm{lbm}}{\mathrm{lbmol}}\right) = 60.09\ \mathrm{lbm/lbmol}$$

The oxygen used with 6.1% by mass silicon is

$$\left(\frac{(2)(16\ \mathrm{lbm})}{28.09\ \mathrm{lbm}}\right)\left(0.061\ \frac{\mathrm{lbm}}{\mathrm{lbm\ coal}}\right) = 0.0695\ \mathrm{lbm/lbm\ coal}$$

Silicon ash produced per lbm of coal is

$$0.061\ \frac{\mathrm{lbm}}{\mathrm{lbm\ coal}} + 0.0695\ \frac{\mathrm{lbm}}{\mathrm{lbm\ coal}} = 0.1305\ \mathrm{lbm/lbm\ coal}$$

Silicon ash produced per hour is

$$\left(0.1305\ \frac{\mathrm{lbm}}{\mathrm{lbm\ coal}}\right)\left(15{,}300\ \frac{\mathrm{lbm\ coal}}{\mathrm{hr}}\right) = 1996.7\ \mathrm{lbm/hr}$$

The silicon in 410 lbm/hr refuse is

$$\left(410\ \frac{\mathrm{lbm}}{\mathrm{hr}}\right)(1 - 0.3) = 287\ \mathrm{lbm/hr}$$

The emission rate is

$$1996.7\ \mathrm{lbm/hr} - 287\ \mathrm{lbm/hr} = \boxed{1709.7\ \mathrm{lbm/hr}}$$

(b) From Table 37.7, the stoichiometric reaction for sulfur is

$$\begin{array}{lccc} & \mathrm{S} & +\ \mathrm{O_2} & \longrightarrow\ \mathrm{SO_2} \\ \mathrm{MW} & 32 & 32 & 64 \end{array}$$

Sulfur dioxide produced for 15,300 lbm/hr of coal feed is

$$\left(15{,}300\ \frac{\mathrm{lbm}}{\mathrm{hr}}\right)(0.0244\ \mathrm{lbm\ S})\left(\frac{64\ \mathrm{lbm\ SO_2}}{32\ \mathrm{lbm\ S}}\right) = \boxed{746.6\ \mathrm{lbm/hr}}$$

Combustion

(c) From Eq. 37.16(b), the heating value of the fuel is

$$\begin{aligned} \text{HHV} &= 14{,}093G_\text{C} + (60{,}958)\left(G_\text{H} - \frac{G_\text{O}}{8}\right) + 3983G_\text{S} \\ &= \left(14{,}093\ \frac{\text{Btu}}{\text{lbm}}\right)(0.7656) \\ &\quad + \left(60{,}958\ \frac{\text{Btu}}{\text{lbm}}\right)\left(0.055 - \frac{0.077}{8}\right) \\ &\quad + \left(3983\ \frac{\text{Btu}}{\text{lbm}}\right)(0.0244) \\ &= 13{,}653\ \text{Btu/lbm} \end{aligned}$$

The gross available combustion power is

$$\begin{aligned} \dot{m}_f(\text{HV}) &= \left(15{,}300\ \frac{\text{lbm}}{\text{hr}}\right)\left(13{,}653\ \frac{\text{Btu}}{\text{lbm}}\right) \\ &= 2.089 \times 10^8\ \text{Btu/hr} \end{aligned}$$

The carbon in 410 lbm/hr refuse is

$$\left(410\ \frac{\text{lbm}}{\text{hr}}\right)(0.3) = 123\ \text{lbm/hr}$$

Power lost in unburned carbon in refuse is $\dot{m}_\text{C}(\text{HV})$.

From App. 37.A, the gross heat of combustion for C is 14,093 Btu/lbm.

$$\left(123\ \frac{\text{lbm}}{\text{hr}}\right)\left(14{,}093\ \frac{\text{Btu}}{\text{lbm}}\right) = 1.733 \times 10^6\ \text{Btu/hr}$$

The remaining combustion power is

$$\begin{aligned} &2.089 \times 10^8\ \frac{\text{Btu}}{\text{hr}} - 1.733 \times 10^6\ \frac{\text{Btu}}{\text{hr}} \\ &\qquad = 2.072 \times 10^8\ \text{Btu/hr} \end{aligned}$$

Losses in the steam generator and electrical generator will further reduce this to

$$(0.86)\left(2.072 \times 10^8\ \frac{\text{Btu}}{\text{hr}}\right) = 1.782 \times 10^8\ \text{Btu/hr}$$

With an electrical output of 17 MW, thermal energy removed by cooling water is

$$\begin{aligned} Q &= 1.782 \times 10^8\ \frac{\text{Btu}}{\text{hr}} - (17\ \text{MW})\left(1000\ \frac{\text{kW}}{\text{MW}}\right) \\ &\quad \times \left(3413\ \frac{\frac{\text{Btu}}{\text{hr}}}{\text{kW}}\right) \\ &= 1.202 \times 10^8\ \text{Btu/hr} \end{aligned}$$

The temperature rise of the cooling water is

$$\Delta T = \frac{Q}{\dot{m}c_p}$$

At 60°F, the specific heat of water is $c_p = 1$ Btu/lbm-°F.

$$\begin{aligned} \Delta T &= \frac{1.202 \times 10^8\ \frac{\text{Btu}}{\text{hr}}}{\left(225\ \frac{\text{ft}^3}{\text{sec}}\right)\left(62.4\ \frac{\text{lbm}}{\text{ft}^3}\right)\left(3600\ \frac{\text{sec}}{\text{hr}}\right)\left(1\ \frac{\text{Btu}}{\text{lbm-°F}}\right)} \\ &= \boxed{2.38°\text{F}} \end{aligned}$$

The electrical generation is not cooled by the cooling water. Therefore, it is not correct to include the generation efficiency in the calculation of losses.

(d) Limiting 0.1 lbm of particulates per million Btu per hour, the allowable emission rate is

$$\begin{aligned} &\left(0.1\ \frac{\text{lbm}}{\text{MBtu}}\right)\left(15{,}300\ \frac{\text{lbm}}{\text{hr}}\right) \\ &\quad \times \left(13{,}653\ \frac{\text{Btu}}{\text{lbm}}\right)\left(\frac{1\ \text{MBtu}}{10^6\ \text{Btu}}\right) = 20.89\ \text{lbm/hr} \end{aligned}$$

The efficiency of the flue gas particulate collectors is

$$\begin{aligned} \eta &= \frac{\text{actual emission rate} - \text{allowable emission rate}}{\text{actual emission rate}} \\ &= \frac{1709.7\ \frac{\text{lbm}}{\text{hr}} - 20.89\ \frac{\text{lbm}}{\text{hr}}}{1709.7\ \frac{\text{lbm}}{\text{hr}}} \\ &= \boxed{0.988\ \ (98.8\%)} \end{aligned}$$

17. *Customary U.S. Solution*

(c) The stoichiometric reaction for propane is given in Table 37.7.

$$\begin{array}{lcccccc} & C_3H_8 + & 5O_2 & \longrightarrow & 3CO_2 & + & 4H_2O \\ \text{MW} & 44.097 & (5)(32.000) & & (3)(44.011) & & (4)(18.016) \end{array}$$

With 40% excess O_2 by weight,

$$\begin{array}{lccc} & C_3H_8 & & (1.4)(5)O_2 \\ \text{MW} & 44.097 & + & (7)(32.000) \\ & 44.097 & & 224 \end{array}$$

$$\begin{array}{ccccccc} & 3CO_2 & & 4H_2O & & 2O_2 \\ \longrightarrow & (3)(44.011) & + & (4)(18.016) & + & (2)(32.000) \\ & 132.033 & & 72.064 & & 64 \end{array}$$

The excess oxygen is

$$(2)(32) = 64 \text{ lbm/lbmol } C_3H_8$$

The mass ratio of nitrogen to oxygen is

$$\frac{G_N}{G_O} = \left(\frac{B_N}{R_N}\right)\left(\frac{R_O}{B_O}\right) = \left(\frac{0.40}{55.16 \, \frac{\text{ft-lbf}}{\text{lbm-°R}}}\right)\left(\frac{48.29 \, \frac{\text{ft-lbf}}{\text{lbm-°R}}}{0.60}\right) = 0.584$$

(Values of R_N and R_O are taken from Table 29.7.)

The nitrogen accompanying the oxygen is

$$(7)(32)(0.584) = 130.8 \text{ lbm}$$

The mass balance per mole of propane is

$$\begin{array}{llllllll} & C_3H_8 + & O_2 + & N_2 \longrightarrow & CO_2 + & H_2O + & O_2 + & N_2 \\ \text{mass per mole} & 44.097 + & 224 + & 130.8 \longrightarrow & 132.033 + & 72.064 + & 64 + & 130.8 \end{array}$$

At standard conditions (60°F, 1 atm), the propane density is given by Eq. 29.50.

$$\rho = \frac{p}{RT}$$

The absolute temperature, T, is

$$60°\text{F} + 460 = 520°\text{R}$$

R for propane from Table 29.7 is 35.04 ft-lbf/lbm-°R.

$$\rho = \frac{\left(14.7 \, \frac{\text{lbf}}{\text{in}^2}\right)\left(144 \, \frac{\text{in}^2}{\text{ft}^2}\right)}{\left(35.04 \, \frac{\text{ft-lbf}}{\text{lbm-°R}}\right)(520°\text{R})} = 0.1162 \text{ lbm/ft}^3$$

The mass flow rate of propane based on 250 SCFM of propane is

$$\left(250 \, \frac{\text{ft}^3}{\text{min}}\right)\left(0.1162 \, \frac{\text{lbm}}{\text{ft}^3}\right) = 29.05 \text{ lbm/min}$$

Scaling the other mass balance factors down by

$$\frac{29.05 \, \frac{\text{lbm}}{\text{min}}}{44.097 \, \frac{\text{lbm}}{\text{lbmol}}} = 0.6588 \text{ lbmol/min}$$

$$\begin{array}{lllll} & C_3H_8 + & O_2 & + N_2 & \\ \frac{\text{lbm}}{\text{min}} & 29.05 + & 147.57 + & 86.17 & \\ \longrightarrow & CO_2 + & H_2O + & O_2 + & N_2 \\ & 86.98 + & 47.48 + & 42.16 + & 86.17 \end{array}$$

The oxygen flow rate is $\boxed{14.76 \text{ lbm/min}}$ [part (c)].

(a) Using R from Table 24.7, the specific volumes of the reactants are given by Eq. 24.50.

$$\upsilon_{C_3H_8} = \frac{RT}{p} = \frac{\left(35.04 \, \frac{\text{ft-lbf}}{\text{lbm-°R}}\right)(80°\text{F} + 460)}{\left(14.7 \, \frac{\text{lbf}}{\text{in}^2}\right)\left(144 \, \frac{\text{in}^2}{\text{ft}^2}\right)} = 8.939 \text{ ft}^3/\text{lbm}$$

$$\upsilon_{O_2} = \frac{\left(48.29 \, \frac{\text{ft-lbf}}{\text{lbm-°R}}\right)(80°\text{F} + 460)}{\left(14.7 \, \frac{\text{lbf}}{\text{in}^2}\right)\left(144 \, \frac{\text{in}^2}{\text{ft}^2}\right)} = 12.319 \text{ ft}^3/\text{lbm}$$

$$\upsilon_{N_2} = \frac{\left(55.16 \, \frac{\text{ft-lbf}}{\text{lbm-°R}}\right)(80°\text{F} + 460)°\text{R}}{\left(14.7 \, \frac{\text{lbf}}{\text{in}^2}\right)\left(144 \, \frac{\text{in}^2}{\text{ft}^2}\right)} = 14.071 \text{ ft}^3/\text{lbm}$$

The total incoming volume is

$$\dot{V} = \left(29.05 \, \frac{\text{lbm}}{\text{min}}\right)\left(8.939 \, \frac{\text{ft}^3}{\text{lbm}}\right) + \left(147.57 \, \frac{\text{lbm}}{\text{min}}\right)\left(12.319 \, \frac{\text{ft}^3}{\text{lbm}}\right) + \left(86.17 \, \frac{\text{lbm}}{\text{min}}\right)\left(14.071 \, \frac{\text{ft}^3}{\text{lbm}}\right) = 3290 \text{ ft}^3/\text{min}$$

Since the velocity must be kept below 400 ft/min, the area of inlet pipe is

$$A_{in} = \frac{\dot{V}}{\text{v}} = \frac{3290 \, \frac{\text{ft}^3}{\text{min}}}{400 \, \frac{\text{ft}}{\text{min}}} = \boxed{8.23 \text{ ft}^2}$$

Combustion

(d) Similarly, the specific volumes of the products are

$$\upsilon_{CO_2} = \frac{\left(35.11\ \frac{\text{ft-lbf}}{\text{lbm-°R}}\right)(460°\text{F} + 460)}{\left(8\ \frac{\text{lbf}}{\text{in}^2}\right)\left(144\ \frac{\text{in}^2}{\text{ft}^2}\right)} = 28.04\ \text{ft}^3/\text{lbm}$$

$$\upsilon_{H_2O} = \frac{\left(85.78\ \frac{\text{ft-lbf}}{\text{lbm-°R}}\right)(460°\text{F} + 460)}{\left(8\ \frac{\text{lbf}}{\text{in}^2}\right)\left(144\ \frac{\text{in}^2}{\text{ft}^2}\right)} = 68.50\ \text{ft}^3/\text{lbm}$$

$$\upsilon_{O_2} = \frac{\left(48.29\ \frac{\text{ft-lbf}}{\text{lbm-°R}}\right)(460°\text{F} + 460)}{\left(8\ \frac{\text{lbf}}{\text{in}^2}\right)\left(144\ \frac{\text{in}^2}{\text{ft}^2}\right)} = 38.56\ \text{ft}^3/\text{lbm}$$

$$\upsilon_{N_2} = \frac{\left(55.16\ \frac{\text{ft-lbf}}{\text{lbm-°R}}\right)(460°\text{F} + 460)}{\left(8\ \frac{\text{lbf}}{\text{in}^2}\right)\left(144\ \frac{\text{in}^2}{\text{ft}^2}\right)} = 44.05\ \text{ft}^3/\text{lbm}$$

The total exhaust volume is

$$\begin{aligned}\dot{V} &= \left(86.98\ \frac{\text{lbm}}{\text{min}}\right)\left(28.04\ \frac{\text{ft}^3}{\text{lbm}}\right) \\ &\quad + \left(47.48\ \frac{\text{lbm}}{\text{min}}\right)\left(68.50\ \frac{\text{ft}^3}{\text{lbm}}\right) \\ &\quad + \left(42.16\ \frac{\text{lbm}}{\text{min}}\right)\left(38.56\ \frac{\text{ft}^3}{\text{lbm}}\right) \\ &\quad + \left(86.98\ \frac{\text{lbm}}{\text{min}}\right)\left(44.05\ \frac{\text{ft}^3}{\text{min}}\right) \\ &= \boxed{11{,}148\ \text{ft}^3/\text{min}}\quad \text{[part (d)]}\end{aligned}$$

(b) Since the velocity of the products must be kept below 800 ft/min, the area of the stack is

$$A_{\text{stack}} = \frac{Q}{\text{v}} = \frac{11{,}148\ \frac{\text{ft}^3}{\text{min}}}{800\ \frac{\text{ft}}{\text{min}}} = \boxed{13.94\ \text{ft}^2}\quad \text{[part (b)]}$$

(e) For ideal gases, the partial pressure is volumetrically weighted. The water vapor partial pressure in the stack is

$$(8\ \text{psia})\left(\frac{\left(47.48\ \frac{\text{lbm}}{\text{min}}\right)\left(68.50\ \frac{\text{ft}^3}{\text{lbm}}\right)}{11{,}148\ \frac{\text{ft}^3}{\text{lbm}}}\right) = 2.33\ \text{psia}$$

The saturation temperature corresponding to 2.33 psia is $T_{\text{dp}} = \boxed{132°\text{F}.}$

SI Solution

(c) Following the procedure for the Customary U.S. Solution, the mass balance per mole of propane is

	C_3H_8 +	O_2 +	N_2 ⟶	CO_2 +	H_2O +	O_2 +	N_2
kg per mole	44.097 +	224 +	130.8 ⟶	132.033 +	72.064 +	64 +	130.8

At standard conditions (16°C, 101.3 kPa), the propane density is given by Eq. 24.50.

$$\rho = \frac{p}{RT}$$

The absolute temperature is

$$T = 16°\text{C} + 273 = 289\text{K}$$

From Table 29.7, R for propane is 188.55 J/kg·K.

$$\rho = \frac{(101.3\ \text{kPa})\left(1000\ \frac{\text{Pa}}{\text{kPa}}\right)}{\left(188.55\ \frac{\text{J}}{\text{kg·K}}\right)(289\text{K})} = 1.86\ \text{kg/m}^3$$

The mass flow rate of propane based on 118 L/s of propane is

$$\left(118\ \frac{\text{L}}{\text{s}}\right)\left(\frac{1\ \text{m}^3}{1000\ \text{L}}\right)\left(1.86\ \frac{\text{kg}}{\text{m}^3}\right) = 0.2195\ \text{kg/s}$$

Scale the other mass balance factors down.

$$\frac{0.2195\ \frac{\text{kg}}{\text{s}}}{44.097\ \frac{\text{kg}}{\text{kmol}}} = 0.004978\ \text{kmol/s}$$

	C_3H_8 +	O_2 +	N_2 ⟶	CO_2 +	H_2O	+ O_2 +	N_2
$\frac{\text{kg}}{\text{s}}$	0.2195	1.115	0.6511	0.6572	0.3587	0.3186	0.6511

The oxygen flow rate is $\boxed{1.115\ \text{kg/s}.}$

(a) Using R from Table 29.7, the specific volumes of the reactants are given by Eq. 29.50.

$$v_{C_3H_8} = \frac{RT}{p} = \frac{\left(188.55\ \frac{J}{kg{\cdot}K}\right)(27^\circ C + 273)}{(101\ kPa)\left(1000\ \frac{Pa}{kPa}\right)}$$
$$= 0.5600\ m^3/kg$$

$$v_{O_2} = \frac{\left(259.82\ \frac{J}{kg{\cdot}K}\right)(27^\circ C + 273)}{(101\ kPa)\left(1000\ \frac{Pa}{kPa}\right)}$$
$$= 0.7717\ m^3/kg$$

$$v_{N_2} = \frac{\left(296.77\ \frac{J}{kg{\cdot}K}\right)(27^\circ C + 273)}{(101\ kPa)\left(1000\ \frac{Pa}{kPa}\right)}$$
$$= 0.8815\ m^3/kg$$

The total incoming volume, $\dot{V}$, is

$$\begin{aligned}\dot{V} &= \left(0.2195\ \frac{kg}{s}\right)\left(0.5600\ \frac{m^3}{kg}\right) \\ &\quad + \left(1.115\ \frac{kg}{s}\right)\left(0.7717\ \frac{m^3}{kg}\right) \\ &\quad + \left(0.6511\ \frac{kg}{s}\right)\left(0.8815\ \frac{m^3}{kg}\right) \\ &= 1.557\ m^3/s\end{aligned}$$

Since the velocity for the reactants must be kept below 2 m/s, the area of inlet pipe is

$$A_{in} = \frac{\dot{V}}{v} = \frac{1.557\ \frac{m^3}{s}}{2\ \frac{m}{s}} = \boxed{0.779\ m^2}$$

(d) Similarly, the specific volumes of the products are

$$v_{CO_2} = \frac{\left(188.92\ \frac{J}{kg}\right)(240^\circ C + 273)}{(55\ kPa)\left(1000\ \frac{Pa}{kPa}\right)}$$
$$= 1.762\ m^3/kg$$

$$v_{H_2O} = \frac{\left(461.5\ \frac{J}{kg}\right)(240^\circ C + 273)}{(55\ kPa)\left(1000\ \frac{Pa}{kPa}\right)}$$
$$= 4.305\ m^3/kg$$

$$v_{O_2} = \frac{\left(259.82\ \frac{J}{kg{\cdot}K}\right)(240^\circ C + 273)}{(55\ kPa)\left(1000\ \frac{Pa}{kPa}\right)}$$
$$= 2.423\ m^3/kg$$

$$v_{N_2} = \frac{\left(296.77\ \frac{J}{kg{\cdot}K}\right)(240^\circ C + 273)}{(55\ kPa)\left(1000\ \frac{Pa}{kPa}\right)}$$
$$= 2.768\ m^3/kg$$

The total exhaust volume is

$$\begin{aligned}\dot{V} &= \left(0.6572\ \frac{kg}{s}\right)\left(1.762\ \frac{m^3}{kg}\right) \\ &\quad + \left(0.3587\ \frac{kg}{s}\right)\left(4.305\ \frac{m^3}{kg}\right) \\ &\quad + \left(0.3186\ \frac{kg}{s}\right)\left(2.423\ \frac{m^3}{kg}\right) \\ &\quad + \left(0.6511\ \frac{kg}{s}\right)\left(2.768\ \frac{m^3}{kg}\right) \\ &= \boxed{5.276\ m^3/s}\end{aligned}$$

(b) Since the velocity of products must be kept below 4 m/s, the area of stack is

$$A_{stack} = \frac{\dot{V}}{v} = \frac{5.276\ \frac{m^3}{s}}{4\ \frac{m}{s}}$$
$$= \boxed{1.319\ m^2}$$

(e) For ideal gases, the partial pressure is volumetrically weighted. The water vapor partial pressure in the stack is

$$(55\ kPa)\left(\frac{\left(0.3584\ \frac{kg}{s}\right)\left(4.305\ \frac{m^3}{kg}\right)}{5.273\ \frac{m^3}{s}}\right)$$
$$= 16.093\ kPa$$

The saturation temperature corresponding to 16.093 kPa is found from App. 29.O to be $T_{dp} = \boxed{54.5^\circ C.}$

18. *Customary U.S. Solution*

(a) Since atmospheric air is not used, the nitrogen and oxygen can be varied independently. Furthermore, since enthalpy increase information is not given for oxygen, a 0% excess oxygen can be assumed.

Combustion

From Table 37.7,

$$\begin{array}{lcccc} & C_3H_8 & + 5O_2 & \longrightarrow 3CO_2 & + 4H_2O \\ \text{moles} & (1) & (5) & (3) & (4) \end{array}$$

Subtract the reactant enthalpies from the product enthalpies to calculate the heat of reaction. The enthalpy of formation of oxygen is zero, since it is an element in its natural state.

$$\begin{aligned} & n_{CO_2}(\Delta H_f)_{CO_2} + n_{H_2O}(\Delta H_f)_{H_2O} \\ & \quad - n_{C_3H_8}(\Delta H_f)_{C_3H_8} - n_{O_2}(\Delta H_f)_{O_2} \\ & \quad = (3 \text{ lbmol})\left(-169{,}300 \ \frac{\text{Btu}}{\text{lbmol}}\right) \\ & \qquad + (4 \text{ lbmol})\left(-104{,}040 \ \frac{\text{Btu}}{\text{lbmol}}\right) \\ & \qquad - (1 \text{ lbmol})\left(28{,}800 \ \frac{\text{Btu}}{\text{lbmol}}\right) \\ & \qquad - (5 \text{ lbmol})(0) \\ & \quad = -952{,}860 \text{ Btu/lbmol of fuel} \end{aligned}$$

The negative sign indicates an exothermal reaction.

Let x be the number of moles of nitrogen per mole of propane. Use the nitrogen to cool the combustion. The above heat of reaction will increase the enthalpy of products from the standard reference temperature to 3600°R. Therefore,

$$\begin{aligned} 952{,}860 \ \frac{\text{Btu}}{\text{lbmol}} &= (3 \text{ lbmol})\left(39{,}791 \ \frac{\text{Btu}}{\text{lbmol}}\right) \\ & \quad + (4 \text{ lbmol})\left(31{,}658 \ \frac{\text{Btu}}{\text{lbmol}}\right) \\ & \quad + x\left(24{,}471 \ \frac{\text{Btu}}{\text{lbmol}}\right) \\ x &= 28.89 \text{ lbmol/lbmol fuel} \end{aligned}$$

The mass of nitrogen per lbmole of propane is

$$\begin{aligned} M_{N_2} &= \left(28.89 \ \frac{\text{lbmol}}{\text{lbmol fuel}}\right)\left(28.016 \ \frac{\text{lbm}}{\text{lbmol}}\right) \\ &= \boxed{809.4 \text{ lbm/lbmol propane}} \end{aligned}$$

The mass of oxygen per lbmole of propane is

$$\begin{aligned} M_{O_2} &= (5 \text{ lbmol})\left(32 \ \frac{\text{lbm}}{\text{lbmol}}\right) \\ &= \boxed{160 \text{ lbm/lbmol fuel}} \end{aligned}$$

(b) The partial pressure is volumetrically weighted. This is the same as molar weighting.

product	lbmol	volumetric fraction
CO_2	3	3/35.89 = 0.0836
H_2O	4	4/35.89 = 0.1115
N_2	28.89	28.89/35.89 = 0.8049
O_2	0	0/35.89 = 0
	35.89 lbmol	1.000

The partial pressure of water vapor is

$$\begin{aligned} p_{H_2O} &= \left(\frac{n_{H_2O}}{n}\right)p = (0.1115)(14.7 \text{ psia}) \\ &= 1.64 \text{ psia} \end{aligned}$$

From App. 29.B, this corresponds to approximately 118°F. Since the stack temperature is 100°F, some of the water will condense. From App. 29.A, the maximum vapor pressure of water is 0.9503 psia. Let n be the number of moles of water in the stack gas.

$$\begin{aligned} n_{H_2O} &= \left(\frac{p_{H_2O}}{p}\right)n = \left(\frac{0.9503 \text{ psia}}{14.7 \text{ psia}}\right)(35.89 \text{ lbmol}) \\ &= \boxed{2.320 \text{ lbmol/lbmol } C_3H_8} \end{aligned}$$

(c) The water removed is

$$\begin{aligned} 4 - n_{H_2O} &= 4 \text{ lbmol} - 2.320 \text{ lbmol} \\ &= 1.680 \text{ lbmol of } H_2O\text{/lbmol of } C_3H_8 \\ m &= (1.680 \text{ lbmol})\left(18.016 \ \frac{\text{lbm}}{\text{lbmol}}\right) \\ &= \boxed{30.27 \text{ lbm } H_2O\text{/lbmol } C_3H_8} \end{aligned}$$

SI Solution

(a) From the customary U.S. solution, the heat of reaction is

$$\begin{aligned} & n_{CO_2}(\Delta H_f)_{CO_2} + n_{H_2O}(\Delta H_f)_{H_2O} \\ & \quad - n_{C_3H_8}(\Delta H_f)_{C_3H_8} - n_{O_2}(\Delta H_f)_{O_2} \\ & \quad = (3 \text{ kmol})\left(-393.8 \ \frac{\text{GJ}}{\text{kmol}}\right) \\ & \qquad + (4 \text{ kmol})\left(-242 \ \frac{\text{GJ}}{\text{kmol}}\right) \\ & \qquad - (1 \text{ kmol})\left(67.0 \ \frac{\text{GJ}}{\text{kmol}}\right) - (5 \text{ kmol})(0) \\ & \quad = -2216.4 \text{ GJ/kmol fuel} \end{aligned}$$

The negative sign indicates an exothermal reaction.

Let x be the number of moles of nitrogen per mole of propane. Use the nitrogen to cool the combustion. The above heat of reaction will increase the enthalpy of products from the standard reference temperature to 1980°C. Therefore,

$$\begin{aligned}2216.4\ \frac{\text{GJ}}{\text{mol}} &= (3\ \text{kmol})\left(92.6\ \frac{\text{GJ}}{\text{kmol}}\right) \\ &\quad + (4\ \text{kmol})\left(73.6\ \frac{\text{GJ}}{\text{kmol}}\right) \\ &\quad + x\left(56.9\ \frac{\text{GJ}}{\text{kmol}}\right) \\ x &= 28.90\ \text{kmol/kmol fuel}\end{aligned}$$

The mass of nitrogen per mole of propane is

$$\begin{aligned}M_{N_2} &= (28.90\ \text{kmol})\left(28.016\ \frac{\text{kg}}{\text{kmol}}\right) \\ &= \boxed{809.7\ \text{kg/kmol fuel}}\end{aligned}$$

The mass of oxygen per kmol of propane is

$$\begin{aligned}M_{O_2} &= (5\ \text{kmol})\left(32\ \frac{\text{kg}}{\text{kmol}}\right) \\ &= \boxed{160\ \text{kg/kmol propane}}\end{aligned}$$

(b) The partial pressure is volumetrically weighted. This is the same as molar weighting.

product	kmol	volumetric fraction
CO_2	3	3/35.90 = 0.0836
H_2O	4	4/35.90 = 0.1114
N_2	28.90	28.90/35.90 = 0.8050
O_2	0	0/35.90 = 0
	35.90 kmol	1.000

The partial pressure of water vapor is

$$\begin{aligned}p_{H_2O} &= \left(\frac{n_{H_2O}}{n}\right)p = (0.1114)(101\ \text{kPa}) \\ &= 11.25\ \text{kPa}\end{aligned}$$

From App. 29.O, this corresponds to approximately 47.6°C. Since the stack temperature is 38°C, some of the water will condense. From App. 29.N, the maximum vapor pressure of water at the stack temperature is 6.632 kPa. Let n be the number of moles of water in the stack gas.

$$\begin{aligned}n_{H_2O} &= \left(\frac{p_{H_2O}}{p}\right)n = \left(\frac{6.632\ \text{kPa}}{101\ \text{kPa}}\right)(35.90\ \text{kmol}) \\ &= \boxed{2.357\ \text{kmol}\ H_2O/\text{kmol}\ C_3H_8}\end{aligned}$$

(c) The liquid water removed is

$$\begin{aligned}4 - n_{H_2O} &= 4\ \text{kmol} - 2.357\ \text{kmol} \\ &= 1.643\ \text{kmol of}\ H_2O/\text{kmol propane} \\ \text{m} &= (1.643\ \text{lbmol})\left(18.016\ \frac{\text{kg}}{\text{kmol}}\right) \\ &= \boxed{29.6\ \text{kg}\ H_2O/\text{kmol}\ C_3H_8}\end{aligned}$$

19. *Customary U.S. Solution*

(a) If the power output is to be unchanged,

$$\begin{aligned}\text{BHP}_1 &= \text{BHP}_2 \\ (\dot{m}_{F,1})(\text{LHV}_1) &= (\dot{m}_{F,2})(\text{LHV}_2)\end{aligned}$$

From Eq. 29.8,

$$\begin{aligned}\dot{m}_F &= (\text{BSFC})(\text{BHP}) \\ (\text{BSFC}_1)(\text{LHV}_1) &= (\text{BSFC}_2)(\text{LHV}_2) \\ \frac{\text{BSFC}_2}{\text{BSFC}_1} &= \frac{\text{LHV}_1}{\text{LHV}_2} = \frac{23{,}200\ \frac{\text{Btu}}{\text{lbm}}}{11{,}930\ \frac{\text{Btu}}{\text{lbm}}} \\ &= 1.945 \\ \frac{\text{BSFC}_2 - \text{BSFC}_1}{\text{BSFC}_1} &= \frac{\text{BSFC}_2}{\text{BSFC}_1} - 1 = 1.945 - 1 \\ &= \boxed{0.945\quad [94.5\%\ \text{increase}]}\end{aligned}$$

(b) If the fuel injection velocity is to be unchanged ($\text{v}_2 = \text{v}_1$),

$$\begin{aligned}\dot{m} &= \rho A \text{v} \\ A &= \frac{\dot{m}}{\rho \text{v}} \\ \dot{m}_2 &= 1.945\dot{m}_1\quad [\text{part (a)}] \\ \frac{A_2 - A_1}{A_1} &= \frac{\frac{\dot{m}_2}{\rho_2 \text{v}_2} - \frac{\dot{m}_1}{\rho_1 \text{v}_1}}{\frac{\dot{m}_1}{\rho_1 \text{v}_1}} = \frac{\frac{\dot{m}_2}{\rho_2} - \frac{\dot{m}_1}{\rho_1}}{\frac{\dot{m}_1}{\rho_1}} \\ &= \frac{\frac{1.945 m_1}{\rho_2} - \frac{\dot{m}_1}{\rho_1}}{\frac{\dot{m}_1}{\rho_1}} = \frac{\frac{1.945}{\rho_2} - \frac{1}{\rho_1}}{\frac{1}{\rho_1}} \\ \frac{A_2 - A_1}{A_1} &= (1.945)\left(\frac{\rho_1}{\rho_2}\right) - 1\end{aligned}$$

Combustion

From Table 37.3, for gasoline, $SG_1 = 0.74$. For ethanol, $SG_2 = 0.794$.

$$\frac{\rho_1}{\rho_2} = \frac{SG_1}{SG_2} = \frac{0.74}{0.794}$$

$$\frac{A_2 - A_1}{A_1} = (1.945)\left(\frac{0.74}{0.794}\right) - 1$$

$$= \boxed{0.810 \quad [81\% \text{ increase}]}$$

(c) If no changes are made to the engine, power output is proportional to the weight flow and heating value.

$$\frac{P_2 - P_1}{P_1} = \frac{\dot{m}_{F2}(\text{LHV}_2) - \dot{m}_{F1}(\text{LHV}_1)}{\dot{m}_{F1}(\text{LHV}_1)}$$

$$\dot{m}_F = \rho A \text{v}$$

$$\frac{P_2 - P_1}{P_1} = \frac{(\rho_2 A_2 \text{v}_2)(\text{LHV}_2) - (\rho_1 A_1 \text{v}_1)(\text{LHV}_1)}{(\rho_1 A_1 \text{v}_1)(\text{LHV}_1)}$$

Since no changes are made to the engine, $A_2 = A_1$ and $\text{v}_2 = \text{v}_1$.

$$\frac{P_2 - P_1}{P_1} = \frac{\rho_2(\text{LHV}_2) - \rho_1(\text{LHV}_1)}{\rho_1(\text{LHV}_1)}$$

$$= \frac{\text{LHV}_2 - \left(\frac{\rho_1}{\rho_2}\right)(\text{LHV}_1)}{\left(\frac{\rho_1}{\rho_2}\right)(\text{LHV}_1)}$$

From part (b),

$$\frac{\rho_1}{\rho_2} = \frac{0.74}{0.794}$$

$$\frac{P_2 - P_1}{P_1} = \frac{11{,}930\ \frac{\text{Btu}}{\text{lbm}} - \left(\frac{0.74}{0.794}\right)\left(23{,}200\ \frac{\text{Btu}}{\text{lbm}}\right)}{\left(\frac{0.74}{0.794}\right)\left(23{,}200\ \frac{\text{Btu}}{\text{lbm}}\right)}$$

$$= \boxed{-0.45 \quad [45\% \text{ decrease}]}$$

SI Solution

(a) From part (a) of the customary U.S. solution,

$$\frac{\text{BSFC}_2}{\text{BSFC}_1} = \frac{\text{LHV}_1}{\text{LHV}_2} = \frac{54\ \frac{\text{MJ}}{\text{kg}}}{27.7\ \frac{\text{MJ}}{\text{kg}}} = 1.949$$

$$\frac{\text{BSFC}_2 - \text{BSFC}_1}{\text{BSFC}_1} = \frac{\text{BSFC}_2}{\text{BSFC}_1} - 1 = 1.949 - 1$$

$$= \boxed{0.949 \quad [94.9\% \text{ increase}]}$$

(b) From part (b) of the customary U.S. solution,

$$\frac{A_2 - A_1}{A_1} = \frac{\frac{\dot{m}_2}{\rho_2} - \frac{\dot{m}_1}{\rho_1}}{\frac{\dot{m}_1}{\rho_1}}$$

$$\dot{m}_2 = 1.949\dot{m}_1 \quad [\text{part (a)}]$$

$$\frac{A_2 - A_1}{A_1} = \frac{\frac{1.949\dot{m}_1}{\rho_2} - \frac{\dot{m}_1}{\rho_1}}{\frac{\dot{m}_1}{\rho_1}}$$

$$= (1.949)\left(\frac{\rho_1}{\rho_2}\right) - 1$$

$$= (1.949)\left(\frac{0.74}{0.794}\right) - 1$$

$$= \boxed{0.816 \quad [81.6\% \text{ increase}]}$$

(c) From part (c) of the customary U.S. solution,

$$\frac{P_2 - P_1}{P_1} = \frac{\text{LHV}_2 - \left(\frac{\rho_1}{\rho_2}\right)(\text{LHV}_1)}{\left(\frac{\rho_1}{\rho_2}\right)(\text{LHV}_1)}$$

$$= \frac{27.7\ \frac{\text{MJ}}{\text{kg}} - \left(\frac{0.74}{0.794}\right)\left(54\ \frac{\text{MJ}}{\text{kg}}\right)}{\left(\frac{0.74}{0.795}\right)\left(54\ \frac{\text{MJ}}{\text{kg}}\right)}$$

$$= \boxed{-0.45 \quad [45\% \text{ decrease}]}$$

38 Air Quality

PRACTICE PROBLEMS

1. The following questions pertain to air pollution regulations of the Clean Air Act (CAA) and the Resource Conservation and Recovery Act (RCRA).

(a) What regulatory standards define acceptable levels for the criteria pollutants?

(A) National Ambient Air Quality Standards (NAAQS)
(B) New Source Performance Standards (NSPS)
(C) National Emission Standards for Hazardous Air Pollutants (NESHAP)
(D) Lowest Achievable Emission Rate (LAER) for Non-attainment Areas (NA)

(b) Which of the following is not included under the CAA?

(A) regional classification as part of Prevention of Significant Deterioration (PSD)
(B) State Implementation Plans (SIP) to define control strategies for criteria pollutants
(C) hazardous air pollutants such as asbestos, mercury, beryllium, and radionuclides
(D) performance standards for principal organic hazardous constituent (POHC) emissions from hazardous waste incinerators

(c) What option is not available under emissions trading?

(A) bubble
(B) emission reduction credit
(C) interregional averaging
(D) offset

(d) What does destruction and removal efficiency (DRE) encompass?

(A) destruction and removal in the incinerator of the principal organic hazardous constituent (POHC) prior to the application of any air pollution control equipment
(B) destruction and removal in the incinerator of all pollutants regulated under the National Emission Standards for Hazardous Air Pollutants (NESHAP) prior to the application of any air pollution control equipment
(C) destruction in the incinerator and removal by any air pollution control equipment of the principal organic hazardous constituent (POHC)
(D) destruction in the incinerator and removal by any air pollution control equipment of all pollutants regulated under the National Emission Standards for Hazardous Air Pollutants (NESHAP)

(e) What is the corrected particulate concentration for particulate at 347 mg/m^3 and 4% O_2?

(A) 58 mg/m^3
(B) 286 mg/m^3
(C) 347 mg/m^3
(D) 421 mg/m^3

2. The questions below address issues related to classification of air pollutants.

(a) What average particle size defines the upper limit of respirable particulate matter?

(A) 0.5 μm
(B) 1 μm
(C) 2.5 μm
(D) 10 μm

(b) What is the settling Reynolds number for a particle with a 5 μm diameter and 1000 kg/m^3 density in air at 60°C and 1 atm?

(A) 7.5×10^{-5}
(B) 1.9×10^{-4}
(C) 3.7×10^{-4}
(D) 1.4×10^{-3}

(c) Historically, what has been the primary source of airborne lead pollution?

(A) automobile exhaust emissions
(B) lead smelting emissions
(C) lead-based paint weathering and flaking
(D) electronic equipment soldering emissions

(d) What are the most common primary forms of nitrogen oxides (NOx) and sulfur oxides (SOx) when emitted?

(A) nitric oxide (NO) and sulfur dioxide (SO_2)
(B) nitric oxide (NO) and sulfur trioxide (SO_3)
(C) nitrogen dioxide (NO_2) and sulfur dioxide (SO_2)
(D) nitrogen dioxide (NO_2) and sulfur trioxide (SO_3)

(e) What condensed chemical formula represents an alkane radical?

(A) $CH_3{-}CH{=}CH_2$
(B) $CH_2{=}O$
(C) $CH_3{-}CH_3$
(D) $CH_2{-}CH_2{-}CH_3$

3. A mixture of 20% methane, 30% oxygen, and 50% nitrogen is stored in a pressurized container at 1.5 atm and 15°C. What is the change in the density of the gas mixture if the temperature is increased to 25°C at constant pressure?

4. Air quality monitoring has produced the following data for particulate matter.

particle size collected	10 μm
clean filter mass	10.1453 g
filter mass at 24 h	10.5818 g
initial air flow	1.6 m^3/min
final air flow	1.58 m^3/min

(a) What is the PM 10 concentration?

(A) 174 $\mu g/m^3$
(B) 191 $\mu g/m^3$
(C) 275 $\mu g/m^3$
(D) 395 $\mu g/m^3$

(b) What is the Pollution Standard Index (PSI) based on PM 10?

(A) 112
(B) 121
(C) 163
(D) 264

(c) What is the air quality descriptor corresponding to the PSI based on PM 10?

(A) moderate
(B) unhealthful
(C) very unhealthful
(D) hazardous

(d) If ozone (O_3) were present along with the PM 10 at a 1 h concentration of 0.42 ppm, what would be the PSI?

(A) 121
(B) 163
(C) 302
(D) 320

(e) What is the air quality descriptor corresponding to the PSI when ozone is included?

(A) moderate
(B) unhealthful
(C) very unhealthful
(D) hazardous

5. An air pollutant is emitted from a stack with an effective height of 185 m at 27 kg/s. The wind speed at 10 m above grade is 5.3 m/s under a clear night sky. (a) What is the ground-level concentration at a point 2000 m downwind and 300 m crosswind of the stack? (b) If the plume temperature at the stack is 30°C and the ambient lapse rate and ground level air temperature are −0.0088°C/m and 18°C, respectively, will the plume likely be at ground level as predicted?

6. An electrostatic precipitator has been selected to treat a particulate-laden air stream characterized as follows.

particulate concentration	11 g/m^3
particulate temperature	175°C
air flow	4 m^3/s
average particle diameter	6 μm
required removal efficiency	95%
applied voltage	80 kV
distance between collectors	36 cm

(a) What is the spacing between the discharge electrodes and the collectors?

(A) 9 cm
(B) 18 cm
(C) 36 cm
(D) 72 cm

(b) What is the average electric field?

(A) 1.0×10^5 N/C
(B) 2.2×10^5 N/C
(C) 4.4×10^5 N/C
(D) 8.9×10^5 N/C

(c) What is the drift velocity?

(A) 0.0071 m/s
(B) 0.034 m/s
(C) 0.14 m/s
(D) 0.56 m/s

(d) What is the required collector area?

(A) 22 m^2
(B) 86 m^2
(C) 350 m^2
(D) 1700 m^2

(e) What is the specific collection area?

(A) 6 $m^2 \cdot s/m^3$
(B) 22 $m^2 \cdot s/m^3$
(C) 88 $m^2 \cdot s/m^3$
(D) 430 $m^2 \cdot s/m^3$

7. The following questions relate to mobile source air pollution.

(a) Which of the following would not contribute to overall improved air quality in urban areas?

(A) Increase access to public transportation.
(B) Increase the proportion of new cars to old cars.
(C) Increase the proportion of cars with diesel engines to those with gasoline engines.
(D) Require routine maintenance of automobile air pollution control equipment.

(b) What strategy does the EPA employ to regulate mobile source emissions?

(A) pre-consumer certification and enforcement
(B) pre-consumer certification, post-manufacture enforcement
(C) pre-consumer enforcement, post-manufacture certification
(D) post-manufacture certification and enforcement

(c) What is the EPA average fuel economy for a car rated at 20 mpg city and 34 mpg highway?

(A) 24.3 mpg
(B) 24.5 mpg
(C) 25.9 mpg
(D) 27.0 mpg

(d) Why are fuel economy ratings important to air quality?

(A) Lower fuel consumption results in lower emissions per mile driven.
(B) Lower fuel consumption reduces fuel production demand, thereby reducing refinery emissions.
(C) Lower fuel consumption reduces fuel tank fill-ups, thereby reducing hydrocarbon emissions.
(D) Lower fuel consumption entices consumers to purchase less-polluting, smaller cars.

(e) What are the primary sources of emissions from automobiles to the air?

(A) ventilation from the fuel tank, exhaust from the tailpipe, and particulates from tire wear and roadway surface abrasion
(B) ventilation from the crankcase, ventilation from the fuel tank, exhaust from the tailpipe, and particulates from tire wear and roadway surface abrasion
(C) volatilization from the carburetor or throttle body, exhaust from the tailpipe, and particulates from tire wear and roadway surface abrasion
(D) ventilation from the crankcase, ventilation from the fuel tank, volatilization from the carburetor or throttle body, and exhaust from the tailpipe

8. Radon gas enters a 1000 m^3 basement room at 2600 pCi/min.

(a) What ventilation flow rate is required to keep the radon concentration in the room below 4 pCi/L?

(A) 0.65 m^3/min
(B) 6.5 m^3/min
(C) 10 m^3/min
(D) 15 m^3/min

(b) At the calculated ventilation rate, how long is required to change out one room-volume of air?

(A) 1.2 h
(B) 1.7 h
(C) 2.5 h
(D) 26 h

(c) What is the primary source of radon in indoor air?

(A) carpeting and furnishings
(B) concrete and masonry
(C) electronic equipment
(D) soil

(d) How does environmental tobacco smoke (ETS) influence radon exposure?

(A) There is no influence.
(B) Radon adsorbs to the ETS particulate, reducing exposure.
(C) Radon adsorbs to the ETS particulate, increasing exposure.
(D) Radon absorbs ETS gases, increasing exposure.

(e) What is the primary health risk associated with radon exposure?

(A) lung cancer
(B) genetic defects
(C) bone decalcification
(D) permanent memory loss

9. Nitrogen dioxide (NO_2) and sulfur trioxide (SO_3) concentrations in the air in an urban area are 254 μg/m^3 and 427 μg/m^3, respectively. Assume that 1 L of rainwater will scrub 10 m^3 of air and that the natural pH of rainwater is 5.5. What will be the rainwater pH after scrubbing the NO_2 and SO_3?

SOLUTIONS

1. (a) The National Ambient Air Quality Standards (NAAQS) define acceptable levels for the criteria pollutants.

The answer is (A).

(b) The CAA does not include performance standards for the principal organic hazardous constituent (POHC) emissions from hazardous waste incinerators. The POHC emissions from hazardous waste incinerators are regulated under RCRA.

The answer is (D).

(c) Interregional averaging is not an option available under emissions trading.

The answer is (C).

(d) Destruction and removal efficiency (DRE) encompasses destruction in the incinerator and removal by any air pollution control equipment of the principal organic hazardous constituent (POHC).

The answer is (C).

(e) From Eq. 38.2,

$$\begin{aligned} P_c &= P_m\left(\frac{14}{21-Y}\right) \\ &= \left(347\ \frac{\text{mg}}{\text{m}^3}\right)\left(\frac{14}{21-4}\right) \\ &= \boxed{286\ \text{mg/m}^3} \end{aligned}$$

The answer is (B).

2. (a) The average particle size that defines the upper limit of respirable particulate matter is 2.5 μm.

The answer is (C).

(b) From App. 38.C, $\mu_g = 1.97 \times 10^{-5}$ kg/m·s and $\rho_g = 1.06$ kg/m^3 at 60°C.

Use Eq. 38.3.

$$\begin{aligned} v_s &= \frac{d_p^2(\rho_p-\rho_g)g}{18\mu_g} \\ &= \frac{\begin{array}{c}(5\ \mu\text{m})^2\left(10^{-6}\ \frac{\text{m}}{\mu\text{m}}\right)^2\left(1000\ \frac{\text{kg}}{\text{m}^3}-1.06\ \frac{\text{kg}}{\text{m}^3}\right) \\ \times\left(9.81\ \frac{\text{m}}{\text{s}^2}\right)\end{array}}{(18)\left(1.97\times10^{-5}\ \frac{\text{kg}}{\text{m·s}}\right)} \\ &= 6.9\times10^{-4}\ \text{m/s} \end{aligned}$$

Use Eq. 38.4.

$$\begin{aligned}\text{Re} &= \frac{\mathrm{v}_s d_p \rho_g}{\mu_g} \\ &= \frac{\left(6.9\times 10^{-4}\ \frac{\text{m}}{\text{s}}\right)(5\ \mu\text{m})\left(10^{-6}\ \frac{\text{m}}{\mu\text{m}}\right)\left(1.06\ \frac{\text{kg}}{\text{m}^3}\right)}{1.97\times 10^{-5}\ \frac{\text{kg}}{\text{m}\cdot\text{s}}} \\ &= \boxed{1.9\times 10^{-4}}\end{aligned}$$

The answer is (B).

(c) Historically, the primary source of airborne lead pollution has been automobile exhaust emissions.

The answer is (A).

(d) The most common primary forms of nitrogen oxides (NO_x) and sulfur oxides (SO_x) when emitted are nitric oxide (NO) and sulfur dioxide (SO_2).

The answer is (A).

(e) The formula representing an alkane radical is CH_2-CH_2-CH_3 (propyl).

The answer is (D).

3. The molecular weight of CH_4 is

$$\left(12\ \frac{\text{g}}{\text{mol}}\right) + (4)\left(1\ \frac{\text{g}}{\text{mol}}\right) = 16\ \text{g/mol}$$

The molecular weight of O_2 is

$$(2)\left(16\ \frac{\text{g}}{\text{mol}}\right) = 32\ \text{g/mol}$$

The molecular weight of N_2 is

$$(2)\left(14\ \frac{\text{g}}{\text{mol}}\right) = 28\ \text{g/mol}$$

From Eq. 38.15 the apparent molecular weight of the gas mixture is

$$\begin{aligned}&(0.20)\left(16\ \frac{\text{g}}{\text{mol}}\right) + (0.30)\left(32\ \frac{\text{g}}{\text{mol}}\right) \\ &\quad +(0.50)\left(28\ \frac{\text{g}}{\text{mol}}\right) = 26.8\ \text{g/mol}\end{aligned}$$

Use Eq. 38.12.

ρ_g at 1.5 atm and 15°C is

$$\begin{aligned}\rho_g &= \frac{p(\text{MW})}{R^*T} \\ &= \frac{(1.5\ \text{atm})\left(26.8\ \frac{\text{g}}{\text{mol}}\right)\left(\frac{1\ \text{kg}}{1000\ \text{g}}\right)}{\left(8.2\times 10^{-5}\ \frac{\text{atm}\cdot\text{m}^3}{\text{mol}\cdot\text{K}}\right)(273+15^\circ\text{C})} \\ &= 1.702\ \text{kg/m}^3\end{aligned}$$

ρ_g at 1.5 atm and 25°C is

$$\begin{aligned}\rho_g &= \frac{(1.5\ \text{atm})\left(26.8\ \frac{\text{g}}{\text{mol}}\right)\left(\frac{1\ \text{kg}}{1000\ \text{g}}\right)}{\left(8.2\times 10^{-5}\ \frac{\text{atm}\cdot\text{m}^3}{\text{mol}\cdot\text{K}}\right)(273+25^\circ\text{C})} \\ &= 1.645\ \text{kg/m}^3\end{aligned}$$

$$\begin{aligned}\Delta\rho_g &= 1.645\ \frac{\text{kg}}{\text{m}^3} - 1.702\ \frac{\text{kg}}{\text{m}^3} \\ &= \boxed{-0.057\ \text{kg/m}^3}\end{aligned}$$

4. (a) The air flow for 24 h PM 10 is

$$\frac{1.6\ \frac{\text{m}^3}{\text{min}} + 1.58\ \frac{\text{m}^3}{\text{min}}}{2} = 1.59\ \text{m}^3/\text{min}$$

$$\begin{aligned}24\ \text{h PM 10} &= \frac{(10.5818\ \text{g} - 10.1453\ \text{g})\left(10^6\ \frac{\mu\text{g}}{\text{g}}\right)}{(24\ \text{h})\left(60\ \frac{\text{min}}{\text{h}}\right)\left(1.59\ \frac{\text{m}^3}{\text{min}}\right)} \\ &= \boxed{191\ \mu\text{g/m}^3}\end{aligned}$$

The answer is (B).

(b) Use Table 38.6. The PSI based on PM 10 is

$$100 + (200-100)\left(\frac{191-150}{350-150}\right) = \boxed{121}$$

The answer is (B).

(c) From Table 38.5, the air quality descriptor corresponding to a PSI of 121 is "unhealthful."

The answer is (B).

(d) Use Table 38.6. The PSI based on O_3 is

$$300 + (300-200)\left(\frac{0.42-0.40}{0.50-0.40}\right) = 320$$

Because 320 is greater than 121, the PSI = 320.

The answer is (D).

(e) From Table 38.5, the air quality descriptor corresponding to a PSI of 320 is "hazardous."

The answer is (D).

Combustion

5. (a) Use Eq. 38.29.

$$C_{x,y,0} = \frac{Q_g \exp\left((-0.5)\left(\frac{y}{\sigma_y}\right)^2\right) \times \exp\left((-0.5)\left(\frac{H}{\sigma_z}\right)^2\right)}{\pi\mu\sigma_y\sigma_z}$$

From Table 38.7, the stability condition is D.

From Figure 38.11, $\sigma_y = 150$ m when $x = 2000$ m for stability condition D.

From Figure 38.12, $\sigma_z = 50$ m when $x = 2000$ m for stability condition D.

$$C_{2000,300,0} = \frac{\left(27\ \frac{\text{kg}}{\text{s}}\right) \exp\left((-0.5)\left(\frac{300\ \text{m}}{150\ \text{m}}\right)^2\right) \times \exp\left((-0.5)\left(\frac{185\ \text{m}}{50\ \text{m}}\right)^2\right)}{\pi\left(5.3\ \frac{\text{m}}{\text{s}}\right)(50\ \text{m})(150\ \text{m})\left(\frac{1\ \text{kg}}{10^9\ \mu\text{g}}\right)}$$
$$= \boxed{30\ \mu\text{g/m}^3}$$

(b) The ambient lapse rate is less than the adiabatic lapse rate, so the resulting plume will be coning, and the pollutant will likely reach ground level as predicted.

6. (a)

$$Z_e = \frac{36\ \text{cm}}{2} = 18\ \text{cm}$$

The answer is (B).

(b) Use Eq. 38.45

$$E_f = \frac{V_d}{Z_e} = \frac{80\,000\ \text{V}}{(18\ \text{cm})\left(\frac{1\ \text{m}}{100\ \text{cm}}\right)} = 4.4 \times 10^5\ \text{V/m} = \boxed{4.4 \times 10^5\ \text{N/C}}$$

The answer is (C).

(c) From App. 38.C, $\mu_g = 2.50 \times 10^{-5}$ kg/m·s at 175°C.

Use Eq. 38.44.

$$w = \frac{8\xi_o E^2 d_p}{24\mu_g} = \frac{(8)\left(8.85 \times 10^{-12}\ \frac{\text{C}^2}{\text{N}\cdot\text{m}^2}\right)\left(4.4 \times 10^5\ \frac{\text{N}}{\text{C}}\right) \times (6\ \mu\text{m})\left(10^{-6}\ \frac{\text{m}}{\mu\text{m}}\right)}{(24)\left(2.50 \times 10^{-5}\ \frac{\text{kg}}{\text{m}\cdot\text{s}}\right)\left(1\ \frac{\text{N}\cdot\text{s}^2}{\text{kg}\cdot\text{m}}\right)} = \boxed{0.14\ \text{m/s}}$$

The answer is (C).

(d) Use Eq. 38.46.

$$A = \frac{-\ln\left(1 - \frac{E}{100\%}\right)Q_g}{w} = \frac{-\ln\left(1 - \frac{95\%}{100\%}\right)\left(4\ \frac{\text{m}^3}{\text{s}}\right)}{0.14\ \text{m/s}} = \boxed{86\ \text{m}^2}$$

The answer is (B).

(e) Use Eq. 38.46.

$$\text{SCA} = \frac{A}{Q_g} = \frac{86\ \text{m}^2}{4\ \frac{\text{m}^3}{\text{s}}} = \boxed{22\ \text{m}^2\cdot\text{s/m}^3}$$

The answer is (B).

7. (a) Increasing the proportion of cars with diesel engines to those with gasoline engines would not contribute to overall improved air quality in urban areas.

The answer is (C).

(b) The EPA employs a pre-consumer certification and enforcement strategy to regulate mobile source emissions.

The answer is (A).

(c)

$$\text{average fuel economy} = \frac{1\ \text{mi}}{\frac{(1\ \text{mi})(0.45)}{20\ \frac{\text{mi}}{\text{gal}}} + \frac{(1\ \text{mi})(0.55)}{34\ \frac{\text{mi}}{\text{gal}}}} = \boxed{25.9\ \text{mpg}}$$

The answer is (C).

(d) Fuel economy ratings are important to air quality because lower fuel consumption results in lower emissions per mile driven.

The answer is (A).

(e) The primary sources of emissions from automobiles to the air are ventilation from the crankcase, ventilation from the fuel tank, volatilization from the carburetor or throttle body, and exhaust from the tailpipe.

The answer is (D).

8. (a) Perform a mass balance with the room as the boundary. Ignore decay, and assume steady state.

$$\text{mass radon in} = \text{mass radon out}$$

The desired radon concentration in the room is 4 pCi/L.

$$2600 \text{ pCi/min} = \left(4 \, \frac{\text{pCi}}{\text{L}}\right)\left(1000 \, \frac{\text{L}}{\text{m}^3}\right)\left(Q_g \, \frac{\text{m}^3}{\text{min}}\right)$$

$$Q_g = \boxed{0.65 \text{ m}^3/\text{min}}$$

The answer is (A).

(b) The time required to change-out 1000 m^3 room air volume is

$$\frac{1000 \text{ m}^3}{\left(0.65 \, \frac{\text{m}^3}{\text{min}}\right)\left(60 \, \frac{\text{min}}{\text{h}}\right)} = \boxed{26 \text{ h}}$$

The answer is (D).

(c) The primary source of radon in indoor air is from soil.

The answer is (D).

(d) Environmental tobacco smoke (ETS) increases radon exposure by adsorption of radon to ETS particulate.

The answer is (C).

(e) The primary health risk associated with radon exposure is lung cancer.

The answer is (A).

9. Since the pH is 5.5, $[\text{H}^+] = 10^{-5.5} = 3.2 \times 10^{-6}$ M. From Eq. 38.25,

$$\text{NO}_2 + \cdot\text{OH} \longrightarrow \text{HNO}_3 \text{ in the atmosphere}$$

$$\text{HNO}_3(aq) \longrightarrow \text{H}^+ + \text{NO}_3^- \text{ in the solution}$$

1 mole NO_2 yields 1 mole H^+.

$$\text{MW NO}_2 = 14 \, \frac{\text{g}}{\text{mol}} + (2)\left(16 \, \frac{\text{g}}{\text{mol}}\right) = 46 \text{ g/mol}$$

$$\begin{aligned}[\text{H}^+] \text{ from NO}_2 &= \frac{\left(254 \, \frac{\mu\text{g}}{\text{m}^3 \text{ air}}\right)\left(\frac{1 \text{ g}}{10^6 \, \mu\text{g}}\right)\left(\frac{1 \text{ mol H}^+}{\text{mol NO}_2}\right)}{\left(\frac{1 \text{ L water}}{10 \text{ m}^3 \text{ air}}\right)\left(46 \, \frac{\text{g}}{\text{mol NO}_2}\right)} \\ &= 5.5 \times 10^{-5} \text{ M}\end{aligned}$$

From Eq. 38.28 and dissociation of H_2SO_4

$$\text{SO}_3 + \text{H}_2\text{O} \longrightarrow \text{H}_2\text{SO}_4(aq) \longrightarrow 2\text{H}^+ + \text{SO}_4^{-2}$$

1 mole SO_3 yields 2 mole H^+.

$$\text{MW SO}_3 = 32 \, \frac{\text{g}}{\text{mol}} + (3)\left(16 \, \frac{\text{g}}{\text{mol}}\right) = 80 \text{ g/mol}$$

$$\begin{aligned}[\text{H}^+] \text{ from SO}_3 &= \frac{\left(427 \, \frac{\mu\text{g}}{\text{m}^3 \text{ air}}\right)\left(\frac{1 \text{ g}}{10^6 \, \mu\text{g}}\right)\left(\frac{2 \text{ mol H}^+}{\text{mol SO}_3}\right)}{\left(\frac{1 \text{ L water}}{10 \text{ m}^3 \text{ air}}\right)\left(80 \, \frac{\text{g}}{\text{mol SO}_3}\right)} \\ &= 1.1 \times 10^{-4} \text{ M}\end{aligned}$$

$$\begin{aligned}\text{total } [\text{H}^+] &= 5.5 \times 10^{-5} \text{ M} + 1.1 \times 10^{-4} \text{ M} \\ &\quad + 3.2 \times 10^{-6} \text{ M} \\ &= 1.7 \times 10^{-4} \text{ M}\end{aligned}$$

$$\begin{aligned}\text{pH} &= -\log[\text{H}^+] = -\log\left(1.7 \times 10^{-4}\right) \\ &= \boxed{3.77}\end{aligned}$$

Combustion

39 Municipal Solid Waste

PRACTICE PROBLEMS

Landfills

1. A town has a current population of 10,000, which is expected to grow linearly, doubling every 15 yr. The town intends to dispose of its municipal solid waste in a 30-acre landfill that will be converted to a park in 20 yr. Solid waste is generated at the rate of 5 lbm/capita-day. The average compacted density in the landfill will be 1000 lbm/yd^3. Disregarding any soil addition for cover and cell construction, how long will it take to fill the landfill to a uniform height of 6 ft?

(A) 2100 days
(B) 4200 days
(C) 6300 days
(D) 9500 days

2. (*Time limit: one hour*) A town of 10,000 people has selected a square landfill site to deposit its solid waste. The minimum unused side borders are 50 ft. The landfill currently consists of a square 50-acre depression with an average depth of 20 ft below the surrounding grade. When the landfill is at final capacity, it will be covered with 10 ft of earth cover. The maximum height of the covered landfill is 20 ft above the surrounding grade. Solid waste is generated at the rate of 5 lbm/capita-day. The average compacted density in the landfill will be 1000 lbm/yd^3.

(a) What is the volumetric capacity of the landfill site?

(A) 1.1×10^6 ft^3
(B) 6.5×10^6 ft^3
(C) 3.1×10^7 ft^3
(D) 5.7×10^7 ft^3

(b) Using a loading factor of 1.25, what is the volume of landfill used each day?

(A) 60 yd^3/day
(B) 120 yd^3/day
(C) 180 yd^3/day
(D) 240 yd^3/day

(c) What is the service life of the landfill site?

(A) 30 yr
(B) 45 yr
(C) 60 yr
(D) 90 yr

SOLUTIONS

1. Find the rate of increase of waste production.

The mass of waste deposited in the first day will be

$$(10{,}000 \text{ people})\left(5\ \frac{\text{lbm}}{\text{person-day}}\right) = 50{,}000 \text{ lbm/day}$$

The mass of waste deposited on the last day will be

$$(20{,}000 \text{ people})\left(5\ \frac{\text{lbm}}{\text{person-day}}\right) = 100{,}000 \text{ lbm/day}$$

The increase in rate is

$$\frac{\Delta m}{\Delta t} = \frac{100{,}000\ \frac{\text{lbm}}{\text{day}} - 50{,}000\ \frac{\text{lbm}}{\text{day}}}{(15 \text{ years})\left(365\ \frac{\text{days}}{\text{year}}\right)} = 9.132 \text{ lbm/day}^2$$

The mass deposited on day D is

$$\begin{aligned} m_D &= 50{,}000 + (9.132)(D-1) \\ &\approx 50{,}000 + 9.132D \end{aligned}$$

The cumulative mass deposited is

$$m_t = \int_0^t m_D\,dt = 50{,}000t + \frac{9.132t^2}{2}$$

With a compacted density of 1000 lbm/yd^3 and a loading factor of 1.00 (no soil cover), the capacity of the site with a 6 ft lift is

$$\begin{aligned} m_{\text{max}} &= \frac{(30 \text{ ac})\left(43{,}560\ \frac{\text{ft}^2}{\text{ac}}\right)(6 \text{ ft})\left(1000\ \frac{\text{lbm}}{\text{yd}^3}\right)}{27\ \frac{\text{ft}^3}{\text{yd}^3}} \\ &= 2.9 \times 10^8 \text{ lbm} \end{aligned}$$

The time to fill is found by solving the quadratic equation.

$$\begin{aligned} 2.9 \times 10^8 &= 50{,}000t + \frac{9.132t^2}{2} \\ t^2 + 10{,}951t &= 6.351 \times 10^7 \\ t &= \boxed{4193 \text{ days}} \end{aligned}$$

The answer is (B).

Solid Waste

2. The side length of the square disposal site is

(a)
$$\text{side} = \sqrt{(50 \text{ ac})\left(43{,}560 \ \frac{\text{ft}^2}{\text{ac}}\right)} = 1476 \text{ ft}$$

With 50 ft borders, the usable area is

$$A = \left(1476 \text{ ft} - (2)(50 \text{ ft})\right)^2 = 1.893 \times 10^6 \text{ ft}^2$$

If the site is excavated 20 ft, 10 ft of soil is used as cover, and the maximum above-ground height is 20 ft, the service capacity of compacted waste is

$$(1.893 \times 10^6 \text{ ft}^2)(30 \text{ ft}) = \boxed{5.68 \times 10^7 \text{ ft}^3}$$

The answer is (D).

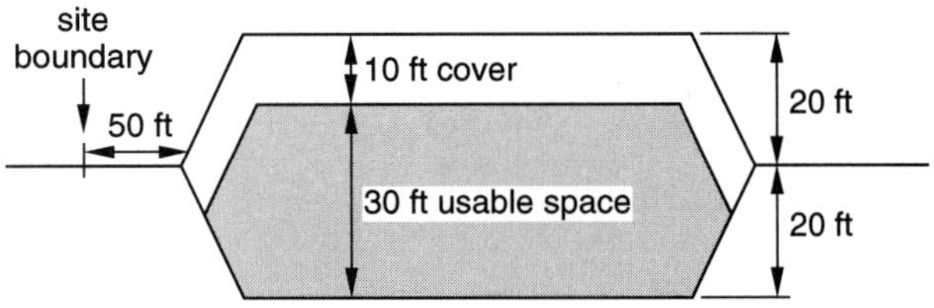

(not to scale or representative of actual construction)

(b) The volume of landfill per day is

$$\frac{(10{,}000 \text{ people})\left(5 \ \frac{\text{lbm}}{\text{day-person}}\right)(1.25)}{1000 \ \frac{\text{lbm}}{\text{yd}^3}} = \boxed{62.5 \text{ yd}^3/\text{day}}$$

The answer is (A).

(c) The service life is

$$\frac{5.68 \times 10^7 \text{ ft}^3}{\left(27 \ \frac{\text{ft}^3}{\text{yd}^3}\right)\left(62.5 \ \frac{\text{yd}^3}{\text{day}}\right)\left(365 \ \frac{\text{days}}{\text{yr}}\right)} = \boxed{92.2 \text{ yr}}$$

The answer is (D).

40 Environmental Pollutants

PRACTICE PROBLEMS

Chapter 40 of the *Environmental Engineering Reference Manual* contains no practice problems.

41 Disposition of Hazardous Materials

PRACTICE PROBLEMS

Chapter 41 of the *Environmental Engineering Reference Manual* contains no practice problems.

42 Environmental Remediation

PRACTICE PROBLEMS

Chapter 42 of the *Environmental Engineering Reference Manual* contains no practice problems.

43 Organic Chemistry

PRACTICE PROBLEMS

Chapter 43 of the *Environmental Engineering Reference Manual* contains no practice problems.

44 Biology and Bacteriology

PRACTICE PROBLEMS

1. (*Time limit: one hour*) A fresh wastewater sample is taken containing nitrate ions, sulfate ions, and dissolved oxygen, and it is placed in a sealed jar absent of air.

(a) What is the sequence of oxidation of the compounds?

(A) nitrate, dissolved oxygen, and then sulfate
(B) sulfate, nitrate, and then dissolved oxygen
(C) dissolved oxygen, nitrate, and then sulfate
(D) none of the above

(b) Obnoxious odors will

(A) appear in the sample when the dissolved oxygen is exhausted.
(B) appear in the sample when the dissolved oxygen and nitrates are exhausted.
(C) appear in the sample when the dissolved oxygen, nitrate, and sulfate are exhausted.
(D) not appear.

(c) Bacteria will convert ammonia to nitrate if the bacteria are

(A) phototropic.
(B) autotropic.
(C) thermophilic.
(D) obligate anaerobes.

(d) Bacteria will reduce nitrate to nitrogen gas if the bacteria are

(A) obligate heterotropic.
(B) obligate anaerobic.
(C) facultative heterotropic.
(D) facultative anaerobic.

(e) A bacteriophage is a

(A) bacterial enzyme.
(B) virus that infects bacteria.
(C) mesophilic organism.
(D) virus that stimulates bacterial growth.

(f) Algal growth in the wastewater

(A) will be inhibited when the dissolved oxygen is exhausted.
(B) will be inhibited when toxins are also found in the solution.
(C) will be inhibited when toxins are found in the solution or when the dissolved oxygen is exhausted.
(D) cannot be inhibited by chemical means.

(g) Algal growth, in the presence of nitrifying bacteria in the wastewater sample,

(A) will occur if the sample is exposed to sunlight.
(B) will occur if inorganic nutrients are absent.
(C) will occur if inorganic nutrients are present.
(D) cannot occur.

(h) The addition of protozoa to the wastewater will

(A) not change the wastewater's biochemical composition.
(B) increase the growth of algae in wastewater.
(C) increase the growth of bacteria in wastewater.
(D) decrease the growth of algae and bacteria in the wastewater.

(i) Coliform bacteria in wastewater from human, animal, or soil sources

(A) can be distinguished in the multiple-tube fermentation test.
(B) can be categorized into only two groups: human/animal and soil.
(C) can be distinguished if multiple-tube fermentation and Eschericheiae coli (EC) tests are both used.
(D) cannot be distinguished.

(j) Under what specific condition would a presence-absence (P-A) coliform test be used on this sample?

(A) if the sample contains known pathogens
(B) when multiple-tube fermentation indicates positive dilution in a presumptive test
(C) when multiple-tube fermentation indicates negative dilution in a presumptive test
(D) if the sample will be used as drinking water

2. A waste stabilization pond will be used in a municipal wastewater treatment system.

(a) Explain the function of bacteria and algae in the stabilization pond's application.

(b) Explain why algae would be a problem if it were present in the discharge from the pond.

(c) Define mechanisms that could be utilized to control algae and that would prove helpful in the development and use of this wastewater treatment system.

(d) Explain what is meant by *facultative pond.*

SOLUTIONS

1. (a) The sequence of oxygen usage reduced by bacteria is dissolved oxygen, nitrate, and then sulfate.

The answer is (C).

(b) Following the sequence of oxidation, obnoxious odors will occur when dissolved oxygen and nitrate are exhausted.

The answer is (B).

(c) Nitrification is performed by autotrophic bacteria to gain energy for growth by synthesis of carbon dioxide in an aerobic environment.

The answer is (B).

(d) Facultative heterotropic bacteria can decompose organic matter to gain energy under anaerobic conditions by removing the oxygen from nitrate, releasing nitrogen gas.

The answer is (C).

(e) A bacteriophage is a virus that infects bacteria.

The answer is (B).

(f) Algae are photosynthetic (gaining energy from light), releasing oxygen during metabolism. They thrive in aerobic environments. The presence of nitrifying bacteria and toxins and the depletion of dissolved oxygen would inhibit the growth process.

The answer is (C).

(g) Nitrifying bacteria are nonphotosynthetic, gaining energy by taking in oxygen to oxidize reduced inorganic nitrogen. The presence of inorganic nutrients would maintain the nitrification process, thereby limiting algal growth.

The answer is (B).

(h) Protozoa consume bacteria and algae in wastewater treatment and in the aquatic food chain.

The answer is (D).

(i) Fecal coliforms from humans and other warm-blooded animals are the same bacterial species. Coliforms originating from the soil can be separated by a confirmatory procedure using EC medium broth incubated at the elevated temperature of 44.5°C (112°F).

The answer is (B).

(j) The presence-absence technique is used for the testing of drinking water.

The answer is (D).

2. (a) A waste stabilization pond's operation is dependent on the reaction of bacteria and algae. Organic matter is metabolized by bacteria to produce the principle products of carbon dioxide, water, and a small amount of ammonia nitrogen. Algae convert sunlight into energy through photosynthesis. They utilize the end products of cell synthesis and other nutrients to synthesize new cells and produce oxygen. The most important role of the algae is in the production of oxygen in the pond for use by aerobic bacteria. In the absence of sunlight, the algae will consume oxygen in the same manner as bacteria. Algae removal is important in producing a high-quality effluent from the pond.

(b) The discharge of algae increases suspended solids in the discharge and may present a problem in meeting water quality criteria. The algae exert an oxygen demand when they settle to the bottom of the stream and undergo respiration.

(c) The following methods have been suggested for control of algae: (a) multiple ponds in series, (2) drawing off of effluent from below the surface by use of a good baffling arrangement to avoid algae concentrations, (3) sand filter or rock filter for algae removal, (4) alum addition and flocculation, (5) microscreening, and (6) chlorination to kill algae. Chlorination may increase BOD loading due to dead algae cells releasing stored organic material.

(d) Facultative ponds have two zones of treatment: an aerobic surface layer in which oxygen is used by aerobic bacteria for waste stabilization and an anaerobic bottom zone in which sludge decomposition occurs. No artificially induced aeration is used.

45 Toxicology

PRACTICE PROBLEMS

1. An environmental engineer has been assigned to manage a laboratory that analyzes hazardous waste samples. She decides to check the workplace air for compliance with the ACGIH values because the ventilation system has been working poorly. She obtained data on possible toxicants in the breathing zone at four workstations.

station	toxicant	concentration for indicated duration (ppm)			
		2 h	4 h	2 h	peak <15 min
1	butane	100	80	120	200
2	pentane	500	70	900	1000
3	hexane	100	200	200	1050
4	mixture of all of above				

The environmental engineer reviewed the ACGIH TLVs for the toxicants identified.

toxicant	TLV-TWA (ppm)	STEL (ppm)	ceiling (ppm)
butane	800	–	–
pentane	600	–	–
hexane	500	1000	–

(a) The TLV-TWA for butane at workstation 1 is most nearly

(A) 80 ppm
(B) 85 ppm
(C) 90 ppm
(D) 95 ppm

(b) The TLV-TWA for pentane at workstation 2 is most nearly

(A) 280 ppm
(B) 325 ppm
(C) 385 ppm
(D) 425 ppm

(c) The TLV-TWA for hexane at workstation 3 is most nearly

(A) 165 ppm
(B) 175 ppm
(C) 185 ppm
(D) 195 ppm

(d) Assuming the effects of the toxicants are additive, the TLV-TWA of the mixture is most nearly

(A) 0.77
(B) 0.94
(C) 1.11
(D) 1.36

(e) Which of the following statements regarding the laboratory are true?

I. Because the concentration of pentane of 900 ppm exceeds the TLV of 600 ppm, workstation 2 exceeds the ACGIH values.
II. Because the concentration of hexane exceeds the STEL, workstation 3 exceeds ACGIH values.
III. The mixture of butane, pentane, and hexane at workstation 4 exceeds ACGIH values.
IV. Even if hexane were not present, workstation 4 would still exceed ACGIH values.
V. If the effects of each toxicant were independent, workstation 4 would not exceed ACGIH values.

(A) I, III, IV, V
(B) I, II, IV
(C) II, IV, V
(D) III, V

2. Two people work side by side at a chemical plant with poor ventilation. They have been exposed to 0.6 ppm of trichloroethylene (TCE) for 20 years. Standard factors for the two are as follows.

parameter	worker A	worker B
body mass (BW)	100 kg	70 kg
respiration rate (CR)	1.5 m^3/h	0.9 m^3/h
absorption factor	100%	90%

The molecular weight of TCE is 131.40 g/mol, and the slope factor-inhalation for TCE is 1.3×10^{-2}

Health, Safety, Welfare

(mg/kg·day)$^{-1}$. Their workstations are at normal temperature and pressure, and they both work 50 weeks a year and 8 h daily.

(a) The chronic daily intake for worker A is most nearly

(A) 2.4×10^{-4} mg/kg·day
(B) 3.8×10^{-3} mg/kg·day
(C) 1.4×10^{-2} mg/kg·day
(D) 7.6×10^{-2} mg/kg·day

(b) The chronic daily intake for worker B is most nearly

(A) 6.8×10^{-3} mg/kg·day
(B) 2.5×10^{-2} mg/kg·day
(C) 3.6×10^{-2} mg/kg·day
(D) 5.8×10^{-2} mg/kg·day

(c) The excess cancer risk for worker A is most nearly

(A) 1.3×10^{-6}
(B) 2.8×10^{-5}
(C) 9.8×10^{-4}
(D) 4.6×10^{-3}

(d) The excess cancer risk for worker B is most nearly

(A) 4.3×10^{-6}
(B) 3.4×10^{-5}
(C) 7.6×10^{-4}
(D) 4.3×10^{-3}

3. Drainage from a former pesticide manufacturing site enters Bear River at river kilometer 240. The drainage contains DDT, and the concentration of DDT after complete mixing is 5 μg/L. The town of Bear Valley, adult population 15,000, uses the river at river kilometer 40 for its water supply with only chlorination. The drainage has occurred for 30 yr. The river velocity is 5 cm/s. Assume the half-life of DDT is 70 days.

(a) The concentration of DDT in the town's drinking water is most nearly

(A) 1.2 μg/L
(B) 2.3 μg/L
(C) 3.2 μg/L
(D) 3.9 μg/L

(b) The lifetime chronic daily intake of DDT for an adult living in Bear Valley is most nearly

(A) 4×10^{-5} mg/kg·day
(B) 8×10^{-4} mg/kg·day
(C) 6×10^{-3} mg/kg·day
(D) 2×10^{-2} mg/kg·day

(c) Assume a slope factor of 0.34 for DDT. The number of excess lifetime cancers that can be expected in the adult population of Bear Valley is most nearly

(A) 0.05
(B) 0.2
(C) 4
(D) 12

4. (a) The major pathways of exposure to toxic agents are

I. inhalation
II. ingestion
III. whole body
IV. skin absorption
V. eyes and hands

(A) I, II, V
(B) I, III, IV
(C) III, IV, V
(D) I, II, IV

(b) The absorption of toxicants into the bloodstream through the lungs depends upon the

I. gradient between the concentration of the toxicant in the blood and its concentration in air
II. solubility of the toxicant in the blood
III. size of the toxicant molecule
IV. respiration rate
V. condition of the lungs

(A) I, II, IV
(B) I, II, IV, V
(C) II, III, IV, V
(D) I, II, III, IV, V

(c) Toxicity effects are exerted on target organs or tissue if the rate of accumulation exceeds the

I. rate of biotransformation
II. rate of elimination, transformation, and storage
III. slope of the dose-response curve
IV. metabolic rate of the exposed individual
V. slope factor

(A) I, II
(B) II
(C) III, IV
(D) V

(d) LD_{50} means the dose in

(A) mg/day at which 50 of the test animals died

(B) mg/day at which 50% of the test animals died

(C) mg/kg body mass at which 50% of the test animals died

(D) mg/kg body mass at which an adverse effect was observed in 50% of the test animals

(e) Which of the following statements are true?

I. Carcinogens have a dose threshold below which no adverse health response can be measured.

II. Exposure to a carcinogen below its threshold has no associated risk.

III. Carcinogens can produce malignant tumors that spread to other parts of the body through the bloodstream.

IV. Carcinogens cause cell mutation through damage to DNA beyond the ability of the body to make repairs.

(A) I, III, IV

(B) II, III

(C) I, II, IV

(D) III, IV

(f) What is the difference between acute and chronic exposure to toxic chemicals?

(A) Acute exposure occurs through the inhalation pathway; chronic exposure occurs through skin absorption.

(B) Acute exposure is a short duration, typically less than 24 hours; chronic exposure is for a long duration, typically 30 days or more.

(C) Acute exposure results in death to 50% of test animals; chronic exposure results in 100% survival of test animals.

(D) Acute exposure occurs above the threshold of the dose-response curve.

(g) Chemical A has an LD_{50} of 3 mg/kg. Chemical B has an LD_{50} of 9 mg/kg. Chemical A is

(A) one-third as toxic as chemical B

(B) three times as toxic as chemical B

(C) not comparable to chemical B because the pathways are not specified

(D) not comparable to chemical B because the target organs are not specified

(h) The law of additive effects

(A) applies to all mixtures of gases and vapors

(B) applies only when the effects from the component gases and vapors occur within the same organs

(C) applies to noncarcinogens less than the TLV

(D) applies to carcinogens with a high slope factor

(i) Biological exposure indices are

(A) intended to replace the ACGIH TLV values

(B) the levels of contaminants that would be observed in specimens from a healthy worker exposed to the same level of TLVs

(C) not suitable for use in calculating the total body burden

(D) limited to use for the dermal pathway

(j) EPA reference doses (RfD) for noncarcinogens are

(A) developed from precise chemical studies and are very accurate

(B) independent of the pathway of exposure

(C) typically based on the dose that results in no observed adverse effect, but the dose associated with the lowest observed adverse effect can be used if necessary

(D) based on the most sensitive end point when there is more than one toxicological effect

Health, Safety, Welfare

SOLUTIONS

1. (a) From Eq. 45.6,

$$E = \frac{1}{8}\sum_{i=1}^{n} C_i T_i$$

$$E_{\text{butane}} = \frac{1}{8\text{ h}}\left(\begin{array}{c}(100\text{ ppm})(2\text{ h}) + (80\text{ ppm})(4\text{ h}) \\ + (120\text{ ppm})(2\text{ h})\end{array}\right) = \boxed{95\text{ ppm}}$$

The answer is (D).

(b) From Eq. 45.6,

$$E_{\text{pentane}} = \frac{1}{8\text{ h}}\left(\begin{array}{c}(500\text{ ppm})(2\text{ h}) + (70\text{ ppm})(4\text{ h}) \\ + (900\text{ ppm})(2\text{ h})\end{array}\right) = \boxed{385\text{ ppm}}$$

The answer is (C).

(c) From Eq. 45.6,

$$E_{\text{hexane}} = \frac{1}{8\text{ h}}\left(\begin{array}{c}(100\text{ ppm})(2\text{ h}) + (200\text{ ppm})(4\text{ h}) \\ + (200\text{ ppm})(2\text{ h})\end{array}\right) = \boxed{175\text{ ppm}}$$

The answer is (B).

(d) From Eq. 45.7,

$$E_m = \sum_{i=1}^{n} \frac{C_i}{L_i}$$

$$\begin{aligned} E_m &= \frac{C_{\text{butane}}}{L_{\text{butane}}} + \frac{C_{\text{pentane}}}{L_{\text{pentane}}} + \frac{C_{\text{hexane}}}{L_{\text{hexane}}} \\ &= \frac{95\text{ ppm}}{800\text{ ppm}} + \frac{385\text{ ppm}}{600\text{ ppm}} + \frac{175\text{ ppm}}{500\text{ ppm}} \\ &= 0.119 + 0.642 + 0.35 \\ &= \boxed{1.11} \end{aligned}$$

The answer is (C).

(e) I. False. The TWA concentration of pentane is 385 ppm and the TLV-TWA is 800 ppm, so the TLV-TWA is not exceeded.

II. False.

III. True.

IV. False. If hexane was not present at workstation 4, the TLV of the mixture would be 0.119 (butane) plus 0.641 (pentane) = 0.760, which is less than 1.0 and would not exceed the ACGIH limit.

V. True. The independent effects are each less than 1.0.

The answer is (D).

2. (a) From Eq. 11.75, the CDI for worker A is

$$\begin{aligned} \text{CDI} &= \frac{(C_{\text{TCE/air}})(\text{CR})(\text{EF})(\text{ED})}{(\text{BW})(\text{AT})} \\ &= \frac{\left(3.22\ \frac{\text{mg}}{\text{m}^3}\right)\left(1.5\ \frac{\text{m}^3}{\text{h}}\right)\left(8\ \frac{\text{h}}{\text{day}}\right)\left(5\ \frac{\text{days}}{\text{wk}}\right) \times \left(50\ \frac{\text{wk}}{\text{yr}}\right)(20\text{ yr})(1)}{(100\text{ kg})(70\text{ yr})\left(365\ \frac{\text{days}}{\text{yr}}\right)} \\ &= \boxed{7.56 \times 10^{-2}\text{ mg/kg·day}} \end{aligned}$$

The answer is (D).

(b) From Eq. 11.75, the CDI for worker B given an absorption of 0.9 is

$$\begin{aligned} \text{CDI} &= \frac{(C_{\text{TCE/air}})(\text{CR})(\text{EF})(\text{ED})}{(\text{BW})(\text{AT})} \\ &= \frac{\left(3.22\ \frac{\text{mg}}{\text{m}^3}\right)\left(0.9\ \frac{\text{m}^3}{\text{h}}\right)\left(8\ \frac{\text{h}}{\text{day}}\right)\left(5\ \frac{\text{days}}{\text{wk}}\right) \times \left(50\ \frac{\text{wk}}{\text{yr}}\right)(20\text{ yr})(0.9)}{(70\text{ kg})(70\text{ yr})\left(365\ \frac{\text{days}}{\text{yr}}\right)} \\ &= \boxed{5.83 \times 10^{-2}\text{ mg/kg·day}} \end{aligned}$$

The answer is (D).

(c) From Eq. 45.4, the risk for worker A is

$$\begin{aligned} R &= (\text{SF})(\text{CDI}) \\ &= \left(1.3 \times 10^{-2}\ \frac{\text{kg·day}}{\text{mg}}\right)\left(7.56 \times 10^{-2}\ \frac{\text{mg}}{\text{kg·day}}\right) \\ &= \boxed{9.8 \times 10^{-4}} \end{aligned}$$

The answer is (C).

(d) From Eq. 45.4, the risk for worker B is

$$\begin{aligned} R &= (\text{SF})(\text{CDI}) \\ &= \left(1.3 \times 10^{-2}\ \frac{\text{kg·day}}{\text{mg}}\right)\left(5.83 \times 10^{-2}\ \frac{\text{mg}}{\text{kg·day}}\right) \\ &= \boxed{7.6 \times 10^{-4}} \end{aligned}$$

The answer is (C).

3. (a) travel distance = 240 km − 40 km = 200 km

river velocity $= 5\ \frac{\text{cm}}{\text{s}}\ \left(5\times10^{-2}\ \frac{\text{m}}{\text{s}}\right)$

$$\begin{aligned}\text{travel time} &= \frac{s}{v}\\ &= \frac{(200\ \text{km})\left(1000\ \frac{\text{m}}{\text{km}}\right)}{\left(5\times10^{-2}\ \frac{\text{m}}{\text{s}}\right)\left(86\,400\ \frac{\text{s}}{\text{day}}\right)}\\ &= 46.3\ \text{days}\end{aligned}$$

Assuming a first-order reaction, $C_t = C_o e^{-kt}$

At half-life,

$$\begin{aligned}C_t &= \frac{C_o}{2}\\ t &= 70\ \text{days}\\ 0.5 &= e^{-k(70\ \text{days})}\\ k &= \frac{-\ln 0.5}{70\ \text{days}} = 9.902\times10^{-3}\ \text{day}^{-1}\\ C_t &= \left(5\ \frac{\mu\text{g}}{\text{L}}\right)e^{-\left(0.009902\ \frac{1}{\text{day}}\right)(46.3\ \text{d})}\\ &= \boxed{3.16\ \mu\text{g/L}}\end{aligned}$$

The answer is (C).

(b) From Table 45.3, the standard factors for adults in Bear Valley are

$$\begin{aligned}C_{\text{DDT/water}} &= 3\ \mu\text{g/L}\\ \text{CR} &= 2\ \text{L/day}\\ \text{BW} &= 70\ \text{kg}\\ \text{ED} &= 30\ \text{yr}\\ \text{AT} &= 70\ \text{yr}\end{aligned}$$

From Eq. 11.75, the CDI is

$$\begin{aligned}\text{CDI} &= \frac{(C_{\text{DDT/water}})(\text{CR})(\text{EF})(\text{ED})}{(\text{BW})(\text{AT})}\\ &= \frac{\left(3\times10^{-6}\ \frac{\text{g}}{\text{L}}\right)\left(2\ \frac{\text{L}}{\text{day}}\right)(30\ \text{yr})\left(1000\ \frac{\text{mg}}{\text{g}}\right)(1)}{(70\ \text{kg})(70\ \text{yr})}\\ &= \boxed{3.67\times10^{-5}\ \text{mg/kg·day}}\end{aligned}$$

The answer is (A).

(c) From Eq. 45.4,

$$\begin{aligned}R &= (\text{SF})(\text{CDI})\\ &= \left(0.34\ \frac{\text{kg·days}}{\text{mg}}\right)\left(3.67\times10^{-5}\ \frac{\text{mg}}{\text{kg·day}}\right)\\ &= 1.25\times10^{-5}\end{aligned}$$

From Eq. 45.3,

$$\begin{aligned}\text{EC} &= (\text{EP})(R)\\ &= (15\,000)(1.25\times10^{-5})\\ &= \boxed{0.188\quad(0.2)}\end{aligned}$$

The answer is (B).

4. (a) The major pathways are inhalation, ingestion, and skin absorption.

The answer is (D).

(b) All statements are true.

The answer is (D).

(c) Only II is true.

The answer is (B).

(d) Only C is true.

The answer is (C).

(e) III and IV are true.

The answer is (D).

(f) Only B is true.

The answer is (B).

(g) Only B is true.

The answer is (B).

(h) Only B is true.

The answer is (B).

(i) Only B is true.

The answer is (B).

(j) Only C is true.

The answer is (C).

46 Industrial Hygiene

PRACTICE PROBLEMS

1. A concentration of 2.2 ppm of methane (CH_4) was measured at a landfill at an air temperature of 25°C. What is the concentration in mg/m^3?

2. A 25 L sample of landfill gas was bubbled through 50 mL of a solution that has an 85% collection efficiency for methane. The sample was analyzed at 25°C and 750 mm Hg pressure. The laboratory reported the result of 25 $\mu g/mL$. What is the concentration of methane in the landfill gas in ppm?

3. During each shift, workers in a sewage treatment plant are exposed to noise from a sludge pump, a blower, and a ventilation fan as indicated.

source	sound pressure level, dB (reference: 20 μPa)	period of operation (h)
sludge pump	96	0800–1200 1600–1700
blower	94	0800–1000 1500–1700
ventilation fan	98	0800–1000 1300–1500

All noise is at 1000 Hz and referenced to the dBA scale.

(a) What is the sound pressure level the workers are exposed to during the period from 0800 to 1000 hours?

- (A) 99 dB
- (B) 100 dB
- (C) 101 dB
- (D) 103 dB

(b) What is the sound pressure level the workers are exposed to during the period from 1600 to 1700 hours?

- (A) 97 dB
- (B) 98 dB
- (C) 99 dB
- (D) 100 dB

(c) What is the time-weighted sound pressure level exposure for one shift?

- (A) 95.2 dB
- (B) 96.4 dB
- (C) 97.8 dB
- (D) 98.3 dB

(d) Relative to OSHA noise exposure standards, which of the following statements is true for this situation?

- (A) Standards are exceeded because the TWA exposure exceeds 85 dBA.
- (B) Standards are not exceeded because none of the permissible noise levels are exceeded for any duration allowed per day.
- (C) Standards are exceeded because the sum of the fractional allowable exposure duration for each sound pressure level exceeds 1.0 (100%).
- (D) Standards are not exceeded because the TWA exposure does not exceed the permissible TWA exposures.

4. Which of the following statements about biological hazards are false?

- I. A bacterium is a prokaryote whose nucleus is not physically separated from its cytoplasm.
- II. Viruses depend on their host cells for reproduction and metabolism.
- III. Exogenous infections are caused by microorganisms normally present in a part of the body that have reached another site in the body.
- IV. Zoonoses are microorganisms capable of infecting both animals and humans.

- (A) I and III
- (B) I, II, and III
- (C) III
- (D) IV

5. Which of the following diseases should be of special concern to the environmental engineer when designing, operating, or managing water supply projects?

- I. acquired immunodeficiency syndrome (AIDS)
- II. Rocky Mountain Spotted Fever
- III. botulism
- IV. tuberculosis
- V. Legionnaires' disease
- VI. hepatitis A

- (A) I, III, and VI
- (B) II, IV, and VI
- (C) V
- (D) V and VI

6. Which of the following statements concerning heat and cold stress are true?

I. Metabolic rate can affect heat gain by one or two orders of magnitude compared to radiation and convection, but affects heat loss to about the same extent as radiation and convection.

II. The NIOSH ceiling limit (CL), recommended exposure limit (REL), and recommended alert level (RAL) are functions of the time-weighted average of the metabolic rate.

III. Heat stress on unacclimatized workers should not exceed the recommended alert level (RAL).

IV. Two physiological functions that can be readily measured to confirm that heat stress is being controlled to allowable limits are core temperature and heart rate.

V. The main hazards from cold stress are hypothermia, tissue damage, and reduced blood circulation.

(A) I, III, IV, V
(B) II, III, IV, V
(C) I, II, V
(D) I, II, III, IV

7. An environmental engineer is the project manager for a project to upgrade two pumps that handle cooling water at a nuclear power plant. The pumps are in a restricted area with ionizing radiation at a rate of 1530 μSv/d. The upgrade will require two workers at all times and is expected to take a total of 6.5 h. The allowable radiation exposure limit for an 8 h workday is 140 μSv/d/person. How should the engineer schedule the work?

8. An environmental engineer must prepare some wastewater samples for analysis using isopropyl alcohol (CH_3CH_3CHOH). Vapor levels over an 8 h workday will be 420 ppm (2 h), 350 ppm (2.5 h), and 390 ppm (3.5 h). The temperature of the lab is 30°C because the air conditioner is broken. Will the engineer's exposure level be in compliance if the threshold limit value (TLV) for isopropyl alcohol is 400 ppm at 25°C?

9. Environmental engineers are testing some air monitoring equipment and need to make a 20 ppm concentration of carbon tetrachloride (CCl_4) in a 200 L drum. How many milliliters of liquid CCl_4 at a liquid density of 1.5 g/mL need to be evaporated at 28°C and 750 mm Hg (0.987 atm, 100 kPa) to make the test drum?

10. Which of the following statements are true about cumulative trauma disorders (CTDs)?

I. Risk of CTDs is associated with highly repetitive use of the same body part.

II. Damage from CTDs is usually noticed by workers within 24 h of the activity.

III. CTDs may affect tendons, tendon sheaths, soft tissue, nerves, bones, and spinal vertebrae.

IV. High repetitiveness is defined as a cycle time of less than one minute.

V. Leaning over a workbench and resting the forearm on a hard edge or surface may lead to carpal tunnel syndrome.

(A) I, III, V
(B) II, III, V
(C) I, III
(D) I, IV, V

SOLUTIONS

1. At STP (0°C and 1 atm), 1 mol of any gas occupies 22.4 L. Calculate its volume at 25°C.

$T_1 = 0°C + 273 = 273K$

$T_2 = 25°C + 273 = 298K$

$$\begin{aligned} V_2 &= V_1\left(\frac{T_2}{T_1}\right) \\ &= \left(22.4\ \frac{\text{L}}{\text{mol}}\right)\left(\frac{298\text{K}}{273\text{K}}\right) \\ &= 24.45\ \text{L/mol} \end{aligned}$$

The molecular weight of methane is

$$\text{MW} = (1)\left(12\ \frac{\text{g}}{\text{mol}}\right) + (4)\left(1\ \frac{\text{g}}{\text{mol}}\right) = 16\ \text{g/mol}$$

Given a methane concentration of 2.2 ppm (2.2×10^{-6} liters of methane per liter of air), calculate the number of moles of methane per unit volume of air.

$$\begin{aligned} \frac{\text{moles methane}}{\text{volume air}} &= \frac{\text{methane concentration}}{\text{methane volume}} \\ &= \frac{2.2 \times 10^{-6}\ \frac{\text{L}}{\text{L}}}{24.45\ \frac{\text{L}}{\text{mol}}} \\ &= 9.0 \times 10^{-8}\ \text{mol/L} \end{aligned}$$

To calculate the concentration in mg/m^3, use the calculation of the number of moles.

$$n = \frac{m}{\text{MW}}$$

$$\begin{aligned} \frac{m}{V_{\text{air}}} &= \left(\frac{n}{V_{\text{air}}}\right)(\text{MW}) \\ &= \left(9.0 \times 10^{-8}\ \frac{\text{mol}}{\text{L}}\right)\left(16\ \frac{\text{g}}{\text{mol}}\right) \\ &\quad \times \left(10^3\ \frac{\text{mg}}{\text{g}}\right)\left(10^3\ \frac{\text{L}}{\text{m}^3}\right) \\ &= \boxed{1.44\ \text{mg/m}^3} \end{aligned}$$

Alternatively, the concentration of methane in mg/m^3 can be calculated as

$$\begin{aligned} C_{m/V} &= C_{\text{ppm}}\left(\frac{\text{MW}}{V_2}\right) \\ &= \left(2.2 \times 10^{-6}\ \frac{\text{L}}{\text{L}}\right)\left(\frac{16\ \frac{\text{g}}{\text{mol}}}{24.45\ \frac{\text{L}}{\text{mol}}}\right) \\ &\quad \times \left(10^3\ \frac{\text{mg}}{\text{g}}\right)\left(10^3\ \frac{\text{L}}{\text{m}^3}\right) \\ &= \boxed{1.44\ \text{mg/m}^3} \end{aligned}$$

2. The total mass of methane in the 25 L sample is

$$\begin{aligned} m &= \frac{\begin{pmatrix}\text{concentration of} \\ \text{sample in solution}\end{pmatrix}\begin{pmatrix}\text{volume of} \\ \text{solution}\end{pmatrix}}{\text{collection efficiency}} \\ &= \frac{\left(25\ \frac{\mu\text{g}}{\text{mL}}\right)(50\ \text{mL})}{0.85} \\ &= 1470.6\ \mu\text{g} \end{aligned}$$

Correct the volume of gas per mole for temperature and pressure.

$T_1 = 0°C + 273 = 273K \qquad p_1 = 760$ mm Hg
$T_2 = 25°C + 273 = 298K \qquad p_2 = 750$ mm Hg

$$\frac{p_1 V_1}{T_1} = \frac{p_2 V_2}{T_2}$$

$$\begin{aligned} V_2 &= V_1\left(\frac{T_2}{T_1}\right)\left(\frac{p_1}{p_2}\right) \\ &= \left(22.4\ \frac{\text{L}}{\text{mol}}\right)\left(\frac{298\text{K}}{273\text{K}}\right)\left(\frac{760\ \text{mm Hg}}{750\ \text{mm Hg}}\right) \\ &= 24.8\ \text{L/mol} \end{aligned}$$

The concentration of methane in ppm is

$$\begin{aligned} C_{\text{ppm}} &= C_{m/V}\left(\frac{V_2}{\text{MW}}\right) \\ &= \left(\frac{1470.6\ \mu\text{g}}{25\ \text{L}}\right)\left(\frac{24.8\ \frac{\text{L}}{\text{mol}}}{16\ \frac{\text{g}}{\text{mol}}}\right) \\ &= 91.2\ \mu\text{g/g} \\ &= \boxed{91.2\ \text{ppm}} \end{aligned}$$

3. (a) Use Eq. 46.8.

$$\begin{aligned} L_{p,\text{total}} &= 10 \log \sum_i^3 10^{\frac{L_i}{10}} \\ &= 10 \log \left(10^{\frac{96}{10}} + 10^{\frac{94}{10}} + 10^{\frac{98}{10}}\right) \\ &= \boxed{101.1\ \text{dB}} \end{aligned}$$

The answer is (C).

(b) Use Eq. 46.8.

$$\begin{aligned} L_{p,\text{total}} &= 10 \log \sum_i^2 10^{\frac{L_i}{10}} \\ &= 10 \log \left(10^{\frac{96}{10}} + 10^{\frac{94}{10}}\right) \\ &= \boxed{98.1\ \text{dB}} \end{aligned}$$

The answer is (B).

(c) Calculate TWA exposure using Eq. 46.8.

period beginning hour	sludge pump (dB)	blower (dB)	ventilation fan (dB)	total exposure (dB)
0800	96	94	98	101.1
0900	96	94	98	101.1
1000	96			96
1100	96			96
1300			98	98
1400			98	98
1500		94		94
1600	96	94		98.1
total TWA				97.8

The answer is (C).

(d)

(A) False; the standards allow 85 dBA to be exceeded for specific durations at various sound pressure levels.

(B) False; the standards allow permissible noise levels to be exceeded for specific durations.

(C) From Eqs. 46.8 and 46.9 and Table 46.3,

total exposure (dB)	exposure duration, C (h)	permissible exposure duration, t (h)	C/t
101.1	2	1.5	2/1.5 = 1.33
96	2	3	2/3 = 0.67
98	2	2	2/2 = 1.00
94	1	4	1/4 = 0.25
98.1	1	2	1/2 = 0.50
Σ	8		3.75

Because the summation of 3.75 is greater than 1.0, the OSHA permissible noise exposure is exceeded. A hearing conservation program is required.

(D) False; the standards are based on an allowable duration at each exposure level, not on the TWA of the permissible exposures.

The answer is (C).

4. Statements I, II, and IV are true. Statement III is false; exogenous infections are caused by microorganisms that enter the body from the environment. Endogenous infections are caused by indigenous microorganisms.

The answer is (C).

5. Statements I through IV are not concerns: AIDS is not a water-borne disease. Rocky Mountain spotted fever is passed to humans by ticks. Botulism is food-borne or can enter the body through contaminated soil or needles. Tuberculosis is spread by inhalation of infectious droplets. Legionnaires' disease, on the other hand, is associated with heat-transfer systems, warm-temperature water, and stagnant water. Hepatitis A can also be spread through contaminated water.

The answer is (D).

6. Statements I, II, III, and IV are true. Statement V is false: Reduced blood circulation is one of the body's mechanisms to reduce heat loss but is not a main hazard.

The answer is (D).

7. Hourly exposure is

$$\left(1530\ \frac{\mu\text{Sv}}{\text{day}}\right)\left(\frac{1\ \text{day}}{24\ \text{h}}\right) = 63.75\ \mu\text{Sv/h}$$

Maximum individual allowable exposure time is

$$\frac{140\ \dfrac{\frac{\mu\text{Sv}}{\text{day}}}{\text{person}}}{63.75\ \dfrac{\mu\text{Sv}}{\text{h}}} = 2.2\ \text{h/day/person}$$

Use a total of four teams, three in 2 h shifts and one in a 1/2 h shift. Other combinations are possible.

8. At STP (0°C and 1 atm), 1 mol of any gas occupies 22.4 L. Calculate its volume at 25°C and 30°C.

$$T_1 = 0°\text{C} + 273 = 273\text{K}$$
$$T_2 = 25°\text{C} + 273 = 298\text{K}$$
$$T_3 = 30°\text{C} + 273 = 303\text{K}$$

At constant pressure (1 atm),

$$\frac{V_1}{T_1} = \frac{V_2}{T_2} = \frac{V_3}{T_3}$$

$$\begin{aligned} V_2 &= V_1\left(\frac{T_2}{T_1}\right) \\ &= \left(22.4\ \frac{\text{L}}{\text{mol}}\right)\left(\frac{298\text{K}}{273\text{K}}\right) \\ &= 24.45\ \text{L/mol} \end{aligned}$$

$$\begin{aligned} V_3 &= V_1\left(\frac{T_3}{T_1}\right) \\ &= \left(22.4\ \frac{\text{L}}{\text{mol}}\right)\left(\frac{303\text{K}}{273\text{K}}\right) \\ &= 24.86\ \text{L/mol} \end{aligned}$$

The molecular weight of isopropyl alcohol is

$$\begin{aligned} \text{MW} &= (3)\left(12\ \frac{\text{g}}{\text{mol}}\right) + (8)\left(1\ \frac{\text{g}}{\text{mol}}\right) + (1)\left(16\ \frac{\text{g}}{\text{mol}}\right) \\ &= 60\ \text{g/mol} \end{aligned}$$

Calculate the concentration of isopropyl alcohol at 25°C for each exposure period. Note that as the temperature increases the volume increases, but the number of moles remains constant, so the concentration decreases.

$$C_2 = C_3\left(\frac{V_2}{V_3}\right) = (420 \text{ ppm})\left(\frac{24.45 \frac{\text{L}}{\text{mol}}}{24.86 \frac{\text{L}}{\text{mol}}}\right)$$
$$= 413 \text{ ppm}$$

Similarly,

$$350 \text{ ppm at } 30°\text{C} = 344 \text{ ppm at } 25°\text{C}$$

$$390 \text{ ppm at } 30°\text{C} = 384 \text{ ppm at } 25°\text{C}$$

Calculate TLV-TWA for the 8 h day.

$$\text{TLV-TWA} = \frac{\left(\begin{array}{c}(2 \text{ h})(413 \text{ ppm}) + (2.5 \text{ h})(344 \text{ ppm}) \\ +(3.5 \text{ h})(384 \text{ ppm})\end{array}\right)}{8 \text{ h}}$$
$$= 379 \text{ ppm}$$

The TLV-TWA of 400 ppm will not be exceeded.

9. At STP (0°C and 1 atm), 1 mol of any gas occupies 22.4 L. Calculate its volume at 28°C.

$$T_1 = 0°\text{C} + 273 = 273\text{K} \qquad p_1 = 760 \text{ mm Hg}$$
$$T_2 = 28°\text{C} + 273 = 301\text{K} \qquad p_2 = 750 \text{ mm Hg}$$

$$\frac{p_1 V_1}{T_1} = \frac{p_2 V_2}{T_2}$$

$$V_2 = V_1\left(\frac{T_2}{T_1}\right)\left(\frac{p_1}{p_2}\right)$$
$$= \left(22.4 \frac{\text{L}}{\text{mol}}\right)\left(\frac{301\text{K}}{273\text{K}}\right)\left(\frac{760 \text{ mm Hg}}{750 \text{ mm Hg}}\right)$$
$$= 25 \text{ L/mol}$$

The molecular weight of carbon tetrachloride is

$$\text{MW} = (1)\left(12 \frac{\text{g}}{\text{mol}}\right) + (4)\left(35.45 \frac{\text{g}}{\text{mol}}\right)$$
$$= 154 \text{ g/mol}$$

Given the drum volume and a carbon tetrachloride concentration of 20 ppm (20×10^{-6} liters CCl_4 per liter of air), calculate the required volume of CCl_4 gas.

$$V_{\text{gas}} = CV_{\text{drum}}$$
$$= \left(20 \times 10^{-6} \frac{\text{L}}{\text{L}}\right)(200 \text{ L})$$
$$= 4 \times 10^{-3} \text{ L}$$

The required number of carbon tetrachloride moles is

$$n = \frac{V_{\text{gas}}}{V_2} = \frac{4 \times 10^{-3} \text{ L}}{25 \frac{\text{L}}{\text{mol}}}$$
$$= 1.6 \times 10^{-4} \text{ mol}$$

$$m = n(\text{MW})$$
$$= (1.6 \times 10^{-4} \text{ mol})\left(154 \frac{\text{g}}{\text{mol}}\right)$$
$$= 2.46 \times 10^{-2} \text{ g}$$

The volume of liquid carbon tetrachloride is

$$V_{\text{liq}} = \frac{m}{\rho} = \frac{2.46 \times 10^{-2} \text{ g}}{1.5 \frac{\text{g}}{\text{mL}}}$$
$$= \boxed{1.64 \times 10^{-2} \text{ mL}}$$

10. Statements I and III are true. Statements II, IV, and V are false: CTDs may not show symptoms for months or years. High repetitiveness is defined as a cycle time of less than thiry seconds. Item V should refer to cubital tunnel syndrome.

The answer is (C).

Health, Safety, Welfare

47 Health and Safety

PRACTICE PROBLEMS

1. Workers at a cleanup site have determined that benzene (C_6H_6) is a contaminant. The equilibrium vapor pressure of benzene in air is 7×10^{-3} mm Hg at NTP.

(a) The concentration of benzene in air at NTP in ppm is most nearly

(A) 4 ppm
(B) 6 ppm
(C) 9 ppm
(D) 12 ppm

(b) The concentration of benzene in air at NTP in mg/m^3 is most nearly

(A) 16 mg/m^3
(B) 20 mg/m^3
(C) 26 mg/m^3
(D) 29 mg/m^3

2. Toluene was evaporated into an airtight room filled with clean air at a temperature of 40°C and a pressure of 760 mm Hg. The specific gravity of toluene is 0.86. The room measures 10 m by 20 m by 10 m.

(a) If the volume of toluene liquid prior to vaporization is 500 mL, the concentration of toluene in the room is most nearly

(A) 60 ppm
(B) 100 ppm
(C) 190 ppm
(D) 240 ppm

(b) If the pressure in the room is increased to 780 mm Hg, the concentration of toluene in the room is most nearly

(A) 35 ppm
(B) 60 ppm
(C) 80 ppm
(D) 100 ppm

3. A supply chamber contains 0.06% ethylene oxide (C_2H_4O) at 10°C and 1 atm. A pump calibrated at 0.4 L/min is to deliver a mixture of clean air and ethylene oxide at a mixed concentration of 100 ppm to a test chamber that is maintained at 45°C and 1 atm.

(a) The concentration of ethylene oxide in the supply chamber is most nearly

(A) 200 ppm
(B) 300 ppm
(C) 500 ppm
(D) 600 ppm

(b) The flow rate of ethylene oxide needed is most nearly

(A) 0.01 L/min
(B) 0.07 L/min
(C) 0.10 L/min
(D) 0.15 L/min

(c) The flow rate of dilution air is most nearly

(A) 0.1 L/min
(B) 0.2 L/min
(C) 0.3 L/min
(D) 0.4 L/min

(d) If the test chamber measures 2 m by 1 m by 1 m, the volume of supply ethylene oxide needed to collect 2 g of ethylene oxide in the test chamber is most nearly

(A) 500 L
(B) 1800 L
(C) 2900 L
(D) 4500 L

4. Workers at a shipyard are exposed to lead while scrapping old navy warships. To determine lead exposure, the air sampling time required to produce sufficient mass for a reliable exposure result must be calculated. The OSHA PEL for lead is 0.050 mg/m^3. The ACGIH TLV is 0.05. NIOSH test method 7082 for inorganic lead metals gives the following sampling information for lead.

range studied	0.13 to 0.4 mg/m^3 or 10 to 200 μg/sample
estimated LOD	2.6 μg/sample
air sampling flow rate	1 to 4 L/min
PEL	0.05 mg/m^3
sampling media	0.8 μm cellulose ester membrane

(a) The sampling technician wants to collect a lead sample mass of 10 times the LOD. He wants to try an

Health, Safety, Welfare

exposure of 25% of the PEL for the first round of preliminary sampling. The maximum air sample he should collect at the field conditions of 30°C and 650 mm Hg is most nearly

(A) 500 L
(B) 1000 L
(C) 2000 L
(D) 2500 L

(b) A sampling pump calibrated at field conditions will deliver 4 L/min. The time to collect the sample is most nearly

(A) 300 min
(B) 400 min
(C) 500 min
(D) 600 min

(c) If the technician collects the minimum sample mass, how will the sampling protocol change?

(A) 2 samples at 4 h each
(B) 1 sample at 8 h each
(C) 4 samples at 2 h each
(D) 2 samples at 3 h each plus 1 sample at 2 h

SOLUTIONS

1. (a) At NTP,

$$T = 25°\text{C} + 273 = 298\text{K}$$

$$p = (760 \text{ mm Hg})\left(\frac{1 \text{ atm}}{760 \text{ mm Hg}}\right) = 1 \text{ atm}$$

Use Eq. 47.16.

$$\begin{aligned} C_{\text{ppm,benzene}} &= \left(\frac{p_{\text{PVP,benzene}}}{p_{\text{total}}}\right)(10^6) \\ &= \left(\frac{7\times10^{-3} \text{ mm Hg}}{760 \text{ mm Hg}}\right)(10^6) \\ &= \boxed{9.21 \text{ ppm at NTP}} \end{aligned}$$

The answer is (C).

(b) The molecular weight of benzene is

$$\begin{aligned} \text{MW}_{\text{benzene}} &= (6)\left(12 \ \frac{\text{g}}{\text{mol}}\right) + (6)\left(1 \ \frac{\text{g}}{\text{mol}}\right) \\ &= 78 \text{ g/mol} \end{aligned}$$

Use Eq. 47.11.

$$\begin{aligned} C_{m/V,\text{benzene}} &= \left(\frac{C_{\text{ppm,benzene}}(\text{MW}_{\text{benzene}})p}{R^*T}\right)(10^{-6}) \\ &= \frac{(9.21 \text{ ppm})\left(78 \ \frac{\text{g}}{\text{mol}}\right)(1 \text{ atm}) \times(10^{-6})\left(10^3 \ \frac{\text{mg}}{\text{g}}\right)\left(10^3 \ \frac{\text{L}}{\text{m}^3}\right)}{\left(0.08205 \ \frac{\text{L·atm}}{\text{mol·K}}\right)(298\text{K})} \\ &= \boxed{29.4 \text{ mg/m}^3} \end{aligned}$$

The answer is (D).

2. (a)

$$T = 40°\text{C} + 273 = 313\text{K}$$

$$p = (760 \text{ mm Hg})\left(\frac{1 \text{ atm}}{760 \text{ mm Hg}}\right) = 1 \text{ atm}$$

From App. 14.B, at 1 atm and 40°C,

$$\begin{aligned} \rho_{\text{water}} &= 992.25 \text{ kg/m}^3 \\ \rho_{\text{toluene}} &= 0.86\rho_{\text{water}} \\ &= (0.86)\left(992.25 \ \frac{\text{kg}}{\text{m}^3}\right) \\ &= 853.34 \text{ kg/m}^3 \end{aligned}$$

From Eq. 47.5, the mass of dissolved toluene is

$$\begin{aligned} m_{\text{toluene}} &= \rho_{\text{toluene}}V_{\text{toluene}} \\ &= \left(853.34 \ \frac{\text{kg}}{\text{m}^3}\right)(0.5 \text{ L})\left(\frac{1 \text{ m}^3}{10^3 \text{ L}}\right) \\ &= 0.427 \text{ kg} \end{aligned}$$

The molecular weight of toluene (C_7H_8) is

$$\begin{aligned} MW_{toluene} &= (7)\left(12\ \frac{g}{mol}\right) + (8)\left(1\ \frac{g}{mol}\right) \\ &= 92\ g/mol \end{aligned}$$

Solve for the volume of the vaporized toluene. From Eqs. 47.3 and 47.4,

$$\begin{aligned} V_{toluene} &= \frac{m_{toluene}R^*T}{(MW_{toluene})p} \\ &= \frac{(427\ g)\left(0.08205\ \frac{L\cdot atm}{mol\cdot K}\right)(313K)}{\left(92\ \frac{g}{mol}\right)(1\ atm)\left(10^3\ \frac{L}{m^3}\right)} \\ &= 0.119\ m^3 \end{aligned}$$

Solve for the volume of air in the room.

$$V_{air} = (10\ m)(20\ m)(10\ m) = 2000\ m^3$$

Solve for the concentration of toluene in the room.

From Eq. 47.8,

$$\begin{aligned} C_{ppm,toluene} &= \left(\frac{V_{toluene}}{V_{air}}\right)(10^6) \\ &= \left(\frac{0.119\ m^3}{2000\ m^3}\right)(10^6) \\ &= \boxed{59.5\ ppm} \end{aligned}$$

The answer is (A).

(b) The concentration in ppm is unchanged by changes in temperature or pressure.

The answer is (B).

3. (a) The concentration of ethylene oxide is given as 0.06%.

$$C_{ppm} = (0.06\%)\left(\frac{10^6\ ppm}{100\%}\right) = \boxed{600\ ppm}$$

The answer is (D).

(b) From the conservation of mass,

$$\begin{aligned} Q_i &= \frac{Q_fC_f}{C_i} \quad [C_d = 0] \\ &= \frac{\left(0.4\ \frac{L}{min}\right)(100\ ppm)}{600\ ppm} \\ &= \boxed{0.0667\ L/min} \end{aligned}$$

The answer is (B).

(c) From the continuity equation,

$$\begin{aligned} Q_d &= Q_f - Q_i = 0.4\ \frac{L}{min} - 0.0667\ \frac{L}{min} \\ &= \boxed{0.333\ L/min} \end{aligned}$$

The answer is (C).

(d) Find the mass concentration of the ethylene oxide (C_2H_4O) in the supply chamber.

$$T = 10°C\ + 273 = 283K$$

$$p = 1\ atm$$

The molecular weight of ethylene oxide is

$$\begin{aligned} MW_{C_2H_4O} &= (2)\left(12\ \frac{g}{mol}\right) + (4)\left(1\ \frac{g}{mol}\right) \\ &\quad + (1)\left(16\ \frac{g}{mol}\right) \\ &= 44\ g/mol \end{aligned}$$

From Eq. 47.11,

$$\begin{aligned} C_{m/V,C_2H_4O} &= \left(\frac{C_{ppm,C_2H_4O}MW_{C_2H_4O}p}{R^*T}\right)(10^{-6}) \\ &= \frac{(600\ ppm)\left(44\ \frac{g}{mol}\right)(1\ atm)(10^{-6})}{\left(0.08205\ \frac{L\cdot atm}{mol\cdot K}\right)(283K)} \\ &= 0.00114\ g/L \end{aligned}$$

From Eq. 47.6, to deliver 2 g,

$$\begin{aligned} V_{supply,C_2H_4O} &= \frac{m_{C_2H_4O}}{C_{m/V,C_2H_4O}} = \frac{2\ g}{0.00114\ \frac{g}{L}} \\ &= \boxed{1750\ L} \end{aligned}$$

The answer is (B).

4. (a) At NTP,

$$T_S = 25°C + 273 = 298K$$

$$p_S = (760\ mm\ Hg)\left(\frac{1\ atm}{760\ mm\ Hg}\right) = 1\ atm$$

At field conditions,

$$T_F = 30°C\ + 273 = 303K$$

$$p_F = (650\ mm\ Hg)\left(\frac{1\ atm}{760\ mm\ Hg}\right) = 0.855\ atm$$

The mass of lead to be collected is 10 times the LOD.

$$m_{lead} = (10)(LOD) = (10)(2.6\ \mu g) = 26\ \mu g$$

The exposure is 25% of the PEL at NTP.

$$\begin{aligned} C_{m/V,\text{lead}} &= (0.25)(\text{PEL}) = (0.25)\left(0.05\ \frac{\text{mg}}{\text{m}^3}\right) \\ &= 0.0125\ \text{mg/m}^3 \end{aligned}$$

From Eq. 47.6,

$$\begin{aligned} V_{s,\text{air}} &= \frac{m_\text{lead}}{C_{m/V,\text{lead}}} \\ &= \frac{(26\ \mu\text{g})\left(\frac{1\ \text{mg}}{10^3\ \mu\text{g}}\right)\left(10^3\ \frac{\text{L}}{\text{m}^3}\right)}{0.0125\ \frac{\text{mg}}{\text{m}^3}} \\ &= 2080\ \text{L} \end{aligned}$$

Correct the sample volume to field conditions. From Eq. 47.1,

$$\begin{aligned} V_{F,\text{air}} &= V_{S,\text{air}}\left(\frac{p_S}{p_F}\right)\left(\frac{T_F}{T_S}\right) \\ &= (2080\ \text{L})\left(\frac{1\ \text{atm}}{0.855\ \text{atm}}\right)\left(\frac{303\text{K}}{298\text{K}}\right) \\ &= \boxed{2470\ \text{L}} \end{aligned}$$

The answer is (D).

(b)

$$\begin{aligned} t &= \frac{V}{Q} = \frac{2470\ \text{L}}{4\ \frac{\text{L}}{\text{min}}} \\ &= \boxed{618\ \text{min}} \end{aligned}$$

The answer is (D).

(c) Assume that the technician collects at least 10 μg/sample.

$$\frac{10\ \mu\text{g}}{2.6\ \mu\text{g}} = 3.85 \text{ times the LOD}$$

From Eq. 47.6,

$$\begin{aligned} V_{s,\text{air}} &= \frac{m_\text{lead}}{C_{m/V,\ \text{lead}}} \\ &= \frac{(10\ \mu\text{g})\left(\frac{1\ \text{mg}}{10^3\ \mu\text{g}}\right)\left(10^3\ \frac{\text{L}}{\text{m}^3}\right)}{0.0125\ \frac{\text{mg}}{\text{m}^3}} \\ &= 800\ \text{L}\quad [\text{at NTP}] \end{aligned}$$

Correct the sample volume to field conditions. From Eq. 47.1,

$$\begin{aligned} V_{F,\text{air}} &= V_{S,\text{air}}\left(\frac{p_S}{p_F}\right)\left(\frac{T_F}{T_S}\right) \\ &= (800\ \text{L})\left(\frac{1\ \text{atm}}{0.855\ \text{atm}}\right)\left(\frac{303\text{K}}{298\text{K}}\right) \\ &= 950\ \text{L} \end{aligned}$$

$$\begin{aligned} t &= \frac{V}{Q} = \frac{(950\ \text{L})\left(\frac{1\ \text{h}}{60\ \text{min}}\right)}{4\ \frac{\text{L}}{\text{min}}} \\ &= \boxed{3.96\ \text{h}} \end{aligned}$$

Collect 2 samples at 4 h each.

The answer is (A).

48 Biological Effects of Radiation

PRACTICE PROBLEMS

1. Which is (are) the most appropriate unit(s) of measurement for use when discussing the biological effects of radiation on a human being?

(A) rad or gray
(B) rem or sievert
(C) roentgen
(D) curie or becquerel

2. What is the annual average effective dose equivalent received from background sources?

(A) 20 mrem
(B) 180 mrem
(C) 360 mrem
(D) 1000 mrem

3. What is the maximum whole body radiation dose allowed radiation workers per year?

(A) 5 mrem
(B) 50 mrem
(C) 5000 mrem
(D) 50 rem

4. Which of the following is false?

(A) Increasing distance from a radiation field decreases radiation exposure.
(B) Increasing shielding from a radiation field decreases radiation exposure.
(C) Decreasing walking speed in a radiation field decreases radiation exposure.
(D) Increasing time spent in a radiation field decreases radiation exposure.

5. A radiation source undergoes 87 039 256 decays per hour. Express this in (a) becquerels and (b) curies.

(A) 2.4×10^{4} Bq; 6.5×10^{-7} Ci
(B) 6.5×10^{-7} Bq; 2.4×10^{4} Ci
(C) 8.9×10^{14} Bq; 6.5×10^{-7} Ci
(D) 6.5×10^{-7} Bq; 8.9×10^{14} Ci

6. How long can a new 15.5 mCi Cs-137 reference source be used if it is to be taken out of service when the activity reaches 10.0 mCi (for Cs-137, half-life = 30.0 yr)?

(A) 1.9 yr
(B) 5.0 yr
(C) 19.0 yr^{-1}
(D) 19.0 yr

7. a) What is the decay constant for a radionuclide that has an initial activity of 225 mCi, and 110 days later has an activity of 81.8 mCi? b) What is the half-life?

(A) 2.44×10^{-3} day^{-1}; 111 days
(B) 6.38×10^{-3} day^{-1}; 108 days
(C) 9.20×10^{-3} yr^{-1}; 75.3 yr
(D) 9.20×10^{-3} day^{-1}; 75.3 days

8. How long could a worker have 2 cm^2 of skin contaminated with 1 MBq carbon-14 and not exceed the SDE occupational exposure limit?

(A) 2.9 h
(B) 2.9 yr
(C) 29 h
(D) 0.5 yr

9. How far would a member of the public have to be kept away from an unshielded 750 mCi chromium-51 source to not exceed the NRC limits?

(A) 18 cm
(B) 180 cm
(C) 400 cm
(D) 4000 cm

10. A worker accidentally ingests 100 μCi of tritiated water. Calculate the committed effective dose equivalent ($H_{E,50}$).

(A) 6.3×10^{-3} rad
(B) 6.3×10^{-3} rem
(C) 4.5×10^{-5} R
(D) 6.3×10^{-3} Sv

11. What is the total effective dose equivalent to a person who inhales $^{11}CO_2$ at an average concentration of 5×10^{-5} μCi/mL for one year?

(A) 2.8 rem
(B) 28 mrem
(C) 1.1 $\times10^{10}$ mrem
(D) 1.1$\times10^{10}$ mSv

12. What is the committed effective dose equivalent to a person whose drinking water was supplied by a facility that releases calcium-45 with an average monthly activity of 3×10^{-8} μCi/mL?

(A) 7.5$\times10^{-7}$ mrem
(B) 7.5$\times10^{-7}$ rem
(C) 7.5$\times10^{-10}$ mSv
(D) 7.5$\times10^{-7}$ Sv

13. How much farther away would one need to move an unshielded gamma source to reduce the radiation intensity to 1% of the original intensity?

(A) 9 times
(B) 10 times
(C) 100 times
(D) 200 times

SOLUTIONS

1. ***The answer is (B).***

2. ***The answer is (C).***

3. ***The answer is (C).***

4. ***The answer is (D).***

5. (a)

$$\left(\frac{87\,039\,256}{\text{h}}\right)\left(\frac{\text{h}}{60\ \text{min}}\right)\left(\frac{\text{min}}{60\ \text{s}}\right)\left(\frac{\frac{1}{\text{s}}}{\text{Bq}}\right) = \boxed{2.42\times10^{4}\ \text{Bq}}$$

(b)

$$(2.42\times10^{4}\ \text{Bq})\left(\frac{1\ \text{Ci}}{3.7\times10^{10}\ \text{Bq}}\right) = \boxed{6.54\times10^{-7}\ \text{Ci}}$$

The answer is (A).

6. From Eq. 48.2,

$$\lambda = \frac{\ln 2}{t_{1/2}} = \frac{0.693}{30.0\ \text{yr}} = 0.0231\ \text{yr}^{-1}$$

From Eq. 48.1,

$$t = \frac{-\ln\left(\frac{A_t}{A_0}\right)}{\lambda} = \frac{-\ln\left(\frac{10.0\ \text{mCi}}{15.5\ \text{mCi}}\right)}{0.0231\ \text{yr}^{-1}} = \boxed{19.0\ \text{yr}}$$

The answer is (D).

7. (a) From Eq. 48.1,

$$\lambda = \frac{-\ln\left(\frac{A_t}{A_0}\right)}{t} = \frac{-\ln\left(\frac{81.8\ \text{mCi}}{225\ \text{mCi}}\right)}{110\ \text{days}} = \boxed{9.20\times10^{-3}\ \text{day}^{-1}}$$

(b) From Eq. 48.2,

$$t_{1/2} = \frac{\ln 2}{\lambda} = \frac{0.693}{9.2\times10^{-3}\ \text{day}^{-1}} = \boxed{75.3\ \text{days}}$$

The answer is (D).

8.

$$A = 1 \text{ MBq} = 1 \times 10^6 \text{ Bq}$$
$$A_s = 2 \text{ cm}^2$$

From Table 48.4 for C-14 at 0.007 cm (where the SDE must be determined),

$$\text{DF} = 2.9 \times 10^{-3} \frac{\frac{\text{Sv}}{\text{yr}}}{\frac{\text{Bq}}{\text{cm}^2}}$$

Use Eq. 48.5.

$$\begin{aligned} X_b &= \text{DF}\left(\frac{A}{A_s}\right) \\ &= \left(2.9 \times 10^{-3} \frac{\text{Sv}\cdot\text{cm}^2}{\text{Bq}\cdot\text{yr}}\right)\left(\frac{1 \times 10^6 \text{ Bq}}{2 \text{ cm}^2}\right) \\ &= 1.45 \times 10^3 \text{ Sv/yr} \end{aligned}$$

The SDE limit for the skin is 50 rem (0.5 Sv) in a year.

$$\begin{aligned} \frac{0.5 \text{ Sv}}{1.45 \times 10^3 \frac{\text{Sv}}{\text{yr}}} &= (3.3 \times 10^{-4} \text{ yr})\left(365 \frac{\text{days}}{\text{yr}}\right)\left(24 \frac{\text{h}}{\text{day}}\right) \\ &= \boxed{3.0 \text{ h}} \end{aligned}$$

The answer is (A).

9. Calculate the annual dose equivalent rate for Cr-51.

$$\begin{aligned} A &= (750 \times 10^{-3} \text{ Ci})\left(3.7 \times 10^{10} \frac{\text{Bq}}{\text{Ci}}\right)\left(\frac{1 \text{ MBq}}{10^6 \text{ Bq}}\right) \\ &= 2.78 \times 10^3 \text{ MBq} \end{aligned}$$

From Table 48.5, for Cr-51 at a distance (d_0) of 1 m,

$$\Gamma = 6.320 \times 10^{-6} \frac{\frac{\text{mSv}}{\text{h}}}{\text{MBq}}$$

From Eq. 48.6,

$$\begin{aligned} X_g &= \Gamma A \\ &= \left(6.320 \times 10^{-6} \frac{\text{mSv}}{\text{MBq}\cdot\text{h}}\right)(2.78 \times 10^4 \text{ MBq}) \\ &\quad \times \left(24 \frac{\text{h}}{\text{day}}\right)\left(365 \frac{\text{days}}{\text{yr}}\right) \\ &= 1.54 \times 10^3 \text{ mSv/yr} \quad [\text{at 1 m}] \end{aligned}$$

Find the distance at which the dose equivalent rate is reduced to 1 mSv/yr, the total effective dose equivalent limit for individual members of the public.

From Eq. 48.15,

$$I_0 {d_0}^2 = I_f {d_f}^2$$

$$\begin{aligned} d_f &= d_0 \sqrt{\frac{I_0}{I_f}} = (1 \text{ m}) \sqrt{\frac{1.6 \times 10^3 \frac{\text{mSv}}{\text{yr}}}{1 \frac{\text{mSv}}{\text{yr}}}} \\ &= \boxed{40 \text{ m}} \end{aligned}$$

The answer is (D).

10. From Table 48.8, part 1, column 1, for oral ingestion of H-3 water, the ALI is 8×10^4 μCi. This would produce 5 rem $H_{E,50}$.

$$5 \text{ rem} = 8 \times 10^4 \ \mu\text{Ci}$$

The $H_{E,50}$ produced by 100 μCi is

$$H_{E,50} = 100 \ \mu\text{Ci}$$

Therefore,

$$\begin{aligned} \frac{H_{E,50}}{100 \ \mu\text{Ci}} &= \frac{5 \text{ rem}}{8 \times 10^4 \ \mu\text{Ci}} \\ H_{E,50} &= \frac{(5 \text{ rem})(100 \ \mu\text{Ci})}{8 \times 10^4 \ \mu\text{Ci}} \\ &= \boxed{6.25 \times 10^{-3} \text{ rem}} \end{aligned}$$

The answer is (B).

11. From Table 48.8, part 2, column 1, for C-11, 9×10^{-7} μCi/mL produces 0.05 rem TEDE, so

$$0.05 \text{ rem} = 9 \times 10^{-7} \ \mu\text{Ci/mL}$$

The TEDE produced by 5×10^{-5} μCi/mL is

$$\text{TEDE} = 5 \times 10^{-5} \ \mu\text{Ci/mL}$$

Therefore,

$$\begin{aligned} \frac{\text{TEDE}}{5 \times 10^{-5} \frac{\mu\text{Ci}}{\text{mL}}} &= \frac{0.05 \text{ rem}}{9 \times 10^{-7} \frac{\mu\text{Ci}}{\text{mL}}} \\ \text{TEDE} &= \frac{\left(5 \times 10^{-5} \frac{\mu\text{Ci}}{\text{mL}}\right)(0.05 \text{ rem})}{9 \times 10^{-7} \frac{\mu\text{Ci}}{\text{mL}}} \\ &= \boxed{2.77 \text{ rem}} \end{aligned}$$

The answer is (A).

Health, Safety, Welfare

12. From Table 48.8, part 3, for Ca-45, 2×10^{-4} μCi/mL produces 0.5 rem $H_{E,50}$, so

$$0.5 \text{ rem} = 2\times10^{-4}\ \mu\text{Ci/mL}$$

The $H_{E,50}$ produced by 3×10^{-8} μCi/mL is

$$H_{E,50} = 3\times10^{-8}\ \mu\text{Ci/mL}$$

Therefore,

$$\frac{H_{E,50}}{3\times10^{-8}\ \frac{\mu\text{Ci}}{\text{mL}}} = \frac{0.5\text{ rem}}{2\times10^{-4}\ \frac{\mu\text{Ci}}{\text{mL}}}$$

$$H_{E,50} = \frac{\left(3\times10^{-8}\ \frac{\mu\text{Ci}}{\text{mL}}\right)(0.5\text{ rem})}{2\times10^{-4}\ \frac{\mu\text{Ci}}{\text{mL}}} = \boxed{7.5\times10^{-5}\text{ rem}}$$

The answer is (D).

13. For the intensity to be 1% of the original

$$\frac{I_2}{I_1} = 0.01$$

From Eq. 48.15,

$$\frac{d_1}{d_2} = \sqrt{\frac{I_2}{I_1}} = \sqrt{0.01} = 0.1$$

$$d_2 = 10d_1 = d_1 + \boxed{9d_1}$$

The answer is (A).

49 Shielding

PRACTICE PROBLEMS

1. Using Einstein's $E = mc^2$ equation, calculate (a) the energy equivalent of a neutron's mass, and (b) the wavelength of a photon that has the equivalent energy.

(A) 3.2×10^{-10} J; 2.6×10^{-20} m
(B) 1.5×10^{-10} J; 1.3×10^{-15} m
(C) 3.2×10^{-10} J; 1.3×10^{-15} m
(D) 1.5×10^{-10} J; 1.3×10^{-15} m

2. What thickness of tin ($\rho = 6.5$ g/cm^3, $A = 118.7$) is needed to shield an isotope that emits a 7.0 MeV ($E_\alpha = 7.0$ MeV) alpha particle?

(A) 9.4×10^{-3} cm
(B) 2.6×10^{-3} cm
(C) 1.7×10^{-2} cm
(D) 5.66 cm

3. What is the density of a 5.0 mm thick shield that shields 100% of the beta particles from indium-118 ($E_{\beta,\max} = 4.250$ MeV)?

(A) 0.31 g/cm^3
(B) 2.9 g/cm^3
(C) 3.4 g/cm^3
(D) 4.3 g/cm^3

4. (a) What is the mass attenuation factor for a shield that is 0.5 in thick, has a density of 10.3 g/cm^3, and reduces the exposure rate of 1.25 MeV gammas from 3.75 R/h to 1.44 mR/h? (b) What is the relaxation length?

(A) 0.5 cm^2/g; 5 cm
(B) 0.60 cm^2/g; 6.19 cm
(C) 1.9 cm^2/g; 5.87 cm
(D) 15 cm^2/g; 0.69 cm

5. A 4 cm thick large iron plate ($\rho_{\text{Fe}} = 7.86$ g/cm^3) is used to shield a high-activity 500 keV gamma ray source. The gamma intensity is measured to be 9.4 mR/h behind the shield. What is the unshielded exposure rate in front of the shield?

(A) 10 mR/h
(B) 84 mR/h
(C) 140 mR/h
(D) 670 mR/h

6. What percent of thermal neutrons will pass through a shield of 1 in of tungsten ($\rho_{\text{W}} = 19.3$ g/cm^3, $A_{\text{W}} = 183.9$)? The σ_a for thermal neutrons in tungsten is 19.2 barns.

(A) 0.01%
(B) 2.5%
(C) 3.7%
(D) 4.6%

7. What is the relaxation length for 500 keV gamma rays in aluminum ($\rho = 2.7$ g/cm^3)?

(A) 4.4 μm
(B) 2.7 cm
(C) 4.4 cm
(D) 4.4 m

SOLUTIONS

1. (a)

$$m = (1.008665 \text{ amu})\left(1.661 \times 10^{-27} \frac{\text{kg}}{\text{amu}}\right) = 1.675 \times 10^{-27} \text{ kg}$$

$$E = mc^2 = (1.675 \times 10^{-27} \text{ kg})\left(3 \times 10^8 \frac{\text{m}}{\text{s}}\right)^2 = \boxed{1.51 \times 10^{-10} \text{ J}}$$

(b) From Eq. 49.4,

$$\lambda = \frac{hc}{E} = \frac{(6.63 \times 10^{-34} \text{ J·s})\left(3 \times 10^8 \frac{\text{m}}{\text{s}}\right)}{1.51 \times 10^{-10} \text{ J}} = \boxed{1.32 \times 10^{-15} \text{ m}}$$

The answer is (D).

2. From Eq. 49.7,

$$R_{\alpha,\text{air}} = k_2E_\alpha - k_3 = \left(1.24 \frac{\text{cm}}{\text{MeV}}\right)(7.0 \text{ MeV}) - 2.62 \text{ cm} = 6.06 \text{ cm}$$

From Eq. 49.8,

$$R_{\alpha,\text{Sn}} = \frac{k_4 (A_{\text{Sn}})^{\frac{1}{3}} R_{\alpha,\text{air}}}{\rho_{\text{Sn}}} = \frac{\left(0.56 \times 10^{-3} \frac{\text{g}}{\text{cm}^3}\right)(118.7)^{\frac{1}{3}}(6.06 \text{ cm})}{6.5 \frac{\text{g}}{\text{cm}^3}} = \boxed{2.57 \times 10^{-3} \text{ cm}}$$

The answer is (B).

3. From Eq. 49.10,

$$\rho_{\text{shield}} = \frac{k_8E_{\beta,\max} - k_9}{R_{\beta,\text{shield}}} = \frac{(0.530 \frac{\text{g}}{\text{cm}^2\text{· MeV}})(4.250 \text{ MeV}) - 0.106 \frac{\text{g}}{\text{cm}^2}}{(5.0 \text{ mm})\left(\frac{1 \text{ cm}}{10 \text{ mm}}\right)} = \boxed{4.29 \text{ g/cm}^3}$$

The answer is (D).

4. (a)

$$x = (0.5 \text{ in})\left(2.54 \frac{\text{cm}}{\text{in}}\right) = 1.27 \text{ cm}$$

From Eq. 49.11,

$$\mu = \frac{-\ln\left(\frac{I_x}{I_0}\right)}{\rho x} = \frac{-\ln\left(\frac{1.44 \frac{\text{mR}}{\text{h}}}{3750 \frac{\text{mR}}{\text{h}}}\right)}{\left(10.3 \frac{\text{g}}{\text{cm}^3}\right)(1.27 \text{ cm})} = \boxed{0.601 \text{ cm}^2/\text{g}}$$

(b) From Eq. 49.17,

$$\lambda = \frac{1}{\mu\rho} = \frac{1}{\left(0.601 \frac{\text{cm}^2}{\text{g}}\right)\left(10.3 \frac{\text{g}}{\text{cm}^3}\right)} = \boxed{6.19 \text{ cm}}$$

The answer is (B).

5. From Table 49.3, for 500 keV gamma radiation in iron, $\mu = 0.0828 \text{ cm}^2/\text{g}$.

$$\mu\rho x = \left(0.0828 \frac{\text{cm}^2}{\text{g}}\right)\left(7.86 \frac{\text{g}}{\text{cm}^3}\right)(4 \text{ cm}) = 2.6$$

From Table 49.4, $B \approx 1.50$.

From Eq. 49.12,

$$I_0 = \frac{I_x}{Be^{-\mu\rho x}} = \frac{9.4 \frac{\text{mR}}{\text{h}}}{1.50e^{-2.6}} = \boxed{84.4 \text{ mR/h } (0.084 \text{ R/h})}$$

The answer is (B).

6. Calculate the number of lead nuclei per cubic centimeter.

From Eq. 49.15,

$$N = \frac{N_A\rho_W}{A_W} = \frac{\left(6.02 \times 10^{23} \frac{\text{nuclei}}{\text{mol}}\right)\left(19.3 \frac{\text{g}}{\text{cm}^3}\right)}{183.9 \frac{\text{g}}{\text{mol}}} = 6.32 \times 10^{22} \text{ nuclei/cm}^3$$

The percentage of neutrons transmitted through the shield (I_x/I_0) is given by Eq. 49.16.

$$\frac{I_x}{I_0} = e^{-N\sigma x}$$

$$= e^{-\left(6.32\times10^{22}\ \frac{\text{nuclei}}{\text{cm}^3}\right)\left(19.2\times10^{-24}\ \frac{\text{cm}^2}{\text{nucleus}}\right)(2.54\ \text{cm})}$$

$$= \boxed{0.0459\ (4.6\%)}$$

The answer is (D).

7. From Table 49.3, for 500 keV gammas in aluminum, $\mu = 0.0840\ \text{cm}^2/\text{g}$.

From Eq. 49.17,

$$\lambda = \frac{1}{\mu\rho_{\text{shield}}}$$

$$= \frac{1}{\left(0.0840\ \frac{\text{cm}^2}{\text{g}}\right)\left(2.7\ \frac{\text{g}}{\text{cm}^3}\right)}$$

$$= \boxed{4.41\ \text{cm}}$$

The answer is (C).

Health, Safety, Welfare

50 Illumination and Sound

PRACTICE PROBLEMS

1. What is the total sound pressure level when a 40 dB source is placed adjacent to a 35 dB source?

2. With no machinery operating, the background noise in a room has a sound pressure level of 43 dB. With the machinery operating, the sound pressure level is 45 dB. What is the sound pressure level due to the machinery alone?

3. An unenclosed source produces a sound pressure level of 100 dB. An enclosure is constructed from a material having a transmission loss of 30 dB. The sound pressure level inside the enclosure from the enclosed source increases to 110 dB. What is the reduction in sound pressure level outside the enclosure?

4. 4 ft (1.2 m) from an isotropic sound source, the sound pressure level is 92 dB. What is the sound pressure level 12 ft (3.6 m) from the source?

5. Octave band measurements of a noise source were made. The measurements were 85, 90, 92, 87, 82, 78, 65, and 54 dB at frequencies of 63, 125, 250, 500, 1000, 2000, 4000, and 8000 Hz, respectively. What is the overall A-weighted sound pressure?

6. What is the maximum possible reduction in sound pressure level if the number of sabins is 50% of the total room area?

7. A room has dimensions of 100 ft × 400 ft × 20 ft (30 m × 120 m × 6 m). All surfaces are precast concrete. 40% of the walls are treated acoustically with a material having a sound absorption coefficient of 0.8. What is the reduction in sound pressure level?

8. An office has dimensions of 20 ft × 50 ft × 10 ft (6 m × 15 m × 3 m). The floor is covered with roll vinyl. The walls and ceiling are sheetrock. 20% of the walls are glass windows. There are 15 desks, 15 occupants, and 5 miscellaneous sabins. After complaints from the occupants, the ceiling is treated with a sound absorbing material with a sound absorption coefficient of 0.7. What is the reduction in sound level?

9. A room has dimensions of 15 ft × 20 ft × 10 ft (4.5 m × 6 m × 3 m). The sound absorption coefficients are 0.03, 0.5, and 0.06 for the floor, ceiling, and walls, respectively. A machine with a sound power level of 65 dB is located at the intersection of the floor and wall, 7.5 ft (2.25 m) from the perpendicular walls. The ambient sound pressure level is 50 dB everywhere in the room. What is the sound pressure level 5 ft (1.5 m) from the machine?

SOLUTIONS

1. Use Eq. 50.16.

$$L = 10 \log \sum 10^{\frac{W_i}{10}} = 10 \log \left(10^{\frac{40}{10}} + 10^{\frac{35}{10}}\right) = \boxed{41.2 \text{ dB}}$$

2. Use Eq. 50.16.

$$L = 10 \log \sum 10^{\frac{W_i}{10}} = 10 \log \left(10^{\frac{43}{10}} + 10^{\frac{W}{10}}\right) = 45 \text{ dB}$$

Solve for the unknown machinery sound pressure level, W.

$$10^{\frac{W}{10}} = 10^{\frac{45}{10}} - 10^{\frac{43}{10}}$$

$$W = 10 \log \left(10^{\frac{45}{10}} - 10^{\frac{43}{10}}\right) = \boxed{40.7 \text{ dB}}$$

3. Define $L_{W,1}$ as the sound pressure level inside the enclosure, and define $L_{W,2}$ as the sound pressure level outside the enclosure. Use Eq. 50.23.

$$L_{W,2} = L_{W,1} - \text{TL} = 110 \text{ dB} - 30 \text{ dB} = 80 \text{ dB}$$

Define $L_{W,1}$ as the sound pressure level for the unenclosed source, and define $L_{W,2}$ as the sound pressure level for the enclosed source.

Use Eq. 50.22 to solve for the insertion loss.

$$\text{IL} = 100 \text{ dB} - 80 \text{ dB} = \boxed{20 \text{ dB}}$$

4. From Eq. 50.15, the free-field sound pressure is inversely proportional to the square of the distance from the source.

$$\frac{p_2}{p_1} = \left(\frac{r_1}{r_2}\right)^2$$

From Eq. 50.13,

$$L_{p,2} = L_{p,1} + 10 \log \left(\frac{r_1}{r_2}\right)^2 = L_{p,1} + 20 \log \left(\frac{r_1}{r_2}\right)$$

Customary U.S. Solution

$$L_{p,2} = 92 \text{ dB} + 20 \log \left(\frac{4 \text{ ft}}{12 \text{ ft}}\right) = 92 \text{ dB} - 9.5 \text{ dB} = \boxed{82.5 \text{ dB}}$$

SI Solution

$$L_{p,2} = 92 \text{ dB} + 20 \log \left(\frac{1.2 \text{ m}}{3.6 \text{ m}}\right) = 92 \text{ dB} - 9.5 \text{ dB} = \boxed{82.5 \text{ dB}}$$

5. Add the corrections from Table 50.5 to the measurements.

frequency	measurement	correction	corrected value
63	85	−26.2	58.8
125	90	−16.1	73.9
250	92	−8.6	83.4
500	87	−3.2	83.8
1000	82	0	82.0
2000	78	+1.2	79.2
4000	65	+1.0	66.0
8000	54	−1.1	52.9

Use Eq. 50.16.

$$L = 10 \log \sum 10^{\frac{W_i}{10}} = 10 \log \left(10^{5.88} + 10^{7.39} + 10^{8.34} + 10^{8.38} + 10^{8.2} + 10^{7.92} + 10^{6.6} + 10^{5.29}\right) = \boxed{88.6 \text{ dBA}}$$

6. Define A as the total room area.

$$\sum S_1 = 0.50A$$

The maximum number of sabins is equal to the room area.

$$\sum S_2 = A$$

Use Eq. 50.21.

$$\text{NR} = 10 \log \left(\frac{\sum S_1}{\sum S_2}\right) = 10 \log \left(\frac{0.50A}{A}\right) = 10 \log (0.50) = \boxed{-3.0 \text{ dB} \quad [\text{decrease}]}$$

7. *Customary U.S. Solution*

The surface area of the room walls is

$$A_1 = \big((2)(100 \text{ ft}) + (2)(400 \text{ ft})\big)(20 \text{ ft}) = 20{,}000 \text{ ft}^2$$

The surface area of the room floor and ceiling is

$$A_2 = (2)(100 \text{ ft})(400 \text{ ft}) = 80{,}000 \text{ ft}^2$$

The sound absorption coefficient of precast concrete is the NRC value of 0.02 from App. 50.A.

Define the sound absorption coefficient of precast concrete as α_{concrete}.

The sabin area of the room with all precast concrete is

$$\begin{aligned}\sum S_2 &= \alpha_{\text{concrete}}(A_1 + A_2)\\ &= (0.02)(20{,}000 \text{ ft}^2 + 80{,}000 \text{ ft}^2)\\ &= 2000 \text{ ft}^2\end{aligned}$$

Define the sound absorption coefficient of the wall acoustical treatment as α_{wall}.

The sabin area of the room with 40% of the walls treated with $\alpha_{\text{wall}} = 0.8$ is

$$\begin{aligned}\sum S_1 &= \alpha_{\text{concrete}} A_2 + \alpha_{\text{concrete}}(0.6A_1) + \alpha_{\text{wall}}(0.4A_1)\\ &= (0.02)(80{,}000 \text{ ft}^2) + (0.02)(0.6)(20{,}000 \text{ ft}^2)\\ &\quad + (0.8)(0.4)(20{,}000 \text{ ft}^2)\\ &= 8240 \text{ ft}^2\end{aligned}$$

Use Eq. 50.21.

$$\begin{aligned}\text{NR} &= 10 \log\left(\frac{\sum S_1}{\sum S_2}\right)\\ &= 10 \log\left(\frac{8240 \text{ ft}^2}{2000 \text{ ft}^2}\right) = \boxed{6.1 \text{ dB}}\end{aligned}$$

SI Solution

The surface area of the room walls is

$$A_1 = \big((2)(30 \text{ m}) + (2)(120 \text{ m})\big)(6 \text{ m}) = 1800 \text{ m}^2$$

The surface area of the room floor and ceiling is

$$A_2 = (2)(30 \text{ m})(120 \text{ m}) = 7200 \text{ m}^2$$

The sound absorption coefficient of precast concrete is the NRC value of 0.02 from App. 50.A.

Define the sound absorption coefficient of precast concrete as α_{concrete}.

The sabin area of the room with all precast concrete is

$$\begin{aligned}\sum S_2 &= \alpha_{\text{concrete}}(A_1 + A_2)\\ &= (0.02)(1800 \text{ m}^2 + 7200 \text{ m}^2)\\ &= 180 \text{ m}^2\end{aligned}$$

Define the sound absorption coefficient of the wall acoustical treatment as α_{wall}.

The sabin area of the room with 40% of the walls treated with $\alpha_{\text{wall}} = 0.8$ is

$$\begin{aligned}\sum S_1 &= \alpha_{\text{concrete}} A_2 + \alpha_{\text{concrete}}(0.6A_1) + \alpha_{\text{wall}}(0.4A_1)\\ &= (0.02)(7200 \text{ m}^2) + (0.02)(0.6)(1800 \text{ m}^2)\\ &\quad + (0.8)(0.4)(1800 \text{ m}^2)\\ &= 741.6 \text{ m}^2\end{aligned}$$

Use Eq. 50.21.

$$\begin{aligned}\text{NR} &= 10 \log\left(\frac{\sum S_1}{\sum S_2}\right)\\ &= 10 \log\left(\frac{741.6 \text{ m}^2}{180 \text{ m}^2}\right) = \boxed{6.1 \text{ dB}}\end{aligned}$$

8. *Customary U.S. Solution*

The area of the walls is

$$A_1 = \big((2)(20 \text{ ft}) + (2)(50 \text{ ft})\big)(10 \text{ ft}) = 1400 \text{ ft}^2$$

From App. 50.A, the sound absorption coefficient of sheetrock and glass is $\alpha_1 = 0.03$.

The area of the floor is

$$A_2 = (20 \text{ ft})(50 \text{ ft}) = 1000 \text{ ft}^2$$

From App. 50.A, the sound absorption coefficient of roll vinyl is $\alpha_2 = 0.03$.

The area of the ceiling is

$$A_3 = (20 \text{ ft})(50 \text{ ft}) = 1000 \text{ ft}^2$$

The sound absorption coefficient of the sheetrock is $\alpha_3 = 0.03$.

From App. 50.A, the desks have approximately 1.5 sabins each and the occupants have approximately 5.0 sabins each.

The total sabin area of the untreated room is

$$\begin{aligned}\sum S_2 &= \alpha_1 S_1 + \alpha_2 S_2 + \alpha_3 S_3 + \text{desks} + \text{occupants}\\ &\quad + \text{miscellaneous}\\ &= (0.03)(1400 \text{ ft}^2) + (0.03)(1000 \text{ ft}^2)\\ &\quad + (0.03)(1000 \text{ ft}^2) + (15)(1.5 \text{ ft}^2)\\ &\quad + (15)(5.0 \text{ ft}^2) + 5.0 \text{ ft}^2\\ &= 204.5 \text{ ft}^2\end{aligned}$$

The total sabin area excluding the ceiling is

$$\begin{aligned}204.5 \text{ ft}^2 - \alpha_3 S_3 &= 204.5 \text{ ft}^2 - (0.03)(1000 \text{ ft}^2)\\ &= 174.5 \text{ ft}^2\end{aligned}$$

The total sabin area of the room with the ceiling treated with $\alpha_3 = 0.7$ sound absorption is

$$\begin{aligned}\sum S_1 &= 174.5\ \text{ft}^2 + \alpha_3 S_3 \\ &= 174.5\ \text{ft}^2 + (0.7)(1000\ \text{ft}^2) \\ &= 874.5\ \text{ft}^2\end{aligned}$$

Use Eq. 50.21.

$$\begin{aligned}\text{NR} &= 10\ \log\left(\frac{\sum S_1}{\sum S_2}\right) \\ &= 10\ \log\left(\frac{874.5\ \text{ft}^2}{204.5\ \text{ft}^2}\right) = \boxed{6.3\ \text{dB}}\end{aligned}$$

SI Solution

The area of the walls is

$$A_1 = \big((2)(6\ \text{m}) + (2)(15\ \text{m})\big)(3\ \text{m}) = 126\ \text{m}^2$$

From App. 50.A, the sound absorption coefficient of sheetrock and glass is $\alpha_1 = 0.03$.

The area of the floor is

$$A_2 = (6\ \text{m})(15\ \text{m}) = 90\ \text{m}^2$$

From App. 50.A, the sound absorption coefficient of roll vinyl is $\alpha_2 = 0.03$.

The area of the ceiling is

$$A_3 = (6\ \text{m})(15\ \text{m}) = 90\ \text{m}^2$$

The sound absorption coefficient of the sheetrock is $\alpha_3 = 0.03$.

From App. 50.A, the desks have approximately 1.5 sabins each and the occupants have approximately 5.0 sabins each.

The total sabin area (m^2) of the untreated room is

$$\begin{aligned}\sum S_2 &= \alpha_1 S_1 + \alpha_2 S_2 + \alpha_3 S_3 + \text{desks} + \text{occupants} \\ &\quad + \text{miscellaneous} \\ &= (0.03)(126\ \text{m}^2) + (0.03)(90\ \text{m}^2) \\ &\quad + (0.03)(90\ \text{m}^2) + (15)(1.5\ \text{ft}^2) \\ &\quad \times \left(\frac{1\ \text{m}}{3.28\ \text{ft}}\right)^2 + (15)(5.0\ \text{ft}^2)\left(\frac{1\ \text{m}}{3.28\ \text{ft}}\right)^2 \\ &\quad + (5.0\ \text{ft}^2)\left(\frac{1\ \text{m}}{3.28\ \text{ft}}\right)^2 \\ &= 18.7\ \text{m}^2\end{aligned}$$

The total sabin area excluding the ceiling is

$$\begin{aligned}18.7\ \text{m}^2 - \alpha_3 S_3 &= 18.7\ \text{m}^2 - (0.03)(90\ \text{m}^2) \\ &= 16.0\ \text{m}^2\end{aligned}$$

The total sabin area of the room with the ceiling treated with $\alpha_3 = 0.7$ sound absorption is

$$\begin{aligned}\sum S_1 &= 16.0\ \text{m}^2 + \alpha_3 S_3 \\ &= 16.0\ \text{m}^2 + (0.7)(90\ \text{m}^2) \\ &= 79.0\ \text{m}^2\end{aligned}$$

Use Eq. 50.21.

$$\text{NR} = 10\ \log\left(\frac{\sum S_1}{\sum S_2}\right) = 10\ \log\left(\frac{79.0\ \text{m}^2}{18.7\ \text{m}^2}\right) = \boxed{6.3\ \text{dB}}$$

9. *Customary U.S. Solution*

Define the sound absorption coefficients of the floor, ceiling, and walls as α_1, α_2, and α_3, respectively.

The floor area is

$$A_1 = (15\ \text{ft})(20\ \text{ft}) = 300\ \text{ft}^2$$

The ceiling area is

$$A_2 = (15\ \text{ft})(20\ \text{ft}) = 300\ \text{ft}^2$$

The area of the walls is

$$A_3 = \big((2)(15\ \text{ft}) + (2)(20\ \text{ft})\big)(10\ \text{ft}) = 700\ \text{ft}^2$$

The total surface area of the room is

$$\begin{aligned}A = \sum A_i &= A_1 + A_2 + A_3 \\ &= 300\ \text{ft}^2 + 300\ \text{ft}^2 + 700\ \text{ft}^2 \\ &= 1300\ \text{ft}^2\end{aligned}$$

From Eq. 50.18, the average sound absorption coefficient of the room is

$$\begin{aligned}\overline{\alpha} &= \frac{\sum S_i}{\sum A_i} = \frac{\alpha_1 A_1 + \alpha_2 A_2 + \alpha_3 A_3}{A} \\ &= \frac{(0.03)(300\ \text{ft}^2) + (0.5)(300\ \text{ft}^2) + (0.06)(700\ \text{ft}^2)}{1300\ \text{ft}^2} \\ &= 0.155\end{aligned}$$

From Eq. 50.19, the room constant is

$$R = \frac{\overline{\alpha} A}{1 - \overline{\alpha}} = \frac{(0.155)(1300\ \text{ft}^2)}{1 - 0.155} = 238.5\ \text{ft}^2$$

From Eq. 50.15, the sound pressure level due to the machine is

$$L_{p,\text{dB}} = 10.5 + L_W + 10 \log\left(\frac{Q}{4\pi r^2} + \frac{4}{R}\right)$$
$$= 10.5 + 65 + 10 \log\left(\frac{4}{4\pi(5 \text{ ft})^2} + \frac{4}{238.5 \text{ ft}^2}\right)$$
$$= 60.2 \text{ dB}$$

Use Eq. 50.16 to combine the machine sound pressure level with the ambient sound pressure level.

$$L = 10 \log \sum 10^{\frac{W_i}{10}} = 10 \log\left(10^{\frac{60.2}{10}} + 10^{\frac{50}{10}}\right)$$
$$= \boxed{60.6 \text{ dB}}$$

SI Solution

Define the sound absorption coefficients of the floor, ceiling, and walls as α_1, α_2, and α_3, respectively.

The floor area is

$$A_1 = (4.5 \text{ m})(6 \text{ m}) = 27 \text{ m}^2$$

The ceiling area is

$$A_2 = (4.5 \text{ m})(6 \text{ m}) = 27 \text{ m}^2$$

The area of the walls is

$$A_3 = \big((2)(4.5 \text{ m}) + (2)(6 \text{ m})\big)(3 \text{ m}) = 63 \text{ m}^2$$

The total surface area of the room is

$$A = \sum A_i = A_1 + A_2 + A_3$$
$$= 27 \text{ m}^2 + 27 \text{ m}^2 + 63 \text{ m}^2$$
$$= 117 \text{ m}^2$$

From Eq. 50.18, the average sound absorption coefficient of the room is

$$\overline{\alpha} = \frac{\sum S_i}{\sum A_i} = \frac{\alpha_1 A_1 + \alpha_2 A_2 + \alpha_3 A_3}{A}$$
$$= \frac{(0.03)(27 \text{ m}^2) + (0.5)(27 \text{ m}^2) + (0.06)(63 \text{ m}^2)}{117 \text{ m}^2}$$
$$= 0.155$$

From Eq. 50.19, the room constant is

$$R = \frac{\overline{\alpha} A}{1 - \overline{\alpha}} = \frac{(0.155)(117 \text{ m}^2)}{1 - 0.155} = 21.5 \text{ m}^2$$

From Eq. 50.15, the sound pressure level due to the machine is

$$L_{p,\text{dB}} = L_W + 10 \log\left(\frac{Q}{4\pi r^2} + \frac{4}{R}\right)$$
$$= 65 + 10 \log\left(\frac{4}{4\pi(1.5 \text{ m})^2} + \frac{4}{21.5 \text{ m}^2}\right)$$
$$= 60.2 \text{ dB}$$

Use Eq. 50.16 to combine the machine sound pressure level with the ambient sound pressure level.

$$L = 10 \log \sum_{10}^{\frac{W_i}{10}} = 10 \log\left(10^{\frac{60.2}{10}} + 10^{\frac{50}{10}}\right) = \boxed{60.6 \text{ dB}}$$

51 Emergency Management

PRACTICE PROBLEMS

1. A spill of DDT from a chemical tank truck accident has permeated a 100 m^2 area of soil next to a stream. The emergency response team is evaluating the risk to a town that uses the stream as a source of water 100 km downstream from the spill site. The instream concentration of DDT just below the spill site is 0.6 μg/L, and the stream velocity is 0.2 m/s. The town's water supply is treated only through disinfection.

(a) If the half-life of DDT ranges from 56 to 110 days and its decay is a first-order reaction, the instream concentration of DDT at the point of use will be most nearly

(A) 0.5 μg/L
(B) 1.0 μg/L
(C) 1.5 μg/L
(D) 2.0 μg/L

(b) If the potency factor for DDT by the oral route is 0.34 $(\text{mg/kg·day})^{-1}$, the lifetime risk to adults in the community will be most nearly

(A) 1 per 10,000
(B) 2.5 per 100,000
(C) 3 per 1,000,000
(D) 5.5 per 1,000,000

2. Workers on a road construction project were exposed to vapors from an area contaminated with an herbicide identified as X49 (a noncarcinogen) for 30 working days (8 hr shifts) before it was recognized that individual protection might be needed. The mean concentration of vapors of X49 in air was 2 $\mu g/m^3$ during this period.

(a) The standard factors for workers at the site are as follows.

air breathed	5.2 m^3/h
retention rate (inhaled air)	100%
absorption rate	100%
adult body weight	70 kg

The chronic daily intake of X49 for each worker is most nearly

(A) 1×10^{-2} mg/kg·day
(B) 2×10^{-2} mg/kg·day
(C) 1×10^{-3} mg/kg·day
(D) 2×10^{-3} mg/kg·day

(b) If the reference dose for X49 for the inhalation route is 1.6×10^{-3} mg/kg·day and the acceptable risk is 1 in 1,000,000, how can exposure of the workers most accurately be characterized?

(A) The exposure was acceptable because the hazard ratio is less than 1.0.
(B) The exposure was not acceptable because the hazard ratio is greater than 1.0.
(C) The exposure was acceptable because the risk is less than 1 in 1,000,000.
(D) The exposure was not acceptable because the risk is greater than 1 in 1,000,000.

3. A rural community water supply was contaminated by wastes resulting from a railroad tank car derailment. The wastes contain the following noncarcinogenic pollutants and have the indicated reference doses.

constituent	concentration	reference dose
chemical A	10 μg/L	5×10^{-3} mg/kg·day
metal B	25 μg/L	1.0×10^{-3} mg/kg·day
chemical C	15 μg/L	3.6×10^{-2} mg/kg·day

Standard factors for the community are as follows.

drinking water use, adult	2.0 L/day
drinking water use, child	1.2 L/day
body weight (mass), adult	70 kg
body weight (mass), child	30 kg

Assume retention and absorption rates of 100% each.

(a) The hazard ratio for adults is most nearly

(A) 0.4
(B) 0.8
(C) 1.0
(D) 1.4

(b) The hazard ratio for children is most nearly

(A) 0.2
(B) 0.5
(C) 1.0
(D) 1.5

(c) How should the exposure to the community be characterized?

(A) There is no significant risk to the community.
(B) There is an unacceptable risk to adults but not to children.
(C) There is an unacceptable risk to children but not to adults.
(D) There is an unacceptable risk to both adults and children.

4. Emergency responders have identified the following flammable liquids stored separately in a warehouse 50 m from a house fire. The atmosphere is at normal conditions.

liquid	auto-ignition temp.	vapor pressure at 20°C	ratio of specific heats	flammable range volume % LFL	UFL
ethane	472°C	3850 kPa	1.206	3.0	12.5
benzene	500°C	10 kPa	1.100	1.2	8.0
n-heptane	285°C	4.6 kPa	1.400	1.1	6.7

(a) If a spill occurs and equilibrium is reached, which chemicals would form a flammable mixture in air?

(A) ethane and n-heptane
(B) benzene only
(C) n-heptane only
(D) ethane and benzene

(b) If the drums are instantly crushed from a collapsing structure and the vapors mix with air and are compressed to 100 atm, which of the vapors would autoignite?

(A) ethane and benzene
(B) benzene
(C) n-heptane
(D) benzene and n-heptane

(c) If the vapors from the three liquids are mixed in air in the amounts given, what would the LFL of the mixture most nearly be?

	no. of moles
ethane	2
benzene	4
n-heptane	3

(A) 0.5%
(B) 1.5%
(C) 2.5%
(D) 3.5%

SOLUTIONS

1. (a) The travel time, t, from the spill to the town is

$$t = \frac{l}{v} = \left(\frac{100 \text{ km}}{0.2 \ \frac{\text{m}}{\text{s}}}\right)\left(10^3 \ \frac{\text{m}}{\text{km}}\right)\left(\frac{1 \text{ min}}{60 \text{ s}}\right)\left(\frac{1 \text{ h}}{60 \text{ min}}\right)\left(\frac{1 \text{ day}}{24 \text{ h}}\right) = 5.79 \text{ days}$$

The half-life, $t_{1/2}$, of DDT ranges from 56 to 110 days. Use the average of 83 days.

The concentration can be found from the first-order reaction equation.

$$C_t = C_o e^{-kt}$$

Find k from the half-concentration and the half-life of DDT.

$$C_t = \frac{C_o}{2} = C_o e^{-k(83 \text{ days})}$$
$$0.5 = e^{-k(83 \text{ days})}$$
$$\ln 0.5 = -k(83 \text{ days})$$
$$k = 8.351 \times 10^{-3}\text{days}^{-1}$$

The concentration of DDT at the point of use is

$$C_t = C_o e^{-kt} = \left(0.6 \ \frac{\mu\text{g}}{\text{L}}\right) e^{-(8.351\times 10^{-3}\text{days}^{-1})(5.79 \text{ days})} = \boxed{0.572 \ \mu\text{g/L}}$$

The answer is (A).

(b) Find the quantity of water ingested daily. Assume a standard adult has a body weight of 70 kg and ingests 2 L of water daily.

Find the chronic daily intake, CDI.

$$\text{CDI} = \frac{C+Q}{m} = \frac{\left(0.572 \ \frac{\mu\text{g}}{\text{L}}\right)\left(2 \ \frac{\text{L}}{\text{day}}\right)\left(10^{-3} \ \frac{\text{mg}}{\mu\text{g}}\right)}{70 \text{ kg}} = 1.634 \times 10^{-5} \text{ mg/kg·day}$$

The potency factor, PF, (also known as slope factor, SF) for DDT by the oral route is 0.34 (mg/kg·day)$^{-1}$.

The lifetime risk is

$$R = (\text{CDI})(\text{PF}) = \left(1.634 \times 10^{-5} \ \frac{\text{mg}}{\text{kg·day}}\right)\left(0.34\left(\frac{\text{mg}}{\text{kg·day}}\right)^{-1}\right)$$

$$R = \boxed{5.556 \times 10^{-6} \quad \text{(5.6 additional cancer cases per million population)}}$$

The answer is (D).

2. (a) The daily intake can be found as follows.

$$I_N = \frac{(C)(\text{CR})(\text{EF})(\text{ED})(\text{RR})(\text{ABS})}{(\text{BW})(\text{AT})}$$

$$C = \frac{(2\ \mu\text{g/m}^3)}{1000\ \frac{\mu\text{g}}{\text{mg}}} = 2\times 10^3\ \frac{\text{mg}}{\text{m}^3}$$

$$\text{CR} = \left(5.2\ \frac{\text{m}^3}{\text{h}}\right)\left(8\ \frac{\text{h}}{\text{day}}\right) = 41.6\ \text{m}^3/\text{day}$$

$$\text{EF} = 30\ \text{days/yr}$$

$$\text{ED} = 1\ \text{yr}$$

$$\text{RR} = 1.0$$

$$\text{ABS} = 1.0$$

$$\text{BW} = 70\ \text{kg}$$

$$\text{AT} = 30\ \text{days}$$

$$I_N = \frac{\left(2\times 10^{-3}\ \frac{\text{mg}}{\text{m}^3}\right)\left(41.6\ \frac{\text{m}^3}{\text{day}}\right)\left(30\ \frac{\text{days}}{\text{yr}}\right)\times(1\ \text{yr})(1.0)(1.0)}{(70\ \text{kg})(30\ \text{days})}$$

$$= \boxed{1.19\times 10^{-3}\ \text{mg/kg·day}}$$

The answer is (C).

(b) For noncarcinogens, use the hazard ratio method.

$$\text{HR} = \frac{I_N}{\text{RfD}} = \frac{1.19\times 10^{-3}\ \frac{\text{mg}}{\text{kg·day}}}{1.6\times 10^{-3}\ \frac{\text{mg}}{\text{kg·day}}}$$

$$= 0.74$$

The HR is less than 1.0, so the exposure was acceptable.

The answer is (A).

3. (a)

$$\text{EF} = 1\ \text{day/yr}$$

$$\text{ED} = 1\ \text{yr}$$

$$\text{RR} = 1.0$$

$$\text{ABS} = 1.0$$

$$\text{AT} = 1\ \text{day}$$

For adults, the dose of chemical A is

$$\text{CR} = 2.0\ \text{L/day}$$

$$\text{BW} = 70\ \text{kg}$$

$$C_{\text{A}} = \left(10\ \frac{\mu\text{g}}{\text{L}}\right)\left(10^{-3}\ \frac{\text{mg}}{\mu\text{g}}\right) = 10\times 10^{-3}\ \text{mg/L}$$

$$I_N = \frac{(C)(\text{CR})(\text{EF})(\text{ED})(\text{RR})(\text{ABS})}{(\text{BW})(\text{AT})}$$

$$= \frac{\left(10\times 10^{-3}\ \frac{\text{mg}}{\text{L}}\right)\left(2.0\ \frac{\text{L}}{\text{day}}\right)\left(1\ \frac{\text{day}}{\text{yr}}\right)\times(1\ \text{yr})(1.0)(1.0)}{(70\ \text{kg})(1\ \text{day})}$$

$$= 2.86\times 10^{-4}\ \text{mg/kg·day}$$

The hazard ratio is

$$\text{HR} = \frac{I_N}{\text{RfD}} = \frac{2.86\times 10^{-4}\ \frac{\text{mg}}{\text{kg·day}}}{5\times 10^{-3}\ \frac{\text{mg}}{\text{kg·day}}}$$

$$= 0.057$$

Similarly, the other pollutants contribute to the hazard ratio for adults.

pollutant	dose $\left(\frac{\text{mg}}{\text{kg·day}}\right)$	RfD $\left(\frac{\text{mg}}{\text{kg·day}}\right)$	hazard ratio
chemical A	2.86×10^{-4}	50×10^{-4}	0.057
metal B	7.14×10^{-4}	10×10^{-4}	0.714
chemical C	4.28×10^{-4}	360×10^{-4}	0.012
total			**0.783**

The answer is (B).

(b) For children, the dose of chemical A is

$$\text{CR} = 1.0\ \text{L/day}$$

$$\text{BW} = 30\ \text{kg}$$

$$I_N = \frac{(C)(\text{CR})(\text{EF})(\text{ED})(\text{RR})(\text{ABS})}{(\text{BW})(\text{AT})}$$

$$= \frac{\left(10\times 10^{-3}\ \frac{\text{mg}}{\text{L}}\right)\left(1.2\ \frac{\text{L}}{\text{day}}\right)\left(1\frac{\text{day}}{\text{yr}}\right)\times(1\ \text{yr})\ (1.0)\ (1.0)}{(30\ \text{kg})(1\ \text{day})}$$

$$= 4\times 10^{-4}\ \text{mg/kg·day}$$

Find the hazard ratio.

$$\text{HR} = \frac{I_N}{\text{RfD}} = \frac{4\times 10^{-4}\ \frac{\text{mg}}{\text{kg·day}}}{5\times 10^{-3}\ \frac{\text{mg}}{\text{kg·day}}}$$

$$= 0.08$$

Similarly, the other pollutants contribute to the hazard ratio for children.

pollutant	dose $\left(\frac{\text{mg}}{\text{kg·day}}\right)$	RfD $\left(\frac{\text{mg}}{\text{kg·day}}\right)$	hazard ratio
chemical A	4×10^{-4}	50×10^{-4}	0.080
metal B	10×10^{-4}	10×10^{-4}	1.000
chemical C	6×10^{-4}	360×10^{-4}	0.017
total			**1.097**

The answer is (C).

(c) Because the hazard ratio is less than 1.0 for adults, there is no unacceptable risk for them. However, there

Health, Safety, Welfare

is an unacceptable risk for children because the hazard ratio for them is greater than 1.0.

The answer is (C).

4. (a) Determine the mole fraction (in vapor phase) of each substance at NTP.

$$T = 25°\text{C} + 273 = 298\text{K}$$

$$p = (1 \text{ atm})\left(\frac{101.325 \text{ kPa}}{1 \text{ atm}}\right) = 101.325 \text{ kPa}$$

$$x_v = \frac{p_{\text{vp, pure}} x_{li}}{p}$$

Because each flammable liquid is stored independently, the mole fraction (in liquid phase) of each ($x_{l,\text{A}}$) is 1.

For ethane,

$$x_{v,\text{ethane}} = \frac{(3850 \text{ kPa})(1)}{101.325 \text{ kPa}} = 38.0 \quad (3800\%)$$

$x_{v,\text{ethane}}$ is greater than ethane's UFL of 12.5% and is too rich to form a flammable mixure in air.

For benzene,

$$x_{v,\text{benzene}} = \frac{(10 \text{ kPa})(1)}{101.325 \text{ kPa}} = 0.0986 \quad (9.86\%)$$

$x_{v,\text{benzene}}$ is greater than benzene's UFL of 8.0% and is too rich to form a flammable mixture in air.

For n-heptane,

$$x_{v,n\text{-heptane}} = \frac{(4.6 \text{ kPa})(1)}{101.325 \text{ kPa}} = 0.0454 \quad (4.54\%)$$

$x_{v,n\text{-heptane}}$ is between n-heptane's LFL and UFL (1.1% and 6.7% respectively) and forms a flammable mixture in air.

The answer is (C).

(b) At NTP,

$$T_1 = 25°\text{C} + 273 = 298\text{K}$$

$$p_1 = (1 \text{ atm})\left(\frac{101.325 \text{ kPa}}{1 \text{ atm}}\right) = 101.325 \text{ kPa}$$

After compression,

$$p_2 = (100 \text{ atm})\left(\frac{101.325 \text{ kPa}}{1 \text{ atm}}\right) = 101.325 \times 10^2 \text{ kPa}$$

Determine the final temperature of each substance. Assume compression occurs adiabatically.

$$T_2 = T_1 \left(\frac{p_2}{p_1}\right)^{\frac{k-1}{k}}$$

For ethane,

$$T_{2,\text{ ethane}} = (298\text{K})\left(\frac{100 \text{ atm}}{1 \text{ atm}}\right)^{\frac{1.206-1}{1.206}} = 654.4\text{K}$$

$$T_{2,\text{ ethane}} = 654.4\text{K} - 273 \approx 381°\text{C}$$

$T_{2,\text{ ethane}}$ is less than 472°C, the autoignition temperature of ethane.

For benzene,

$$T_{2,\text{ benzene}} = (298\text{K})\left(\frac{100 \text{ atm}}{1 \text{ atm}}\right)^{\frac{1.100-1}{1.100}} = 452.9\text{K}$$

$$T_{2,\text{ benzene}} = 452.9\text{K} - 273 \approx 180°\text{C}$$

$T_{2,\text{ benzene}}$ is less than 500°C, the autoignition temperature of benzene.

For n-heptane,

$$T_{2,n\text{-heptane}} = (298\text{K})\left(\frac{100 \text{ atm}}{1 \text{ atm}}\right)^{\frac{1.4-1}{1.4}} = 1111\text{K}$$

$$T_{2,n\text{-heptane}} = 1111\text{K} - 273 = 838°\text{C}$$

$T_{2,n\text{-heptane}}$ is greater than 285°C, the autoignition temperature of n-heptane.

The answer is (C).

(c)

For ethane,

$$x_{v,\text{ ethane}} = \frac{n_{\text{ethane}}}{n_{\text{total}}} = \frac{2 \text{ mol}}{2 \text{ mol} + 4 \text{ mol} + 3 \text{ mol}} = 0.22$$

For benzene,

$$x_{v,\text{ benzene}} = \frac{n_{\text{benzene}}}{n_{\text{total}}} = \frac{4 \text{ mol}}{9 \text{ mol}} = 0.44$$

For n-heptane,

$$x_{v,n\text{-heptane}} = \frac{n_{n\text{-heptane}}}{n_{\text{total}}} = \frac{3 \text{ mol}}{9 \text{ mol}} = 0.33$$

$$\text{LFL}_{\text{mix}} = \frac{1}{\sum\left(\frac{x_{v,i}}{\text{LFL}_i}\right)} = \frac{1}{\frac{0.22}{3.0\%} + \frac{0.44}{1.2\%} + \frac{0.33}{1.1\%}} = \boxed{1.35\%}$$

The answer is (B).

52 Electrical Systems and Equipment

PRACTICE PROBLEMS

Chapter 52 of the *Environmental Engineering Reference Manual* contains no practice problems.

53 Instrumentation and Digital Numbering Systems

PRACTICE PROBLEMS

Chapter 53 of the *Environmental Engineering Reference Manual* contains no practice problems.

54 Engineering Economic Analysis

PRACTICE PROBLEMS

1. At 6% effective annual interest, how much will be accumulated if \$1000 is invested for ten years?

2. At 6% effective annual interest, what is the present worth of \$2000 that becomes available in four years?

3. At 6% effective annual interest, how much should be invested to accumulate \$2000 in 20 years?

4. At 6% effective annual interest, what year-end annual amount deposited over seven years is equivalent to \$500 invested now?

5. At 6% effective annual interest, what will be the accumulated amount at the end of ten years if \$50 is invested at the end of each year for ten years?

6. At 6% effective annual interest, how much should be deposited at the start of each year for ten years (a total of 10 deposits) in order to empty the fund by drawing out \$200 at the end of each year for ten years (a total of 10 withdrawals)?

7. At 6% effective annual interest, how much should be deposited at the start of each year for five years to accumulate \$2000 on the date of the last deposit?

8. At 6% effective annual interest, how much will be accumulated in ten years if three payments of \$100 are deposited every other year for four years, with the first payment occurring at $t = 0$?

9. \$500 is compounded monthly at a 6% nominal annual interest rate. How much will have accumulated in five years?

10. What is the effective annual rate of return on an \$80 investment that pays back \$120 in seven years?

11. A new machine will cost \$17,000 and will have a resale value of \$14,000 after five years. Special tooling will cost \$5000. The tooling will have a resale value of \$2500 after five years. Maintenance will be \$2000 per year. The effective annual interest rate is 6%. What will be the average annual cost of ownership during the next five years?

12. An old covered wooden bridge can be strengthened at a cost of \$9000, or it can be replaced for \$40,000. The present salvage value of the old bridge is \$13,000. It is estimated that the reinforced bridge will last for 20 years, will have an annual cost of \$500, and will have a salvage value of \$10,000 at the end of 20 years. The estimated salvage value of the new bridge after 25 years is \$15,000. Maintenance for the new bridge would cost \$100 annually. The effective annual interest rate is 8%. Which is the best alternative?

13. A firm expects to receive \$32,000 each year for 15 years from sales of a product. An initial investment of \$150,000 will be required to manufacture the product. Expenses will run \$7530 per year. Salvage value is zero, and straight-line depreciation is used. The income tax rate is 48%. What is the after-tax rate of return?

14. A public works project has initial costs of \$1,000,000, benefits of \$1,500,000, and disbenefits of \$300,000. (a) What is the Winfrey benefit/cost ratio? (b) What is the excess of benefits over costs?

15. A speculator in land pays \$14,000 for property that he expects to hold for ten years. \$1000 is spent in renovation, and a monthly rent of \$75 is collected from the tenants. (Use the year-end convention.) Taxes are \$150 per year, and maintenance costs are \$250 per year. What must be the sale price in ten years to realize a 10% rate of return?

16. What is the effective annual interest rate for a payment plan of 30 equal payments of \$89.30 per month when a lump sum payment of \$2000 would have been an outright purchase?

17. A depreciable item is purchased for \$500,000. The salvage value at the end of 25 years is estimated at \$100,000. What is the depreciation in each of the first three years using the (a) straight line, (b) sum-of-the-years' digits, and (c) double-declining balance methods?

18. Equipment that is purchased for $12,000 now is expected to be sold after ten years for $2000. The estimated maintenance is $1000 for the first year, but it is expected to increase $200 each year thereafter. The effective annual interest rate is 10%. What are the (a) present worth and (b) annual cost?

19. A new grain combine with a 20-year life can remove seven pounds of rocks from its harvest per hour. Any rocks left in its output hopper will cause $25,000 damage in subsequent processes. Several investments are available to increase the rock-removal capacity, as listed in the table. The effective annual interest rate is 10%. What should be done?

rock removal rate	probability of exceeding rock removal rate	required investment to achieve removal rate
7	0.15	0
8	0.10	$15,000
9	0.07	$20,000
10	0.03	$30,000

20. (*Time limit: one hour*) A mechanism that costs $10,000 has operating costs and salvage values as given. An effective annual interest rate of 20% is to be used.

year	operating cost	salvage value
1	$2000	$8000
2	$3000	$7000
3	$4000	$6000
4	$5000	$5000
5	$6000	$4000

(a) What is the economic life of the mechanism? (b) Assuming that the mechanism has been owned and operated for four years already, what is the cost of owning and operating the mechanism for one more year?

21. (*Time limit: one hour*) A salesperson intends to purchase a car for $50,000 for personal use, driving 15,000 miles per year. Insurance for personal use costs $2000 per year, and maintenance costs $1500 per year. The car gets 15 miles per gallon, and gasoline costs $1.50 per gallon. The resale value after five years will be $10,000. The salesperson's employer has asked that the car be used for business driving of 50,000 miles per year and has offered a reimbursement of $0.30 per mile. Using the car for business would increase the insurance cost to $3000 per year and maintenance to $2000 per year. The salvage value after five years would be reduced to $5000. If the employer purchased a car for the salesperson to use, the initial cost would be the same, but insurance, maintenance, and salvage would be $2500, $2000, and $8000, respectively. The salesperson's effective annual interest rate is 10%. (a) Is the reimbursement offer adequate? (b) With a reimbursement of $0.30 per mile, how many miles must the car be driven per year to justify the employer buying the car for the salesperson to use?

22. (*Time limit: one hour*) Alternatives A and B are being evaluated. The effective annual interest rate is 10%. Which alternative is economically superior?

	alternative A	alternative B
first cost	$80,000	$35,000
life	20 years	10 years
salvage value	$7000	0
annual costs		
years 1–5	$1000	$3000
years 6–10	$1500	$4000
years 11–20	$2000	0
additional cost		
year 10	$5000	0

23. (*Time limit: one hour*) A car is needed for three years. Plans A and B for acquiring the car are being evaluated. An effective annual interest rate of 10% is to be used. Which plan is economically superior?

plan A: lease the car for $0.25/mile (all inclusive)

plan B: purchase the car for $30,000
keep the car for three years
sell the car after three years for $7200
pay $0.14 per mile for oil and gas
pay other costs of $500 per year

24. (*Time limit: one hour*) Two methods are being considered to meet strict air pollution control requirements over the next ten years. Method A uses equipment with a life of ten years. Method B uses equipment with a life of five years that will be replaced with new equipment with an additional life of five years. Capacities of the two methods are different, but operating costs do not depend on the throughput. Operation is 24 hours per day, 365 days per year. The effective annual interest rate for this evaluation is 7%.

	method A	method B	
	years 1–10	years 1–5	years 6–10
installation cost	$13,000	$6000	$7000
equipment cost	$10,000	$2000	$2200
operating cost per hour	$10.50	$8.00	$8.00
salvage value	$5000	$2000	$2000
capacity (tons/yr)	50	20	20
life	10 years	5 years	5 years

(a) What is the uniform annual cost per ton for each method? (b) Over what range of throughput (in tons/yr) does each method have the minimum cost?

25. (*Time limit: one hour*) A transit district has asked for your assistance in determining the proper fare for its bus system. An effective annual interest rate of 7% is to be used. The following additional information was compiled for your study.

cost per bus	\$60,000
bus life	20 years
salvage value	\$10,000
miles driven per year	37,440
number of passengers per year	80,000
operating cost	\$1.00 per mile in the first year, increasing \$0.10 per mile each year thereafter

(a) If the fare is to remain constant for the next 20 years, what is the break-even fare per passenger? (b) If the transit district decides to set the per-passenger fare at \$0.35 for the first year, by what amount should the per-passenger fare go up each year thereafter so that the district can break even in 20 years? (c) If the transit district decides to set the per-passenger fare at \$0.35 for the first year and the per-passenger fare goes up \$0.05 each year thereafter, what additional governmental subsidy (per passenger) is needed for the district to break even in 20 years?

26. Make a recommendation to your client to accept one of the following alternatives. Use the present worth comparison method. (Initial costs are the same.)

A: a 25 year annuity paying \$4800 at the end of each year, where the interest rate is a nominal 12% per annum

B: a 25 year annuity paying \$1200 every quarter at 12% nominal annual interest

27. A firm has two alternatives for improvement of its existing production line. The data are as follows.

	alternative A	alternative B
initial installment cost	\$1500	\$2500
annual operating cost	\$800	\$650
service life	5 years	8 years
salvage value	0	0

Determine the best alternative using an interest rate of 15%.

28. Two mutually exclusive alternatives requiring different investments are being considered. The life of both alternatives is estimated at 20 years with no salvage values. The minimum rate of return that is considered acceptable is 4%. Which alternative is best?

	alternative A	alternative B
investment required	\$70,000	\$40,000
net income per year	\$5620	\$4075
rate of return on total investment	5%	8%

29. Compare the costs of two plant renovation schemes, A and B. Assume equal lives of 25 years, no salvage values, and interest at 25%. Make the comparison on the basis of (a) present worth, (b) capitalized cost, and (c) annual cost.

	alternative A	alternative B
first cost	\$20,000	\$25,000
annual expenditure	\$3000	\$2500

30. With interest at 8%, obtain the solutions to the following to the nearest dollar. (a) A machine costs \$18,000 and has a salvage value of \$2000. It has a useful life of 8 years. What is its book value at the end of 5 years using straight line depreciation? (b) Using data from part (a), find the depreciation in the first three years using the sinking fund method. (c) Repeat part (a) using double declining balance depreciation to find the first five years' depreciation.

31. A chemical pump motor unit is purchased for \$14,000. The estimated life is 8 years, after which it will be sold for \$1800. Find the depreciation in the first two years by the sum-of-the-years' digits method. Calculate the after-tax depreciation recovery using 15% interest with 52% income tax.

32. A soda ash plant has the water effluent from processing equipment treated in a large settling basin. The settling basin eventually discharges into a river that runs alongside the basin. Recently enacted environmental regulations require all rainfall on the plant to be diverted and treated in the settling basin. A heavy rainfall will cause the entire basin to overflow. An uncontrolled overflow will cause environmental damage and heavy fines. The construction of additional height on the existing basic walls is under consideration.

Data on the costs of construction and expected costs for environmental cleanup and fines are shown. Data on 50 typical winters have been collected. The soda ash plant management considers 12% to be their minimum rate of return, and it is felt that after 15 years the plant will be closed. The company wants to select the alternative that minimizes its total expected costs.

additional basin height (ft)	number of winters with basin overflow	expense for environmental clean up per year	construction cost
0	24		0
5	14	\$600,000	\$600,000
10	8	\$650,000	\$710,000
15	3	\$700,000	\$900,000
20	1	\$800,000	\$1,000,000
	50		

Systems, Mgmt., Professional

33. A wood processing plant installed a waste gas scrubber at a cost of $30,000 to remove pollutants from the exhaust discharged into the atmosphere. The scrubber has no salvage value and will cost $18,700 to operate next year, with operating costs expected to increase at the rate of $1200 per year thereafter. When should the company consider replacing the scrubber? Money can be borrowed at 12%.

34. Two alternative piping schemes are being considered by a water treatment facility. On the basis of a 10-year life and an interest rate of 12%, determine the number of hours of operation for which the two installations will break even.

	alternative A	alternative B
pipe diameter	4 in	6 in
head loss for required flow	48 ft	26 ft
size motor required	20 hp	7 hp
energy cost per hour of operation	$ 0.30	$ 0.10
cost of motor installed	$3600	$2800
cost of pipes and fittings	$3050	$5010
salvage value at end of 10 years	$200	$280

35. An 88% learning curve is used with an item whose first production time was 6 weeks. How long will it take to produce the fourth item? How long will it take to produce the sixth through fourteenth items?

36. *(Time limit: one hour)* A company is considering two alternatives, only one of which can be selected.

alternative	initial investment	salvage value	annual net profit	life
A	$120,000	$15,000	$57,000	5 yr
B	$170,000	$20,000	$67,000	5 yr

The net profit is after operating and maintenance costs, but before taxes. The company pays 45% of its year-end profit as income taxes. Use straight line depreciation. Do not use investment tax credit. Find the best alternative if the company's minimum attractive rate of return is 15%.

37. *(Time limit: one hour)* A company is considering the purchase of equipment to expand its capacity. The equipment cost is $300,000. The equipment is needed for 5 years, after which it will be sold for $50,000. The company's before-tax cash flow will be improved $90,000 annually by the purchase of the asset.

The corporate tax rate if 48%, and straight line depreciation will be used. The company will take an investment tax credit of 6.67%. What is the after-tax rate of return associated with this equipment purchase?

38. *(Time limit: one hour)* A 120-room hotel is purchased for $2,500,000. A 25-year loan is available for 12%. A study was conducted to determine the various occupancy rates.

occupancy	probability
65% full	0.40
70%	0.30
75%	0.20
80%	0.10

The operating costs of the hotel are as follows.

taxes and insurance	$20,000 annually
maintenance	$50,000 annually
operating	$200,000 annually

The life of the hotel is figured to be 25 years when operating 365 days per year. The salvage value after 25 years is $500,000.

Neglect tax credit and income taxes. Determine the average rate that should be charged per room per night to return 15% of the initial cost each year.

39. *(Time limit: one hour)* A company is insured for $3,500,000 against fire. The insurance rate is $0.69/1000. The insurance company will decrease the rate to $0.47/1000 if fire sprinklers are installed. The initial cost of the sprinklers is $7500. Annual costs are $200; additional taxes are $100 annually. The system life is 25 years. What is the rate of return on this investment?

40. *(Time limit: one hour)* Heat losses through the walls in an existing building cost a company $1,300,000 per year. This amount is considered excessive, and two alternatives are being evaluated. Neither of the alternatives will increase the life of the existing building beyond the current expected life of 6 years, and neither of the alternatives will produce a salvage value.

Alternative A: Do nothing. Continue with current losses.

Alternative B: Spend $2,000,000 immediately to upgrade the building and reduce the loss by 80%. This alternative will require annual maintenance of $150,000.

Alternative C: Spend $1,200,000 immediately. Repeat the $1,200,000 expenditure 3 years from now. Heat loss the first year will be reduced 80%. Due to deterioration, the reduction will be 55% and 20% in the second and third years. (The pattern is repeated starting after the second expenditure.) There are no maintenance costs.

All energy and maintenance costs are considered expenses for tax purposes. The company's tax rate is 48%, and straight line depreciation is used. 15% is regarded as the effective annual interest rate. Evaluate each alternative on an after-tax basis, and recommend the best alternative.

41. *(Time limit: one hour)* You have been asked to determine if a 7-year-old machine should be replaced. Give a full explanation for your recommendation. Base your decision on a before-tax interest rate of 15%.

The existing machine is presumed to have a 10-year life. It has been depreciated on a straight line basis from its original value of $1,250,000 to a current book value of $620,000. Its ultimate salvage value was assumed to be $350,000 for purposes of depreciation. Its present salvage value is estimated at $400,000, and this is not expected to change over the next 3 years. The current operating costs are not expected to change from $200,000 per year.

A new machine costs $800,000, with operating costs of $40,000 the first year, and increasing by $30,000 each year thereafter. The new machine has an expected life of 10 years. The salvage value depends on the year the new machine is retired.

year retired	salvage
1	$600,000
2	$500,000
3	$450,000
4	$400,000
5	$350,000
6	$300,000
7	$250,000
8	$200,000
9	$150,000
10	$100,000

SOLUTIONS

1.

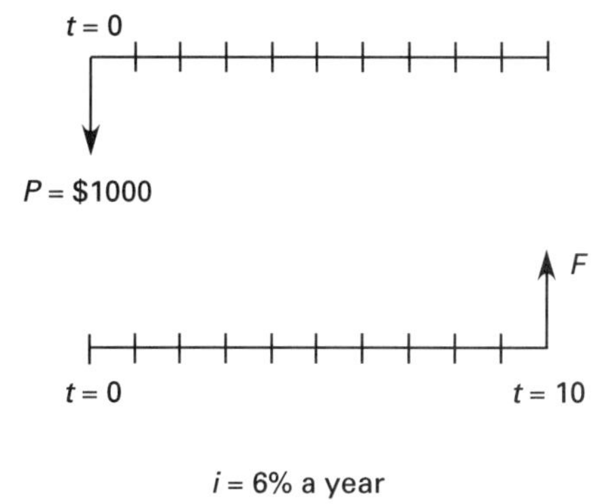

By the formula from Table 54.1,

$$F = P(1+i)^n = (\$1000)(1+0.06)^{10} = \boxed{\$1790.85}$$

By the factor converting P to F, $(F/P, i, n) = 1.7908$ for $i = 6\%$ a year and $n = 10$ years.

$$F = P(F/P, 6\%, 10) = (\$1000)(1.7908) = \boxed{\$1790.80}$$

2.

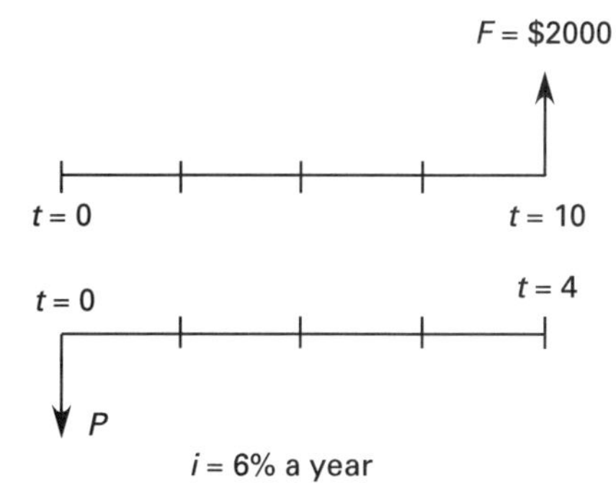

By the formula from Table 54.1,

$$P = \frac{F}{(1+i)^n} = \frac{\$2000}{(1+0.06)^4} = \boxed{\$1584.19}$$

From the factor converting F to P, $(P/F, i, n) = 0.7921$ for $i = 6\%$ a year and $n = 4$ years.

$$\text{P} = F(P/F, 6\%, 4) = (\$2000)(0.7921) = \boxed{\$1584.20}$$

3.

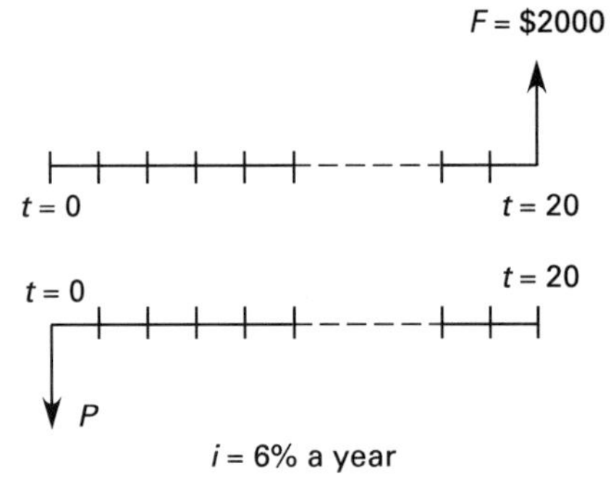

By the formula from Table 54.1,

$$P = \frac{F}{(1+i)^n} = \frac{\$2000}{(1+0.06)^{20}} = \boxed{\$623.61}$$

From the factor converting F to P, $(P/F, i, n) = 0.3118$ for $i = 6\%$ a year and $n = 20$ years.

$$P = F(P/F, 6\%, 20) = (\$2000)(0.3118) = \boxed{\$623.60}$$

4.

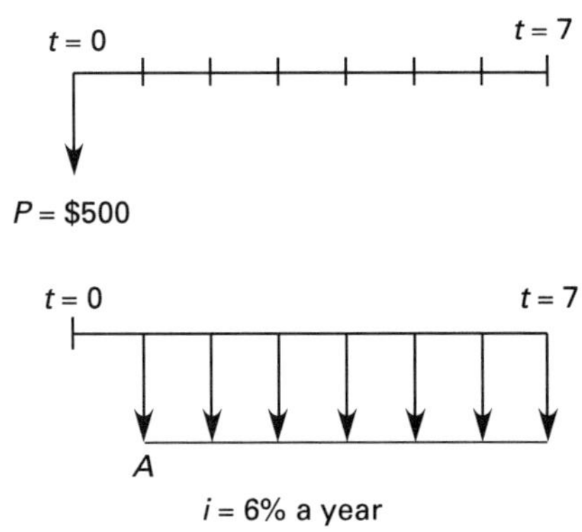

By the formula from Table 54.1,

$$A = P\left(\frac{i(1+i)^n}{(1+i)^n - 1}\right) = (\$500)\left(\frac{(0.06)(1+0.06)^7}{(1+0.06)^7 - 1}\right) = \boxed{\$89.57}$$

By the factor converting P to A, $(A/P, i, n) = 0.17914$ for $i = 6\%$ a year and $n = 7$ years.

$$A = P(A/P, 6\%, 7) = (\$500)(0.17914) = \boxed{\$89.57}$$

5.

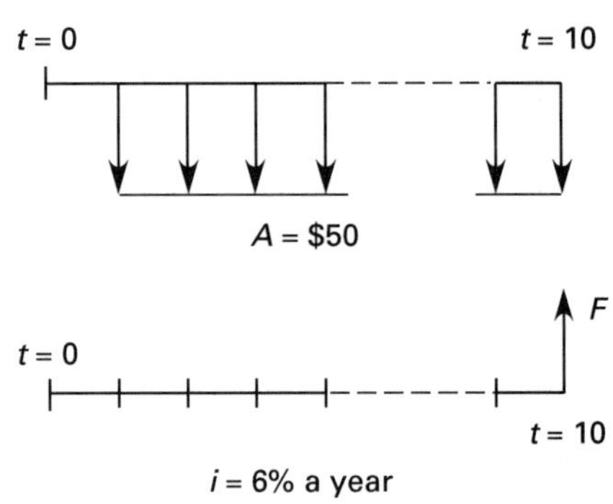

By the formula from Table 54.1,

$$F = A\left(\frac{(1+i)^n - 1}{i}\right) = (\$50)\left(\frac{(1+0.06)^{10} - 1}{0.06}\right) = \boxed{\$659.04}$$

By the factor converting A to F, $(F/A, i, n) = 13.181$ for $i = 6\%$ a year and $n = 10$ years.

$$F = A(F/A, 6\%, 10) = (\$50)(13.181) = \boxed{\$659.05}$$

6.

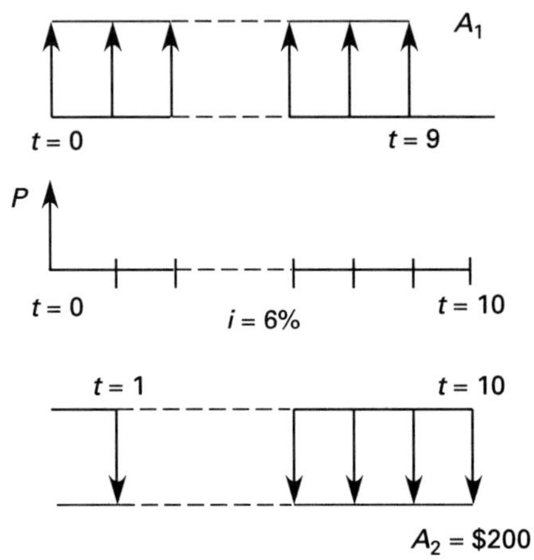

By the formula from Table 54.1, for each cash flow diagram,

$$\begin{aligned} P &= A_1 + (A_1)\left(\frac{(1+0.06)^9 - 1}{(0.06)(1+0.06)^9}\right) \\ &= A_2\left(\frac{(1+0.06)^{10} - 1}{(0.06)(1+0.06)^{10}}\right) \end{aligned}$$

Therefore for $A_2 = \$200$,

$$\begin{aligned} A_1 + (A_1)\left(\frac{(1+0.06)^9 - 1}{(0.06)(1+0.06)^9}\right) &= (\$200)\left(\frac{(1+0.06)^{10} - 1}{(0.06)(1+0.06)^{10}}\right) \\ 7.80A_1 &= \$1472.02 \\ A_1 &= \boxed{\$188.72} \end{aligned}$$

By the factor converting A to P,

$$\begin{aligned} (P/A, 6\%, 9) &= 6.8017 \\ (P/A, 6\%, 10) &= 7.3601 \\ A_1 + A_1(6.8017) &= (\$200)(7.3601) \\ 7.8017A_1 &= \$1472.02 \\ A_1 &= \frac{\$1472.02}{7.8017} = \boxed{\$188.68} \end{aligned}$$

7.

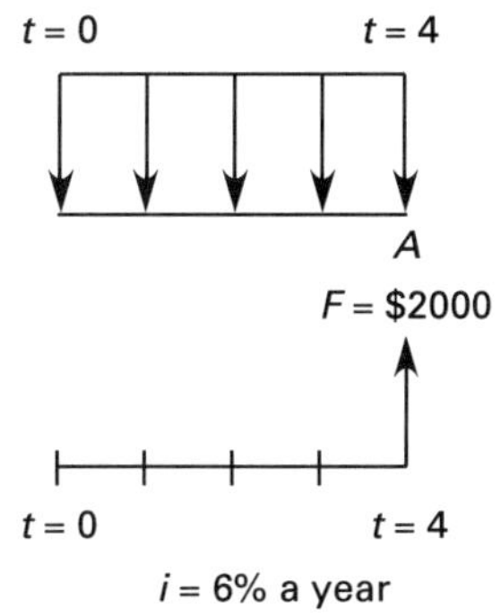

By the formula from Table 54.1,

$$F = A\left(\frac{(1+i)^n - 1}{i}\right)$$

Since the deposits start at the start of each year, $n = 4 + 1$, for a total of five deposits.

$$\begin{aligned} F &= A\left(\frac{(1+i)^{n+1} - 1}{i}\right) = \$2000 \\ &= A\left(\frac{(1+0.06)^5 - 1}{0.06}\right) \\ \$2000 &= 5.6371A \\ A &= \frac{\$2000}{5.6371} = \boxed{\$354.79} \end{aligned}$$

By the factor converting P and A to F,

$$\begin{aligned} F &= A\big((F/P, 6\%, 4) + (F/A, 6\%, 4)\big) \\ \$2000 &= A(1.2625 + 4.3746) \\ A &= \boxed{\$354.79} \end{aligned}$$

8.

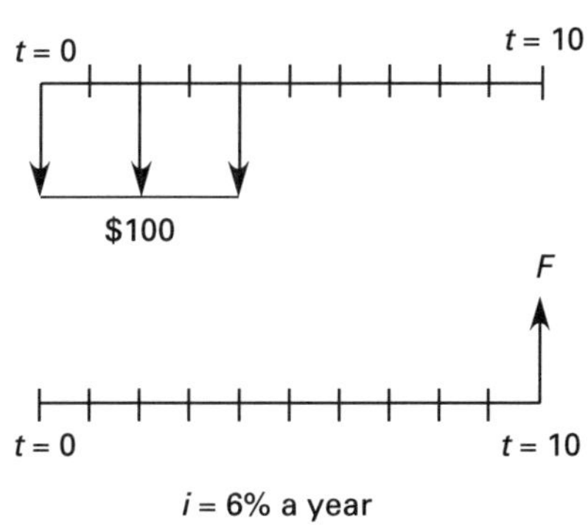

By the formula from Table 54.1, $F = P(1+i)^n$. If each deposit is considered as P, each will accumulate interest for periods of 10, 8, and 6 years.

Therefore,

$$\begin{aligned} F &= (\$100)(1+0.06)^{10} + (\$100)(1+0.06)^8 \\ &\quad + (\$100)(1+0.06)^6 \\ &= (\$100)(1.7908 + 1.5938 + 1.4185) \\ &= (\$100)(4.8031) \\ &= \boxed{\$480.31} \end{aligned}$$

By the factor converting P to F,

$$\begin{aligned} (F/P, i, n) &= 1.7908 \text{ for } i = 6\% \text{ and } n = 10 \\ &= 1.5938 \text{ for } i = 6\% \text{ and } n = 8 \\ &= 1.4185 \text{ for } i = 6\% \text{ and } n = 6 \end{aligned}$$

By summation,

$$\begin{aligned} F &= (\$100)(1.7908 + 1.5938 + 1.4185) \\ &= (\$100)(4.8031) \\ &= \boxed{\$480.31} \end{aligned}$$

9.

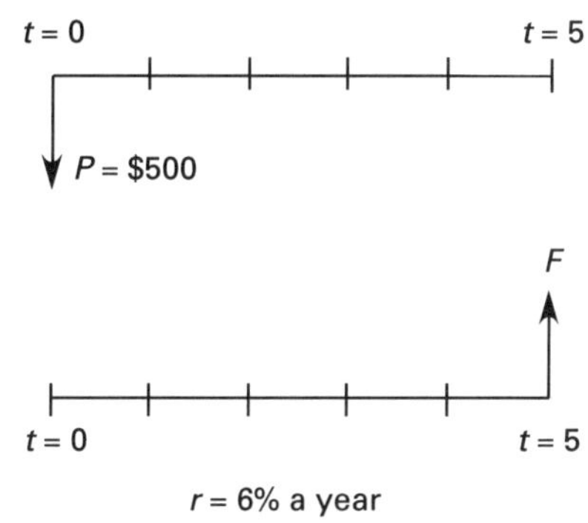

Since the deposit is compounded monthly, the effective interest rate should be calculated as shown by Eq. 54.54.

$$\begin{aligned} i &= \left(1 + \frac{r}{k}\right)^k - 1 = \left(1 + \frac{0.06}{12}\right)^{12} - 1 \\ &= 0.061677 \quad (6.1677\%) \end{aligned}$$

By the formula from Table 54.1,

$$F = P(1+i)^n = (\$500)(1 + 0.061677)^5 = \boxed{\$674.42}$$

To use a table of factors, interpolation is required.

i%	factor F/P
6	1.3382
6.1677	desired
7	1.4026

$$i = \left(\frac{6.1677 - 6}{7 - 6}\right)(1.4026 - 1.3382)$$
$$= 0.0108$$

Therefore,

$$F/P = 1.3382 + 0.0108 = 1.3490$$
$$F = P(F/P, 6.1677\%, 5) = (\$500)(1.3490)$$
$$= \boxed{\$674.50}$$

10.

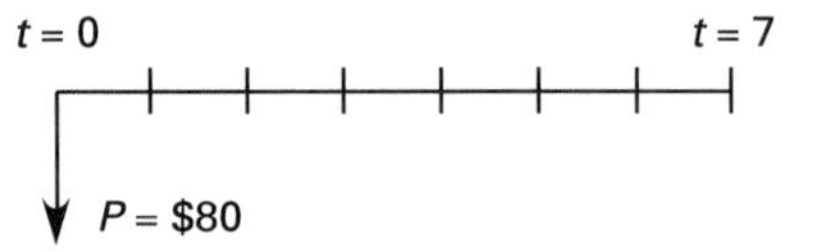

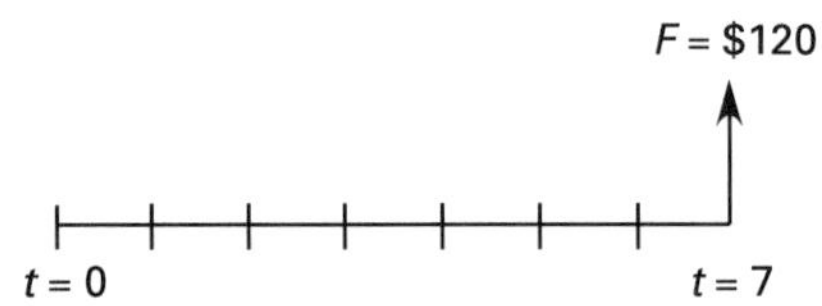

By the formula from Table 54.1,

$$F = P(1 + i)^n$$

Therefore,

$$(1 + i)^n = F/P$$
$$i = (F/P)^{\frac{1}{n}} - 1 = \left(\frac{\$120}{\$80}\right)^{\frac{1}{7}} - 1$$
$$= 0.059 \approx \boxed{6\%}$$

By the factor coverting P to F,

$$F = P(F/P, i\%, 7)$$
$$(F/P, i\%, 7) = F/P = \frac{\$120}{\$80} = 1.5$$

By checking the interest tables,

$$(F/P, i\%, 7) = 1.4071 \text{ for } i = 5\%$$
$$= 1.5036 \text{ for } i = 6\%$$
$$= 1.6058 \text{ for } i = 7\%$$

Therefore, $i \approx \boxed{6\%.}$

11.

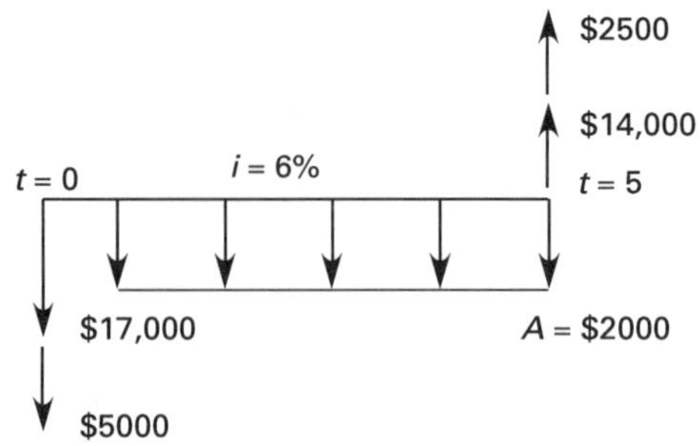

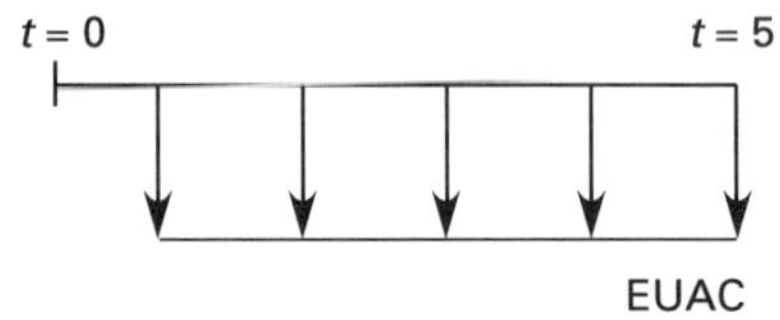

Annual cost of ownership, EUAC, can be obtained by the factors converting P to A and F to A.

$$P = \$17{,}000 + \$5000$$
$$= \$22{,}000$$
$$F = \$14{,}000 + \$2500$$
$$= \$16{,}500$$
$$\text{EUAC} = A + P(A/P, 6\%, 5) - F(A/F, 6\%, 5)$$
$$(A/P, 6\%, 5) = 0.23740$$
$$(A/F, 6\%, 5) = 0.17740$$
$$\text{EUAC} = \$2000 + (\$22{,}000)(0.23740) - (\$16{,}500)(0.17740)$$
$$= \boxed{\$4295.70}$$

12.

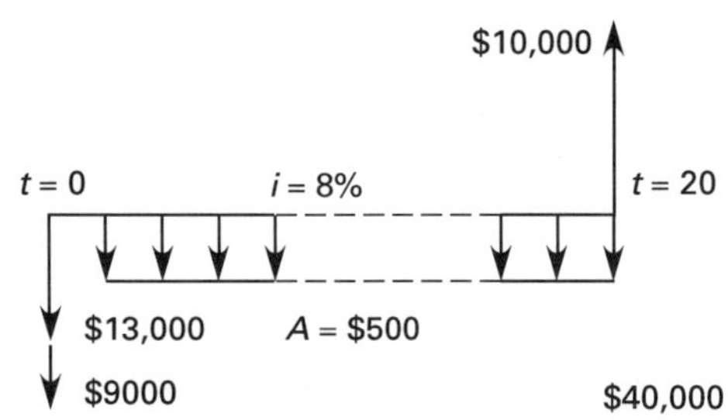

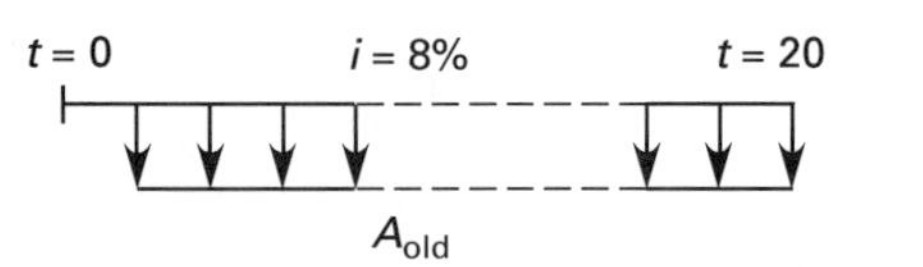

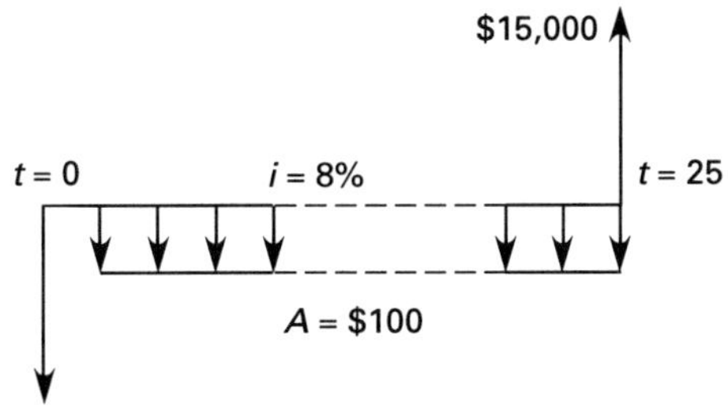

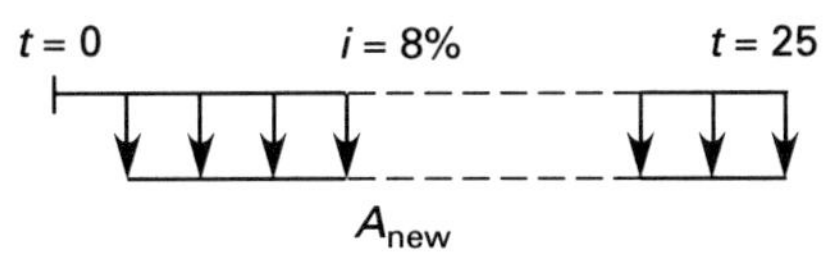

Consider the salvage value as a benefit lost (cost).

$$\begin{aligned}\text{EUAC}_{\text{old}} &= \$500 + (\$22{,}000)(A/P, 8\%, 20) \\ &\quad - (\$10{,}000)(A/F, 8\%, 20) \\ (A/P, 8\%, 20) &= 0.1019 \\ (A/F, 8\%, 20) &= 0.0219 \\ \text{EUAC}_{\text{old}} &= \$500 + (\$22{,}000)(0.1019) \\ &\quad - (\$10{,}000)(0.0219) \\ &= \$2522.80\end{aligned}$$

Similarly,

$$\begin{aligned}\text{EUAC}_{\text{new}} &= \$100 + (\$40{,}000)(A/P, 8\%, 25) \\ &\quad - (\$15{,}000)(A/F, 8\%, 25) \\ (A/P, 8\%, 25) &= 0.0937 \\ (A/F, 8\%, 25) &= 0.0137 \\ \text{EUAC}_{\text{new}} &= \$100 + (\$40{,}000)(0.0937) \\ &\quad - (\$15{,}000)(0.0137) \\ &= \$3642.50\end{aligned}$$

Therefore, the new bridge is going to be more costly.

The best alternative is to strengthen the old bridge.

13.

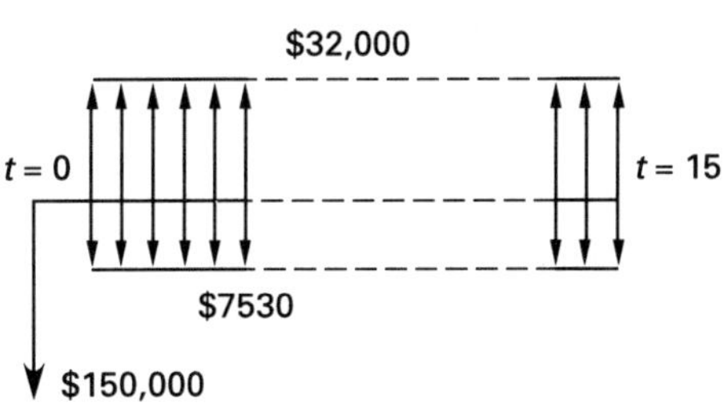

The annual depreciation is

$$\begin{aligned}D &= \frac{C - S_n}{n} = \frac{\$150{,}000}{15} \\ &= \$10{,}000/\text{year}\end{aligned}$$

The taxable income is

$$\$32{,}000 - \$7530 - \$10{,}000 = \$14{,}470/\text{year}$$

Taxes paid are

$$(\$14{,}470)(0.48) = \$6945.60/\text{year}$$

The after-tax cash flow is

$$\$24{,}470 - \$6945.60 = \$17{,}524.40$$

The present worth of the alternate is zero when evaluated at its ROR.

$$0 = -\$150{,}000 + (\$17{,}524.40)(P/A, i\%, 15)$$

Therefore,

$$(P/A, i\%, 15) = \frac{\$150{,}000}{\$17{,}524.40} = 8.55949$$

By checking the tables, this factor matches $i = 8\%$.

$$\boxed{\text{ROR} = 8\%}$$

14. The Winfrey benefit/cost ratio is

$$B/C = \frac{B - D}{C}$$

(a) The benefit/cost ratio will be

$$B/C = \frac{\$1{,}500{,}000 - \$300{,}000}{\$1{,}000{,}000} = \boxed{1.2}$$

(b) The excess of benefits over cost is **$200,000.**

15.

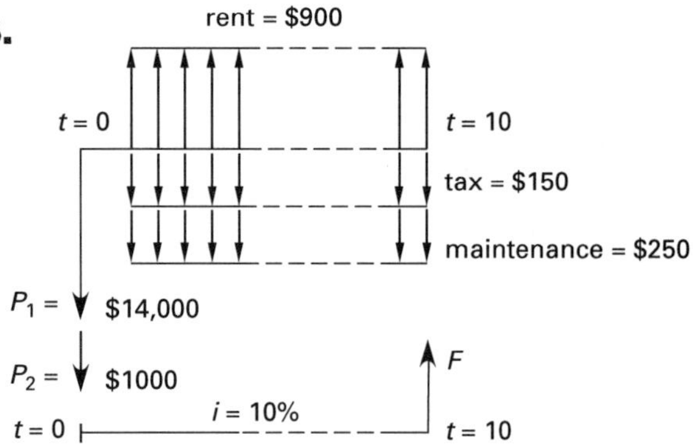

The annual rent is

$$(\$75)\left(12\ \frac{\text{months}}{\text{year}}\right) = \$900$$

$$P = P_1 + P_2 = \$15{,}000$$
$$A_1 = -\$900$$
$$A_2 = \$250 + \$150 = \$400$$

By the factors converting P to F and A to F,

$$\begin{aligned} F &= (\$15{,}000)(F/P, 10\%, 10) \\ &\quad + (\$400)(F/A, 10\%, 10) \\ &\quad - (\$900)(F/A, 10\%, 10) \end{aligned}$$

$$(F/P, 10\%, 10) = 2.5937$$
$$(F/A, 10\%, 10) = 15.937$$

$$\begin{aligned} F &= (\$15{,}000)(2.5937) + (\$400)(15.937) \\ &\quad - (\$900)(15.937) \\ &= \boxed{\$30{,}937} \end{aligned}$$

16.

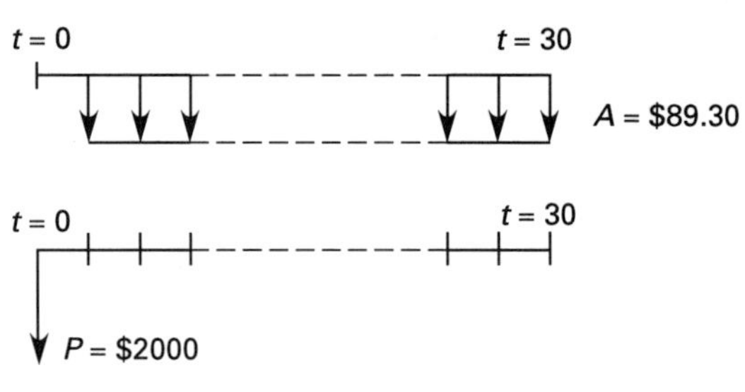

By the formula relating P to A,

$$P = A\left(\frac{(1+i)^n - 1}{i(1+i)^n}\right)$$

$$\frac{(1+i)^{30} - 1}{i(1+i)^{30}} = \frac{\$2000}{\$89.30} = 22.30$$

By trial and error,

$i(\%)$	$(1+i)^{30}$	$\frac{(1+i)^{30}-1}{i(1+i)^{30}}$
10	17.45	9.42
6	5.74	13.76
4	3.24	17.28
2	1.81	22.37

2% per month is close.

$$i = (1 + 0.02)^{12} - 1 = \boxed{0.2682\ \ (26.82\%)}$$

17. (a) Use the straight line method, Eq. 54.25.

$$D = \frac{C - S_n}{n}$$

Each year depreciation will remain the same.

$$D = \frac{\$500{,}000 - \$100{,}000}{25} = \boxed{\$16{,}000}$$

(b) Sum-of-the-years' digits (SOYD) can be calculated as shown by Eq. 54.28.

$$D_j = \frac{(C - S_n)(n - j + 1)}{T}$$

Use Eq. 54.27.

$$T = \tfrac{1}{2}n(n+1) = \left(\tfrac{1}{2}\right)(25)(25+1) = 325$$

$$D_1 = \frac{(\$500{,}000 - \$100{,}000)(25 - 1 + 1)}{325} = \boxed{\$30{,}769}$$

$$D_2 = \frac{(\$500{,}000 - \$100{,}000)(25 - 2 + 1)}{325} = \boxed{\$29{,}538}$$

$$D_3 = \frac{(\$500{,}000 - \$100{,}000)(25 - 3 + 1)}{325} = \boxed{\$28{,}308}$$

(c) The double declining balance (DDB) method can be used. By Eq. 54.32,

$$D_j = dC(1 - d)^{j-1}$$

Use Eq. 54.31.

$$d = \frac{2}{n}$$
$$= \tfrac{2}{25}$$

$$D_1 = \left(\tfrac{2}{25}\right)(\$500{,}000)\left(1 - \tfrac{2}{25}\right)^0 = \boxed{\$40{,}000}$$

$$D_2 = \left(\tfrac{2}{25}\right)(\$500{,}000)\left(1 - \tfrac{2}{25}\right)^1 = \boxed{\$36{,}800}$$

$$D_3 = \left(\tfrac{2}{25}\right)(\$500{,}000)\left(1 - \tfrac{2}{25}\right)^2 = \boxed{\$33{,}856}$$

18.

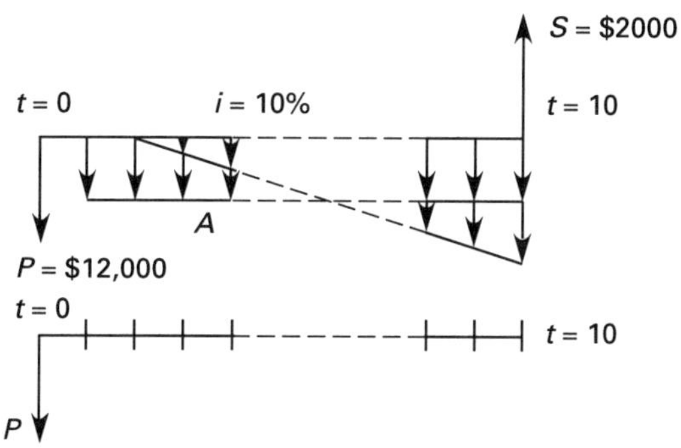

(a) $A = \$1000$ and $G = \$200$ for $t = n - 1 = 9$ years.

$$F = S = \$2000$$
$$P = \$12{,}000 + A(P/A, 10\%, 10) + G(P/G, 10\%, 10) - F(P/F, 10\%, 10)$$
$$= \$12{,}000 + (\$1000)(6.1446) + (\$200)(22.8913) - (\$2000)(0.3855)$$
$$= \boxed{\$21{,}952}$$

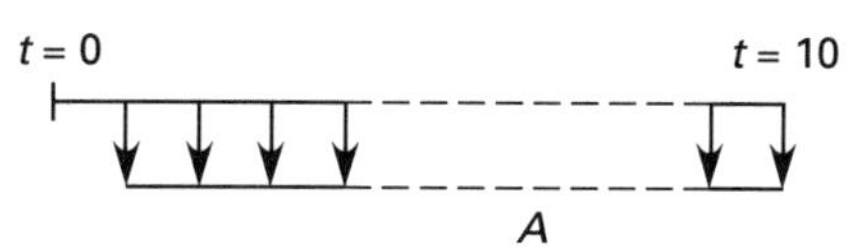

(b) $A = (\$12{,}000)(A/P, 10\%, 10) + \$1000 + (\$200)(A/G, 10\%, 10) - (\$2000)(A/F, 10\%, 10)$

$$= (\$12{,}000)(0.1627) + \$1000 + (\$200)(3.7255) - (\$2000)(0.0627)$$
$$= \boxed{\$3572.70}$$

19. An increase in rock removal capacity can be achieved by a 20-year loan (investment). Different cases available can be compared by equivalent uniform annual cost (EUAC).

$$\text{EUAC} = \text{annual loan cost} + \text{expected annual damage}$$
$$= \text{cost }(A/P, 10\%, 20) + (\$25{,}000)(\text{probability})$$
$$(A/P, 10\%, 20) = 0.1175$$

A table can be prepared for different cases.

rock removal rate	cost ($)	annual loan cost ($)	expected annual damage ($)	EUAC ($)
7	0	0	3750	3750.00
8	15,000	1761.90	2500	4261.90
9	20,000	2349.20	1750	4099.20
10	30,000	3523.80	750	4273.80

It is cheapest to do nothing.

20. Calculate the cost of owning and operating for each year.

$$A_1 = (\$10{,}000)(A/P, 20\%, 1) + \$2000 - (\$8000)(A/F, 20\%, 1)$$
$$(A/P, 20\%, 1) = 1.2$$
$$(A/F, 20\%, 1) = 1.0$$
$$A_1 = (\$10{,}000)(1.2) + \$2000 - (\$8000)(1.0)$$
$$= \$6000$$
$$A_2 = (\$10{,}000)(A/P, 20\%, 2) + \$2000 + (\$1000)(A/G, 20\%, 2) - (\$7000)(A/F, 20\%, 2)$$
$$(A/P, 20\%, 2) = 0.6545$$
$$(A/G, 20\%, 2) = 0.4545$$
$$(A/F, 20\%, 2) = 0.4545$$
$$A_2 = (\$10{,}000)(0.6545) + \$2000 + (\$1000)(0.4545) - (\$7000)(0.4545)$$
$$= \$5818$$
$$A_3 = (\$10{,}000)(A/P, 20\%, 3) + \$2000 + (\$1000)(A/G, 20\%, 3) - (\$6000)(A/F, 20\%, 3)$$
$$(A/P, 20\%, 3) = 0.4747$$
$$(A/G, 20\%, 3) = 0.8791$$
$$(A/F, 20\%, 3) = 0.2747$$

$$\begin{aligned} A_3 &= (\$10{,}000)(0.4747) + \$2000 \\ &\quad + (\$1000)(0.8791) \\ &\quad - (\$6000)(0.2747) \\ &= \$5977.90 \end{aligned}$$

$$\begin{aligned} A_4 &= (\$10{,}000)(A/P, 20\%, 4) \\ &\quad + \$2000 + (\$1000)(A/G, 20\%, 4) \\ &\quad - (\$5000)(A/F, 20\%, 4) \end{aligned}$$

$$(A/P, 20\%, 4) = 0.3863$$
$$(A/G, 20\%, 4) = 1.2762$$
$$(A/F, 20\%, 4) = 0.1863$$

$$\begin{aligned} A_4 &= (\$10{,}000)(0.3863) + \$2000 \\ &\quad + (\$1000)(1.2762) - (\$5000)(0.1863) \\ &= \$6207.70 \end{aligned}$$

$$\begin{aligned} A_5 &= (\$10{,}000)(A/P, 20\%, 5) + \$2000 \\ &\quad + (\$1000)(A/G, 20\%, 5) \\ &\quad - (\$4000)(A/F, 20\%, 5) \end{aligned}$$

$$(A/P, 20\%, 5) = 0.3344$$
$$(A/G, 20\%, 5) = 1.6405$$
$$(A/F, 20\%, 5) = 0.1344$$

$$\begin{aligned} A_5 &= (\$10{,}000)(0.3344) + \$2000 \\ &\quad + (\$1000)(1.6405) - (\$4000)(0.1344) \\ &= \$6446.90 \end{aligned}$$

(a) Since the annual owning and operating cost is smallest after two years of operation, it is advantageous to sell the mechanism after the second year.

The economic life is two years.

(b) After four years of operation, the owning and operating cost of the mechanism for one more year will be

$$A = \$6000 + (\$5000)(1+i) - \$4000$$
$$i = 0.2 \quad (20\%)$$
$$A = \$6000 + (\$5000)(1.2) - \$4000$$
$$= \boxed{\$8000}$$

21. To find out if the reimbursement is adequate, calculate the business-related expense.

Charge the company for business travel.

$$\begin{aligned} \text{insurance: } & \$3000 - \$2000 = \$1000 \\ \text{maintenance: } & \$2000 - \$1500 = \$500 \\ \text{drop in salvage value: } & \$10{,}000 - \$5000 = \$5000 \end{aligned}$$

The annual portion of the drop in salvage value is

$$A = (\$5000)(A/F, 10\%, 5)$$
$$(A/F, 10\%, 5) = 0.1638$$
$$A = (\$5000)(0.1638) = \$819/\text{year}$$

(a) The annual cost of gas is

$$\left(\frac{50{,}000 \text{ mi}}{15 \ \frac{\text{mi}}{\text{gal}}}\right)\left(\frac{\$1.50}{\text{gal}}\right) = \$5000$$

$$\begin{aligned} \text{EUAC per mile} &= \frac{\$1000 + \$500 + \$819 + \$5000}{50{,}000 \text{ mi}} \\ &= \boxed{\$0.14638/\text{mi}} \end{aligned}$$

Since the reimbursement per mile was \$0.30 and since $\$0.30 > \$0.14638$, the reimbursement is adequate.

(b) Next, determine (with reimbursement) how many miles the car must be driven to break even.

If the car is driven M miles per year,

$$\begin{aligned} \left(\frac{\$0.30}{1 \text{ mi}}\right) M &= (\$50{,}000)(A/P, 10\%, 5) + \$2500 \\ &\quad + \$2000 - (\$8000)(A/F, 10\%, 5) \\ &\quad + \left(\frac{M}{15 \ \frac{\text{mi}}{\text{gal}}}\right)(\$1.50) \end{aligned}$$

$$(A/P, 10\%, 5) = 0.2638$$
$$(A/F, 10\%, 5) = 0.1638$$

$$\begin{aligned} 0.3M &= (\$50{,}000)(0.2638) + \$2500 + \$2000 \\ &\quad - (\$8000)(0.1638) + 0.1M \\ 0.2M &= \$16{,}379.60 \\ M &= \frac{\$16{,}379.60}{0.2 \ \frac{\$}{\text{mi}}} = \boxed{81{,}898 \text{ mi}} \end{aligned}$$

22.

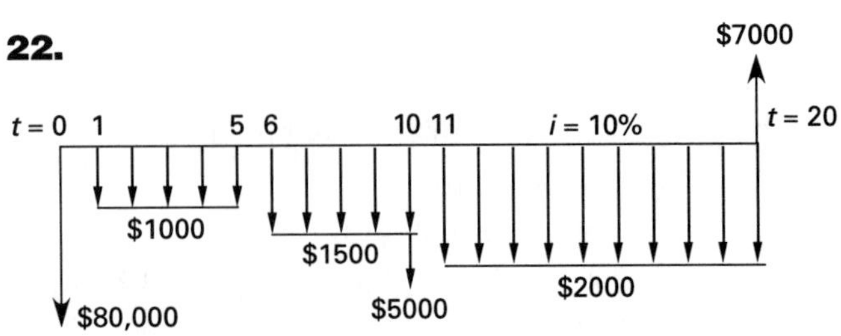

$$\begin{aligned} P_{\text{A}} = {} & \$80{,}000 + (\$1000)(P/A, 10\%, 5) \\ & + (\$1500)(P/A, 10\%, 5)(P/F, 10\%, 5) \\ & + (\$2000)(P/A, 10\%, 10)(P/F, 10\%, 10) \\ & + (\$5000)(P/F, 10\%, 10) \\ & - (\$7000)(P/F, 10\%, 20) \end{aligned}$$

$$\begin{aligned} (P/A, 10\%, 5) &= 3.7908 \\ (P/F, 10\%, 5) &= 0.6209 \\ (P/A, 10\%, 10) &= 6.1446 \\ (P/F, 10\%, 10) &= 0.3855 \\ (P/F, 10\%, 20) &= 0.1486 \\ P_{\text{A}} &= \$80{,}000 + (\$1000)(3.7908) \\ &\quad + (\$1500)(3.7908)(0.6209) \\ &\quad + (\$2000)(6.1446)(0.3855) \\ &\quad + (\$5000)(0.3855) - (\$7000)(0.1486) \\ &= \$92{,}946.15 \end{aligned}$$

Since the lives are different, compare by EUAC.

$$\begin{aligned} \text{EUAC}(A) &= (\$92{,}946.14)(A/P, 10\%, 20) \\ &= (\$92{,}946.14)(0.1175) = \$10{,}921 \end{aligned}$$

Similarly, evaluate alternative B.

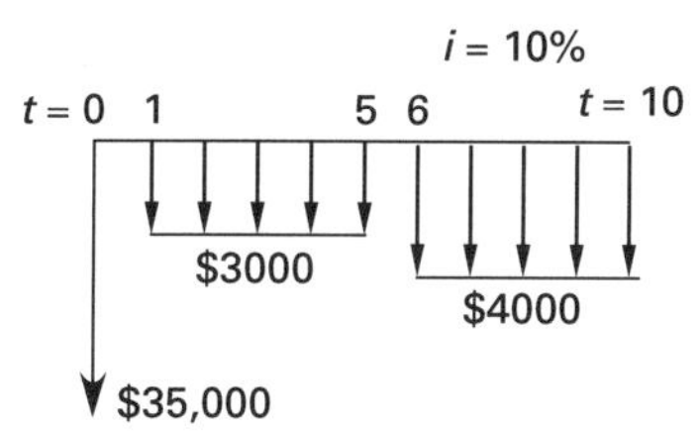

$$\begin{aligned} P_{\text{B}} &= \$35{,}000 + (\$3000)(P/A, 10\%, 5) \\ &\quad + (\$4000)(P/A, 10\%, 5)(P/F, 10\%, 5) \\ (P/A, 10\%, 5) &= 3.7908 \\ (P/F, 10\%, 5) &= 0.6209 \\ P_{\text{B}} &= \$35{,}000 + (\$3000)(3.7908) \\ &\quad + (\$4000)(3.7908)(0.6209) \\ &= \$55{,}787.23 \\ \text{EUAC}(B) &= (\$55{,}787.23)(A/P, 10\%, 10) \\ &= (\$55{,}787.23)(0.1627) = \$9077 \end{aligned}$$

Since $\text{EUAC}(B) < \text{EUAC}(A)$,

> Alternative B is economically superior.

23. For both cases, if the annual cost is compared with a total annual mileage of M,

$$\begin{aligned} A_{\text{A}} &= \$0.25M \\ A_{\text{B}} &= (\$30{,}000)(A/P, 10\%, 3) + \$0.14M \\ &\quad + \$500 - (\$7200)(A/F, 10\%, 3) \\ (A/P, 10\%, 3) &= 0.4021 \\ (A/F, 10\%, 3) &= 0.3021 \\ A_{\text{B}} &= (\$30{,}000)(0.4021) + \$0.14M + \$500 \\ &\quad - (\$7200)(0.3021) \\ &= \$12{,}063 + \$0.14M \\ &\quad + \$500 - \$2175.12 \\ &= \$10{,}387.88 + \$0.14M \end{aligned}$$

For an equal annual cost $A_{\text{A}} = A_{\text{B}}$,

$$\$0.25M = \$10{,}387.88 + \$0.14M$$

An annual mileage would be $M = 94{,}435$ mi.

For an annual mileage less than that, $A_{\text{A}} < A_{\text{B}}$.

> Plan A is economically superior until that mileage is exceeded.

24. (a) Method A:

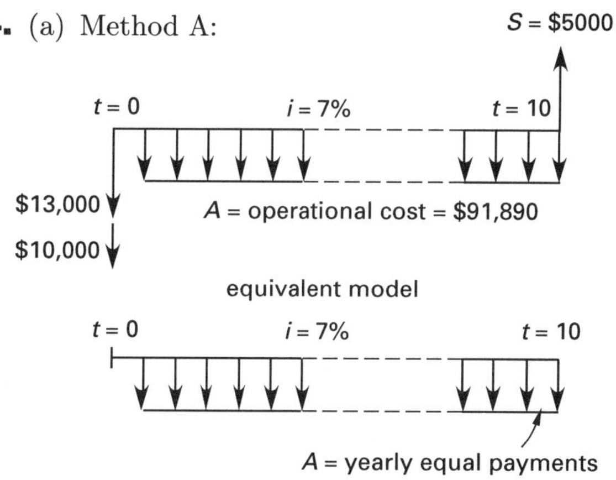

24 hr/day
365 days/yr
total of (24)(365) = 8760 hr/yr
$10.50 operational cost/hr
total of (8760)($10.50) = $91,890 operational cost/yr

$$\begin{aligned} A &= \$91{,}980 + (\$23{,}000)(A/P, 7\%, 10) \\ &\quad - (\$5000)(A/F, 7\%, 10) \\ (A/P, 7\%, 10) &= 0.1424 \\ (A/F, 7\%, 10) &= 0.0724 \\ A &= \$91{,}980 + (\$23{,}000)(0.1424) \\ &\quad - (\$5000)(0.0724) \\ &= \$94{,}893.30/\text{yr} \end{aligned}$$

Systems, Mgmt., Professional

Therefore, the uniform annual cost per ton each year will be

$$\frac{\$94{,}893.30}{50 \text{ ton}} = \boxed{\$1897.87}$$

Method B:

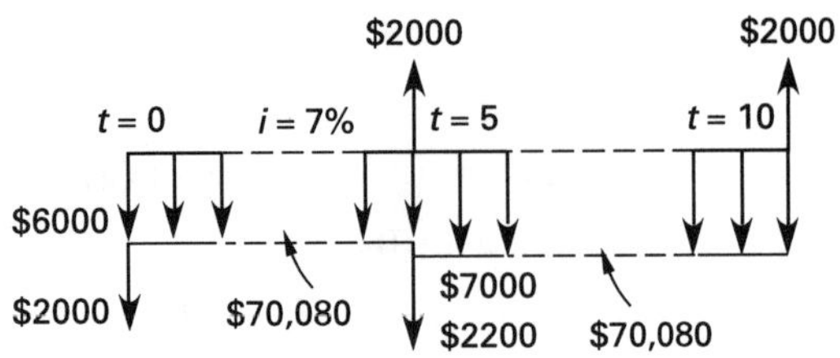

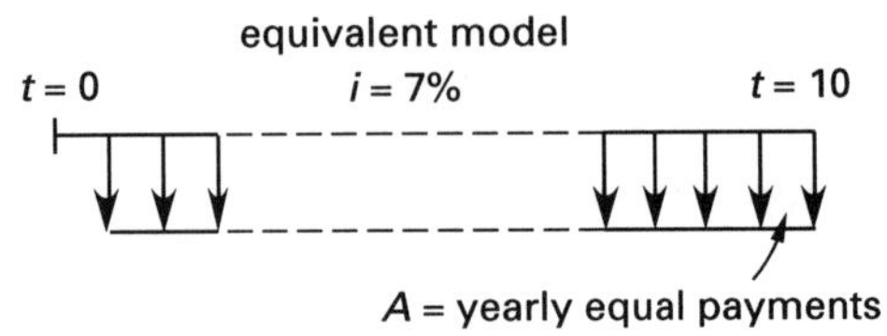

8760 hr/yr
$8 operational cost/hr
total of $70,080 operational cost/yr

$$\begin{aligned} A &= \$70{,}080 + (\$6000 + \$2000)(A/P, 7\%, 10) \\ &\quad + (\$7000 + \$2200 - \$2000)(P/F, 7\%, 5) \\ &\quad \times (A/P, 7\%, 10) \\ &\quad - (\$2000)(A/F, 7\%, 10) \end{aligned}$$

$$(A/P, 7\%, 10) = 0.1424$$

$$(A/F, 7\%, 10) = 0.07238$$

$$(P/F, 7\%, 5) = 0.7130$$

$$\begin{aligned} A &= \$70{,}080 + (\$8000)(0.1424) \\ &\quad + (\$7200)(0.7130)(0.1424) \\ &\quad - (\$2000)(0.0724) \\ &= \$71{,}805.42/\text{yr} \end{aligned}$$

Therefore, the uniform annual cost per ton each year will be

$$\frac{\$71{,}805.42}{20 \text{ ton}} = \boxed{\$3590.27}$$

(b)

tons/yr	cost of using A	cost of using B	cheapest
0–20	$94,893 (1x)	$71,805 (1x)	B
20–40	$94,893 (1x)	$143,610 (2x)	A
40–50	$94,893 (1x)	$215,415 (3x)	A
50–60	$189,786 (2x)	$215,415 (3x)	A
60–80	$189,786 (2x)	$287,220 (4x)	A

25.

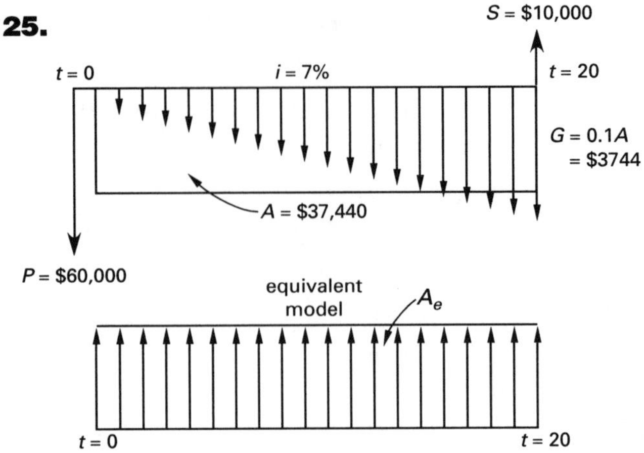

$$\begin{aligned} A_e &= (\$60{,}000)(A/P, 7\%, 20) + A \\ &\quad + G(P/G, 7\%, 20)(A/P, 7\%, 20) \\ &\quad - (\$10{,}000)(A/F, 7\%, 20) \end{aligned}$$

$$(A/P, 7\%, 20) = 0.0944$$

$$A = (37{,}440 \text{ mi})\left(\frac{\$1.0}{\text{mi}}\right) = \$37{,}440$$

$$G = 0.1A = (0.1)(\$37{,}440) = \$3744$$

$$(P/G, 7\%, 20) = 77.5091$$

$$(A/F, 7\%, 20) = 0.0244$$

$$\begin{aligned} A_e &= (\$60{,}000)(0.0944) + \$37{,}440 \\ &\quad + (\$3744)(77.509)(0.0944) \\ &\quad - (\$10{,}000)(0.0244) \\ &= \$70{,}253.78 \end{aligned}$$

(a) With 80,000 passengers a year, the break-even fare per passenger would be

$$\text{fare} = \frac{A_e}{80{,}000} = \frac{\$70{,}253.78}{80{,}000} = \boxed{\$0.878/\text{passenger}}$$

(b)

$$\begin{aligned} \$0.878 &= \$0.35 + G(A/G, 7\%, 20) \\ G &= \frac{\$0.878 - \$0.35}{7.3163} \\ &= \boxed{\$0.072 \text{ increase per year}} \end{aligned}$$

(c) As in part (b), the subsidy should be

$$\begin{aligned} \text{subsidy} &= \text{cost} - \text{revenue} \\ P &= \$0.878 - \big(\$0.35 + (\$0.05)(A/G, 7\%, 20)\big) \\ &= \$0.878 - \big(\$0.35 + (\$0.05)(7.3163)\big) \\ &= \boxed{\$0.162} \end{aligned}$$

26.

$$\begin{aligned} P(\text{A}) &= (\$4800)(P/A, 12\%, 25) \\ &= (\$4800)(7.8431) \\ &= \$37{,}646.88 \end{aligned}$$

(4 quarters)(25 years) = 100 compounding periods

$$\begin{aligned} P(\text{B}) &= (\$1200)(P/A, 3\%, 100) \\ &= (\$1200)(31.5989) \\ &= \$37{,}918.68 \end{aligned}$$

Alternative B is economically superior.

27.

$$\begin{aligned} \text{EUAC(A)} &= (\$1500)(A/P, 15\%, 5) + \$800 \\ &= (\$1500)(0.2983) + \$800 \\ &= \$1247.45 \\ \text{EUAC(B)} &= (\$2500)(A/P, 15\%, 8) + \$650 \\ &= (\$2500)(0.2229) + \$650 \\ &= \$1207.25 \end{aligned}$$

Alternative B is economically superior.

28. The data given imply that both investments return 4% or more. However, the increased investment of \$30,000 may not be cost effective. Do an incremental analysis.

incremental cost = \$70,000 – \$40,000 = \$30,000

incremental income = \$5620 – \$4075 = \$1545

$$0 = -\$30{,}000 + (\$1545)(P/A, i\%, 20)$$

$$(P/A, i\%, 20) = 19.417$$

$$i \approx 0.25\% < 4\%$$

Alternative B is economically superior.

(The same conclusion could be reached by taking the present worths of both alternatives at 4%.)

29. (a)

$$\begin{aligned} P(\text{A}) &= (-\$3000)(P/A, 25\%, 25) - \$20{,}000 \\ &= (-\$3000)(3.9849) - \$20{,}000 \\ &= -\$31{,}954.70 \\ P(\text{B}) &= (-\$2500)(3.9849) - \$25{,}000 \\ &= -\$34{,}962.25 \end{aligned}$$

A is better.

(b)

$$\text{CC(A)} = \$20{,}000 + \frac{\$3000}{0.25} = \$32{,}000$$

$$\text{CC(B)} = \$25{,}000 + \frac{\$2500}{0.25} = \$35{,}000$$

A is better.

(c)

$$\begin{aligned} \text{EUAC(A)} &= (\$20{,}000)(A/P, 25\%, 25) + \$3000 \\ &= (\$20{,}000)(0.2509) + \$3000 \\ &= \$8018.00 \\ \text{EUAC(B)} &= (\$25{,}000)(0.2509) + \$2500 \\ &= \$8772.50 \end{aligned}$$

A is better.

30. (a)

$$\text{BV} = \$18{,}000 - (5)\left(\frac{\$18{,}000 - \$2000}{8}\right) = \boxed{\$8000}$$

(b) With the sinking fund method, the basis is

$$\begin{aligned} &(\$18{,}000 - \$2000)(A/F, 8\%, 8) \\ &= (\$18{,}000 - \$2000)(0.0940) \\ &= \$1504 \end{aligned}$$

$$D_1 = (\$1504)(1.000) = \boxed{\$1504}$$

$$D_2 = (\$1504)(1.0800) = \boxed{\$1624}$$

$$D_3 = (\$1504)(1.0800)^2 = (\$1504)(1.1664) = \boxed{\$1754}$$

(c)

$$D_1 = \left(\tfrac{2}{8}\right)(\$18{,}000) = \boxed{\$4500}$$

$$D_2 = \left(\tfrac{2}{8}\right)(\$18{,}000 - \$4500) = \boxed{\$3375}$$

$$D_3 = \left(\tfrac{2}{8}\right)(\$18{,}000 - \$4500 - \$3375) = \boxed{\$2531}$$

$$D_4 = \left(\tfrac{2}{8}\right)(\$18{,}000 - \$4500 - \$3375 - \$2531) = \boxed{\$1898}$$

$$D_5 = \left(\tfrac{2}{8}\right)(\$18{,}000 - \$4500 - \$3375 - \$2531 - \$1898) = \boxed{\$1424}$$

$$\text{BV} = \$18{,}000 - \$4500 - \$3375 - \$2531 - \$1898 - \$1424 = \boxed{\$4272}$$

Systems, Mgmt., Professional

31.

$$
\begin{aligned}
T &= \left(\tfrac{1}{2}\right)(8)(9) = 36 \\
D_1 &= \left(\tfrac{8}{36}\right)(\$14{,}000 - \$1800) = \$2711 \\
\Delta D &= \left(\tfrac{1}{36}\right)(\$14{,}000 - \$1800) = \$339 \\
D_2 &= \$2711 - \$339 = \$2372 \\
\text{DR} &= (0.52)(\$2711)(P/A, 15\%, 8) \\
&\quad - (0.52)(\$339)(P/G, 15\%, 8) \\
&= (0.52)(\$2711)(4.4873) \\
&\quad - (0.52)(\$339)(12.4807) \\
&= \boxed{\$4125.74}
\end{aligned}
$$

32.

$$
\begin{aligned}
\text{EUAC}_{5\text{ ft}} &= (\$600{,}000)(A/P, 12\%, 15) \\
&\quad + \left(\tfrac{14}{50}\right)(\$600{,}000) + \left(\tfrac{8}{50}\right)(\$650{,}000) \\
&\quad + \left(\tfrac{3}{50}\right)(\$700{,}000) + \left(\tfrac{1}{50}\right)(\$800{,}000) \\
&= \$418{,}080 \\
\text{EUAC}_{10\text{ ft}} &= (\$710{,}000)(A/P, 12\%, 15) \\
&\quad + \left(\tfrac{8}{50}\right)(\$650{,}000) + \left(\tfrac{3}{50}\right)(\$700{,}000) \\
&\quad + \left(\tfrac{1}{50}\right)(\$800{,}000) \\
&= \$266{,}228 \\
\text{EUAC}_{15\text{ ft}} &= (\$900{,}000)(A/P, 12\%, 15) \\
&\quad + \left(\tfrac{3}{50}\right)(\$700{,}000) + \left(\tfrac{1}{50}\right)(\$800{,}000) \\
&= \$190{,}120 \\
\text{EUAC}_{20\text{ ft}} &= (\$1{,}000{,}000)(A/P, 12\%, 15) \\
&\quad + \left(\tfrac{1}{50}\right)(\$800{,}000) \\
&= \$162{,}800
\end{aligned}
$$

$$\boxed{\text{Build to 20 ft.}}$$

33. Assume replacement after 1 year.

$$
\begin{aligned}
\text{EUAC}(1) &= (\$30{,}000)(A/P, 12\%, 1) + \$18{,}700 \\
&= (\$30{,}000)(1.12) + \$18{,}700 = \$52{,}300
\end{aligned}
$$

Assume replacement after 2 years.

$$
\begin{aligned}
\text{EUAC}(2) &= (\$30{,}000)(A/P, 12\%, 2) \\
&\quad + \$18{,}700 + (\$1200)(A/G, 12\%, 2) \\
&= (\$30{,}000)(0.5917) + \$18{,}700 \\
&\quad + (\$1200)(0.4717) = \$37{,}017
\end{aligned}
$$

Assume replacement after 3 years.

$$
\begin{aligned}
\text{EUAC}(3) &= (\$30{,}000)(A/P, 12\%, 3) \\
&\quad + \$18{,}700 + (\$1200)(A/G, 12\%, 3) \\
&= (\$30{,}000)(0.4163) + \$18{,}700 \\
&\quad + (\$1200)(0.9246) = \$32{,}299
\end{aligned}
$$

Similarly, calculate to obtain the numbers in the following table.

years in service	EUAC
1	\$52,300
2	\$37,017
3	\$32,299
4	\$30,207
5	\$29,152
6	\$28,602
7	\$28,335
8	\$28,234
9	\$28,240
10	\$28,312

$$\boxed{\text{Replace after 8 yr.}}$$

34. Assume the head and horsepower data are already reflected in the hourly operating costs.

Let N = no. of hours operated each year.

$$
\begin{aligned}
\text{EUAC}(A) &= (\$3600 + \$3050)(A/P, 12\%, 10) \\
&\quad - (\$200)(A/F, 12\%, 10) + 0.30N \\
&= (\$6650)(0.1770) - (\$200)(0.0570) + 0.30N \\
&= 1165.65 + 0.30N \\
\text{EUAC}(B) &= (\$2800 + \$5010)(A/P, 12\%, 10) \\
&\quad - (\$280)(A/F, 12\%, 10) + 0.10N \\
&= (\$7810)(0.1770) - (\$280)(0.0570) + 0.10N \\
&= 1366.41 + 0.10N
\end{aligned}
$$

$$
\begin{aligned}
\text{EUAC(A)} &= \text{EUAC(B)} \\
1165.65 + 0.30N &= 1366.41 + 0.10N
\end{aligned}
$$

$$N = \boxed{1003.8 \text{ hr}}$$

35. (a) From Eq. 54.78,

$$
\begin{aligned}
\frac{T_2}{T_1} &= 0.88 = 2^{-b} \\
\log 0.88 &= -b \log 2 \\
-0.0555 &= -(0.3010)b \\
b &= 0.1843
\end{aligned}
$$

$$T_4 = (6)(4)^{-0.1843} = \boxed{4.65 \text{ wk}}$$

(b) From Eq. 54.79,

$$\begin{aligned} T_{6\text{–}14} &= \left(\frac{6}{1-0.1843}\right) \\ &\quad \times \left(\left(14+\tfrac{1}{2}\right)^{1-0.1843} - \left(6-\tfrac{1}{2}\right)^{1-0.1843}\right) \\ &= \left(\frac{6}{0.8157}\right)(8.857-4.017) \\ &= \boxed{35.6 \text{ wk}} \end{aligned}$$

36. First check that both alternatives have an ROR greater than the MARR. Work in thousands of dollars.

Evaluate alternative A.

$$\begin{aligned} P(\text{A}) &= -\$120 + (\$15)(P/F, i\%, 5) \\ &\quad + (\$57)(P/A, i\%, 5)(1-0.45) \\ &\quad + \left(\frac{\$120-\$15}{5}\right)(P/A, i\%, 5)(0.45) \\ &= -\$120 + (\$15)(P/F, i\%, 5) \\ &\quad + (\$40.8)(P/A, i\%, 5) \end{aligned}$$

Try 15%.

$$\begin{aligned} P(\text{A}) &= -\$120 + (\$15)(0.4972) + (\$40.8)(3.3522) \\ &= \$24.23 \end{aligned}$$

Try 25%.

$$\begin{aligned} P(\text{A}) &= -\$120 + (\$15)(0.3277) + (\$40.8)(2.6893) \\ &= -\$5.36 \end{aligned}$$

Since $P(\text{A})$ goes through 0,

$$(\text{ROR})_\text{A} > \text{MARR} = 15\%$$

Next, evaluate alternative B.

$$\begin{aligned} P(\text{B}) &= -\$170 + (\$20)(P/F, i\%, 5) \\ &\quad + (\$67)(P/A, i\%, 5)(1-0.45) \\ &\quad + \left(\frac{\$170-\$20}{5}\right)(P/A, i\%, 5)(0.45) \\ &= -\$170 + (\$20)(P/F, i\%, 5) \\ &\quad + (\$50.35)(P/A, i\%, 5) \end{aligned}$$

Try 15%.

$$\begin{aligned} P(\text{B}) &= -\$170 + (\$20)(0.4972) + (\$50.35)(3.352) \\ &= \$8.72 \end{aligned}$$

Since $P(\text{B}) > 0$ and will decrease as i increases,

$$(\text{ROR})_\text{B} > 15\%$$

ROR > MARR for both alternatives.

Do an incremental analysis to see if it is worthwhile to invest the extra $170 – $120 = $50.

$$\begin{aligned} P(\text{B}-\text{A}) &= -\$50 + (\$20-\$15)(P/F, i\%, 5) \\ &\quad + (\$50.35-\$40.8)(P/A, i\%, 5) \end{aligned}$$

Try 15%.

$$\begin{aligned} P(\text{B}-\text{A}) &= -\$50 + (\$5)(0.4972) \\ &\quad + (\$9.55)(3.3522) \\ &= -\$15.50 \end{aligned}$$

Since $P(\text{B}-\text{A}) < 0$ and would become more negative as i increases, the ROR of the added investment is < 15%.

$$\boxed{\text{Alternative A is superior.}}$$

37. Use the year-end convention with the tax credit. The purchase is made at $t = 0$. However, the tax credit is received at $t = 1$ and must be multiplied by $(P/F, i\%, 1)$.

(Note that 0.0667 is actually 2/3 of 10%.)

$$\begin{aligned} P &= -\$300{,}000 + (0.0667)(\$300{,}000)(P/F, i\%, 1) \\ &\quad + (\$90{,}000)(P/A, i\%, 5)(1-0.48) \\ &\quad + \left(\frac{\$300{,}000-\$50{,}000}{5}\right)(P/A, i\%, 5)(0.48) \\ &\quad + (\$50{,}000)(P/F, i\%, 5) \\ &= -\$300{,}000 + (\$20{,}000)(P/F, i\%, 1) \\ &\quad + (\$46{,}800)(P/A, i\%, 5) \\ &\quad + (\$24{,}000)(P/A, i\%, 5) \\ &\quad + (\$50{,}000)(P/F, i\%, 5) \end{aligned}$$

By trial and error,

i	P
10%	$17,616
15%	–$20,412
12%	$1448
13%	–$6142
$12\frac{1}{4}$%	–$479

$$\boxed{i \text{ is between } 12\% \text{ and } 12^1/_4\%.}$$

38. Assume loan payments are made at the end of each year. Find the annual payment.

$$\begin{aligned} \text{payment} &= (\$2{,}500{,}000)(A/P, 12\%, 25) \\ &= (\$2{,}500{,}000)(0.1275) \\ &= \$318{,}750 \\ \text{distributed profit} &= (0.15)(\$2{,}500{,}000) \\ &= \$375{,}000 \end{aligned}$$

After paying all expenses and distributing the 15% profit, the remainder should be 0.

$$\begin{aligned} 0 = \text{EUAC} &= \$20{,}000 + \$50{,}000 + \$200{,}000 \\ &\quad + \$375{,}000 + \$318{,}750 - \text{annual receipts} \\ &\quad - (\$500{,}000)(A/F, 15\%, 25) \\ &= \$963{,}750 - \text{annual receipts} \\ &\quad - (\$500{,}000)(0.0047) \end{aligned}$$

This calculation assumes $i = 15\%$, which equals the desired return. However, this assumption only affects the salvage calculation, and since the number is so small, the analysis is not sensitive to the assumption.

$$\text{annual receipts} = \$961{,}400$$

The average daily receipts are

$$\frac{\$961{,}400}{365} = \$2634$$

Use the expected value approach. The average occupancy is

$$\begin{aligned} &(0.40)(0.65) + (0.30)(0.70) + (0.20)(0.75) \\ &\quad + (0.10)(0.80) = 0.70 \end{aligned}$$

The average number of rooms occupied each night is

$$(0.70)(120 \text{ rooms}) = 84 \text{ rooms}$$

The minimum required average daily rate per room is

$$\frac{\$2634}{84} = \boxed{\$31.36}$$

39.
$$\begin{aligned} \text{annual savings} &= \left(\frac{0.69 - 0.47}{1000}\right)(\$3{,}500{,}000) = \$770 \\ P &= -\$7500 + (\$770 - \$200 - \$100) \\ &\quad \times (P/A, i\%, 25) = 0 \\ (P/A, i\%, 25) &= 15.957 \end{aligned}$$

Searching the tables and interpolating,

$$i \approx \boxed{3.75\%}$$

40. Work in millions of dollars.

$$\begin{aligned} P(\text{A}) &= -(\$1.3)(1 - 0.48)(P/A, 15\%, 6) \\ &= (\$1.3)(0.52)(3.7845) \\ &= -\$2.56 \quad [\text{millions}] \end{aligned}$$

Since this is an after-tax analysis and since the salvage value was mentioned, assume that the improvements can be depreciated.

Use straight line depreciation.

Evaluate alternative B.

$$\begin{aligned} D_j &= \tfrac{2}{6} = 0.333 \\ P(\text{B}) &= -\$2 - (\$0.20)(\$1.3)(1 - 0.48)(P/A, 15\%, 6) \\ &\quad - (\$0.15)(1 - 0.48)(P/A, 15\%, 6) \\ &\quad + (\$0.333)(0.48)(P/A, 15\%, 6) \\ &= -\$2 - (\$0.20)(\$1.3)(0.52)(3.7845) \\ &\quad - (\$0.15)(0.52)(3.7845) \\ &\quad + (\$0.333)(0.48)(3.7845) \\ &= -\$2.206 \quad [\text{millions}] \end{aligned}$$

Next, evaluate alternative C.

$$\begin{aligned} D_j &= \frac{1.2}{3} = 0.4 \\ P(\text{C}) &= -(\$1.2)\big(1 + (P/F, 15\%, 3)\big) \\ &\quad - (\$0.20)(\$1.3)(1 - 0.48) \\ &\quad \times \big((P/F, 15\%, 1) + (P/F, 15\%, 4)\big) \\ &\quad - (\$0.45)(\$1.3)(1 - 0.48) \\ &\quad \times \big((P/F, 15\%, 2) + (P/F, 15\%, 5)\big) \\ &\quad - (\$0.80)(\$1.3)(1 - 0.48) \\ &\quad \times \big((P/F, 15\%, 3) + (P/F, 15\%, 6)\big) \\ &\quad + (\$0.4)(\$0.48)(P/A, 15\%, 6) \\ &= -(\$1.2)(1.6575) \\ &\quad - (\$0.20)(\$1.3)(0.52)(0.8696 + 0.5718) \\ &\quad - (\$0.45)(\$1.3)(0.52)(0.7561 + 0.4972) \\ &\quad - (\$0.80)(\$1.3)(0.52)(0.6575 + 0.4323) \\ &\quad + (\$0.4)(0.48)(3.7845) \\ &= -\$2.436 \quad [\text{millions}] \end{aligned}$$

$$\boxed{\text{Alternative B is superior.}}$$

41. This is a replacement study. Since production capacity and efficiency are not a problem with the defender, the only question is when to bring in the challenger.

Since this is a before-tax problem, depreciation is not a factor, nor is book value.

The cost of keeping the defender one more year is

$$\begin{aligned} \text{EUAC(defender)} &= \$200{,}000 + (0.15)(\$400{,}000) \\ &= \$260{,}000 \end{aligned}$$

For the challenger,

$$\begin{aligned}\text{EUAC(challenger)} &= (\$800{,}000)(A/P, 15\%, 10) + \$40{,}000 \\ &\quad + (\$30{,}000)(A/G, 15\%, 10) \\ &\quad - (\$100{,}000)(A/F, 15\%, 10) \\ &= (\$800{,}000)(0.1993) + \$40{,}000 \\ &\quad + (\$30{,}000)(3.3832) \\ &\quad - (\$100{,}000)(0.0493) \\ &= \$296{,}006\end{aligned}$$

Since the defender is cheaper, keep it. The same analysis next year will give identical answers. Therefore, keep the defender for the next 3 years, at which time the decision to buy the challenger will be automatic.

Having determined that it is less expensive to keep the defender than to maintain the challenger for 10 years, determine whether the challenger is less expensive if retired before 10 years.

If retired in 9 years,

$$\begin{aligned}\text{EUAC(challenger)} &= (\$800{,}000)(A/P, 15\%, 9) + \$40{,}000 \\ &\quad + (\$30{,}000)(A/G, 15\%, 9) \\ &\quad - (\$150{,}000)(A/F, 15\%, 9) \\ &= (\$800{,}000)(0.2096) \\ &\quad + \$40{,}000 + (\$30{,}000)(3.0922) \\ &\quad - (\$150{,}000)(0.0596) \\ &= \$291{,}506\end{aligned}$$

Similar calculations yield the following results for all the retirement dates.

n	EUAC
10	\$296,000
9	\$291,506
8	\$287,179
7	\$283,214
6	\$280,016
5	\$278,419
4	\$279,909
3	\$288,013
2	\$313,483
1	\$360,000

Since none of these equivalent uniform annual costs is less than that of the defender, it is not economical to buy and keep the challenger for any length of time.

Keep the defender.

Systems, Mgmt., Professional

55 Project Management

PRACTICE PROBLEMS

1. The activities that constitute a project are listed. The project starts at $t = 0$. (a) Draw the critical path network. (b) Indicate the critical path. (c) What is the earliest finish? (d) What is the latest finish? (e) What is the slack along the critical path? (f) What is the float along the critical path?

activity	predecessors	successors	duration
start	–	A	0
A	start	B,C,D	7
B	A	G	6
C	A	E,F	5
D	A	G	2
E	C	H	13
F	C	H,I	4
G	D,B	I	18
H	E,F	finish	7
I	F,G	finish	5
finish	H,I	–	0

2. PERT activities constituting a short project are listed with their characteristic completion times. If the project starts at $t = 15$, what is the probability that the project will be completed on or before $t = 42$?

activity	predecessors	successors	t_{min}	t_{likely}	t_{max}
start	–	A	0	0	0
A	start	B,D	1	2	5
B	A	C	7	9	20
C	B	D	5	12	18
D	A,C	finish	2	4	7
finish	D	–	0	0	0

3. Listed is a set of activities and sequence requirements to start a warehouse construction project. Prepare a CPM project diagram.

activity	letter code	code of immediate predecessor
move-in	A	
job layout	B	A
excavations	C	B
make-up forms	D	A
shop drawing, order rebar	E	A
erect forms	F	C,D
rough in plumbing	G	F
install rebar	H	E,F
pour, finish concrete	I	G,H

4. Activities constituting a bridge construction project are listed. (a) Draw the CPM network showing the critical path. (b) Compute ES, EF, LS, and LF for each of the activities. Assume a target time for completing the project that, for the bridge, is 3 days after the EF time.

job number		immediate predecessors	time (days)
a	start		0
b		a	4
c		b	2
d		c	4
e		d	6
f		c	1
g		f	2
h		f	3
i		d	2
j		d,g	4
k		i,j,h	10
l		k	3
m		l	1
n		l	2
o		l	3
p		e	2
q		p	1
r		c	1
s		o,t	2
t		m,n	3
u		t	1
v		q,r	2
w		v	5
x	finish	s,u,w	0

5. Prepare a PERT diagram for the activities and sequence requirements given in Prob. 4.

6. Listed is a set of activities, sequence requirements, and estimated activity times required for the renewal of a pipeline.

(a) Prepare a PERT project diagram.

activity	letter code	code of immediate predecessor	activity time requirement (days)
assemble crew for job	A		10
use old line to build inventory	B	D	28
measure and sketch old line	C	A	2
develop materials list	D	C	1
erect scaffold	E	D	2
procure pipe	F	D	30
procure valves	G	D	45
deactivate old line	H	B	1
remove old line	I	E,H	6
prefabricate new pipe	J	F	5
place valves	K	I,G	1
place new pipe	L	I,J	6
weld pipe	M	L	2
connect valves	N	K,M	1
insulate	O	N	4
pressure test	P	K	1
remove scaffold	Q	O,P	1
clean up and turn over to operating crew	R	O,Q	1

(b) There is additional information in the form of optimistic, most likely, and pessimistic time estimates for the project. Compute the expected mean time, t_m, and the variance, σ^2, for the activities. Which activities have the greatest uncertainty in their completion schedules?

activity code	optimistic time (t_{min})	most likely time (t_{ml})	pessimistic time (t_{max})
A	8	10	12
B	26	26.5	36
C	1	2	3
D	0.5	1	1.5
E	1.5	1.63	4
F	28	28	40
G	40	42.5	60
H	1	1	1
I	4	6	8
J	4	4.5	8
K	0.5	0.9	2
L	5	5.25	10
M	1	2	3
N	0.5	1	1.5
O	3	3.75	6
P	1	1	1
Q	1	1	1
R	1	1	1

(c) Suppose that, due to penalties in the contract, each day the pipeline renewal project can be shortened is worth \$100. Which of the following possibilities would you follow and why?

- Shorten t_m of activity B by 4 days at a cost of \$100.
- Shorten t_m of activity G by 5 days at a cost of \$50.
- Shorten t_m of activity O by 2 days at a cost of \$150.
- Shorten t_m of activity O by 2 days by drawing resources from activity N, thereby lengthening its t_e by 2 days.

SOLUTIONS

1. (a) The critical path network diagram is as follows.

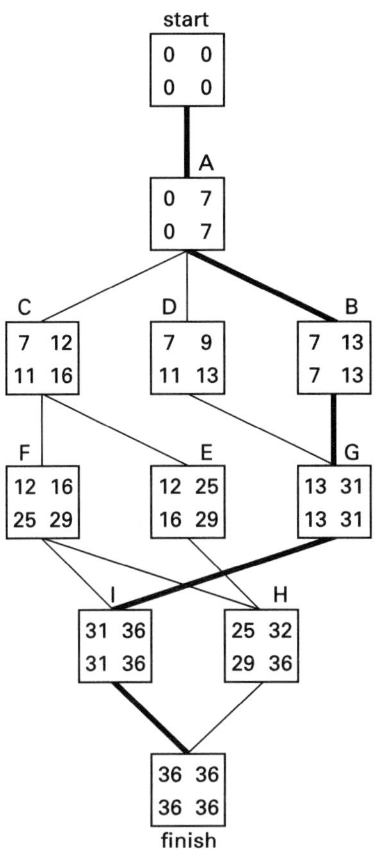

ES (earliest start) Rule: The earliest start time for an activity leaving a particular node is equal to the largest of the earliest finish times for all activities entering the node.

LF (latest finish) Rule: The latest finish time for an activity entering a particular node is equal to the smallest of the latest start times for all activities leaving the node.

The activity is critical if the earliest start equals the latest start.

(b) The critical path is A-B-G-I.

(c) The earliest finish is 36.

(d) The latest finish is 36.

(e) The slack along the critical path is 0.

(f) The float along the critical path is 0.

2. From Eq. 55.1,

$$\mu = \frac{t_{\text{minimum}} + 4t_{\text{likely}} + t_{\text{maximum}}}{6}$$

From Eq. 55.2,

$$\sigma^2 = \left(\frac{t_{\text{maximum}} - t_{\text{minimum}}}{6}\right)^2$$

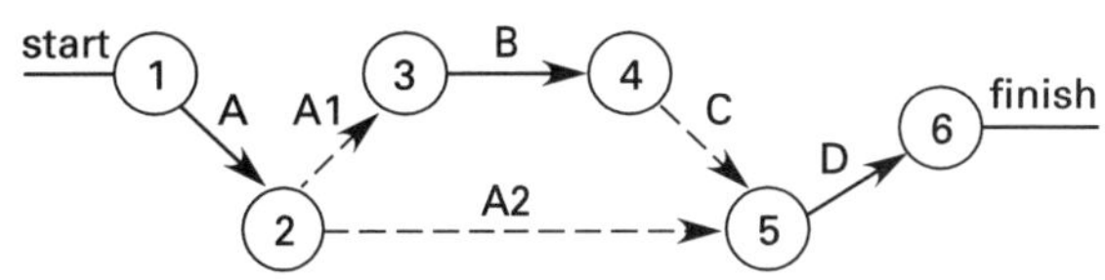

The critical path is A-A1-B-C-D.

The following probability calculations assume that all activities are independent. Use the following theorems for the sum of independent random variables and use the normal distribution for T (project time).

$$\mu_{\text{total}} = t_A + t_B + t_C + t_D$$
$$\sigma^2_{\text{total}} = \sigma^2_A + \sigma^2_B + \sigma^2_C + \sigma^2_D$$

The variance is 10.52778, and the standard deviation is 3.244654. (See table below.)

PERT Analysis for Prob. 2

activity				earliest	latest	earliest	latest	slack
no.	name	exp. time	variance	start	start	finish	finish	LS – ES
1	A	+2.33333	+0.44444	15	15	+17.3333	+17.3333	0
2	A1	0	0	+17.3333	+17.3333	+17.3333	+17.3333	0
3	A2	0	0	+17.3333	+39.6667	+17.3333	+39.6667	+22.3333
4	B	+10.5000	+4.69444	+17.3333	+17.3333	+27.8333	+27.8333	0
5	C	+11.8333	+4.69444	+27.8333	+27.8333	+39.6667	+39.6667	0
6	D	+4.16667	+0.69444	+39.6667	+39.6667	+43.8333	+43.8333	0
expected completion time = 43.83333								

Systems, Mgmt., Professional

$$\mu_{\text{total}} = 43.83333$$
$$\sigma^2_{\text{total}} = 10.52778$$
$$\sigma_{\text{total}} = 3.244654$$
$$z = \frac{t - \mu_{\text{total}}}{\sigma} = \left|\frac{42 - 43.83333}{3.244654}\right| = 0.565$$

From the normal table, the probability of finishing for $T \leq 42$ is 0.286037 (28.6%).

3.

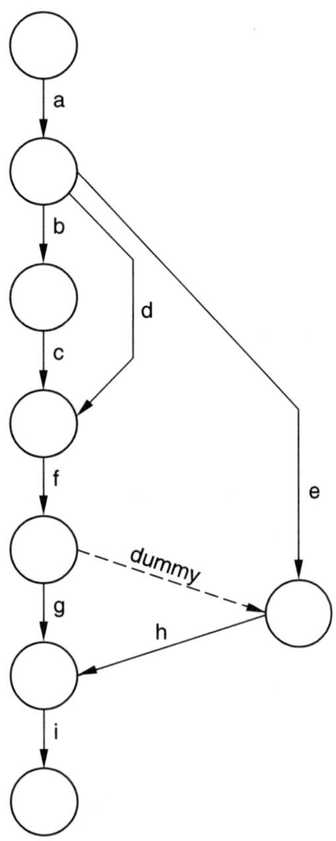

(Notice that a dummy node cannot be replaced with any other activity letter without duplicating that activity.)

4.

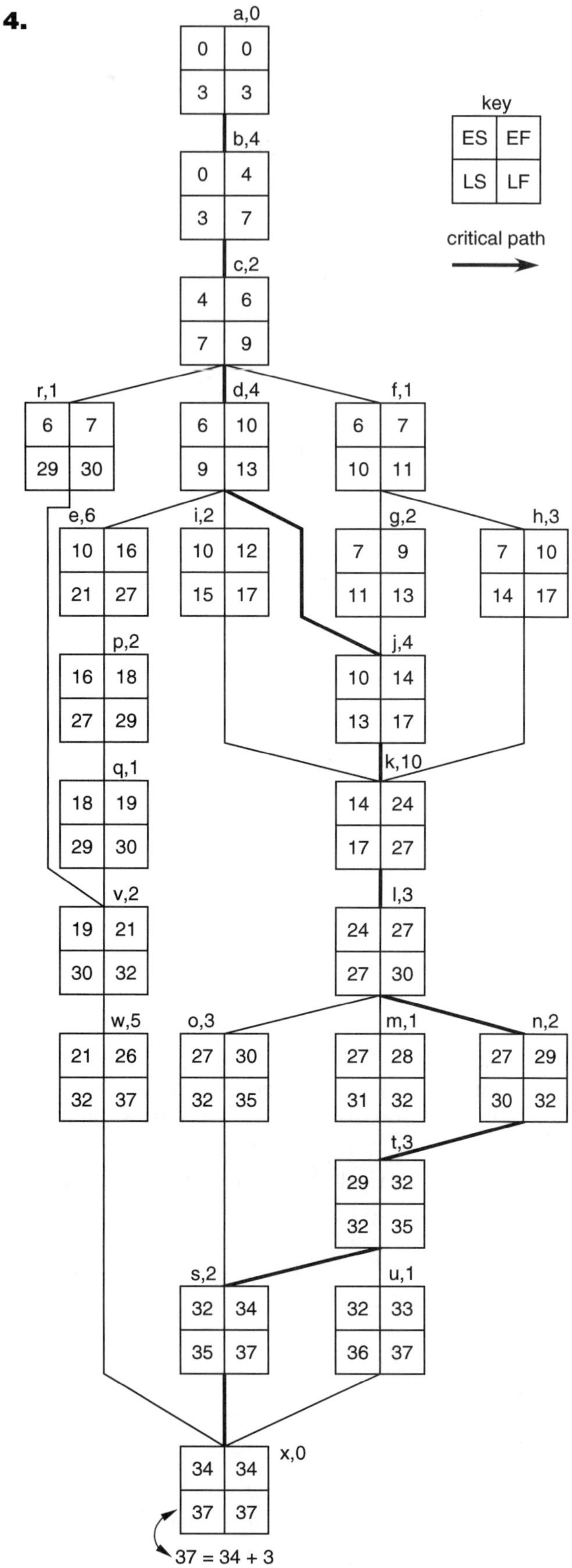

5.

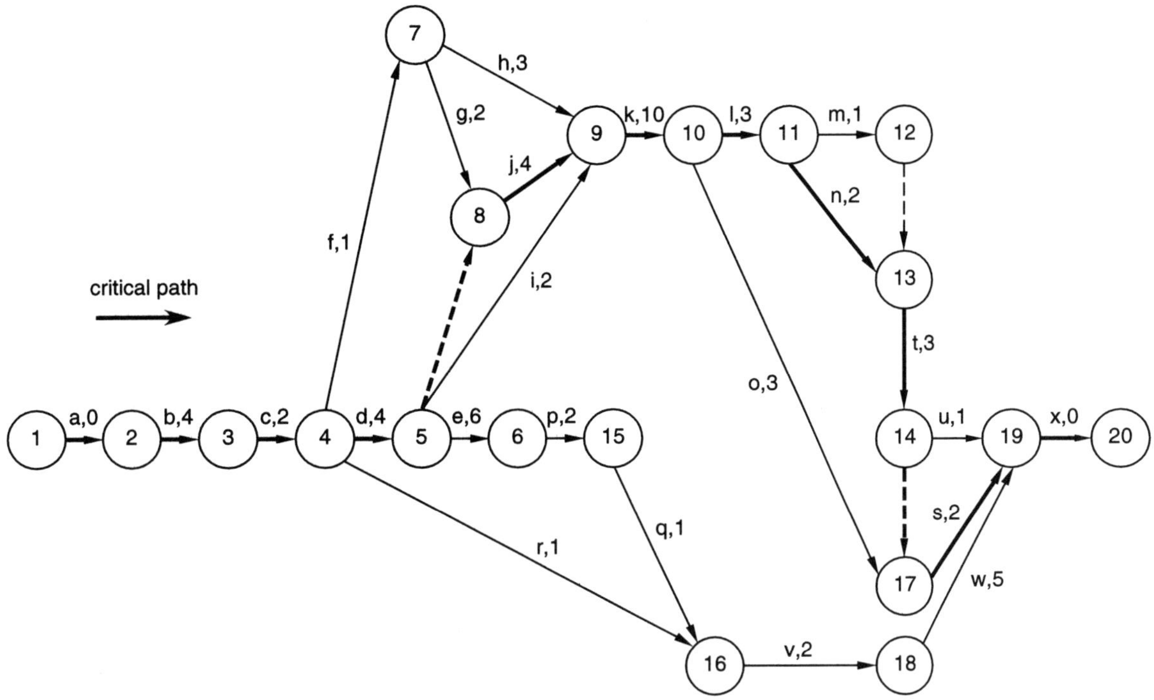

6. (a)

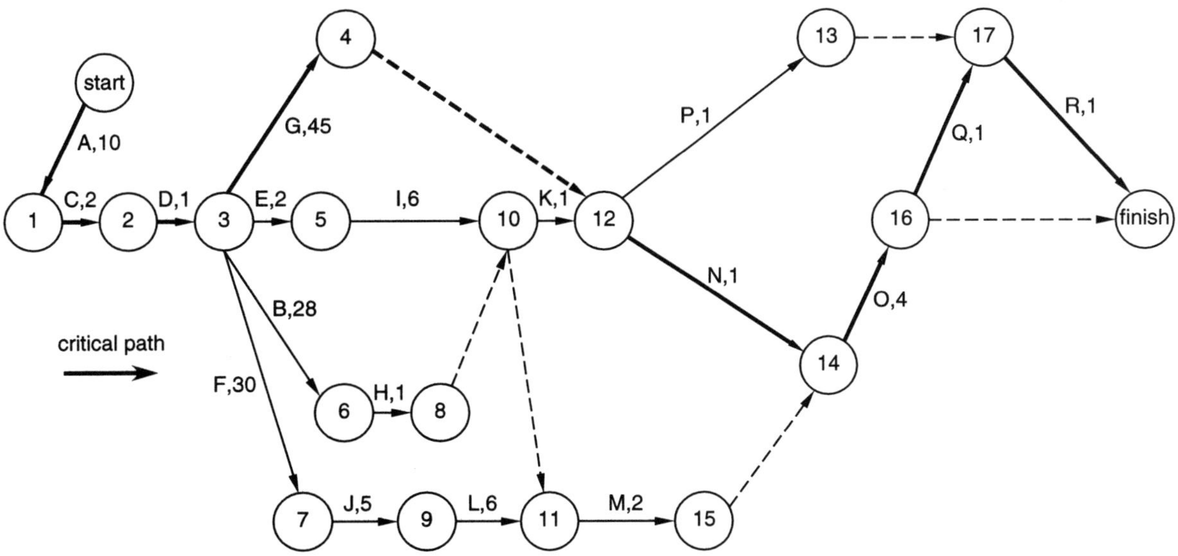

Systems, Mgmt., Professional

(b)

activity	variance	meantime
A	0.44	10
B	2.77	28
C	0.11	2
D	0.027	1
E	0.17	2
F	4.0	30
G	11.11	45
H	0	1
I	0.44	6
J	0.44	5
K	0.6225	1
L	0.69	6
M	0.11	2
N	0.027	1
O	0.25	4
P	0	1
Q	0	1
R	0	1

Activities B, F, and G have the three largest variances, so these activities have the greatest uncertainties.

(c)

- Activity B is not on the critical path, so shortening it will not shorten the project.
- Activity G is on the critical path, so it should be shortened. Shortening it 5 days may change the critical path, however.

  $$\text{path D-G-K: } 1 + 45 = 46 \text{ days}$$

  $$\text{path D-E-I-K: } 1 + 2 + 6 = 9 \text{ days}$$

  $$\text{path D-B-H-I-K: } 1 + 28 + 1 + 6 = 36 \text{ days}$$

  Since the shortened path D-G-K has a length of 46 days − 5 days = 41 days and is still the longest path from D to K, it should be shortened.
- Activity O is on the critical path, but the cost of the crash schedule exceeds the benefit.
- The current path length is

  $$\text{N-O-Q-R: } 1 + 4 + 1 = 6 \text{ days}$$

  The new paths would be

  $$\text{N-P-Q-R: } 3 + 1 + 1 = 5 \text{ days}$$

  $$\text{N-O-Q-R: } 3 + 2 + 1 = 6 \text{ days}$$

  $$\text{N-O-R: } (3 + 2) + (4 - 2) = 7 \text{ days}$$

  N-O-R would become part of the critical path and would be longer than the original critical path. Therefore, it is not acceptable.

56 Engineering Law

PRACTICE PROBLEMS

Chapter 56 of the *Environmental Engineering Reference Manual* contains no practice problems.

57 Engineering Ethics

PRACTICE PROBLEMS

Chapter 57 of the *Environmental Engineering Reference Manual* contains no practice problems.

58 Engineering Licensing in the United States

PRACTICE PROBLEMS

Chapter 58 of the *Environmental Engineering Reference Manual* contains no practice problems.

NO POSTAGE
NECESSARY
IF MAILED
IN THE
UNITED STATES

BUSINESS REPLY MAIL

FIRST CLASS MAIL PERMIT NO. 33 BELMONT, CA

POSTAGE WILL BE PAID BY ADDRESSEE

PROFESSIONAL PUBLICATIONS INC
1250 FIFTH AVE
BELMONT CA 94002-9979

NO POSTAGE
NECESSARY
IF MAILED
IN THE
UNITED STATES

BUSINESS REPLY MAIL

FIRST CLASS MAIL PERMIT NO. 33 BELMONT, CA

POSTAGE WILL BE PAID BY ADDRESSEE

PROFESSIONAL PUBLICATIONS INC
1250 FIFTH AVE
BELMONT CA 94002-9979

NO POSTAGE
NECESSARY
IF MAILED
IN THE
UNITED STATES

BUSINESS REPLY MAIL

FIRST CLASS MAIL PERMIT NO. 33 BELMONT, CA

POSTAGE WILL BE PAID BY ADDRESSEE

PROFESSIONAL PUBLICATIONS INC
1250 FIFTH AVE
BELMONT CA 94002-9979